T0305935

DESIGN AND CONSTRUCTION OF COORDINATION POLYMERS

DESIGN AND CONSTRUCTION OF COORDINATION POLYMERS

Edited by

MAO-CHUN HONG
LING CHEN

A JOHN WILEY & SONS, INC., PUBLICATION

Published by John Wiley & Sons, Inc., Hoboken, New Jersey.
Published simultaneously in Canada.

Library of Congress Cataloging-in-Publication Data:

Design and construction of coordination polymers/
[edited by] Mao-Chun Hong and Ling Chen

Includes bibliographic references and index

ISBN 978-0-470-29450-5 (cloth)

Printed in the United States of America

10 9 8 7 6 5 4 3 2 1

Dedicated to

the Fujian Institute of Research on the Structure of Matter,
Chinese Academy of Sciences, on the occasion of
the fiftieth anniversary of its founding
and
Professor Xin-Tao Wu, one of the pioneers in the
development of the Fujian Institute, on the
occasion of his seventieth birthday

CONTENTS

CONTRIBUTORS

Xian-He Bu, Department of Chemistry, Nankai University, Tianjin, People's Republic of China

Chun-Long Chen, MOE Laboratory of Bioinorganic and Synthetic Chemistry, State Key Laboratory of Optoelectronic Materials and Technologies, School of Chemistry and Chemical Engineering, Sun Yat-Sen University, Guangzhou, People's Republic of China

Lian Chen, State Key Laboratory of Structural Chemistry, Fujian Institute of Research on the Structure of Matter, Chinese Academy of Sciences, Fuzhou, Fujian, People's Republic of China

Ling Chen, State Key Laboratory of Structural Chemistry, Fujian Institute of Research on the Structure of Matter, Chinese Academy of Sciences, Fuzhou, Fujian, People's Republic of China

Li-Juan Chen, State Key Laboratory of Structural Chemistry, Fujian Institute of Research on the Structure of Matter, Chinese Academy of Sciences, Fuzhou, Fujian, People's Republic of China

Shu-Mei Chen, State Key Laboratory of Structural Chemistry, Fujian Institute of Research on the Structure of Matter, Chinese Academy of Sciences, Fuzhou, Fujian, People's Republic of China

Xiao-Ming Chen, MOE Laboratory of Bioinorganic and Synthetic Chemistry, School of Chemistry and Chemical Engineering, Sun Yat-Sen University, Guangzhou, People's Republic of China

Christopher D. Collier, Department of Chemistry and Biochemistry, Miami University, Oxford, Ohio

Miao Du, College of Chemistry and Life Science, Tianjin Normal University, Tianjin, People's Republic of China

Song Gao, State Key Laboratory of Rare Earth Materials and Applications, College of Chemistry and Molecular Engineering, Peking University, Beijing, People's Republic of China

Mao-Chun Hong, State Key Laboratory of Structural Chemistry, Fujian Institute of Research on the Structure of Matter, Chinese Academy of Sciences, Fuzhou, Fujian, People's Republic of China

Ping Hu, MOE Laboratory of Bioinorganic and Synthetic Chemistry, School of Chemistry and Chemical Engineering, Sun Yat-Sen University, Guangzhou, People's Republic of China

Fei-Long Jiang, State Key Laboratory of Structural Chemistry, Fujian Institute of Research on the Structure of Matter, Chinese Academy of Sciences, Fuzhou, Fujian, People's Republic of China

Xiao-Yu Jiang, State Key Laboratory of Structural Chemistry, Fujian Institute of Research on the Structure of Matter, Chinese Academy of Sciences, Fuzhou, Fujian, People's Republic of China

Bei-Sheng Kang, MOE Laboratory of Bioinorganic and Synthetic Chemistry, State Key Laboratory of Optoelectronic Materials and Technologies, School of Chemistry and Chemical Engineering, Sun Yat-Sen University, Guangzhou, People's Republic of China

Jian-Ping Lang, College of Chemistry, Chemical Engineering and Materials Science, Suzhou University, Suzhou, Jiangsu, People's Republic of China

JeongYong Lee, Department of Chemistry and Chemical Biology, Rutgers University, Piscataway, New Jersey

Hong-Xi Li, College of Chemistry, Chemical Engineering and Materials Science, Suzhou University, Suzhou, Jiangsu, People's Republic of China

Jing Li, Department of Chemistry and Chemical Biology, Rutgers University, Piscataway, New Jersey

Kun-Hao Li, Department of Chemistry and Chemical Biology, Rutgers University, Piscataway, New Jersey

Zheng-Zhong Lin, State Key Laboratory of Structural Chemistry, Fujian Institute of Research on the Structure of Matter, Chinese Academy of Sciences, Fuzhou, Fujian, People's Republic of China

Can-Zhong Lu, State Key Laboratory of Structural Chemistry, Fujian Institute of Research on the Structure of Matter, Chinese Academy of Sciences, Fuzhou, Fujian, People's Republic of China

Sheng-Qian Ma, Department of Chemistry and Biochemistry, Miami University, Oxford, Ohio

Zong-Wan Mao, MOE Laboratory of Bioinorganic and Synthetic Chemistry, School of Chemistry and Chemical Engineering, Sun Yat-Sen University, Guangzhou, People's Republic of China

Gerd Meyer, Institut für Anorganische Chemie, Universität zu Köln, Köln, Germany

David H. Olson, Department of Chemistry and Chemical Biology, Rutgers University, Piscataway, New Jersey

Long Pan, Department of Chemistry and Chemical Biology, Rutgers University, Piscataway, New Jersey

Ingo Pantenburg, Institut für Anorganische Chemie, Universität zu Köln, Köln, Germany

Zhi-Gang Ren, College of Chemistry, Chemical Engineering and Materials Science, Suzhou University, Suzhou, Jiangsu, People's Republic of China

Muhamet Sehabi, Institut für Anorganische Chemie, Universität zu Köln, Köln, Germany

Cheng-Yong Su, MOE Laboratory of Bioinorganic and Synthetic Chemistry, State Key Laboratory of Optoelectronic Materials and Technologies, School of Chemistry and Chemical Engineering, Sun Yat-Sen University, Guangzhou, People's Republic of China

Jin-Tao Wang, MOE Laboratory of Bioinorganic and Synthetic Chemistry, School of Chemistry and Chemical Engineering, Sun Yat-Sen University, Guangzhou, People's Republic of China

Xin-Yi Wang, State Key Laboratory of Rare Earth Materials and Applications, College of Chemistry and Molecular Engineering, Peking University, Beijing, People's Republic of China

Yu-Jia Wang, MOE Laboratory of Bioinorganic and Synthetic Chemistry, School of Chemistry and Chemical Engineering, Sun Yat-Sen University, Guangzhou, People's Republic of China

Zhe-Ming Wang, State Key Laboratory of Rare Earth Materials and Applications, College of Chemistry and Molecular Engineering, Peking University, Beijing, People's Republic of China

Li-Ming Wu, State Key Laboratory of Structural Chemistry, Fujian Institute of Research on the Structure of Matter, Chinese Academy of Sciences, Fuzhou, Fujian, People's Republic of China

Xiao-Yuan Wu, State Key Laboratory of Structural Chemistry, Fujian Institute of Research on the Structure of Matter, Chinese Academy of Sciences, Fuzhou, Fujian, People's Republic of China

Ren-Gen Xiong, Ordered Matter Science Research Center, Southeast University, Nanjing, People's Republic of China

Heng-Yun Ye, Ordered Matter Science Research Center, Southeast University, Nanjing, People's Republic of China

Quan-Guo Zhai, State Key Laboratory of Structural Chemistry, Fujian Institute of Research on the Structure of Matter, Chinese Academy of Sciences, Fuzhou, Fujian, People's Republic of China

Jian-Yong Zhang, MOE Laboratory of Bioinorganic and Synthetic Chemistry, State Key Laboratory of Optoelectronic Materials and Technologies, School of Chemistry and Chemical Engineering, Sun Yat-Sen University, Guangzhou, People's Republic of China

Jie-Peng Zhang, MOE Laboratory of Bioinorganic and Synthetic Chemistry, School of Chemistry and Chemical Engineering, Sun Yat-Sen University, Guangzhou, People's Republic of China

Wen Zhang, Ordered Matter Science Research Center, Southeast University, Nanjing, People's Republic of China

Wen-Hua Zhang, College of Chemistry, Chemical Engineering and Materials Science, Suzhou University, Suzhou, Jiangsu, People's Republic of China

Zhen-Guo Zhao, State Key Laboratory of Structural Chemistry, Fujian Institute of Research on the Structure of Matter, Chinese Academy of Sciences, Fuzhou, Fujian, People's Republic of China

Hong-Cai Zhou, Department of Chemistry and Biochemistry, Miami University, Oxford, Ohio; now at the Department of Chemistry, Texas A&M University, College Station, Texas

PREFACE

This volume, *Design and Construction of Coordination Polymers*, is published in recognition of the fiftieth anniversary of the beginning of the Fujian Institute of Research on the Structure of Matter, Chinese Academy of Science, and is dedicated to Prof. Xin-Tao Wu, on the occasion of his seventieth birthday. The volume contains 13 chapters that focus on this theme, contributed by former graduates and friends of this institution. They span a wide range of topics, systems, and approaches to coordination polymers. In Chapter 1, G. Meyer and co-workers describe a variety of AgL_1L_2 coordination complexes achieved with mixed N-containing ligands, some of which exhibit particularly short Ag–N distances. A series of indium(III)–carboxylate coordination polymers with a variety of dimensionalities are reported in Chapter 2 by M.-C. Hong and colleagues. Following this, in Chapter 3, X.-M. Chen and co-workers examine an interesting series of coordination polymers that are formed by solvothermal reactions between diverse metals and ligands, covering a variety of functionalities and reaction types. In Chapter 4, C.-Z. Lu and co-workers report on a variety of hybrid polyoxometallates built of oxo-Mo, oxo-Cu, and oxo-Ag centers and 1,2,4-triazolate or pyridine derivatives. In Chapter 5, C.-Y. Su and others describe the range of Ag(I) coordination environments that can be obtained with various ligand types as well as interpenetrating polymers and their diverse properties and applications. A different type of variety results when coordination polymers are tuned via systematic variations in the bridging organic backbones within multidentate ligands, as related by X.-H. Bu and others in Chapter 6.

Chapter 7, by R.-G. Xiong and co-workers, features a different direction, the design and synthesis of ferroelectric coordination compounds or polymers that are constructed of monochiral ligands and occur in polar space groups. The generation and properties of paramagnetic solids of different dimensionalities from three-atom bridging ligands and 3d metal ions are described by S. Gao and colleagues in Chapter 8. In Chapter 9, L.-M. Wu and L. Chen describe and explain structural, optical, and thermal properties of various iodometallic phases of Pb or Bi with different dimensionalities and motifs. J.-P. Lang and co-workers relate in Chapter 10 the design, construction, and optical properties of supramolecular cluster-based phases of Mo(W)/Cu/S. Microporous metal–organic frameworks as materials for gas storage and separation are addressed in Chapter 11 by J. Li and colleagues. In Chapter 12, H.-C. Zhou and co-workers enumerate the design, construction, and properties of various metal–organic frameworks for storage and separation of gases, H_2 in particular. Finally, in Chapter 13, Z.-W. Mao and colleagues outline the application of some bioinorganic coordination compounds as mimics or models for a range of biological functions or therapeutic agents. Enjoy!

We are grateful for all the authors who unselfishly spent their most precious time in writing contributions for the volume and meeting deadlines. Special thanks are given to Prof. Thomas C. W. Mak at the Chinese University of Hong Kong, Prof. Bei-Sheng Kang at Sun Yat-Sen University, Guangzhou and Prof. Qiu -Tian Liu at FIRSM for their kind help, and A. Lekhwani and R. Amos of John Wiley & Sons, Inc. for editorial assistance.

MAO-CHUN HONG
LING CHEN

Fuzhou
August 2008

1

COORDINATIVE FLEXIBILITY OF MONOVALENT SILVER IN $[Ag^{I} \leftarrow L1]L2$ COMPLEXES

GERD MEYER, MUHAMET SEHABI, AND INGO PANTENBURG
Institut für Anorganische Chemie, Universität zu Köln, Köln, Germany

1.1 INTRODUCTION

Monovalent silver, Ag^+, is a fifth-period closed-shell d^{10}-ion. It is therefore often considered a pseudo-alkali-metal cation with an ionic radius close to that of Na^+, 114 versus 113 pm for coordination number (CN) 4 [1]. Indeed, AgCl crystallizes with an NaCl type of structure, with $a = 554.9$ versus $a = 563.9$ pm [2]. On the other hand, many physical properties are quite different. One striking example is that of AgCl and NaCl solubilities in water, 1.88×10^{-3} g/L versus 358 g/L [3]. The solubility of AgCl is enhanced dramatically through the addition of aqueous ammonia, and linear $[Ag(NH_3)_2]^+$ cations are formed. Quite obviously, there is a much larger affinity of Ag^+ toward the N-donor ligand ammonia than toward water or chloride as competing ligands.

There are many complex salts of the $[Ag(NH_3)_2]^+$ cation with common inorganic cations, NO_3^- or ClO_4^- (see, e.g., ref. 4). In $[Ag(NH_3)_2](ClO_4)$, the $[H_3N\text{–}Ag\text{–}NH_3]^+$ complex cation is linear by symmetry in both the high- and low-temperature modifications [4d]. The Ag–N distances are 214 pm, on average, at 170 K. The cations are arranged such that the shortest Ag–Ag contacts are 302 pm, well above the Ag–Ag distance of 288.9 pm in metallic silver but below the sum of the

Design and Construction of Coordination Polymers, Edited by Mao-Chun Hong and Ling Chen

van der Waals radii of 344 pm. However, *argentophilicity* [5] (i.e., d^{10}–d^{10} bonding interactions at a level of weak hydrogen bonds) may be associated with these short distances. Argentophilicity appears to be a slightly smaller effect than *aurophilicity*, judging by the Ag^+–Ag^+ and Au^+–Au^+ distances in the isostructural compounds [**Ag**$(NH_3)_2$](ClO_4) [4d] and [**Au**$(NH_3)_2$](ClO_4) [6] at the same temperature (170 K): 302.0(2) and 299.0(1) pm, respectively. Much shorter Ag–Ag distances may be seen in constrained systems, of which the dimeric structure of the simple silver acetate is perhaps the most spectacular example, with d(Ag–Ag) = 279.4(4) to 280.9(3) pm [7].

Hg^{2+} as a diagonally related d^{10}-ion forms analogous linear complexes, $[Hg(NH_3)_2]^{2+}$, with Hg–N distances of 207.2(16) pm in $[Hg(NH_3)_2][HgCl_3]_2$ [8], shorter than d(Ag–N) = 212.9(11) to 216.0(12) pm in $[Ag(NH_3)_2](ClO_4)$ [4d] but larger than d(Au–N) = 205.2(2) in $[Au(NH_3)_2](ClO_4)$ [6]. Mercuriphilic effects have not been observed, perhaps due to the higher charge. On the other hand, relativistic effects are, for Hg^{II} as for Au^I, much more pronounced than for Ag^I. One evidence of these effects in mercuric chemistry is the pronounced preference for linear two-coordinate complexes [9], also termed as *characteristic coordination number* (CCN) 2 [10].

As relativistic effects are much less important in silver chemistry, Ag^I exhibits a much larger coordinative flexibility, apparent in coordination numbers between 2 and 6 and in typical closed-shell ion coordination polyhedra (as closed as possible). Other than the hard (Pearson acid) alkali-metal ions, the Ag^+ ion is much more polarizable: thus is a much softer Pearson acid with a higher tendency to coordinate to softer Pearson bases, hence with higher covalent bonding contributions.

In this chapter we report on a number of recently discovered Ag^I complexes with multi-N donor ligands [11] but do not consider this work a comprehensive review. The fact that silver coordination chemistry [12] is presently a rather hot topic may also be seen from a series of leading-edge research papers which have recently been published in the *Australian Journal of Chemistry* [13]. Silver complexes may have a number of functionalities; they may conduct electric current or luminesce, or they may have antimicrobial activity [14], to name only two.

As ammonia (NH_3) is the parent of all N-donor ligands, Ag–N distances as short as, say, 210 pm are considered the landmark for the strongest $Ag^+ \leftarrow N$ interactions ("bonds") possible in $[Ag^I \leftarrow L1]L2$ coordination compounds. L1 is usually a neutral N-donor ligand and L2 is an auxiliary ligand (co-ligand) with a negative charge competing with the L1 ligand for space in the coordination sphere of Ag^+.

1.2 LIGANDS L1 WITH 1,2 N-DONOR FUNCTIONS

1,2-Pyrazole has two directly neighboring nitrogen functions. As competing L2 ligands, triangular (NO_3^-), tetrahedral (BF_4^-, ClO_4^-, SO_4^{2-}), and octahedral anions (PF_6^-) as well as trifluoroacetate ($CF_3COO^- = Tfa^-$) were attempted. In all of these, pyrazole coordinates as a neutral ligand with N2, which bears no hydrogen atom.

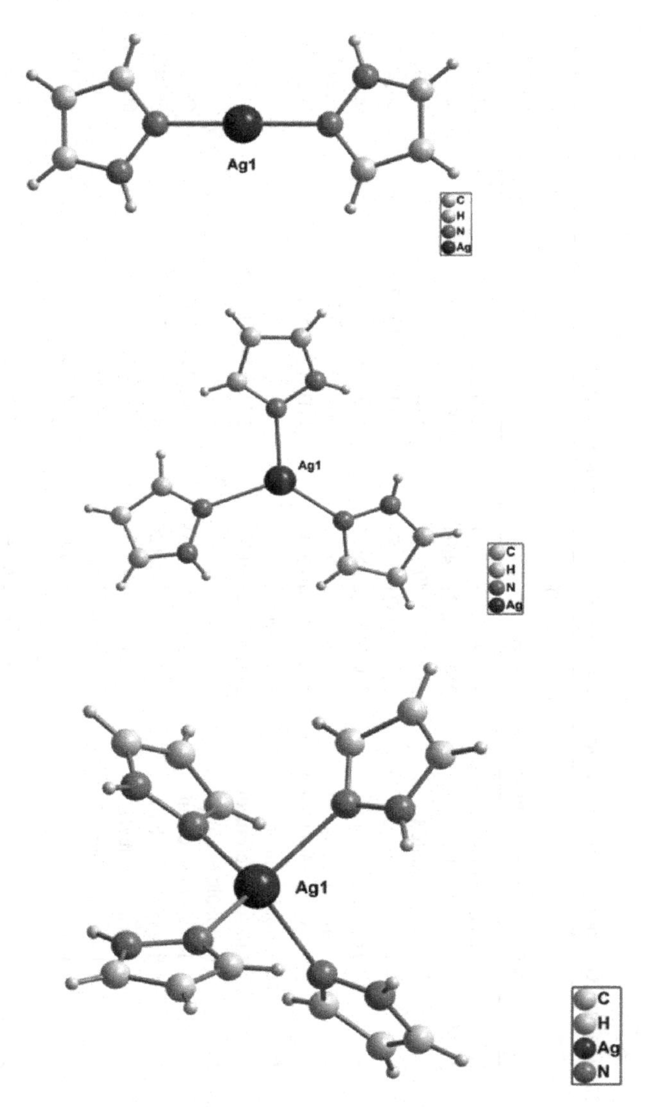

FIGURE 1.1 Linear, triangular, and tetrahedral environments of Ag^+ in $[Ag(Pyz)_2](BF_4)$ (**1**), $[Ag(Pyz)_3](Tfa)$ (**5**), and $[Ag(Pyz)_4]_2(SO_4)$ (**6**).

In $[Ag(Pyz)_2](BF_4)$ (**1**), the isotypic $[Ag(Pyz)_2](ClO_4)$ (**2**), and in $Ag(Pyz)(NO_3)$ (**3**), we observe truly linear N–Ag–N environments (Fig. 1.1), whereas in $[Ag(Pyz)_2]_2(PF_6)_2(Pyz)$ (**4**) the N–Ag–N angle is bent by a few degrees (see Table 1.1). Cations and anions seem to be independent of each other in **1**, **2**, and **4**, which all have 1 : 2 : 1 (Ag : L1 : L2) stoichiometries. In the nitrate (**3**), with its unusual 1 : 1 : 1

TABLE 1.1 Coordination Numbers and Geometry, Distances, and Angles of the Ag^{I}–N Donor Complexes

Compound	CN	Coordination Geometry	d(Ag–N)/pm, *d(Ag–O)/pm*	<(N–Ag–N)/°	CCDC No.
$[Ag(Pyz)_2](BF_4)$ (**1**)	2	Linear, isolated	212.0(3), 213.1(3)	180	662474
$[Ag(Pyz)_2](ClO_4)$ (**2**)	2	Linear, isolated	214.4(7), 216.4(7)	180	662475
$Ag(Pyz)(NO_3)$ (**3**)	2	Linear, isolated	212.6(3), *240.3(4)*	180	662476
$[Ag(Pyz)_2](PF_6)(Pyz)$ (**4**)	2	Linear, isolated	211.7(5)–212.8(5)	174.7(2), 177.3(2)	662477
$[Ag(Pyz)_3](Tfa)$ (**5**)	3	Trigonal, isolated	220.1(3)–227.5(3)	107.4(1)–142.7(1)	662478
$[Ag(Pyz)_4]_2(SO_4)$ (**6**)	4	Tetrahedral, isolated	223.9(1)–239.9(1)	96.5(3)–125.5(3)	662479
Ag(Dcp)(OAc) (**7**)	2 4	Linear + tetrahedral, double chain	218.4(3), 249.2(4), *227.6(3)*	170.1(2), 95.0(1)	662480
Ag(Dcp)(Tfa) (**8**)	2 4	Linear + tetrahedral, double chain	218.2(4), 244.1(4), *228.9(4)*	172.2(2), 93.8(2)	662481
$Ag(Dcp)_2(BF_4)$ (**9**)	4	Tetrahedral, chain	233.3(3), 250.7(3)	85.2(2)–137.6(1)	662482
$Ag(Dcp)_2(ClO_4)$ (**10**)	4	Tetrahedral, chain	233.4(2), 251.2(2)	83.3(1)–137.6(1)	662483
$[Ag(Mel)_2](BF_4)$ (**11**)	2	Linear, isolated	215.4(3), 216.8(3)	170.7(1)	662484
$[Ag(Mel)_2](BF_4)(Mel)_2$ (**12**)	2	Linear, isolated	213.1(2), 213.6(2)	176.5(1)	662485
$Ag(Mel)_2(Tfa)(H_2O)$ (**13**)	4 + 1	Pyramidal, chain	229.7(6), 232.8(6), *245.4(6)*	172.2(2)	662486
$Ag(Mel)(NO_3)$ (**14**)	2 + 2	Bent, chain	225.3(5), 229.0(4), *257.6(6)*	126.6(2)	662487
$[Ag(Dpt)_2(NO_3)]_2$ (**15**)	2 + 2	Bent, dimer	220.6(4), 221.9(6), *282.6(4)*	145.4(2)	662488

$[Ag(Dpt)_2(OAc)]_2$ (**16**)	2 + 2	Tetrahedral, dimer	230.1(4), 235.3(5), *237.4(3)*	122.8(2)	662489
$Ag(Dpt)(H_2O)(ClO_4)$ (**17**)	2 + 1	Bent, chain	228.2(5), 240.2(5), *240.1(7)*	111.2(2)	662490
$Ag(Bpy)_2(ClO_4)$ (**18**)	4	Square, monomer	227.6(3)–243.0(3)	71.4(1)–159.2(1)	662491
$Ag(Bpy)(Tfa)$ (**19**)	2 + 1	Trigonal, weak dimer	229.7(4), 232.2(4), *218.2(3)*	72.0(1)	662492
$Ag(Bpq)_2(Tfa)$ (**20**)	4 + 2	Trigonal prism, monomer	240.5(12)–248.7(13), *233.9(10)–267.8(13)*	68.2(4)–153.0(4)	662493
$Ag_2(Bpq)(NO_3)_2$ (**21**)	1 + 3	Chain	227.9(3), *238.2(2)–252.2(2)*	70.3(1)	662494
	2 + 2		236.4(3), 243.5(2), *240.5(2), 242.0(2)*		
$Ag(Tpt)(NO_3)$ (**22**)	3 + 1	Winding chain	242.7(4)–261.3(4), 236.1(4)	63.9(1)–136.0(2)	662495
$Ag_4(Tpt)_2(Tfa)_4(H_2O)$ (**23**)	3 + 2	Tetramers, stairs	238.6(7)–253.1(9), *225.6(7), 276.0(2)*	65.9(3)–133.8(2)	662496
	2 + 2		222.3(7), 264.7(5), *220.9(6), 274.0(3)*		
$[Ag(Pip)](BF_4)$ (**24**)	2	Linear, isolated	216.6(7), 217.4(9)	174.1(2)	662497
$[Ag_2(Pip)_3](Tfa)_2(H_2O)_6$ (**25**)	3	Trigonal, layer	226.3(2), 229.1(2), 239.2(2)	104.6(1), 116.9(1), 137.6(1)	662498
$[Ag(Pip)_2](NO_3)$ (**26**)	4	Tetrahedral, waved layer	238.6(4), 239.4(4), 245.2(4), 245.6(5)	89.9(2)–116.6(2)	662499

stoichiometry, however, there are two symmetrically and functionally independent Ag^+ ions. One is, as above in **1**, **2**, and **4**, incorporated in the linear $[Ag(Pyz)_2]^+$ cation. The second Ag^+ ion has four contacts to C3 and C4 of pyrazole with d(Ag–C) = 249.4(4) and 281.8(5) pm and to two oxygen atoms of two nitrate ions at d(Ag–O) = 240.3(4) pm, such that a chain of the stoichiometry $Ag(Pyz)_2Ag(NO_3)_2$ is created (see Fig. 1.2). $[Ag(Pyz)_3](Tfa)$ (**5**) and $[Ag(Pyz)_4]_2(SO_4)$ (**6**) exhibit triangular and tetrahedral environments, respectively, as the 1 : 3 and 1 : 4 stoichiometries of

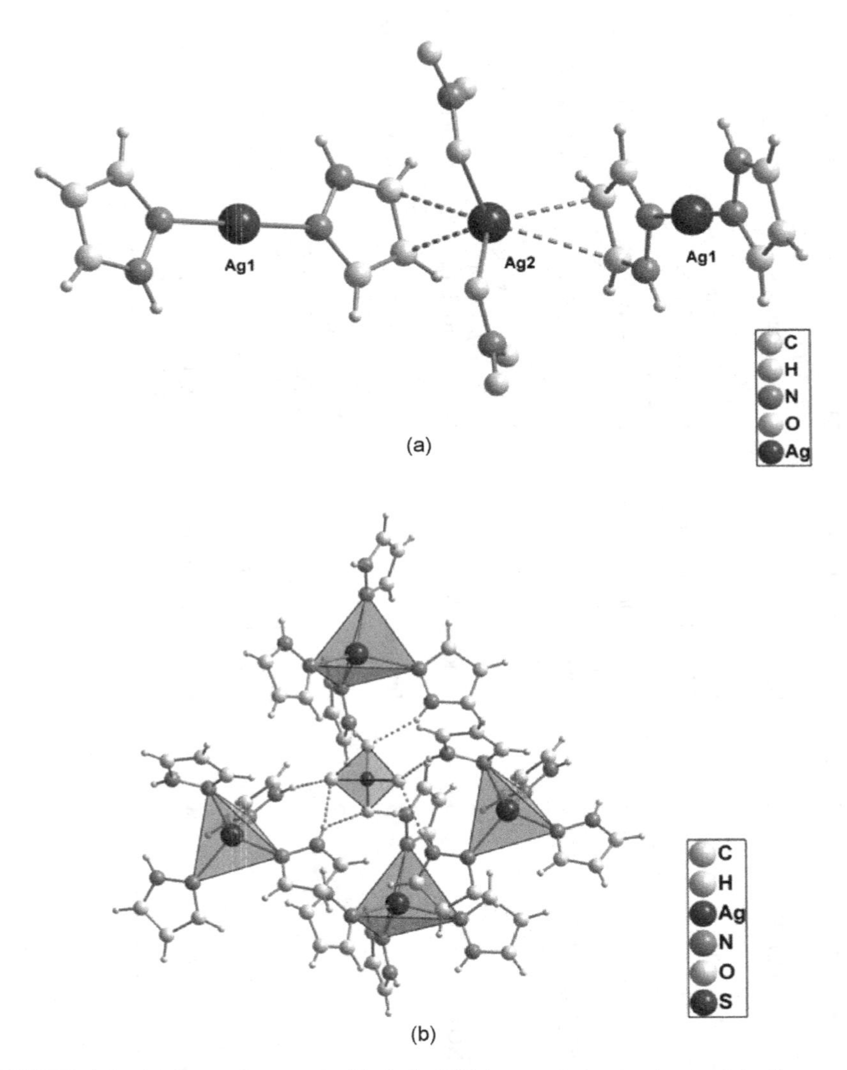

FIGURE 1.2 (a) Part of a $\cdots Ag(Pyz)_2Ag(NO_3)_2\cdots$ chain as observed in the crystal structure of $Ag(Pyz)(NO_3)$ (**3**); (b) the surrounding of $(SO_4)^{2-}$ by four $[Ag(Pyz)_4]^+$ cations stabilized by N–H$\cdots$O hydrogen bonding in the crystal structure of $[Ag(Pyz)_4]_2(SO_4)$ (**6**).

Ag : L1 suggest (see Fig. 1.1). Distances increase steadily with coordination numbers (see Table 1.1). Angles in the AgN_3 triangles and AgN_4 tetrahedra of **5** and **6** cover a rather wide range (Table 1.1); hence, these coordination polyhedra are rather distorted. The sulfate ion in **6** is surrounded by four $[Ag(Pyz)_4]^+$ cations and appears to form hydrogen bonds to N–H hydrogen atoms with d(N–H · · · O) between 193 and 207 pm (see Fig. 1.2).

1.3 LIGANDS L1 WITH 1,3 N-DONOR FUNCTIONS

Although pyrazole has two possible nitrogen functions, –N| and –N–H, the free pair of electrons of the –N| function binds only to Ag^+ with distances dependent on the coordination number.

Pyrimidine (1,3-diazine, Pym) has two –N| functions in the 1,3 positions. Both functions are used to form coordination polymers in the pyrimidine derivative 2,4-diamino-6-chloropyrimidine (Dcp) when acetate (OAc^-) or trifluoroacetate (Tfa^-) are the competing L2 ligands [15]. As chelating anions, these are incorporated as bridging ligands in the essentially isostructural compounds Ag(Dcp)(OAc) (**7**) and Ag(Dcp)(Tfa) (**8**). There are two crystallographically and functionally independent Ag^+ cations in the structures, both of which are coordinated by two –N| functions of Dcp. These –N| functions are different, as one (N3) is neighboring the C4–Cl function and the other one is opposite it (N1). Ag1 is almost linearly coordinated by two N3 atoms at d(Ag–N) = 218.4(4) pm and <(N–Ag–N) = 170.08(2)°. Ag2 has two N1 atoms as neighbors, although at a distance of 249.1(4) pm with two oxygen atoms of two acetate anions at 227.5(3) pm; N–Ag2–N and O–Ag2–O are 94.95(9)° and 154.34(1)°, respectively. Thus, dimers $Ag_2(Dcp)_4(OAc)_2$ are formed with rather close Ag–Ag contacts of 321.33(8) pm. These dimers can, however, not be isolated [note that the stoichiometry is Ag(Dcp)(OAc)!]. Rather, they are bridged via N1 and N3 functions to zigzag chains (see Fig. 1.3).

In Ag(Dcp)(OAc) (**7**) the amino groups are not involved in the coordination sphere of Ag^I. This is, however, the case in the isostructural compounds $Ag(Dcp)_2(BF_4)$ (**9**) and $Ag(Dcp)_2(ClO_4)$ (**10**). Each Ag^+ has two nitrogen atoms of pyrimidine functions (N3) and two amino-nitrogen atoms (N2′) as nearest neighbors at distances of, for **9**, 233.3(3) and 250.7(3) pm, respectively. This leads to a chain with Ag–Ag distances of 485.8(1) pm (see Fig. 1.3).

Melamine (2,4,6-triamino-1,2,3-triazine; Mel) is a highly symmetrical ligand with three –N| and three $-NH_2$ functions. There are a small but increasing number of transition metal–melamine complexes. A molecular, zwitterionic complex has been found with divalent mercury in $[MelH^+HgCl_3^-](Mel)$ [16]. The Hg–N separation in this molecule is 233.6(6) pm, pretty close to that in the Hg^{II}–pyridine complex $Hg(Pyr)_2Cl_2$ [9,17].

With Ag^I, two sorts of melamine complexes have been observed with a variety of competing ligands, with CN 2 and higher, respectively [18]. In $Ag(Mel)_2(BF_4)$ (**11**) and in the isostructural $Ag(Mel)_2(ClO_4)$ (**11a**) [19], melamine also acts as a monodentate μ_1 ligand with one of the three –N| functions forming somewhat bent

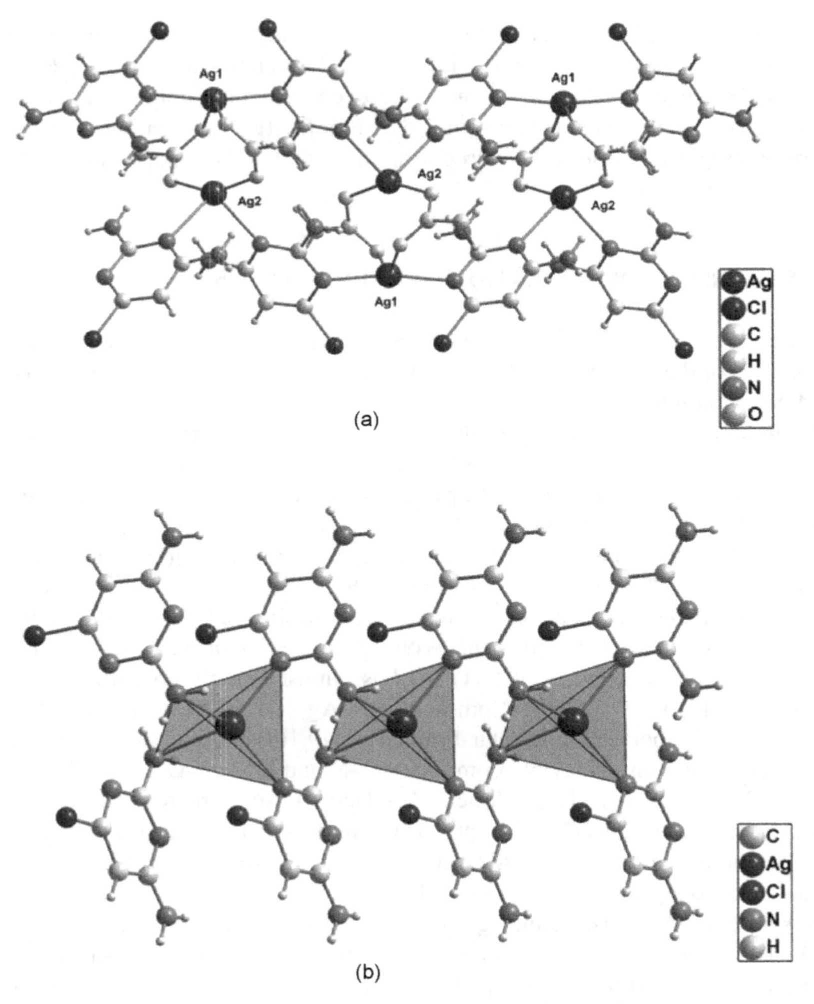

FIGURE 1.3 (a) Part of the $Ag_2(Dcp)_{4/2}(OAc)_2$ chain as observed in the crystal structure of Ag(Dcp)(OAc) (**7**) and analogously in the crystal structure of Ag(Dcp)(Tfa) (**8**); (b) part of the $[Ag(Dcp)_2]^+$ chain as observed in the crystal structures $Ag(Dcp)_2(BF_4)$ (**9**) and $Ag(Dcp)_2(ClO_4)$ (**10**).

cations with Ag–N distances of 215.4(3) pm and an N–Ag–N angle of 170.7(1)°. This angle might be due to a weak attraction to one fluorine atom of the $(BF_4)^-$ anion, d(Ag–F) = 284.4(3) pm (Fig. 1.4). An analogous almost linear cation $[Ag(Mel)_2]^+$ with even slightly smaller Ag–N distances [213.1(2), 213.6(2) pm] is seen in $[Ag(Mel)_2](BF_4)(Mel)_2$ (**12**). Two solvent molecules of melamine are very loosely attached to the $[Ag(Mel)_2]^+$ cations with amino functions at distances of 287.3(3) and 302.0(3) pm (see Fig. 1.4).

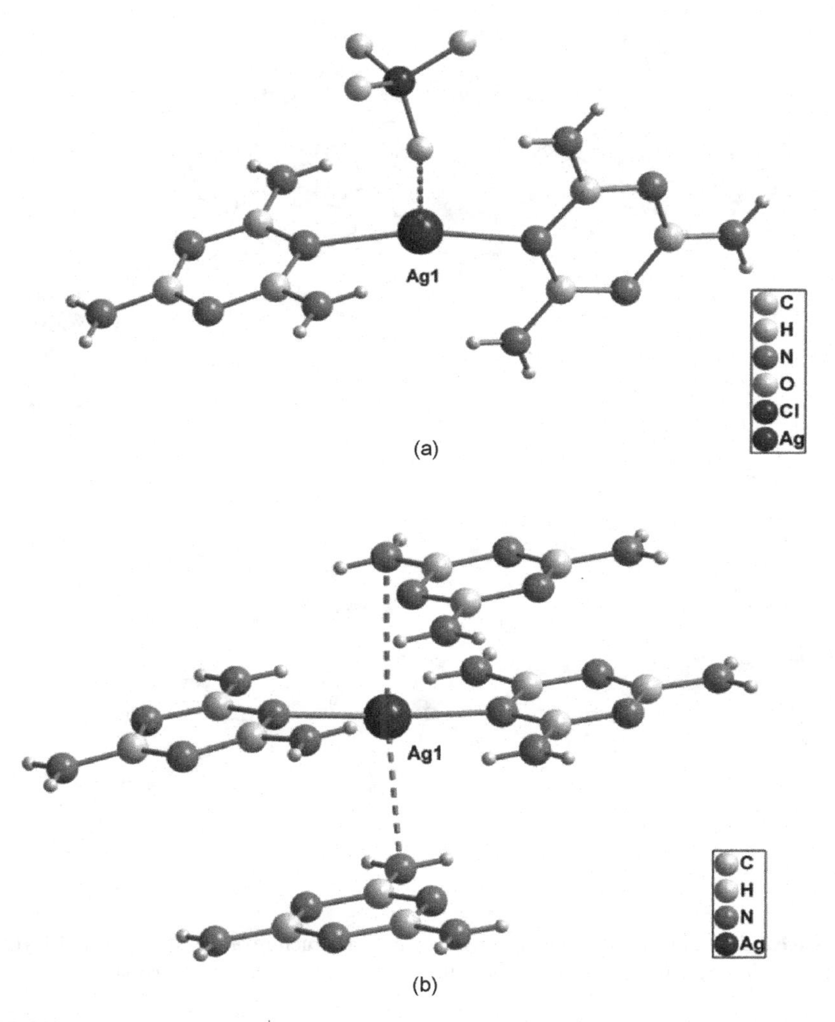

FIGURE 1.4 (a) $[Ag(Mel)_2]^+$ cations to which X = $(BF_4)^-/(ClO_4)^-$ are loosely attached in the crystal structure of $[Ag(Mel)_2]X$ (**11**); (b) the $[Ag(Mel)_2]^+$ cation to which two melamine molecules are loosely attached in $[Ag(Mel)_2](BF_4)(Mel)_2$ (**12**).

With the somewhat stronger ligands trifluoroacetate and nitrate, chain structures are formed with melamine acting as a bridging μ_2 ligand. The coordination environment around Ag^I in $Ag(Mel)_2(Tfa)(H_2O)$ (**13**) is a (distorted) square pyramid with four –N| functions of four melamine molecules forming the base at Ag–N distances of 229.7(6) and 232.8(6) pm and an apical oxygen atom of one Tfa^- anion at d(Ag–O) = 245.4(6) pm. The $Ag(Mel)_4(Tfa)$ units bridge via neighboring –N| functions to chains with the AgN_4O pyramids directed alternately up and down

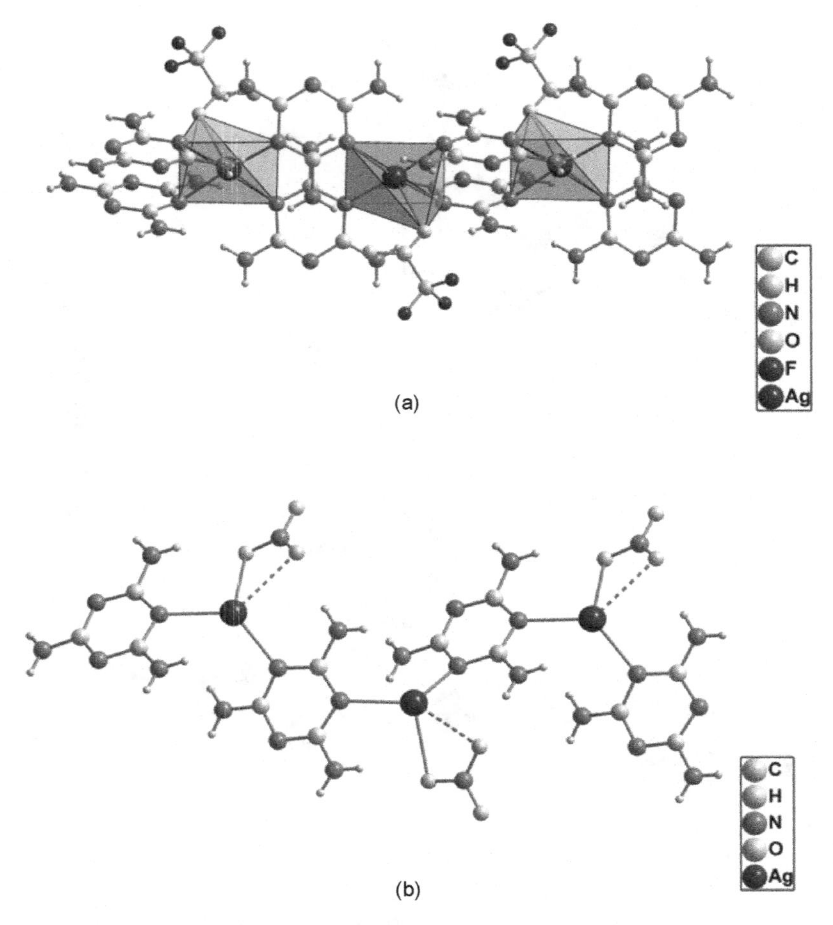

FIGURE 1.5 (a) $Ag(Mel)_{4/2}(Tfa)$ chains in the crystal structure of $Ag(Mel)_2(Tfa)(H_2O)$ (**13**); (b) part of the $Ag(Mel)_{2/2}(NO_3)$ chains in the crystal structure of $Ag(Mel)(NO_3)$ (**14**).

(Fig. 1.5). The nitrate ligand influences the coordination sphere around Ag^I much more than does trifluoroacetate. This is documented more clearly by the much more bent N–Ag–N part [126.6(2)°] than by the Ag–N distances (see Table 1.1 and ref. 20). The 1 : 1 : 1 stoichiometry also affords a chain structure with neighboring –N| functions of melamine ligand bridging (see Fig. 1.5).

A similar μ_2 bridging coordination mode was recently found in $Cu_2(Mel)Br_2$. In $Cu_3(Mel)Cl_3$ melamine even acts as a μ_3 bridging ligand using all three –N| functions [21]. Hydrogen bonding between the amino-H atoms and the halogen atoms is made responsible for the formation of two- and three-dimensionally extended structures.

In the ligand 2,4-diamino-6-phenyl-1,3,5-triazine (Dpt), one amino ligand of melamine is substituted by a phenyl ring. With nitrate as L2 and the stoichiometry

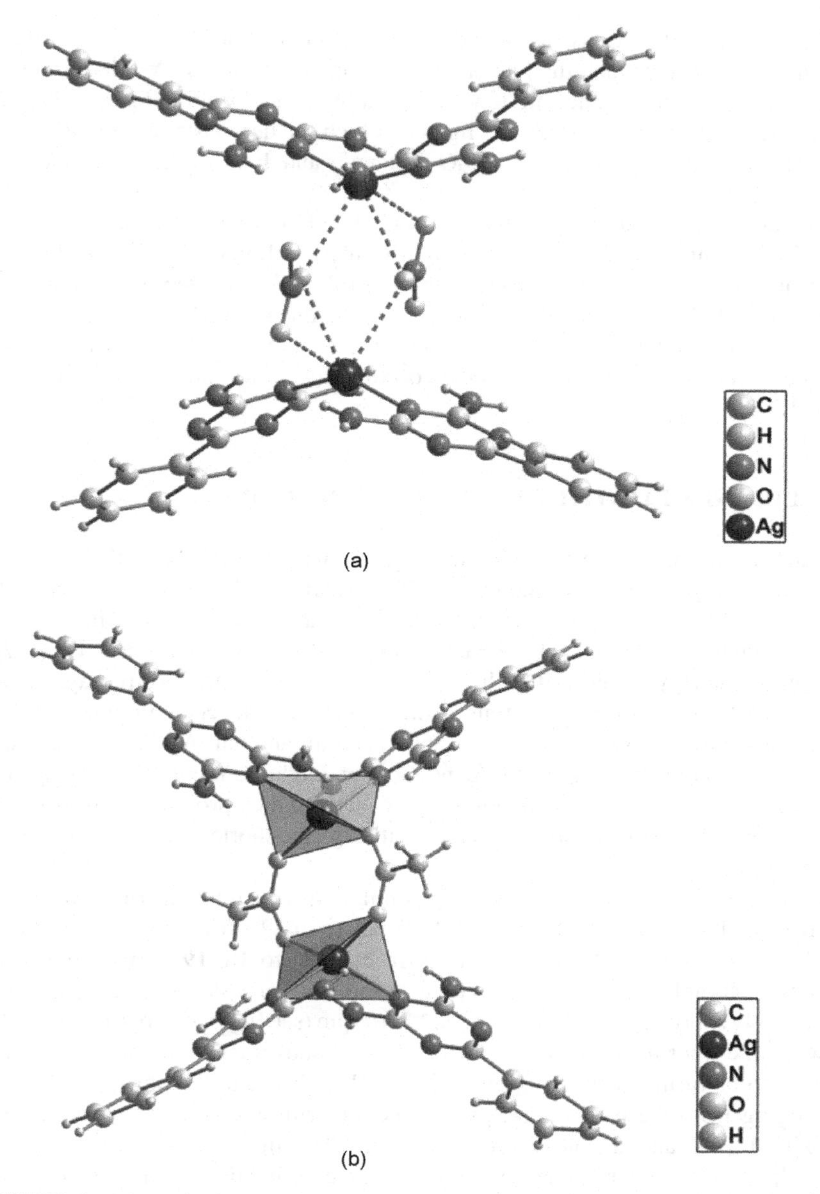

FIGURE 1.6 Dimers in the crystal structures of (a) $[Ag(Dpt)_2(NO_3)]_2$ (**15**) and (b) $[Ag(Dpt)_2(OAc)]_2$ (**16**).

1 : 2 : 1, a dinuclear complex $[Ag(Dpt)_2(NO_3)]_2$ is formed in (**15**) with bent $[Ag(Dpt)_2]^+$ cations [<(N–Ag–N) = 145.4(2)°] and still rather short Ag–N distances of 220.6(4) and 221.9(6) pm, respectively. The cations are bridged by two nitrate ions with rather long Ag–O distances of 282.2 (14) pm (Fig. 1.6).

Acetate also functions as a bridging ligand in the crystal structure of $Ag(Dpt)_2(OAc)$ (**16**), such that, again, dimeric units are formed. While nitrate acts as a monodentate ligand, acetate functions as a bidentate bridging ligand (Fig. 1.6). Acetate acts as a much stronger ligand, which is demonstrated by the much smaller N–Ag–N angle, shorter Ag–O distance, and longer Ag–N distance (see Table 1.1).

The crystal structure of $Ag(Dpt)(H_2O)(ClO_4)$ (**17**) is special in that Ag^+ has a coordination number of 1 + 1 + 1 with bridging Dpt ligands forming a chain and one water molecule coordinating with the central Ag^+ ion. One Ag–N distance is rather short [228.2(5) pm]; the other is equal to the $Ag–OH_2$ distance [240.1(7) pm]. The chains run down the *b*-axis and are arranged such that layers appear between which the ClO_4^- ions and the coordinating water molecules are included (Fig. 1.7).

1.4 LIGANDS L1 WITH 1,4 N-DONOR FUNCTIONS

2,2′-Bipyridine (Bpy) is one of simplest ligands, with a *cisoid* N–C–C–N function; hence it usually acts as a bidentate ligand. In $Ag(Bpy)_2(ClO_4)$ (**18**) [22], Ag^+ has an unusually squarelike environment with rather strong deviations of the N–Ag–N angles from 90° (see Table 1.1), subject mainly to the bite distance, d(N–N) = 271.4 (13) pm in the Bpy ligand and to the torsion angle between the two pyridyl rings of 12.6(5)°. The $[Ag(Bpy)_2]^+$ cations form a chain in the crystallographic [0 1 0] direction with Ag–Ag distances of 366.6(1) pm and an Ag–Ag–Ag angle of 152.4(1)°, where some π–π stacking interactions might be involved [23] (Fig. 1.8). The $(ClO_4)^-$ anions act as "noncoordinating" anions; they provide electroneutrality and are not really involved in coordination with Ag^I, the shortest Ag–O distance being 496.8(5) pm.

In Ag(Bpy)Tfa (**19**) [22], a very similar five-membered ring to that of $Ag(Bpy)_2(ClO_4)$ is observed, with Ag–N distances of 229.7(4) and 232.2(4) pm and an N–Ag–N angle of 72.0(1)°. However, in contrast to **18**, **19** forms a molecular, triangular complex Ag(Bpy)Tfa. These complexes are stacked to dimers with surprisingly short Ag–Ag distances of 303.73(9) pm (see Fig. 1.8). Hydrogen bonding between the fluorine atoms of the trifluormethyl substituents and phenyl-H atoms appears to be further stabilizing, with d(F···H) = 294 and 299 pm.

Judging from these two Ag–Bpy complexes, with a very weakly coordinating $(ClO_4)^-$ ligand and a considerably stronger Tfa^- ligand competing with 2,2′-bipyridine, Ag^I appears to prefer rather small coordination numbers with rather strong Ag–N bonding interactions. Recently, we have also been investigating Mn^{II} complexes with bypyridine and a variety of other N-donor ligands [24]. In all of these, Mn^{II} prefers an octahedral coordination environment: for example, in the simple $Mn(Bpy)_2$*cis*-Br_2, with Mn–N distances around 230 pm. With a 1 : 1 : 2 stoichiometry in $Mn(Bpy)Br_2$, zigzag chains of edge-sharing MnN_2Br_4 octahedra are formed with d(Mn–N) = 226.3(3) pm. However, despite the different coordination numbers, the

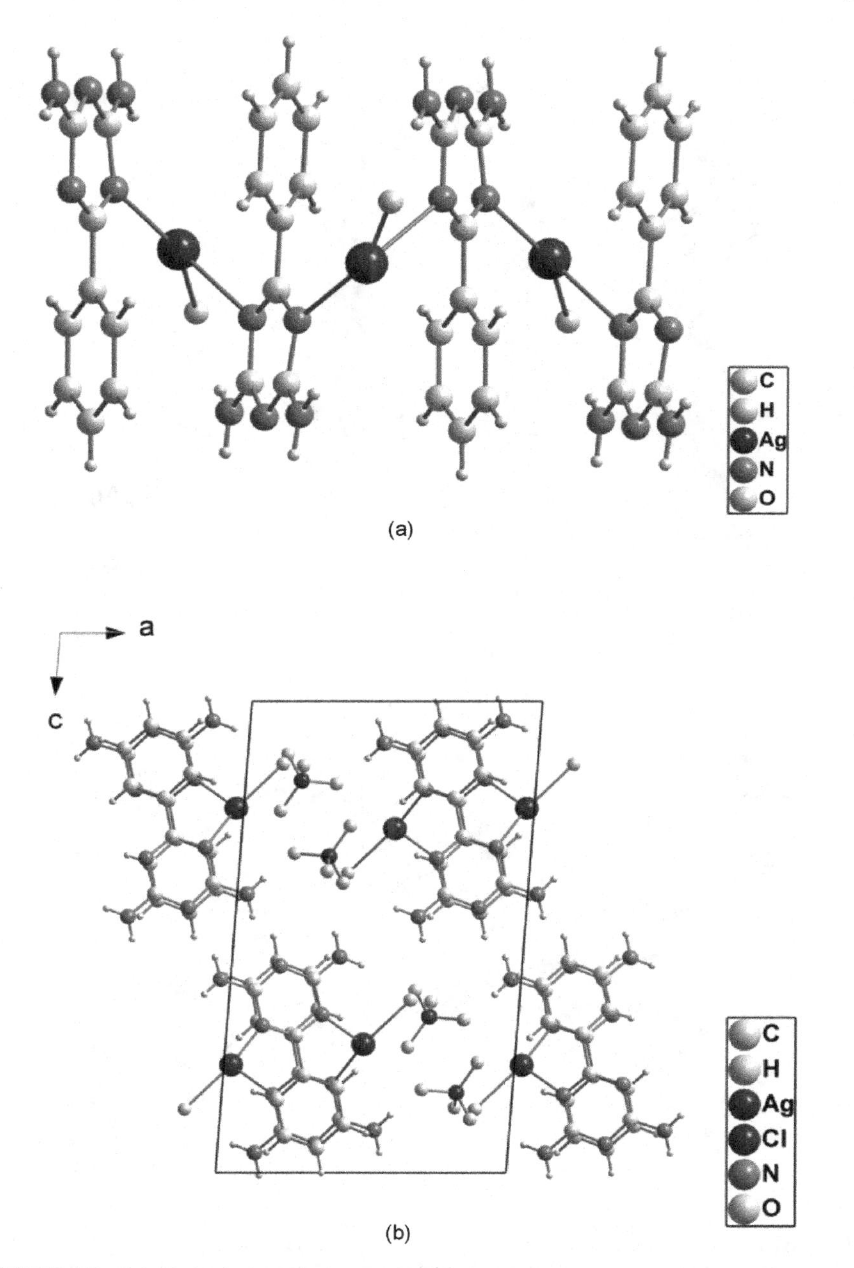

FIGURE 1.7 (a) Cationic $[Ag(Dpt)_{2/2}(H_2O)]^+$ chains in the crystal structure of Ag(Dpt) $(H_2O)(ClO_4)$ (**17**); (b) a projection of the structure onto (010).

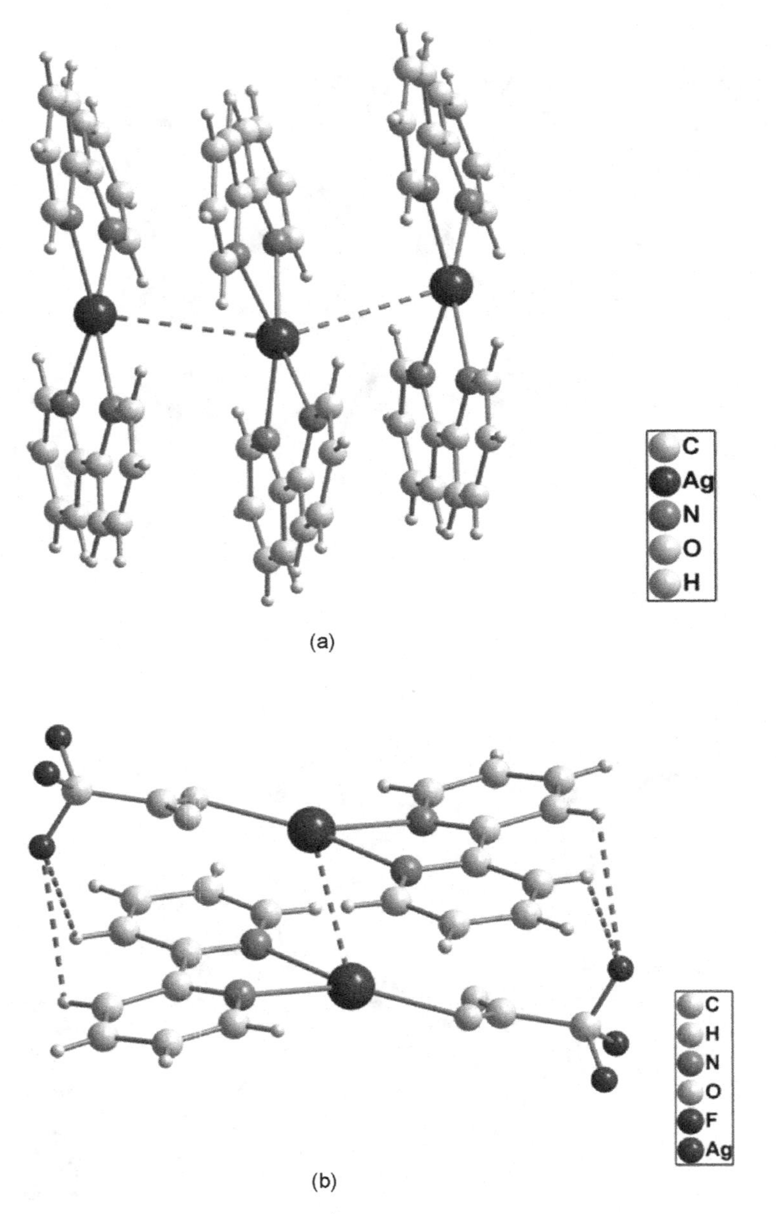

FIGURE 1.8 (a) Cationic $[Ag(Bpy)_2]^+$ chains in the crystal structure of $Ag(Bpy)_2(ClO_4)$ (**18**); (b) connection between Ag(Bpy)Tfa molecules and dimers in Ag(Bpy)Tfa (**19**).

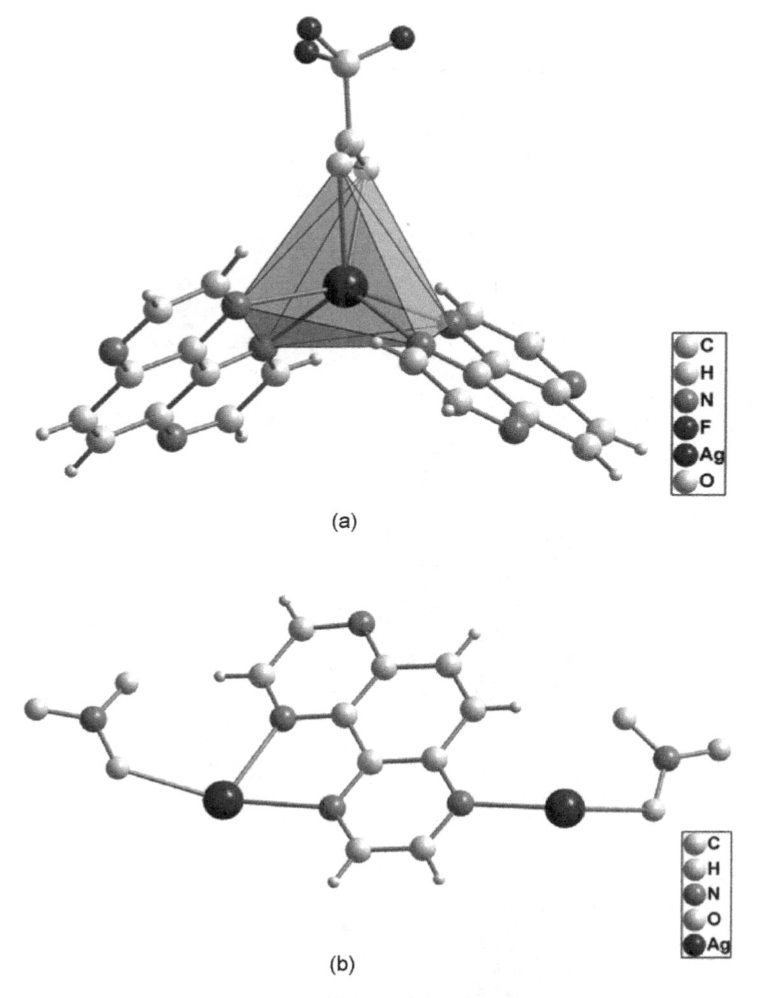

FIGURE 1.9 (a) The complex molecule $Ag(Bpq)_2(Tfa)$ in **20**; (b) the connection of Ag^I via bridging Bpq ligands in $Ag_2(Bpq)(NO_3)_2$ (**21**).

very similar Ag–N and Mn–N distances in the five-membered C–N–M–N–C rings (M = Ag, Mn) attest to the fact that it is mainly the ring geometry and the similar ionic radii (Ag^+ : 129, Mn^{2+} : 97 pm for CN 6 [1]) that is responsible for the respective M–N distances.

Another ligand with two –N| functions in the 1,4 positions as in 2,2′-bipyridine and with two further –N| functions in the pyrazine rings is bis(pyrazino)[2,3-*f*]quinoxaline (Bpq). In $Ag(Bpq)_2(Tfa)$ (**20**), two Bpq ligands and the two oxygen atoms of the Tfa^- carboxylate group build up a distorted trigonal prism (Fig. 1.9a). These complex

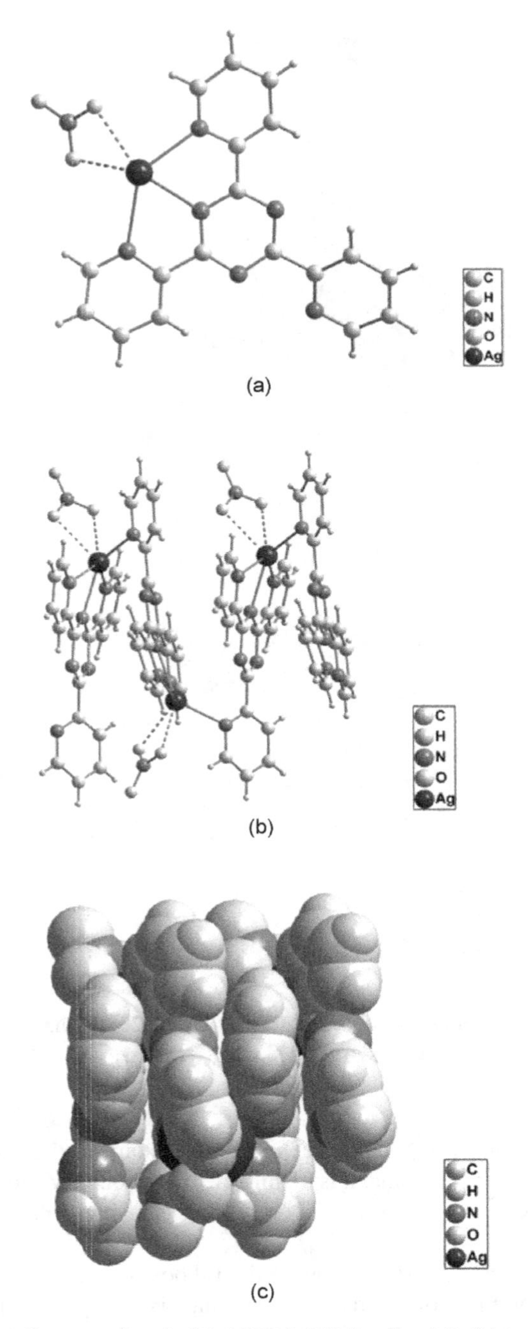

FIGURE 1.10 (a) One complex Ag(Tpt)(NO_3) (**22**) "molecule"; (b) its connection through pyridyl-N atoms to a winding chain, with (c) rather effective space filling.

molecules are arranged such that, again, hydrogen bonds between the phenyl-H and the trifluormethyl-F atoms stabilize the three-dimensional structure.

With nitrate as the auxiliary ligand, Bpq acts as a bridging ligand in $Ag_2(Bpq)(NO_3)_2$ (**21**), in contrast to $Ag(Bpq)_2(NO_3)$ [25], where molecules of this composition are formed. In **21**, one type of Ag^I is coordinated by Bpq in the same manner as in **20**, with two oxygen atoms of nitrate adding to the coordination sphere, while the second type of Ag^I is coordinated by an opposite pyrazine-N with a rather short Ag–N bond [227.9(3) pm]. Three oxygen atoms of nitrate ions add to a heavily distorted tetrahedral coordination. The $Ag_2(Bpq)(NO_3)_2$ chains (Fig. 1.9b) are connected further to layers via nitrate-O.

2,4,6-Tris-(2-pyridyl)-1,3,5-triazine (Tpt) is a flexible 6N ligand with three 1,4-bipyridyl functions and the ability to act as a tridentate ligand. This is observed in $Ag(Tpt)(NO_3)$ (**22**), where nitrate coordinates bidentately, although with rather long Ag–O distances. More important, Ag^I is also coordinated by one neighboring complex with a pyridyl-N, with the shortest Ag–N distance in this compound, only 236.1(4) pm. Thus, through the bridging Tpt ligand, an unusual winding chain is created with rather efficient space filling (Fig. 1.10).

In $Ag_4(Tpt)_2(Tfa)_4(H_2O)$ (**23**), tetrameric complexes are formed with two functionally distinct Ag^I atoms. The two inner silver atoms are three-coordinate by Tpt in the same manner as in **22** and are connected by two bridging Tfa^- anions (Fig. 1.11). The two outer silver atoms are included in a bipyridyl-like five-membered ring and are further coordinated by water and Tfa^- anions, which bridge to further tetramers such that a staircase-like coordination polymer is created (Fig. 1.11).

Again, Ag^I shows a high level of coordinational flexibility in these Tpt complexes, in fair contrast to Mn^{II}. Here we always observe octahedral MnN_3L_3 environments with the Tpt ligand in meridional conformation [24]; for example, in $Mn(Tpt)Cl_2(H_2O)$ or in the trimeric $Mn_3(Tpt)_2(OAc)_6$ (see Fig. 1.12).

Piperazine (Pip) is a nonaromatic 1,4 *transoid* N–C–C–N ligand, as the chair conformation of Pip is more stable. It therefore also tends to form coordination polymers with Ag^I. With the weakly coordinating (BF_4^-) ion, zigzag chains are built in $[Ag(Pip)](BF_4)$ (**24**). Both opposite N–H functions of Pip coordinate to Ag^I, which exhibits strong, almost linear N–Ag–N bonding with d(Ag–N) = 216.6(7) and 217.4 (9) pm and <(N–Ag–N) = 174.1(2)° (Fig. 1.13). Tetrahedral coordination is seen in $[Ag(Pip)_2](BF_4)$ [26], with d(Ag–N) ranging from 231.6(8) to 243.9(6) pm and with the Pip molecules bridging to "undulated" layers.

Trigonal coordination is observed in $[Ag_2(Pip)_3](Tfa)_2(H_2O)_6$ (**25**), with Ag–N distances ranging from 226.3(2) to 239.2(2) pm and N–Ag–N angles between 104.62(9) and 137.58(9)°. As the composition Ag : Pip = 2 : 3 suggests, Pip acts as a μ_2 bridging ligand, and a layer structure with 6 + 6-membered rings is formed (Fig. 1.14). $(Tfa)^-$ and noncoordinating water molecules lie between these layers. The anhydrous compound $Ag_2(Pip)_4(Tfa)_2$ [27] is built from dimeric complexes of that composition, with the Tfa^- anions bridging with d(Ag–Ag) = 400.1(2) pm.

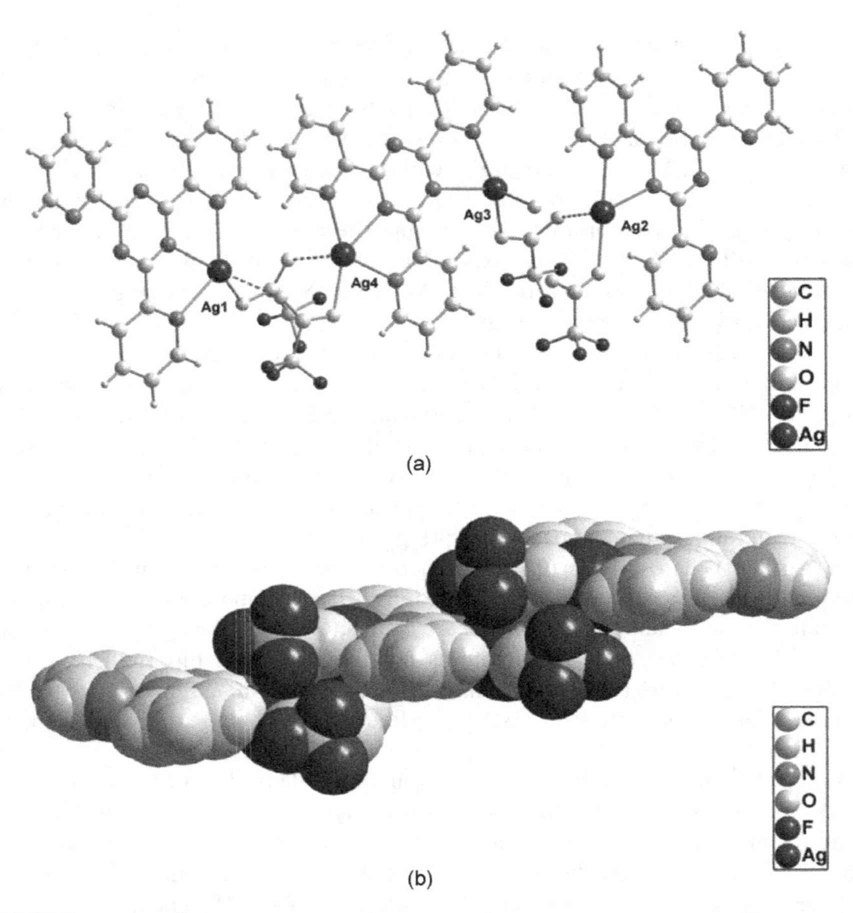

FIGURE 1.11 (a) The tetrameric complexes of $Ag_4(Tpt)_2(Tfa)_4(H_2O)$ (**23**); (b) their arrangement as a staircase coordination oligomer.

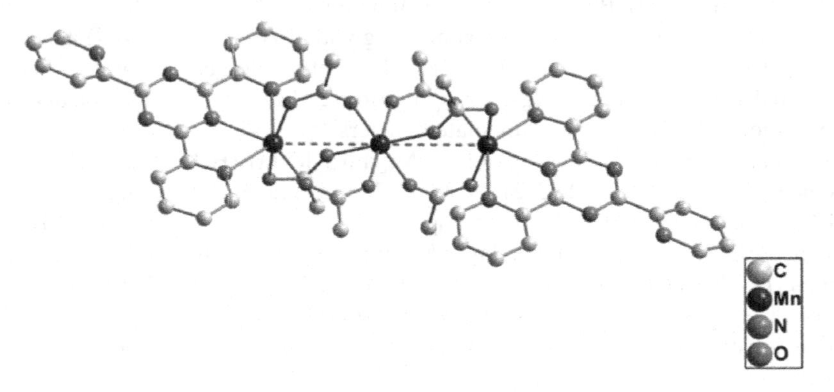

FIGURE 1.12 Acetate-bridged trimers in the crystal structure of $Mn_3(Tpt)_2(OAc)_6$.

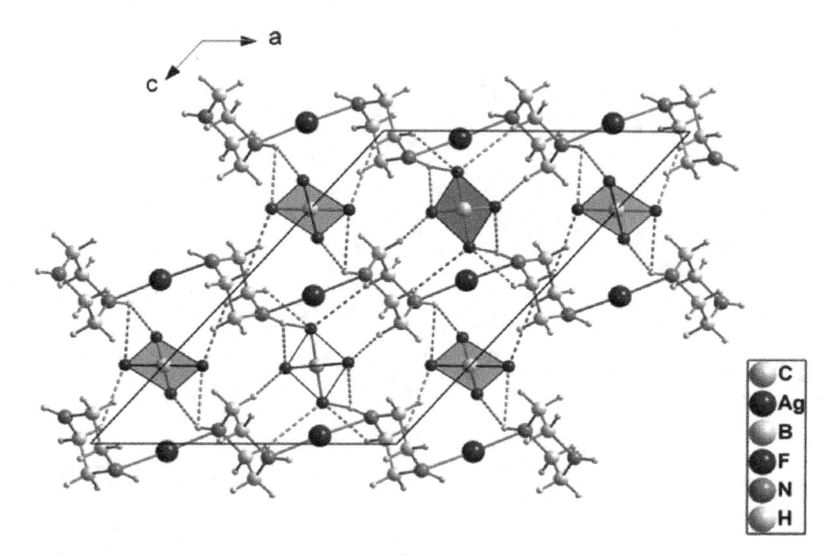

FIGURE 1.13 Projection of part of the crystal structure of $[Ag(Pip)](BF_4)$ (**24**) exhibiting the $[Ag(Pip)_{2/2}]^+$ chains and $(BF_4)^-$ with the hydrogen-bonding system shown.

In the crystal structure of $[Ag(Pip)_2](NO_3)$ (**26**), Ag^+ is tetrahedrally coordinated by piperazine molecules which produce heavily waved layers through μ_2 bridging (see Fig. 1.15). Their stacking in the third direction, [001], produces channels that are occupied by nitrate anions. Ag–N distances of 238.6(4)/239.4(4) and 245.2(4)/245.6(5) pm attest to rather strong bonding within the AgN_4 tetrahedra such that the nitrate anions cannot act as ligands in $[Ag(Pip)_2](NO_3)$, much the same as in $[Ag(Pip)](BF_4)$ and $[Ag_2(Pip)_3](Tfa)_2(H_2O)_6$.

1.5 CONCLUSIONS

In $[Ag^I \leftarrow L1]L2$ complexes, L1 as a (neutral) (multi-)N-donor ligand competes with an auxiliary negatively charged ligand (co-ligand) L2 for space in the coordination sphere of Ag^+. As a closed-shell d^{10}-cation, Ag^+ exhibits a rather strong tendency for small coordination numbers and low-dimensional structures. This tendency is, however, not as pronounced as in Hg^{II} chemistry, where relativistic effects have a much greater influence on coordinative bonding. Therefore, coordination geometries surrounding Ag^+ are more flexible and range from isolated complex cations via oligomers to chains and layers. Strong N-donor ligands such as pyrazole, melamine, or piperazine coordinate Ag^+ linearly in cationic complexes, or with trigonal or tetrahedral environments, depending on the strength of the competing ligand and upon stoichiometry. Multi-N donor ligands lead to chelating coordinations [e.g., with 2,2′-bipyridine (Bpy) or with 2,4,6-tris-(2-pyridyl)-1,3,5-triazine (Tpt)]. Ag^+–Ag^+

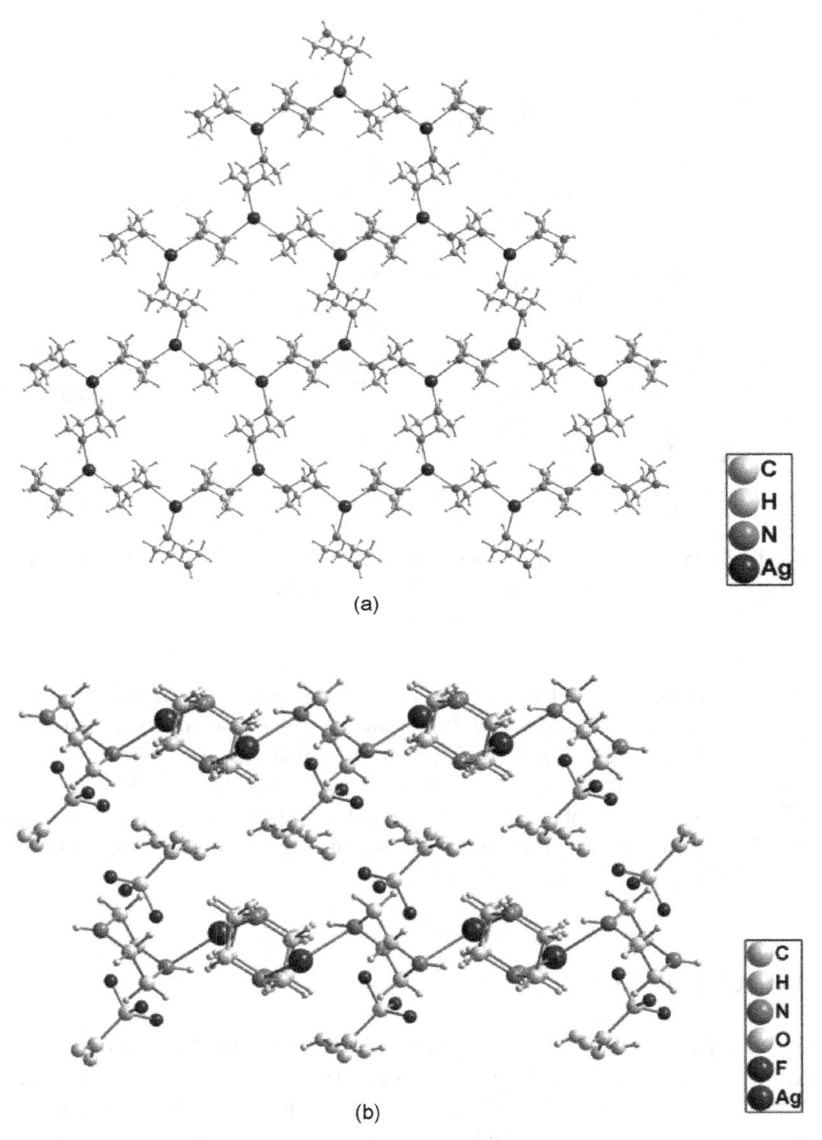

FIGURE 1.14 (a) $[Ag_2(Pip)_3]^{2+}$ layers in the crystal structure of $[Ag_2(Pip)_3](Tfa)_2(H_2O)_6$ (**25**); (b) their stacking and separation by $(Tfa)^-$ anions and water molecules.

interactions ("argentophilicity") seem to play only a minor role, comparable to hydrogen bonding. Chelating ligands such as Bpy seem to be necessary for special coordinations, such as (almost) square planar in $Ag(Bpy)_2(ClO_4)$, which bring the Ag^+ cations close together.

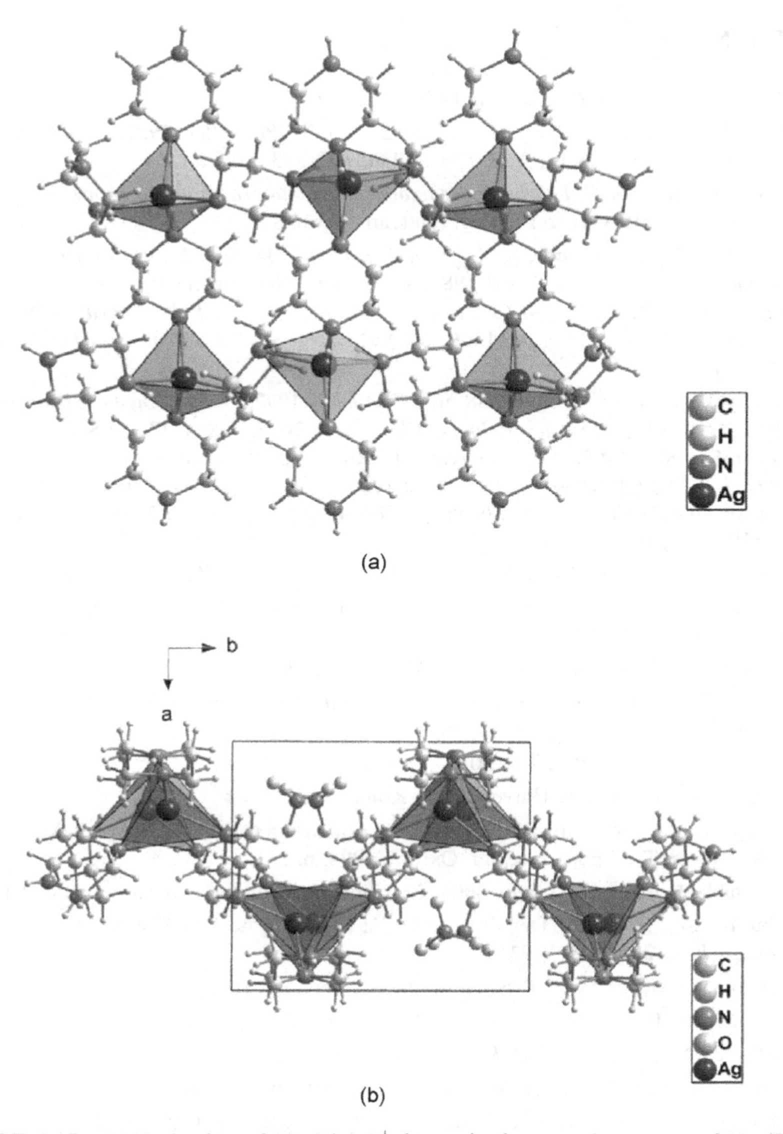

FIGURE 1.15 (a) Projection of $[Ag(Pip)_2]^+$ layers in the crystal structure of $[Ag(Pip)_2](NO_3)$ (**26**); (b) the stacking of these waved layers with $(NO_3)^-$ anions in between.

Acknowledgments

This work was made possible exclusively through the support of the Universität zu Köln, for which the authors are exceedingly grateful.

REFERENCES

1. Shannon, R. D. *Acta Crystallogr.* **1976**, *A32*, 751.
2. AgCl: Swanson, H. E.; Fuyat, R. K.; Ugrinic, G. M. *Phys. Rev.* **1922**, *19*, 248. NaCl: Straumanis, M. E.; Jevins, A. *Z. Phys.* **1936**, *102*, 353.
3. Jander, G.; Blasius, E. *Lehrbuch der analytischen und präparativen anorganischen Chemie*, 16th ed., S. Hirzel Verlag, Stuttgart, Germany, 2006.
4. (a) Corey, R. B.; Pestrecov, K. Z. *Kristallografiya* **1934**, *89*, 528. (b) Yamaguchi, T.; Lindquist, O. *Acta Chem. Scand.* **1983**, *A37*, 685. (c) Nockemann, P.; Meyer, G. Z. *Anorg. Allg. Chem.* **2002**, *628*, 1636. (d) Zachwieja, U.; Jacobs, H. Z. *Kristallografiya* **1992**, *201*, 207. (e) Zeng, S.-L.; Tong, M.-L.; Chen, X.-M.; Ng, S. W. *J. Chem. Soc., Dalton Trans.* **2002**, 360.
5. (a) Pyykkö, P. *Chem. Rev.* **1988**, *88*, 563; *Chem. Rev.* **1997**, *97*, 597; *Angew. Chem., Int. Ed.* **2004**, *43*, 4412. (b) Pyykkö, P. *Inorg. Chim. Acta* **2005**, *358*, 4113. (c) Schmidbaur, H. *Chem. Soc. Rev.* **1995**, 391. (d) Pyykkö, P.; Runeberg, N.; Mendizabal, F. *Chem. Eur. J.* **1997**, *3*, 1451. (e) Pyykkö, P.; Mendizabal, F. *Inorg. Chem.* **1998**, *37*, 3018. (f) O'Grady, E.; Kaltsoyannis, N. *Phys. Chem. Chem. Phys.* **2004**, *6*, 680. (g) Jansen, M. *Angew. Chem., Int. Ed.* **1987**, *26*, 1098.
6. Zheng, S. L.; Nygren, C. L.; Messerschmidt, M.; Coppens, P. *Chem. Commun.* **2006**, 3711.
7. Olson, L. P.; Whitcomb, D. R.; Rajeswaran, M.; Blanton, T. N.; Stwertka, B. J. *Chem. Mater.* **2006**, *18*, 1667.
8. Nockemann, P.; Meyer, G. *Z. Anorg. Allg. Chem.* **2003**, *629*, 123.
9. Meyer, G.; Nockemann, P. Z. *Anorg. Allg. Chem.* **2003**, *629*, 1447, and literature cited therein.
10. Grdenic, D. *Q. Rev.* **1965**, *19*, 303.
11. Sehabi, M. Dissertation, Universität zu Köln, 2006.
12. Kitagawa, S.; Noro, S. In *Comprehensive Coordination Chemistry II*, Vol. 7, McCleverty, J. A., Meyer, T. J., Eds., Elsevier, Oxford, **2004**, p. 231.
13. Constable, E. C. *Aust. J. Chem.* **2006**, *59*, 1, and the seven following articles cited therein.
14. Abu-Youssef, M. A. M.; Dey, R.; Gohar, Y.; Massoud, A. A.; Öhrström, L.; Langer, V. *Inorg. Chem.* **2007**, *46*, 5893.
15. Sehabi, M.; Meyer, G. Z. *Kristallogr.* **2005**, Suppl. 21, 167.
16. Nockemann, P.; Meyer, G. Z. *Anorg. Allg. Chem.* **2004**, *630*, 2571.
17. Grdenic, D.; Krstanovic, I. *Ark. Kemi* **1955**, *27*, 143.
18. Sehabi, M.; Meyer, G. Z. *Anorg. Allg. Chem.* **2004**, *630*, 1758.
19. Zhu, H.; Yu, Z.; Yu, X.; Hu, H.; Huang, X. *J. Chem. Cryst.* **1999**, *29*, 239.
20. Sivashankar, K.; Ranganathan, A.; Pedireddi, V. R.; Rao, C. N. R. *J. Mol. Struct.* **2001**, *559*, 41.
21. Zhang, L.; Zhang, J.; Li, Z.-J.; Cheng, J.-K.; Yin, P.-X.; Yao, Y.-G. *Inorg. Chem.* **2007**, *46*, 5838.
22. Bowmaker, G. A.; Effendy; Marfuah, S.; Skelton, B. W.; White, A. H. *Inorg. Chim. Acta* **2005**, *358*, 4371.
23. Hunter, C. A.; Sanders, J. K. M. *J. Am. Chem. Soc.* **1990**, *12*, 5525.
24. Cesur, N. Dissertation, Universität zu Köln, 2006.

25. Nasielski, J.; Nasielski-Hinkens, R.; Heliporn, S.; Rypens, C.; Declercq, J. P. *Bull. Soc. Chim. Belg.* **1988**, *97*, 983. Marsh, R. E. *Acta Crystallogr.* **1997**, *B53*, 317.
26. Carlucci, L.; Ciani, G.; Proserpio, D. M.; Sironi, A. *Inorg. Chem.* **1995**, *34*, 5698.
27. Brammer, L.; Burgard, M. D.; Eddleston, M. D.; Rodger, C. S.; Rath, N. P.; Adams, H. *CrystEngComm* **2002**, *4*, 239.

2

INDIUM(III)–ORGANIC COORDINATION POLYMERS WITH VERSATILE TOPOLOGICAL STRUCTURES BASED ON MULTICARBOXYLATE LIGANDS

LIAN CHEN, FEI-LONG JIANG, ZHENG-ZHONG LIN, AND MAO-CHUN HONG

State Key Laboratory of Structural Chemistry, Fujian Institute of Research on the Structure of Matter, Chinese Academy of Sciences, Fuzhou, Fujian, People's Republic of China

2.1 INTRODUCTION

There is currently considerable interest in metal–organic coordination polymers (MOCPs) constructed through the deliberate selection of metal ions and multifunctional exodentate ligands, motivated by their intriguing structural diversity and potential functions as microporous solids for molecular adsorption, ion exchange, and heterogeneous catalysis [1–12]. The benzenemulticarboxylic acids (BMCs) are a class of promising ligands that provide high symmetries, diverse charge forms, and multiconnecting ability, and have been used in the design of metal–organic coordination complexes that exploit both the diversity of metal coordination geometries and weak intermolecular forces such as π–π interactions and hydrogen bondings. In addition, the rigid conformation and strong coordinating ability of the carboxylate groups of BMC ligands bestow excellent thermal stabilities on the resulting metal–organic coordination complexes. The utilization of benzenemulticarboxylate ligands, notably benzenedicarboxylic acid (H_2BDC), 1,3,5-benzenetricarboxylic acid (H_3BTC), and 1,2,4,5-benzenetetracarboxylic acid (H_4BTEC) (Fig. 2.1) and divalent

Design and Construction of Coordination Polymers, Edited by Mao-Chun Hong and Ling Chen

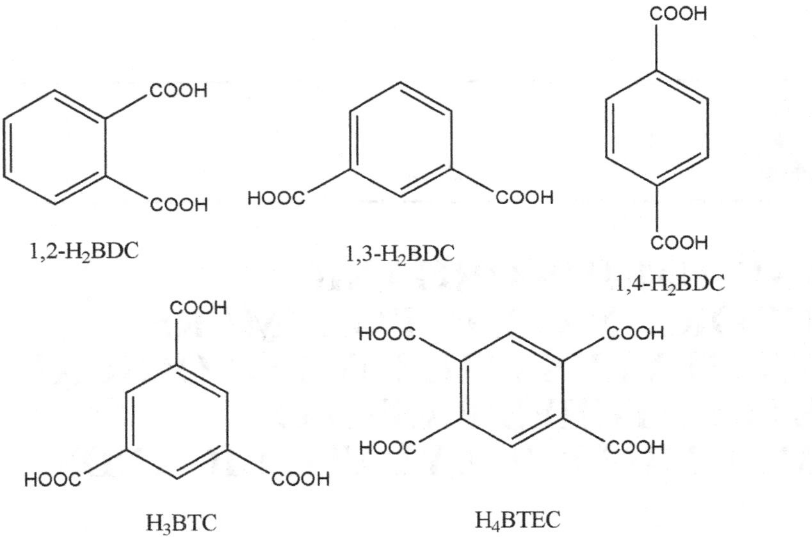

FIGURE 2.1 Schematic diagrams of some benzenemulticarboxylic acids: 1,2-benzenedicarboxylic acid (1,2-H_2BDC), 1,3-benzenedicarboxylic acid (1,3-H_2BDC), 1,4-benzenedicarboxylic acid (1,4-H_2BDC), 1,3,5-benzenetricarboxylic acid (H_3BTC), and 1,2,4,5-benzenetetracarboxylic acid (H_4BTEC).

metal ions (M^{II}) has led to the generation of diverse networks consisting of one-dimensional (1D) chains, two-dimensional (2D) layers, and three-dimensional (3D) structural frameworks [13–24]. Compared with the systematic and extensive studies on M^{II}–BMC complexes, however, relatively few efforts have been made to investigate M^{III}–BMC complexes [25,26]. It has been postulated that the incorporation of trivalent metal ions might create structures different from those containing divalent metal ions because of the increased valence charge. To investigate the influence of a change in the metal center on the coordination architecture during the course of the assembly of the metal ions with benzenemulticarboxylates, among the trivalent metal ions a particular emphasis is placed on indium(III). In contrast to the traditional divalent transition metals, which are usually four- or six-coordinated, indium ions are typical of the +3 oxidation state and have the ability to adopt MO_6 [27,28], MO_7 [29–31], or even MO_8 [32] coordination numbers [33,34].

Due to their unusual valence state and rich coordination modes, indium(III) ions have been widely used in several corner-linked MS_4 tetrahedral metal–sulfide compounds with large pores and channels known as the *supertetrahedra* [35–37]. Nevertheless, indium(III) ions are liable to hydrolyze, which limits their use in the construction of MOCPs. We find that adding appropriate basic reagents to deprotonate the benzenemulticarboxylates and controlling the reaction conditions very carefully may be important in the assembly of In^{III}–BMC complexes.

Pyridine and its derivatives (pyds) (Fig. 2.2) are a kind of basic reagent that can be employed to deprotonate carboxyl groups of BMC ligands and control the

pyridine 4-picoline 2-picoline imidazole

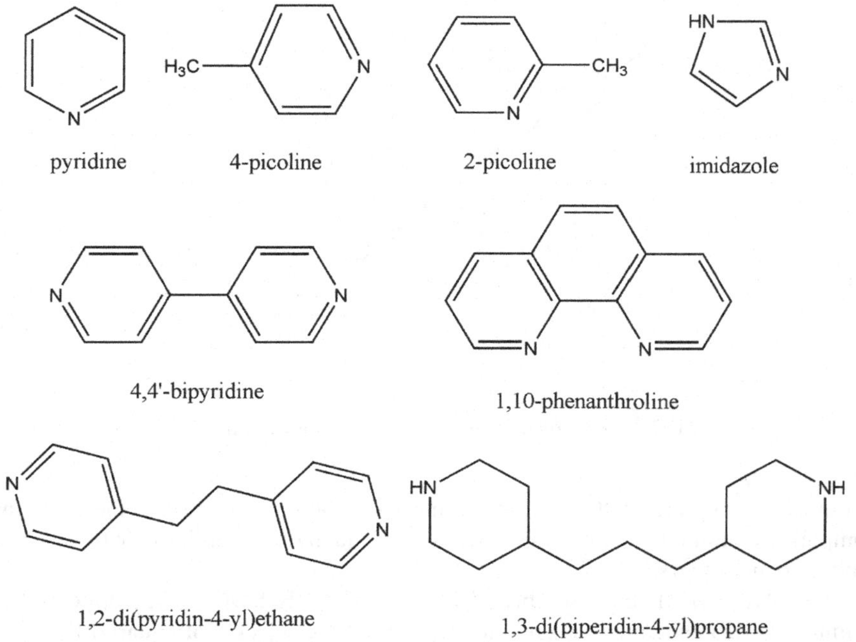

4,4'-bipyridine 1,10-phenanthroline

1,2-di(pyridin-4-yl)ethane 1,3-di(piperidin-4-yl)propane

FIGURE 2.2 Schematic diagrams of pyridine and its derivatives cited in the text.

reaction conditions. They usually act as guests to fill in the cavities of host frameworks or as terminal ligands to complete metal coordination environments [38–40]. Furthermore, interactions including hydrogen-bond contacts and π–π stacking are usually found in a pyds-containing system and play an important role in the assembly process. This chapter is concerned mainly with our recent investigations of metal–organic coordination polymers (MOCPs) assembled by indium(III) and benzenemulticarboxylates.

2.2 ARCHITECTURES CONSTRUCTED BY In(III) AND BENZENEDICARBOXYLATES

Benzenedicarboxylate (BDC) ligands have been found to be useful building blocks in the construction of organic–inorganic materials with desired topologies, owing to their rich coordination modes. However, MOCPs constructed by main-group metal ion, indium(III), and BDC are limited. As far as we know, there are only several examples: $[InH(1,4\text{-}BDC)_2]_n$ (**1**) [32], $[In_2(OH)_3(1,4\text{-}BDC)_{1.5}]_n$ (**2**) [28], $[In(OH)(1,4\text{-}BDC)]_n \cdot 0.75n(1,4\text{-}H_2BDC)$ (**3**) [41], $[In(1,4\text{-}BDC)(OH)]_n \cdot 7nH_2O$ (**4**) [42], $[In(1,4\text{-}BDC)_{1.5}(bpy)]_n$ (**5**), $[In_2(1,4\text{-}BDC)_2(OH)_2(phen)_2]_n$ (**6**) [43], $[In(1,2\text{-}BDC)(OH)(H_2O)]_n$ (**7**) [44], and $[In_3O(1,3\text{-}BDC)_3(H_2O)_{1.5}(C_3N_2H_3)(C_3N_2H_4)_{0.5}]_n \cdot n(DMF) \cdot 0.5n(CH_3CN)$ (**8**) [32]. Although all of them were aggregated by the

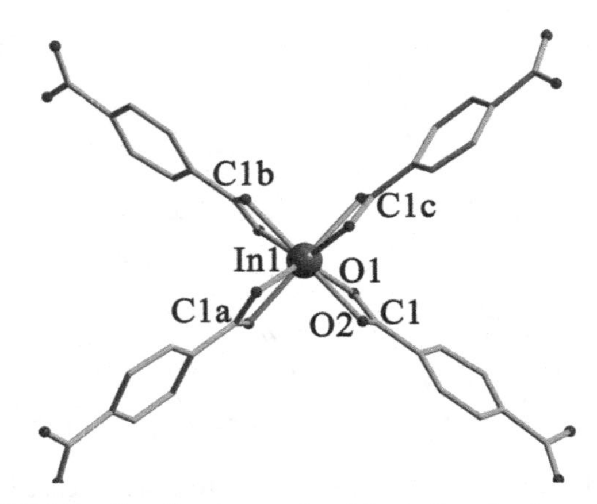

FIGURE 2.3 Coordination mode of the indium ion in complex **1**.

indium(III) ions and BDC, by controlling the reaction conditions, the resulting complexes exhibited diverse frameworks such as quartzlike, vernier structures, CaB_6 topology, and others.

The solvothermal self-assembly of $InCl_3$ with 1,4-H_2BDC in *N,N*-dimethylformamide (DMF) generated the complex $[InH(1,4\text{-BDC})_2]_n$ (**1**) with quartzlike topology. All the indium centers in **1** adopt triangulated dodecahedral geometry by the chelation of four carboxylate groups from four terephthalate anions. The molecular structure shows that **1** is a complex anion with a proton to balance the charge [45]. If the carboxylate groups are considered as the connecting points [46], the T building blocks comprising carbon atoms around the indium centers in **1** show a highly distorted pseudotetrahedral geometry (Fig. 2.3), with the angles of C–In–C being 91.5°, 143.1°, and 100.08°. Using an indium–carboxylate unit $[In(O_2C)_4]^-$ as a four-connected node and terephthalate as a linear rod, the anion-type quartzlike network can be assembled as described in Figure 2.4, resulting in the high symmetry of β-quartz, $P6_422$. Topological analysis indicates that complex **1** is characteristic of twofold interpenetrated 6^48^2-b nets, in which the polyhedron centers in one net occupy the centers of the six-membered ring sets in another network, and organic linkers are interpenetrated through a six-membered ring in the up–down–up–down form. Similar to β-quartz silica, the structure of **1** exhibits a sixfold screw axis that runs in the *c*-direction. Along this sixfold axis, the 3D view of **1** shows a right-handed channel (Fig. 2.5), whose diameter, calculated from the distance between diagonally opposite indium atoms, is about 15.0 Å. If van der Waals radii are considered, the diameter of the channel is about 7.8 Å.

Introducing the deprotonation reagent Et_3N (triethylamine) to the system, the reaction of $InCl_3$ with 1,4-H_2BDC under hydrothermal conditions gave an inorganic–organic hybrid compound $[In_2(OH)_3(1,4\text{-BDC})_{1.5}]_n$ (**2**) with scalelike inorganic In–O sheets. The asymmetric unit of **2** contains two independent indium atoms. Both of them are six-coordinated by three oxygen atoms from three different 1,4-BDC ligands and three μ_2-OH groups displaying the distorted InO_6 octahedral geometry.

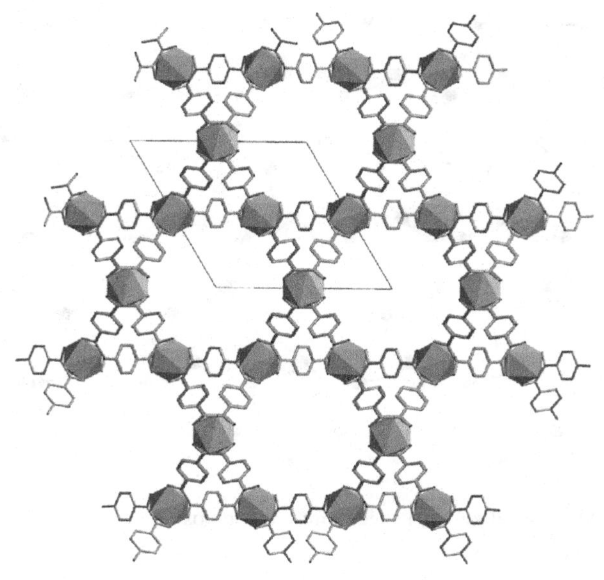

FIGURE 2.4 Single quartz net of **1** in the *c*-direction.

Using the oxygen atoms of the OH groups as bridges, every two neighboring indium atoms are connected by one μ_2-oxygen atom to form inorganic 12-membered In_6O_6 rings. As shown in Figure 2.6, these 12-membered In–O rings form an infinite scalelike (6,3) sheet with the composition $[In_2(OH)_3]^{3+}$. The average distance

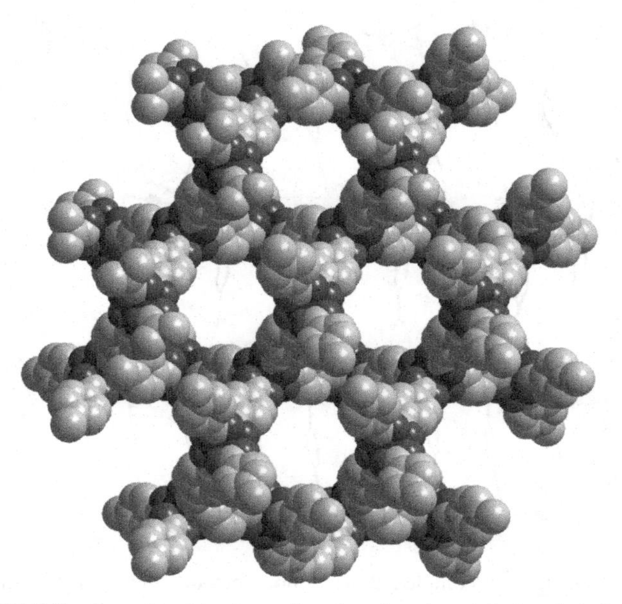

FIGURE 2.5 Complex **1** in the *c*-direction showing right-handed channels.

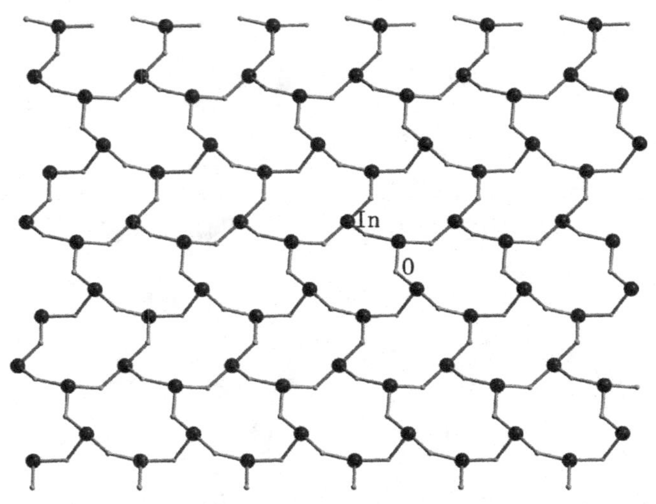

FIGURE 2.6 Scalelike (6,3) sheet with the composition of $[In_2(OH)_3]^{3+}$ in **2**.

between the indium atoms is about 3.77 Å, which is longer than those of purely inorganic structures such as the corundumlike In_2O_3 structure [47] ($d = 3.34$ Å). The layer is repeated along the c stacking direction at a distance of about 10 Å between two layers. As illustrated in Figure 2.7, each of the 1,4-BDC^{2-} counterions coordinates to

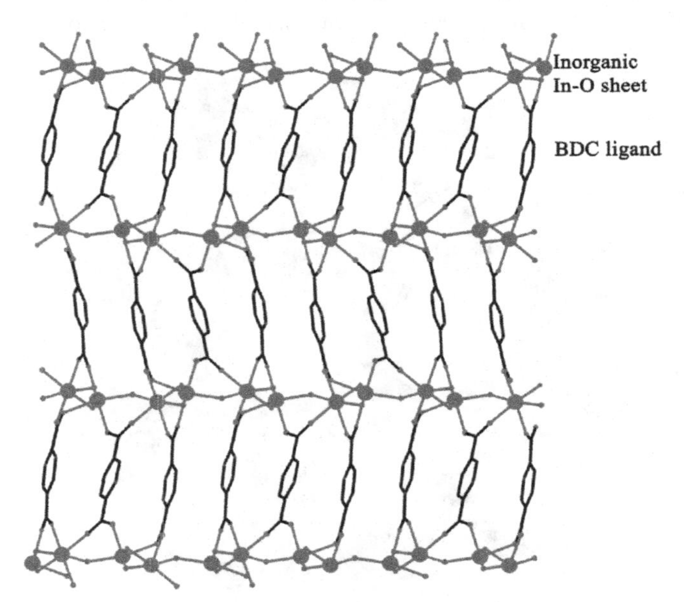

FIGURE 2.7 3D view of complex **2** with inorganic–organic hybrid framework in the bc-plane.

four different indium atoms of the neighboring scalelike inorganic In–O sheets. Thus, the 1,4-BDC ligands, acting as organic linkers, connect the inorganic $[In_2(OH)_3]^{3+}$ layers, giving rise to a 3D inorganic–organic hybrid framework.

The hydrothermal self-assembly of In powder with 1,4-H_2BDC at 220°C in the presence of HF (48%) yielded complex $[In(OH)(1,4\text{-BDC})]_n \cdot 0.75n(1,4\text{-}H_2BDC)$ (**3**). The structure of **3** can be described as a hybrid inorganic coordination polymer–organic vernier structure [48–51], where the two sublattices are formed by a covalently linked In(OH)(1,4-BDC) lattice and ordered chains of hydrogen-bonded 1,4-H_2BDC molecules (Fig. 2.8). In every asymmetric unit of **3**, there are five independent indium atoms, with each coordinated by six oxygen atoms, exhibiting a slightly compressed octahedral geometry with axial In–O bond lengths of 2.077(3) to 2.084(3) Å and equatorial In–O bond lengths of 2.133(4) to 2.154(3) Å. The neighboring InO_6 octahedra share corners through the oxygen atoms in the axial positions, forming a zigzag $\cdots$-OH–In–OH–In–$\cdots$ backbone with In–OH–In angles of 118.6(2) to 120.6(2)°. The four equatorial oxygen atoms come from four different 1,4-BDC^{2-} anions which cross-link the octahedral chains into a 3D framework. The framework shows large channels located by oriented 1,4-H_2BDC guests which are bonded to the host 3D framework through relatively weak interactions: hydrogen bonds and π–π interactions. The O$\cdots$O distances of the hydrogen bonds between the guest 1,4-H_2BDC molecules and the host framework range from 2.5 to 2.7 Å. Each corner-shared InO_6 octahedron of the zigzag chain has a length of 3.6 Å along the chain axis (measured from In$\cdots$In distance), whereas each 1,4-H_2BDC molecule has a length of 9.6 Å along the guest acid column axis. Therefore, the length of eight InO_6 octahedra in the zigzag chain is equivalent to that of three 1,4-H_2BDC molecules of the guest column, with a lattice parameter $c(\sin\beta) = 28.76$ Å. Considering that there are two guest columns for each octahedral chain, there is $\frac{3}{4}$ of a 1,4-H_2BDC guest molecule for each framework indium atom.

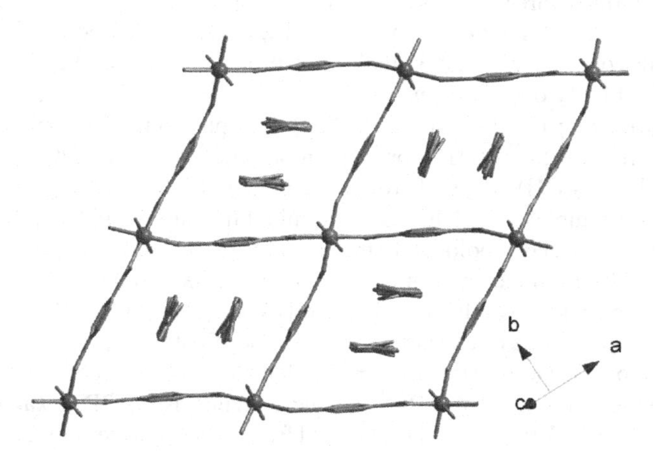

FIGURE 2.8 Structure of **3**, showing the orientation of the guest 1,4-H_2BDC molecules within the In(OH)(1,4-BDC) framework.

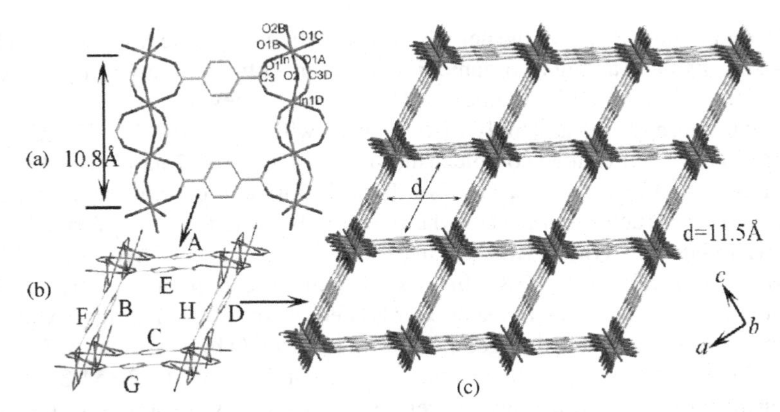

FIGURE 2.9 (a) Two chains are bridged by two 1,4-BDC ligands into a lateral side of a prism for complex **4**. (b) A building unit conceptually imaged as a right prism is constructed by four chains and eight 1,4-BDC ligands. (c) Packed prisms generate a 3D network with four different tunnels, and the largest one is shown here.

Using 2-picoline to adjust the reaction condition, hydrothermal reaction of $InCl_3 \cdot 4H_2O$ and 1,4-H_2BDC gave $[In(1,4\text{-}BDC)(OH)]_n \cdot 7nH_2O$ (**4**). As shown in Figure 2.9, the structure of **4** can be considered as a framework built up by a prism-building unit, which contains 16 indium atoms bridged by 16 μ-OH groups and eight 1,4-BDC bridged ligands. In **4**, each indium atom is octahedrally coordinated by four oxygen atoms from four different 1,4-BDC ligands and two axial μ-OH groups. Each μ-OH group bridges two indium(III) centers, giving rise to infinite indium–oxygen chains, which are further bridged by carboxylate groups of 1,4-BDC ligands to generate a 3D framework with four different tunnels. The biggest of the tunnels is a kind of rhombic channel along the [0 1 0] direction with dimensions 11.5 × 11.5 Å (Fig. 2.9). The other three are along the [1 1 1], [1 0 0], and [0 1 1] directions, with dimensions of 9.9 × 4.8 Å, 9.9 × 4.8 Å, and 11.5 × 4.8 Å, respectively. Based on Platon calculation, the channel volume constitutes 50.8% of the total volume, indicative of a highly open framework.

When introducing the 2,2′-bipyridyl (bpy) or *o*-phenanthroline (phen) into the system, two different In^{III}–BDT coordination polymers, $[In(1,4\text{-}BDC)_{1.5}(bpy)]_n$ (**5**) and $[In_2(1,4\text{-}BDC)_2(OH)_2(phen)_2]_n$ (**6**), were obtained [43]. Although the bpy and phen in the two complexes both act as the terminal ligands, **5** and **6** exhibit different coordination modes and topological structures. X-ray crystallographic analysis of **5** reveals that the asymmetric unit (Fig. 2.10) consists of one indium atom, 1.5, bridging 1,4-benzendicarboxylic ligands as well as one terminal ligand, 2,2′-bipyridyl. This compound presents a rare example of eight-coordinated indium ions [54], where each indium ion is coordinated by three bidentate chelating carboxylate groups of three fully deprotonated 1,4-BDC^{2-} ligands, giving rise to 2D hexagonal layers which further interlock to form a 3D entangled framework. Furthermore, each indium center is bidentately coordinated by two nitrogen atoms of one bpy molecule to form eight-coordinated metal ions. The bond distances of the carboxylic O–In and the bpy's

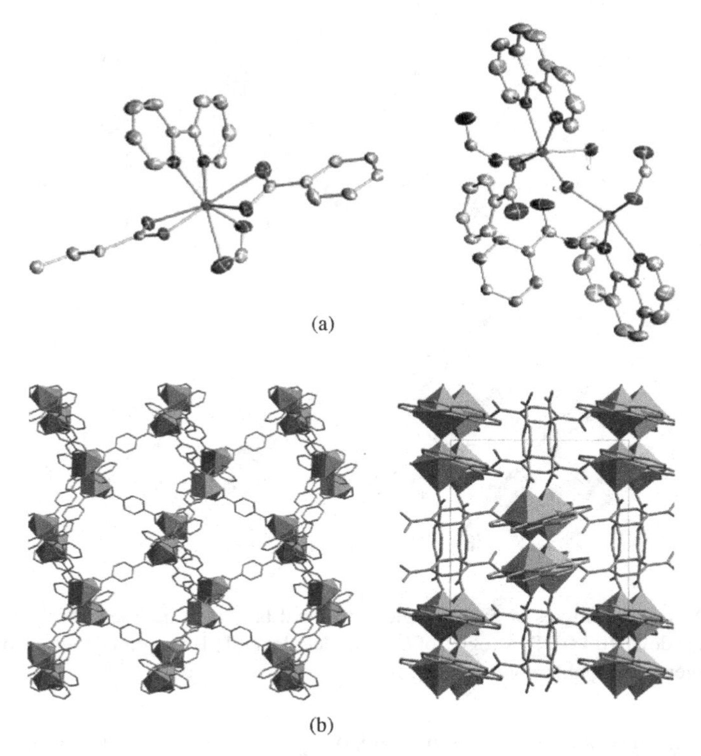

FIGURE 2.10 (a) Ortep drawing with ellipsoid probability 50% of $In(BDC)_{1.5}(bpy)$ (**5**) and $In_2(BDC)_2(OH)_2$ $(phen)_2$ (**6**); (b) view of structures showing the In polyhedra for **5** and **6**.

N–In, which range from 2.204(3) to 2.551(4) Å and 2.295(3) to 2.349(3) Å, respectively, are all comparable to those usually encountered for indium–oxygen and indium–nitrogen coordinations in the literature [52,53]. In the structure, the connectors (1,4-BDC) can be divided into two types, *L* and *L′*, according to their different roles. L, which is situated in the general position, connects the indium atoms along the *b*-direction, generating a 1D infinite zigzag chain with the composition $[In(bpy)(L)]_{\infty}{}^{+}$. The other type of 1.4-BDC molecule L′, with its centroid situated on an inversion center, links the aforementioned zigzag chains along the [1 0 0] direction, forming a (6,3) hexagonal layer with the formula $[In(bpy)(L - L')]$. Stacking at two different directions, these (6,3) hexagonal layers are interlocked in "inclined" mode and form a 3D polycatenation architecture, as Figure 2.9 shows. The six-membered rings in two distinct inclined layers are interlocked with a DOC (degree of catenation) value of 2/2 [55].

X-ray crystallographic analysis of **6** reveals that the asymmetric unit (Fig. 2.11) consists of two independent indium atoms, two bridging 1,4-benzendicarboxylic ligands, two phenantroline molecules, and μ_2-OH groups. In **6**, each independent indium ion is six-coordinated by two nitrogen atoms from a terminal phen ligand,

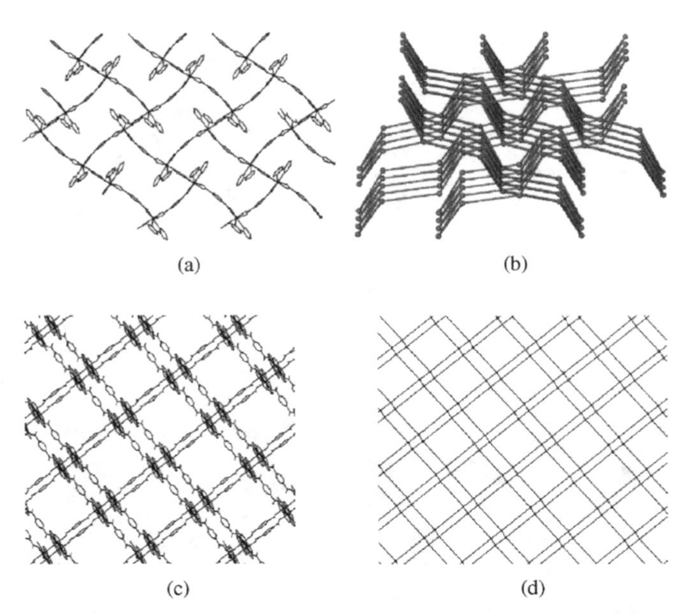

FIGURE 2.11 Topologies of **5**: (a) single hexagonal layer; (b) inclined interlocking mode of (6,3) layers down [0 1 0]. Topologies of **6**: (c) molecular and (d) simplified views along [0 0 1] of the 4^4 layer with double connections.

two μ_2-OH groups, and two oxygen atoms of two carboxylate groups from different 1,4-BDC connectors, exhibiting slightly distorted octahedral coordination geometry (Fig. 2.10). Every four indium atoms are connected by four μ_2-OH groups to form tetrameric $[In_4(OH)_4(phen)_4]_\infty^{8+}$ units, which are linked further by the 1,4-BDC connectors along the *a* and *b* directions, giving rise to infinite layers perpendicular in the (001) direction with the composition $[In(1,4\text{-}BDC)(OH)(phen)]_\infty$. The topological analysis shows that complex **6** exhibits a new 2D layer of square type (4^4) with "double" connections in Figure 2.11 [55]. It is worth mentioning that in every layer, each side of the square is doubled, which makes it different from the general 4^4 square topology. These double-sided square layers arrange themselves in parallel along the [0 1 0] direction with the {ABAB} stacking mode.

Reports about metal–organic coordination polymers based on indium and 1,2-BDC and 1,3-BDC are comparatively rare. Following are two examples. $[In(1,2\text{-}BDC)(OH)(H_2O)]_n$ (**7**) was synthesized by the hydrothermal reaction of $InCl_3$ and 1,2-benzenedicarboxylic acid with 2-picoline as the deprontonation reagent. As depicted in Figure 2.12, the asymmetric unit of **7** consists of one indium center, one 1,2-benzendicarboxylic ligand, one μ_2-OH group, and a coordinated water molecule. The indium atom is six-coordinated by four oxygen atoms from three 1,2-BDC^{2-} ligands, one coordinated water molecule in the square-planar position, and two oxygen atoms from two hydroxyl groups in the axial position. The axial oxygen atom corners are shared by neighboring octahedra to form a zigzag ···OH–In–OH–In··· chain propagating along the *c*-axis, with an

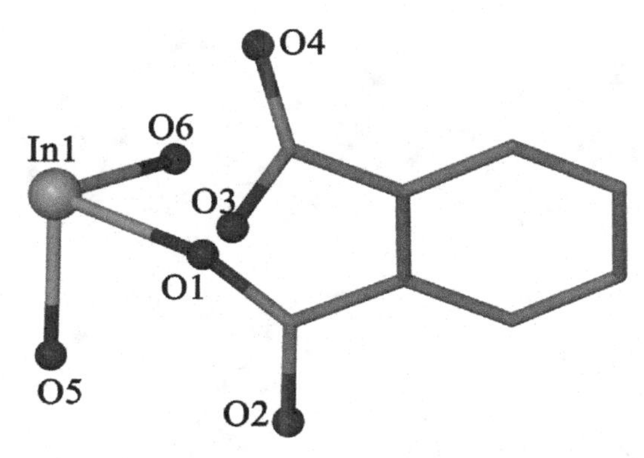

FIGURE 2.12 Asymmetric unit of **7**, showing the atom-numbering scheme. (Hydrogen atoms have been omitted for clarity.)

In–OH–In angle of 126.4(1)° (Fig. 2.12). The 1,2-BDC^{2-} ligand, which acts as a tridentate bridging linker with a bidentate and a monodentate carboxylate group, links the indium atoms of the neighboring zigzag ···OH–In–OH–In··· chains, giving rise to the 2D layered structure described in Figure 2.13. These layers, with the benzene rings located on both sides of the plane, stack parallelly and form the 3D supramolecular framework (Fig. 2.14) by hydrogen-bonding interactions and π–π stackings.

Treated with imidazole, reaction between 1,3-H_2BDC and $In(NO_3)_3{\cdot}2H_2O$ in a DMF/CH_3CN solution yielded complex **8** with CaB_6 topology [56–58]. Crystallographic analysis shows that compound **8** contains oxygen-centered indium–carboxylate clusters $[In_3O(CO_2)_6]$, which can be viewed as six-connected nodes with trigonal–prismatic geometry (Fig. 2.15a). Each cluster comprises three indium-centered octahedra by sharing one central μ_3-oxygen atom located on a threefold axis, with the three In–(μ_3-O)–In angles approximate to 120°. The equatorial positions of the indium centers in the $[In_3O(CO_2)_6]$ building unit are occupied by four oxygen atoms from four organic ligands. One end of the apical positions is occupied by the central μ_3-O atom, and the other is occupied statistically by a water molecule or a neutral imidazole molecule. Each six-connected node $[In_3O(CO_2)_6]$ unit is linked by six separate organic linkers 1,3-H_2BDC, forming a 3D structure, including cuboidal cages, illustrated in Figure 2.15b. The framework of **8** exhibits cuboidal cages and CaB_6 topology similar to those reported in the literature [55–58]. Determined by Platon software, the total solvent-accessible volume for **8** is estimated to be 18.9% (the coordinated imidazolate ligands are not omitted in this calculation).

2.3 ARCHITECTURES CONSTRUCTED BY In(III) AND BENZENETRICARBOXYLATES

With high symmetry, diverse charge, and multiconnecting ability, 1,3,5-benzenetricarboxylic acid (H_3BTC) has been a promising ligand in the design of hybrid

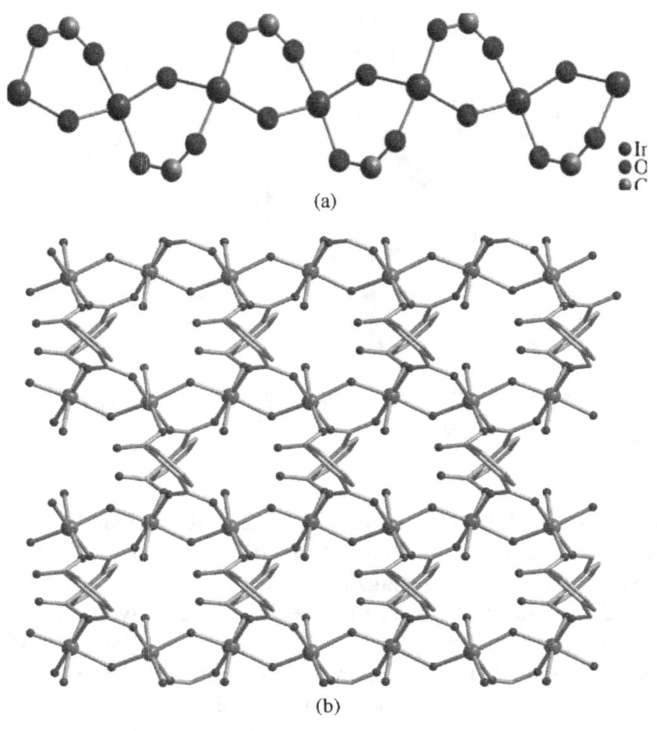

FIGURE 2.13 (a) Zigzag ···OH–In–OH–In··· chain and the six-membered ring A; (b) perspective view of the 2D layered structure of **7**.

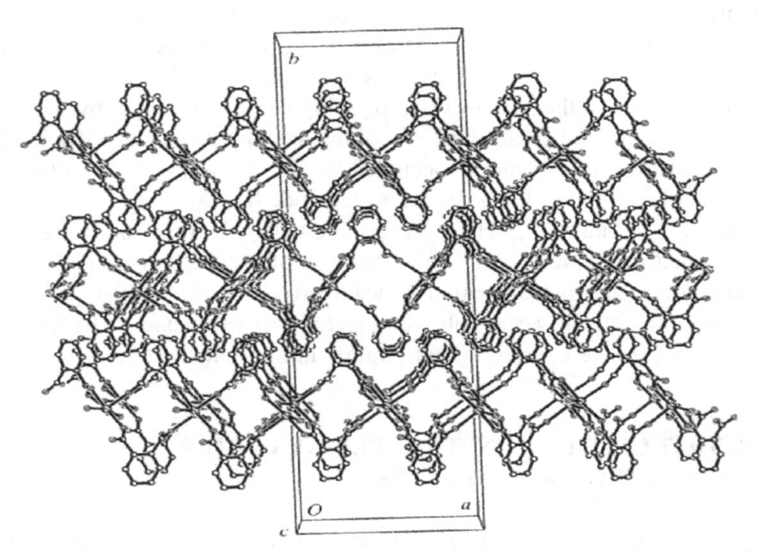

FIGURE 2.14 3D view of the packing of **7** in the *c*-direction.

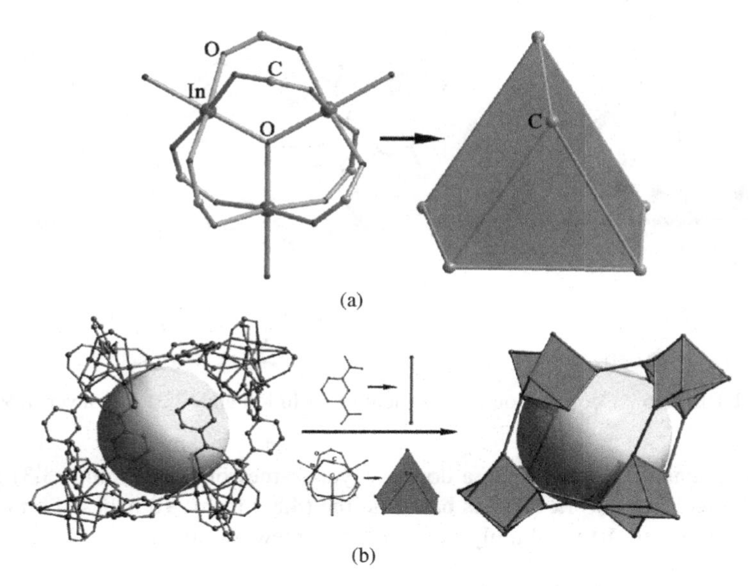

FIGURE 2.15 X-ray crystal structure of **8**: (a) oxygen-centered indium–carboxylate $[In_3O(CO_2)_6]$ building unit; (b) ball-and-stick and polyhedral representations of the cuboidal cage of **8** constructed by the organic linker 1,3-benzenedicarboxylate (1,3-BDC) and $[In_3O(CO_2)_6]$ building unit. The cavity size is indicated by the large spheres. Hydrogen atoms, axial coordinating ligands, and solvent molecules are omitted for clarity.

complexes. Much current research is focused on the syntheses and structural characterization of coordination polymers assembled from H_3BTC and divalent metal ions [59–62]. Trivalent metal ions, however, have been confined to the field of rare earth metals, which are typical of high-coordination-number conformation [63–65]. The main group element, indium(III), can present rich coordination modes with a six-, seven-, or eight-coordination number. Based on this idea, using In^{III} as a metal precursor and H_3BTC as an organic ligand with control of the reaction conditions, a series of In^{III}–BTC complexes with a 1D chain, 2D layer, and 3D network have been synthesized successfully: $[In(BTC)(H_2O)]_n{\cdot}nH_2O$ (**9**) [31], $[In_2(BTC)_2(OH)_2]_n{\cdot}2n$Hpy (**10**) [30], $\{[In_4(BTC)_3(OH)_4]{\cdot}3H_2O{\cdot}$Hpy·py$\}_n$ (**11**) [66], $\{[In_4(BTC)_3(OH)_4]{\cdot}3H_2O{\cdot}$Hapic·apic$\}_n$ (**12**) [66], $\{[In_4(BTC)_3(OH)_4]{\cdot}3H_2O{\cdot}$Hgpic·gpic$\}_n$ (**13**) [66], $\{[In_4(BTC)_3(OH)_4]{\cdot}3H_2O{\cdot}$Hdpea$\}_n$ (**14**) [66], $[In(BTC)(2,2'\text{-bpy})]_n{\cdot}2nH_2O$ (**15**) [67], $[In(BTC)(2,2'\text{-bpy})(H_2O)]_n{\cdot}nH_2O$ (**16**) [68], $[In(BTC)(H_2O)(\text{phen})]_n$ (**17**) [43], $\{[In(HBTC)_2(4,4'\text{-bpy})](4,4'\text{-Hbpy}){\cdot}0.5H_2O\}_n$ (**18**) [26], and $\{[In_6(BTC)_8]{\cdot}3H_2\text{tmdp}{\cdot}40H_2O\}_n$ (**19**) [69] (py = pyridine, apic = 2-picoline, gpic = 4-picoline, dpea = 1,2-di(4-pyridyl)ethane, 2,2′-bpy = 2,2′-bipyridne, phen = *o*-phenanthroline, 4,4′-bpy = 4,4′-bipyridne, and tmdp = 4,4′-trimethylenedipiperidine). Among them, **9** is characteristic of the double-layer motif containing (3,3) honeycomb grids; **10** can be described as a 3D framework containing 1D chiral tunnels; **11–14** have similar frameworks, which grow as 3D host–guest architectures containing 1D tunnels; **15** combines the motifs of double-chain and neutral

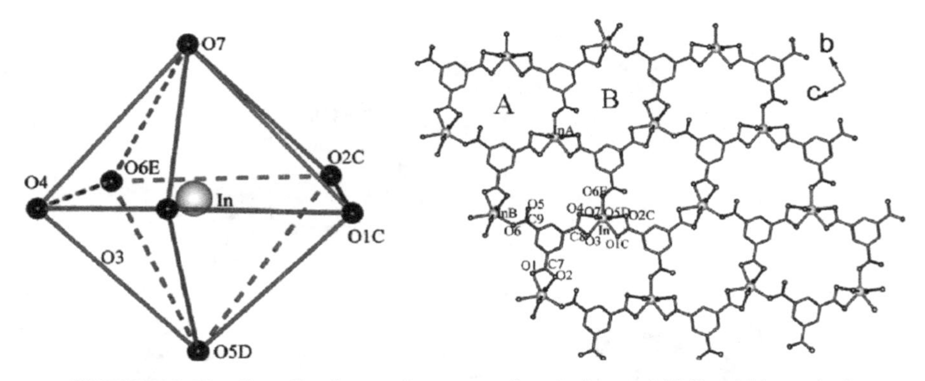

FIGURE 2.16 Coordination environment of an In ion and (3,3) grid layer in **9**.

rectangular tunnel; **16** represents a double-layer framework containing (3,3) grids; the topological framework of **17** is based on the (4.8^2) layer; **18** is based on a (4,2) rhombic grid layer; **19** is a highly open cubic framework containing tunnels along four directions.

Compound **9** exhibits a double-layer sheet configuration generated from the interconnection of two single layers which are produced from the extension of honeycomblike grids along the (1 0 0) plane. Three In^{III} centers and three BTC anions form such a six-membered grid with dimensions of about 7.4×9.2 Å (Fig. 2.16). Inside the grid, each indium atom binds to six carboxylate oxygen atoms from four BTC groups and one water oxygen atom to form a slightly distorted pentagonal bipyramidal motif. A bidentate carboxylate arm lies out of the BTC benzene ring plane and acts as a bridge linking two single layers into a double layer (Fig. 2.17). The other two carboxylate arms bond in a chelating bidentate mode and participate in the propagation of the grids. The two single layers are related to each other through a twofold screw symmetry operation, and the perpendicular distance between them

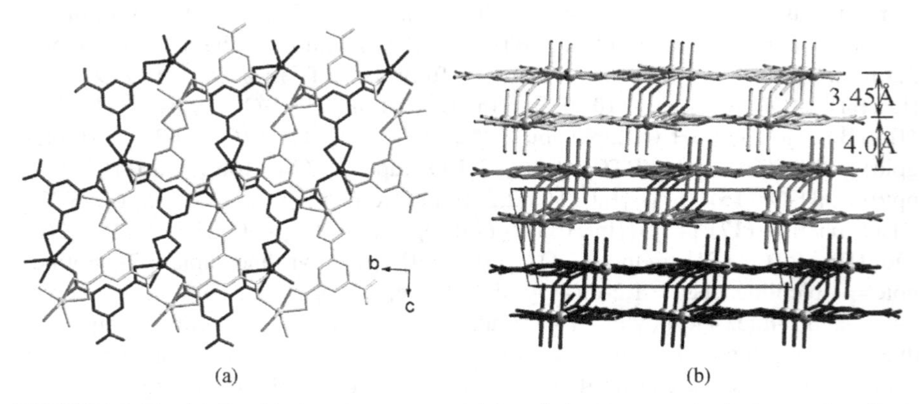

FIGURE 2.17 (a) Double-layer structure of **9** interlinked by two single layers described in Figure 2.16; (b) arrangement of the double layers in the spaces.

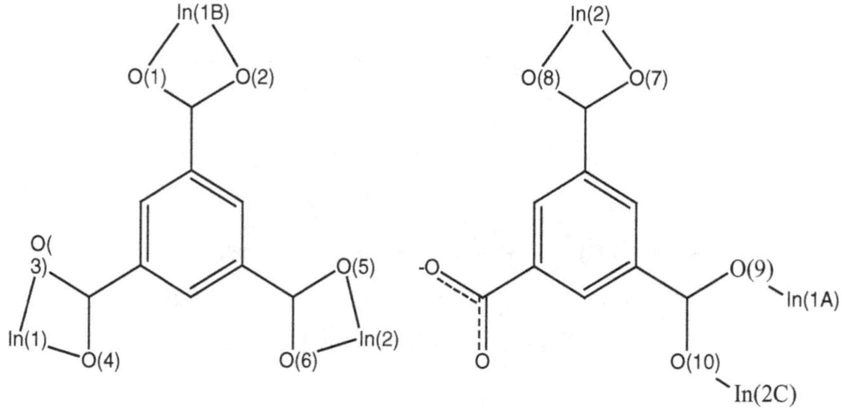

FIGURE 2.18 Schematic diagrams of BTC groups: (a) BTC-A; (b) BTC-B.

is about 3.8 Å. Because of the translation operation of the screw axis, the available sizes of honeycomb-like grids are reduced significantly. Furthermore, the double layers are stacked in parallel by separation of about 4 Å (Fig. 2.17).

$[In_2(BTC)_2(\mu\text{-}OH)_2]_n \cdot 2n$Hpy (**10**) possesses a noninterpenetrated 3D structure with a 1D chiral tunnel. In **10**, each of the two indium(III) ions binds to five oxygen atoms from three BTC ligands and two μ-OH groups to form a capped octahedral coordination geometry for In(1) or a slightly distorted pentagonal bipyramid for In(2). The two different coordination modes of the BTC ligands, BTC-A and BTC-B, are illustrated in Figure 2.18. Indium(III) atoms are linked together by BTC ligands to form a 3D framework possessing a 1D tunnel along the [1 0 0] direction. Figure 2.19 shows a cross-sectional layer of the framework with nonclose rings extended to the entire layer. Each ring consists of three indium(III) atoms, two BTC-A groups, and one BTC-B group. Rings can be divided into two categories according to the proportion of In(1) and In(2) atoms in the rings being either 2 or $\frac{1}{2}$.

The rings parallel to each other further link along the *a*-axis into 1D chiral tunnels. The two neighboring rings are supported by three pillars, and the distance between two neighboring rings is about 3.8 Å. These pillars can also be divided into two categories according to their contents: One includes only a μ-OH group, and the other can be regarded as a combination of a μ-OH group and a carboxylate arm of BTC-B. These two types of pillars are arranged along the *a*-axis in an alternative way to interlink the rings. The stacking of pillars creates infinite In–O–In–O– chains extended throughout the tunnels, as shown in Figure 2.20. It is the inequivalent connections between inequivalent rings that result in the chirality of the tunnels, which are of the same handedness. Using the Platon program, the accessible volume within the crystal is calculated to be 31.8% [70].

$\{[In_4(BTC)_3(OH)_4] \cdot 3H_2O \cdot Hpy \cdot py\}_n$ (**11**) presents a noninterpenetrated 3D structure with channels located by protonated pyridine molecules as organic guests. Three independent indium atoms in **11** are each bonded by six oxygen atoms from four carboxyl and two μ-OH groups to form a distorted octahedral geometry. The BTC

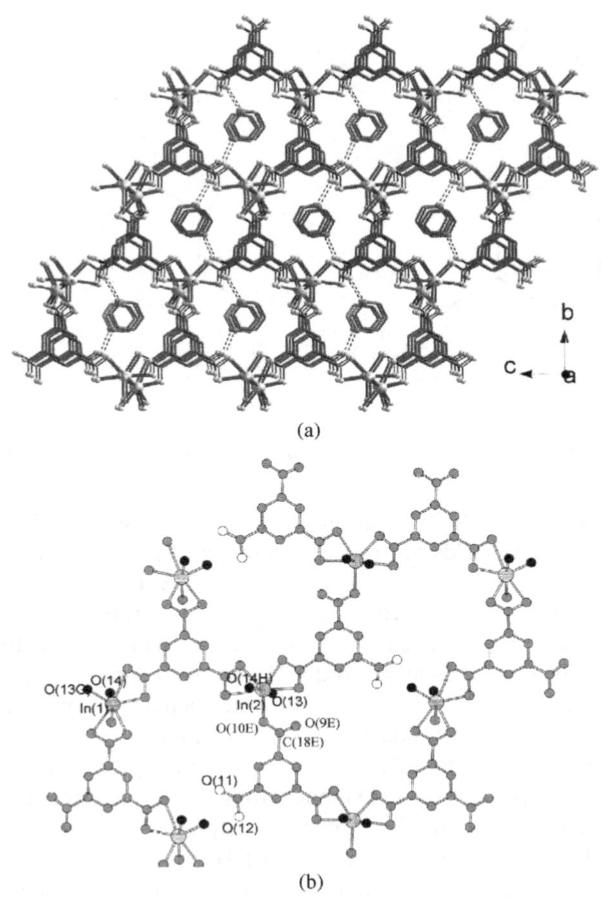

FIGURE 2.19 (a) Packed structure of complex **10**, showing 1D tunnels parallel to the [1 0 0] direction; (b) cross-sectional layer, where the pyridine cations are not shown for clarity.

ligands in complex **11** are divided into two types: BTC-A and BTC-B (Fig. 2.19). Each InO_6 octahedron is connected to two adjacent InO_6 octahedra by μ-OH groups through corner sharing, resulting in an infinite –In–O–In–O– chain like a sinusoidal curve (Fig. 2.21). Arrangement of 12 consecutive InO_6 octahedra completes a period of the sinusoidal curve. The two bidentate carboxyl units of BTC-B are bonded to two indium atoms belonging to the neighboring chains. Thus, BTC-B ligands act as bridges connecting chains into a layered substructure as shown in Figure 2.19. These layers pack in the way of ABAB that neighboring layers are related by the half-unit-cell translation along the crystallographic *b*-axis direction and interlink by the other type of BTC ligand, BTC-A (Fig. 2.22a). The interaction between layers and BTC-A groups results in a 3D framework containing 1D tunnels of size $9.797 \times 8.917\,Å^2$ calculated from the centroid-to-centroid distances between

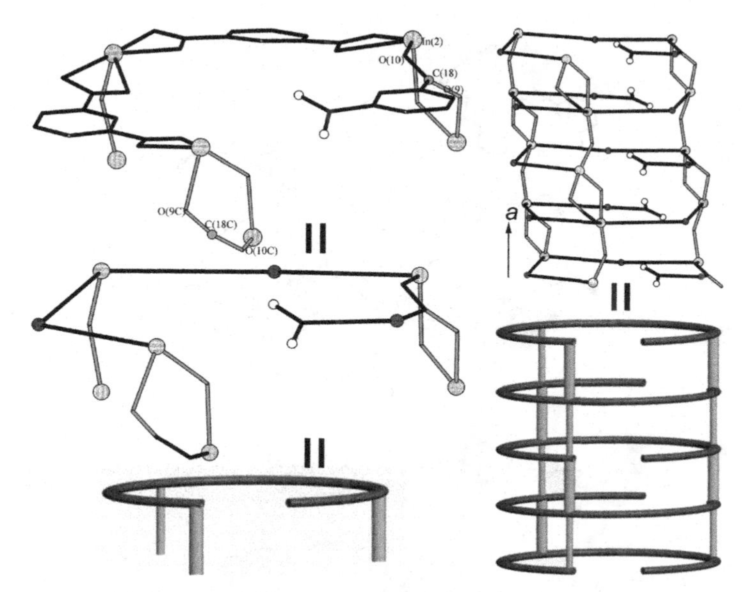

FIGURE 2.20 Conceptual diagram of **10** showing a chiral tunnel.

the benzene rings from opposite BTC^{3-} groups. The length 9.797 Å is also the distance between two neighboring layers. Protonated organic guests and free water molecules are located in the tunnels and form hydrogen bonds with the host framework. The accessible volume within the crystal is more than 36% as

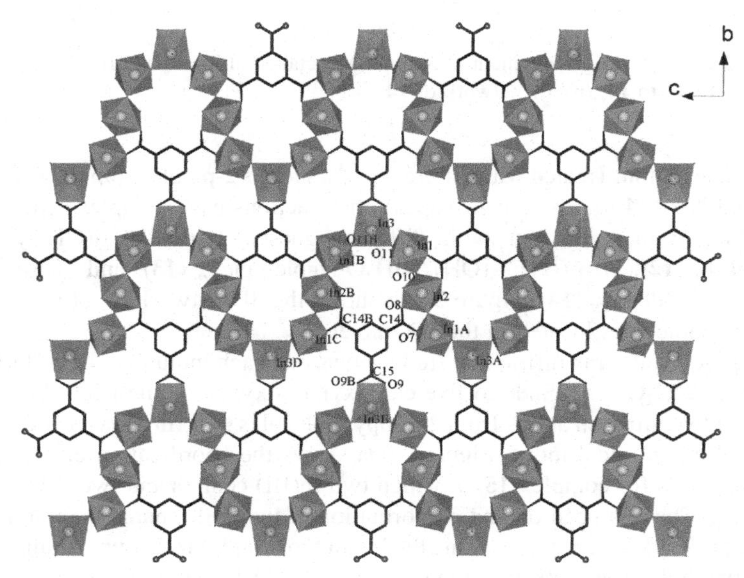

FIGURE 2.21 Layer formed through octahedral chains shaped like a sinusoidal curve in **14**. The coordination environments of BTC-B ligands can be seen clearly.

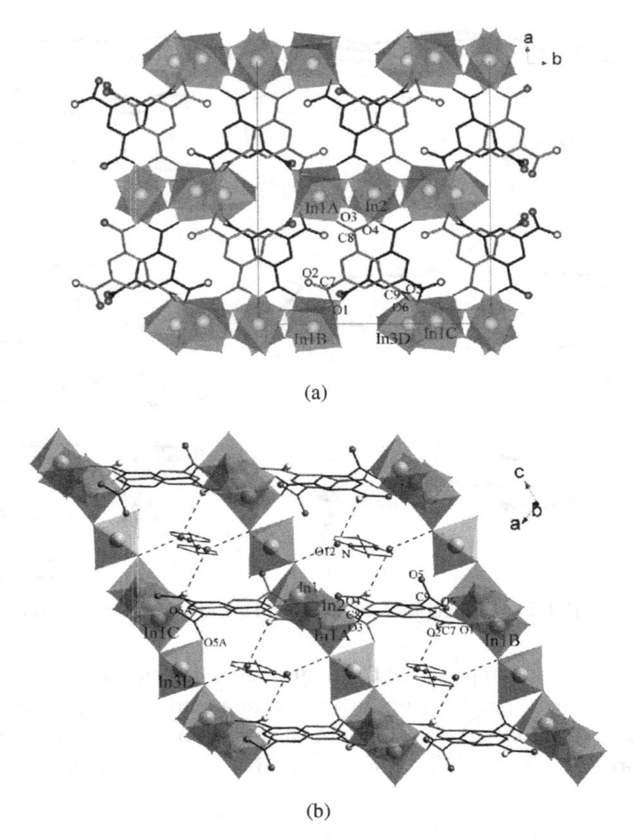

FIGURE 2.22 (a) The arrangement of alternating octahedral layers and BTC-A give a 3D framework for **11** to **14**; (b) 3D network of **14**.

calculated by Platon. Treated with different pyds, such as 2-picoline (apic), 4-picoline (gpic), and 1,2-di(4-pyridyl)ethane (dpea), the reactions gave complexes of similar In(III)–BTC frameworks but with different guests: $\{[In_4(BTC)_3(OH)_4]\cdot 3H_2O\cdot$ Hapic·apic$\}_n$ (**12**), $\{[In_4(BTC)_3(OH)_4]\cdot 3H_2O\cdot$Hgpic·gpic$\}_n$ (**13**), and $\{[In_4(BTC)_3$ $(OH)_4]\cdot 3H_2O\cdot$Hdpea$\}_n$ (**14**). Figure 2.22b shows the 3D network of **14** with protonated dpea molecules located in the 1D tunnels.

A single-crystal x-ray diffraction study shows that each indium atom in [In(BTC) (2,2′-bpy)]$_n\cdot 2n$H$_2$O (**15**) binds to five carboxylate oxygen atoms from three BTC groups and two nitrogen atoms from 2,2′-bpy molecules to form a severely distorted pentagonal bipyramidal motif. Figure 2.20a shows the coordination environments of indium atoms for complex **15**, in which two In(III) centers and two BTC anions form a grid. The composition and conformation of the grids enable them to extend along the [1 −1 0] direction by sharing the indium ions and BTC groups, resulting in a double-stranded chain motif when viewed from the [1 1 −1] direction (Fig. 2.23a). However, viewed along [1 −1 0] direction, the double-stranded chains look like

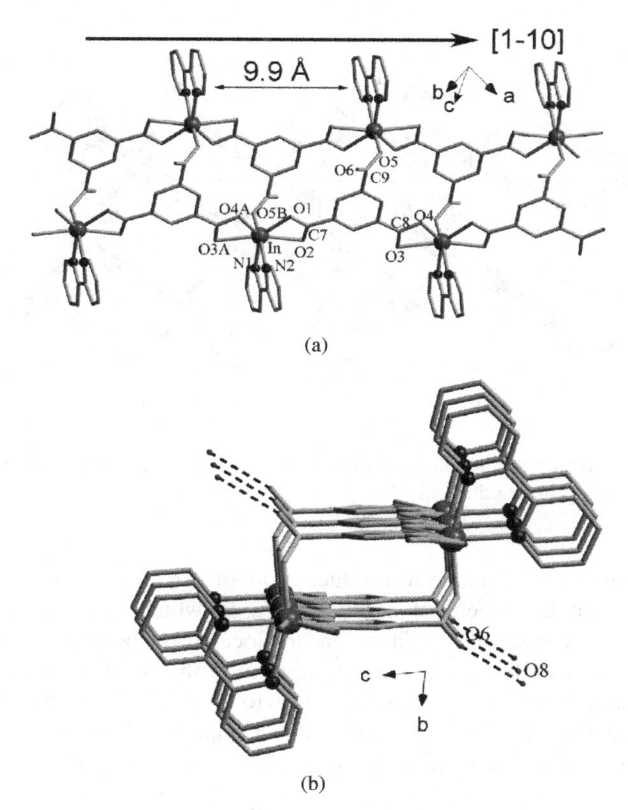

FIGURE 2.23 (a) Double-stranded chain of **15** extend along [1 1 −1] direction; (b) along the [1 −1 0] direction, the double-stranded chains can be viewed as a rectangular tunnel with 2,2′-bpy molecules chelated to the opposite indium vertexes.

rectangular tunnels (Fig. 2.23b). The tunnels in **15** with diagonal lengths of about $8.5 \times 8.5\,\text{Å}^2$ are empty since all free water molecules exist only in the interchain regions. Thus, **15** represents an unprecedented structure with double-stranded chain framework [16,71,72] and neutral independent rectangular tunnels. The 2,2′-bpy ligands link to the metal atoms from two opposite sides of the double-stranded chain, and the distance between the neighboring 2,2′-bpy ligands in the same side of the chain is 9.9 Å. These chains are parallel to each other and form a 3D supramolecular framework by hydrogen bonds between uncoordinated carboxylate oxygen atoms and free water oxygen atoms (Fig. 2.24).

In $[\text{In(BTC)(2,2'-bpy)(H}_2\text{O)}]_n \cdot n\text{H}_2\text{O}$ (**16**), each indium ion is surrounded by three oxygen atoms from three BTC groups, two nitrogen atoms from one 2,2′-bpy molecule, and one terminal water molecule to form a slightly distorted octahedral motif. Each of the three carboxylate groups of the BTC ligand binds to one indium(III) center via a monodentate mode. In this way, three indium ions and three BTC groups are connected to generate a severely distorted (3,3) grid (Fig. 2.25). These grids are

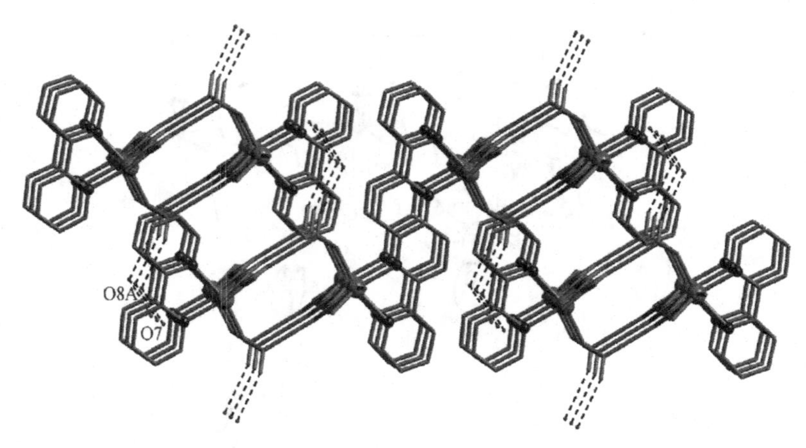

FIGURE 2.24 The tunnels are parallel to each other and form 3D supramolecular structures by hydrogen-bonding interactions in **15**.

extended through a (110) plane to produce a double-layer sheet (Fig. 2.26a), with distances between the layers of about 6.86 Å, in which the chelating 2,2′-bpy molecules coordinated to the indium ion are located on both sides of the sheet. The terminal water molecules stretch slightly out of the sheet, interacting with the carboxylate oxygen atoms in the same sheet to form intrasheet hydrogen bonds and with the carboxylate oxygen atoms in the adjacent sheet to form intersheet hydrogen bonds. The free water molecules are located inside the sheet and form

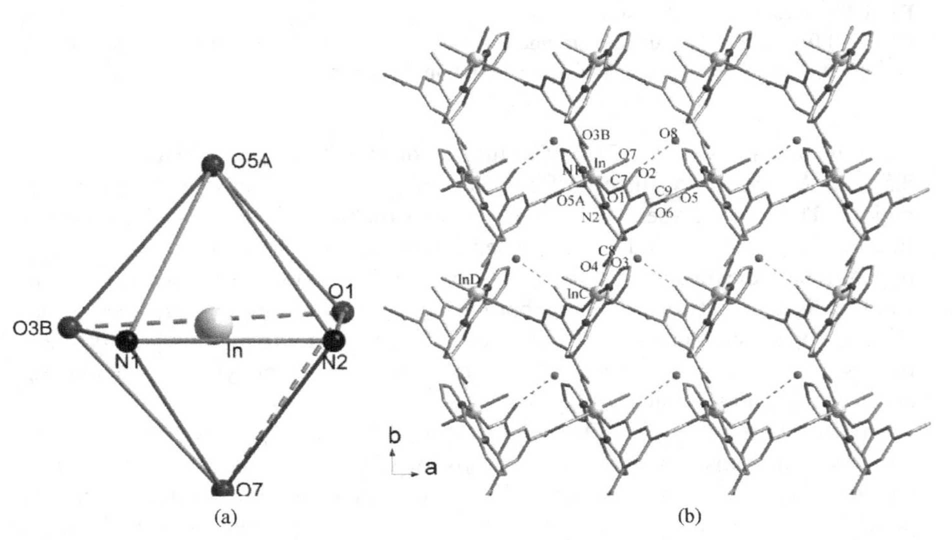

FIGURE 2.25 (a) Coordination mode of In atoms; (b) (3,3) grid single layer with hydrogen bonds in **16**.

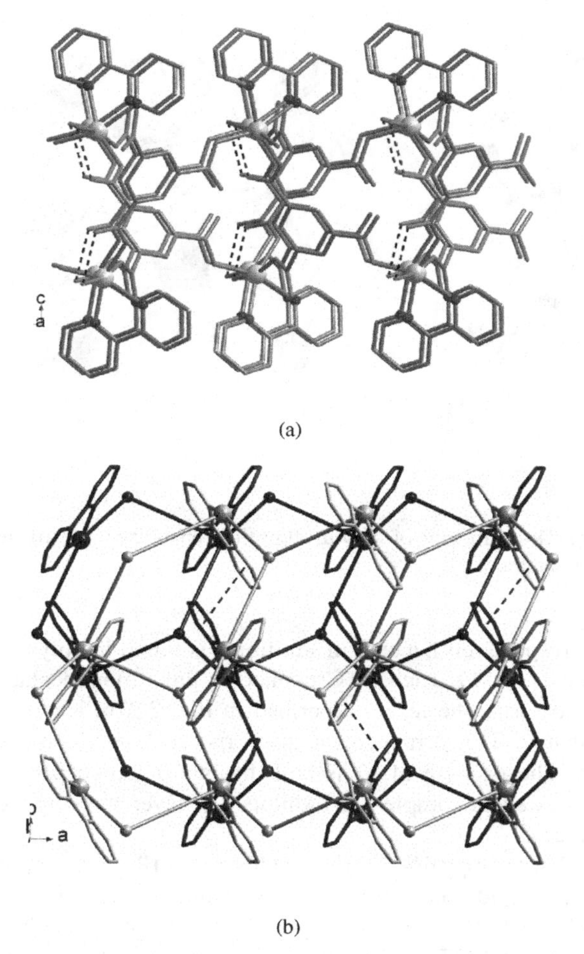

FIGURE 2.26 (a) Double-layer sheet and intersheet hydrogen bonds in **16**; (b) packing scheme and π–π interactions (dashed lines) between two sheets in **16**.

weaker hydrogen bonds with the sheet. Additionally, sheets stacked in the order ABAB have a short distance of 2.49 Å between them, which forces the 2,2′-bpy molecules from one sheet to protrude into the neighboring sheet, arousing π–π contacts in 2,2′-bpy molecules, with centroid-to-centroid distances of 3.90 to 3.75 Å (Fig. 2.26b). In this way, a 3D supramolecular network is constructed via the weak interactions mentioned above.

In $[In(BTC)(H_2O)(phen)]_n$ (**17**), the local coordination geometry around each indium atom can be described as a slightly distorted octahedron formed by two nitrogen atoms of phenanthroline molecule, three oxygen atoms of three carboxylate groups from different BTC connectors, and one water molecule (Fig. 2.27a). Despite having different crystal systems for **16** and **17** (orthorhombic

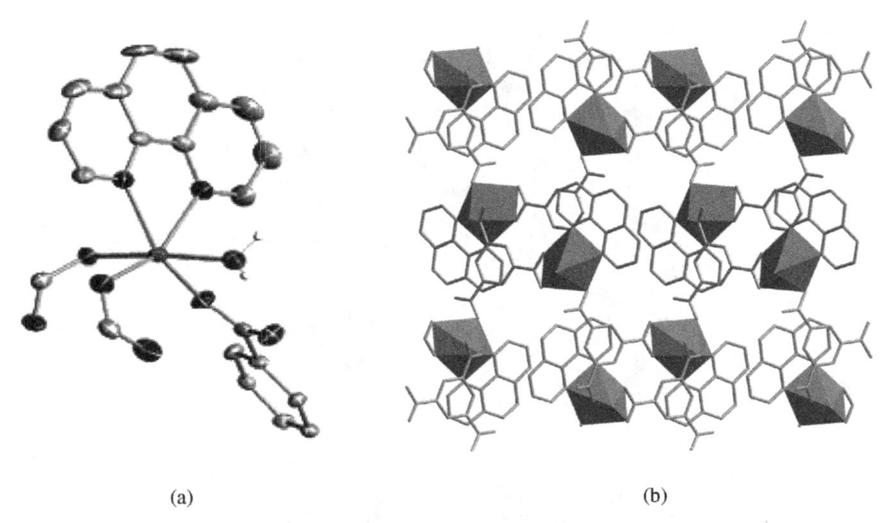

FIGURE 2.27 (a) Ortep drawing of **17** with ellipsoid probability 50%; (b) view of structures showing the In polyhedra.

and monoclinic, respectively), a certain similitude is found in polymeric packing of the two compounds. In **17**, each BTC connector links monodentately with three indium atoms and forms the layers described in Fig. 2.27b, leaving three oxygen atoms uncoordinated. This arrangement gives rise to layers perpendicular to the [1 0 0] direction. The packing of **17** is built by interdigitated parallel layers, and topology study shows that complex **17** Exhibits 4.8^2 layer stacking along the [1 0 0] direction (Fig. 2.28).

In $\{[In(HBTC)_2(4,4'\text{-bpy})](4,4'\text{-Hbpy})\cdot 0.5H_2O\}_n$ (**18**), each indium(III) center is hepta-coordinated and bonded by six oxygen and one nitrogen atom from four HBTC units and one 4,4′-bpy to form a pentangular bipyramidal geometry. Three carboxylate groups of HBTC adopt three different coordination modes; one is uncoordinated, and the other two bond in a mono- and a bidentate fashion, respectively. Thus, the HBTC acts as a bridge and connects the indium(III) centers to form a 2D network with rhombic grids. As shown in Figure 2.29a, each rhombus grid unit is formed from four indium atoms and four HBTC units with the dimensions $10.611(1) \times 10.526(1)$ Å. The grids are extended along [1 0 0] and [0 1 0] to generate a 2D layer structure. The adjacent layers are parallel to each other and pack in a staggered way with grid shifts of $a/2$ and $b/2$, as shown in Figure 2.29b. One type of 4,4′-bpy molecule in **18** acts as a terminal group and coordinates with indium vertexes of grids, and each is opposite the neighboring 4,4′-bpy ligands, which undulates the 2D layers slightly. The other type of 4,4′-bpy molecule is singly protonated and acts as a countercation. The isolated 4,4′-Hbpy molecules fill in the void space of the crystal, with the pyridyl rings being within the grid or within the layer region. The pyridyl rings of 4,4′-bpy molecules are twisted by 37.86° and 9.81° for the coordinated and solvated molecules, respectively. The

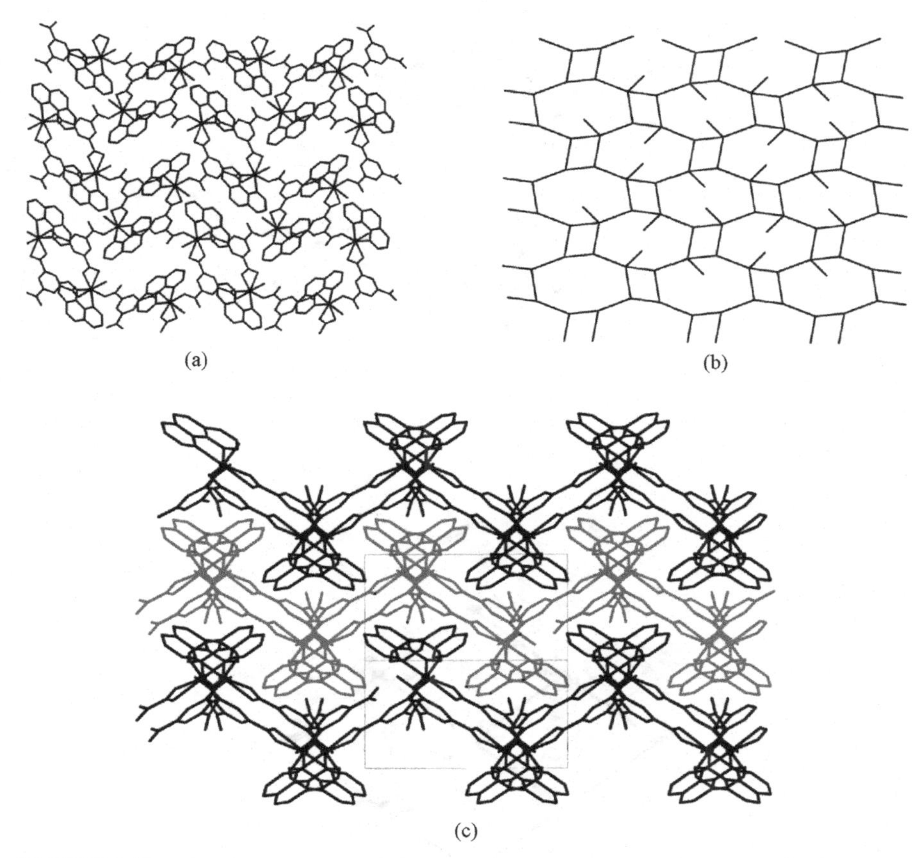

FIGURE 2.28 Topology structure of **17:** (a) molecular and (b) simplified of (4.8^2) layer down [1 0 0]; (c) stacking of layers down [1 0 1].

distance between neighboring layers amounts to approximately 8.7 Å, and they interact with hydrogen bonds.

The structure of $\{[In_6(BTC)_8]\cdot 3H_2\cdot tmdp\cdot 40H_2O\}_n$ (**19**) is a 3D coordination framework consisting of four independent intersecting channels. Each indium ion in **19** is coordinated by eight oxygen atoms from four BTC ligands to form a decahedron motif, which exhibits the excellent coordinating ability of main group elements. This therefore defines a three-connecting node within the structure. Each of the three carboxylate arms of the BTC ligand binds to an indium atom in chelating bidentate mode, thus defining a four-connecting node. These joints give rise to an $In_9(BTC)_{11}$ cagelike unit (Fig. 2.30) in which two parallel basal BTC groups are located at the top and bottom of the cage, respectively. The basal BTC groups together with nine lateral BTC groups and nine indium ions encapsulate a large space. The gigantic pores in the cage can be identified by the following data: 20.1 Å, the edge length of the triangle In1A–In1B–In1C in the middle of the cage, and 8.8 Å,

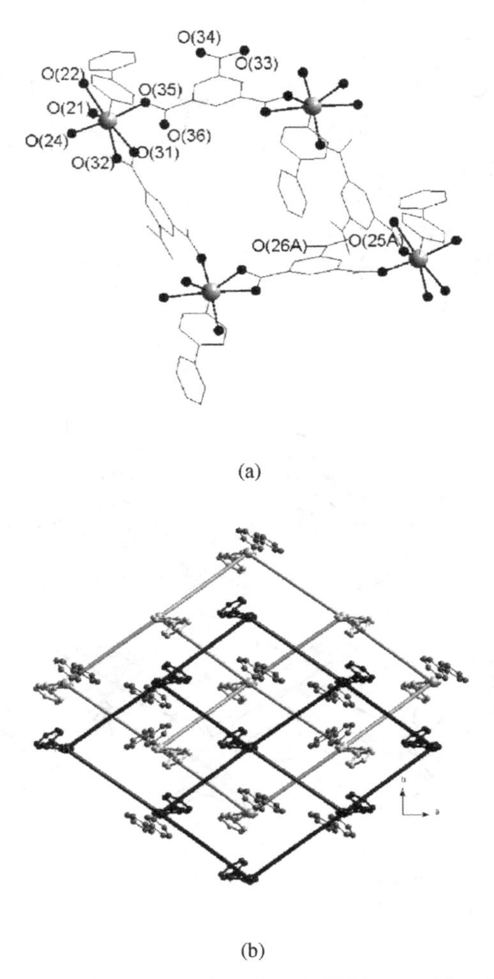

FIGURE 2.29 (a) Rhombic grid formed from four In(III) ions and four HBTCs; (b) staggered layers and the location of 4,4′-bpy molecules.

the distance between the top and bottom basal BTC groups. A threefold rotational axis passes through the centroids of the basal BTC benzene rings and thus gives a C_3 molecular symmetry. The cages aggregate to give a 3D architecture with four channels running along the diagonal body directions of the cubic lattice. These four independent sets of channels having the same shape and size and intersecting at the aforementioned void spaces are embraced by the $In_9(BTC)_{11}$ cages, resulting in a highly porous open framework (Fig. 2.31). Better insight into the nature of this complicated framework can be achieved using a topological approach. As discussed above, the In atoms act as four-coordinate centers and the BTC ligands act as four-coordinate nodes. The framework can be represented topologically as (8,3/4)—a net. The long Schläfli symbol is $8_3{\cdot}8_3{\cdot}8_3{\cdot}8_3{\cdot}8_4{\cdot}8_4$ for the In node and $8_5{\cdot}8_5{\cdot}8_5$ for the BTC

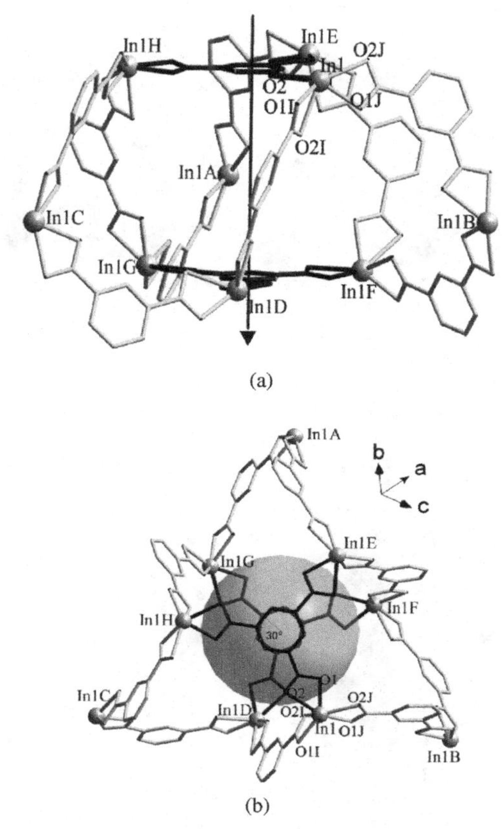

FIGURE 2.30 (a) $In_9(BTC)_{11}$ unit showing the C_3-axis; (b) another view of an $In_9(BTC)_{11}$ cage.

ligand node, giving the net the symbol $(8_5{\cdot}8_5{\cdot}8_5)_4(8_3{\cdot}8_3{\cdot}8_3{\cdot}8_3{\cdot}8_4{\cdot}8_4)_3$, which can be represented by the shorthand symbol $(8^3)_4(8^6)_3$. As estimated by the Platon program, the extra framework volume within the crystal is calculated to be 65.4%, which is 5527 Å^3 per unit cell.

2.4 ARCHITECTURES CONSTRUCTED BY In(III) AND OTHER BENZENEMULTICARBOXYLATES

In addition to the aforementioned indium complexes constructed by benzenedicarboxylate and benzenetricarboxylate ligands, there are still some examples of MOCPs based on In(III) and other benzenemulticarboxylic acids, such as 1,2,4,5-benzenetetracarboxylic acid (H_4BTEC).

Using pyridine as a base reagent to adjust the pH value in the reaction system, $[In_2(H_2BTEC)_2(OH)_2]_n{\cdot}2nH_2O$ (**20**) [31] was obtained under hydrothermal

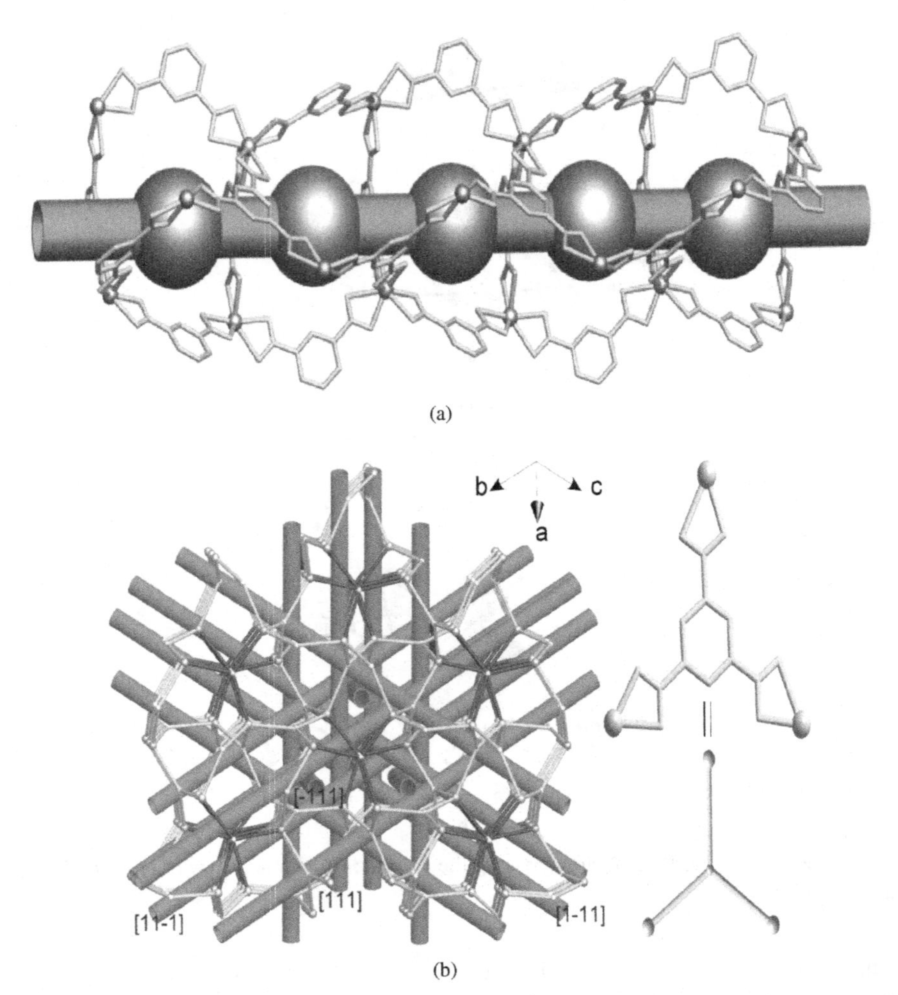

FIGURE 2.31 (a) Channel (represented by the hollow cylinders) and voids (represented by the balls) encapsulated by $In_9(BTC)_{11}$ units; (b) conceptual presentation of the porous network of **19**. The hollow cylinders represent channels. The sticks correspond to the counterparts in Figure 2.30.

conditions and its structure shows characteristic $In^{III}O_4(OH)_2$ octahedral chains. In **20**, each indium atom binds to four carboxylate oxygen atoms from four H_2BTEC groups and two other bridging oxygen atoms to form an octahedron. The octahedra are linked into infinite chains running along the [0 1 0] direction via μ-OH functions. Furthermore, two adjacent octahedra in one chain are bridged by two carboxylate arms from two different organic linkers. Each H_2BTEC group binds to four indium atoms, in a bidentate mode, situated in two different chains through the

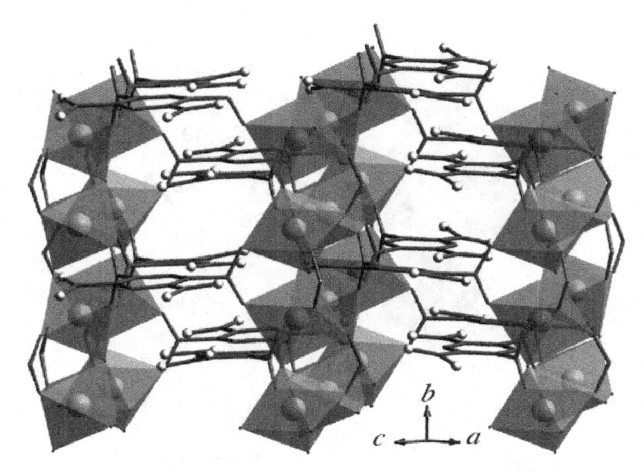

FIGURE 2.32 Interconnection between polyhedral chains and H_2BTEC groups in **20**. For clarity, free water oxygen atoms are not shown.

1,5-dicarboxylate arms. The other two carboxylate arms are uncoordinated and protonated. Geometrically, the latter two carboxylate arms are coplanar with the phenyl ring, whereas the other two are perpendicular. All phenyl rings are perpendicular to the directions of the chains, and the chains are interweaved by organic linkers [73]. These types of connections between the organic and inorganic moieties result in 1D polygonal tunnels (Fig. 2.32). Water molecules reside in the tunnels and give rise to hydrogen bonds with uncoordinated carboxylic functions.

Using 2-picoline (apic), 4-picoline (gpic), and 1,2-bis(4-pyridyl)ethane (dpea) as templates and counter anions, $\{[In_3(BTEC)_2(OH)_2]\cdot Hapic\cdot apic\}_n$ (**21**), $\{[In_3(BTEC)_2(OH)_2]\cdot Hapic\cdot apic\}_n$ (**22**), and $\{(Hdpea)[In_3(BTEC)_2(OH)_2]\}_n$ (**23**) were obtained [74]. Since they possess the same framework structure, complex **22** is selected to represent their structures. In **22**, the four carboxylate arms of the BTEC ligand are all deprotonated; two of them bind bidentately to four indium ions, and the other two adopt the chelating bidentate and monodentate modes. Two μ_3-OH groups bridge the indium ions to form an In_4O_{22} tetranuclear cluster in which four indium ions are in a plane (Fig. 2.33a). If four tetranuclear clusters and eight BTEC ligands are viewed conceptually as edges, a subunit similar to a right prism (Fig. 2.33b) is obtained. The prisms illustrated in Figure 2.33b are extended along the [1 0 0], [0 1 1], and [0 −1 1] directions to generate a 3D network (Fig. 2.34) containing the planar layers defined by In1 ions. By sharing In1 vertexes, tetranuclear clusters are extended through the crystallographic *a*-axis to give infinite chains which act as pillars to support the planar layer. The shortest contact between the pillars is 5.496 Å. The BTEC groups reproduced along the [0 1 1] axis are arranged alternately above and below the In1 planar layer. This also applies to the BTEC groups reproduced along the [0 −1 1] direction. In such an arrangement, the BTEC groups fill in the space between the pillars and therefore prevent the framework from

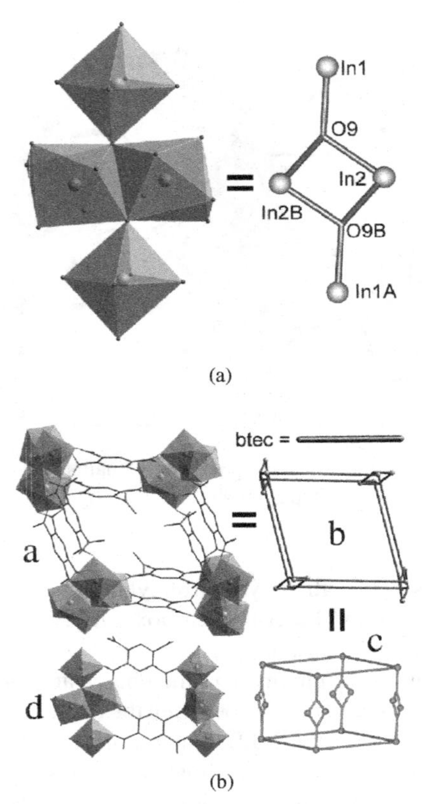

FIGURE 2.33 (a) In_4O_{22} tetranuclear cluster and its conceptual representation; (b) right prism subunit.

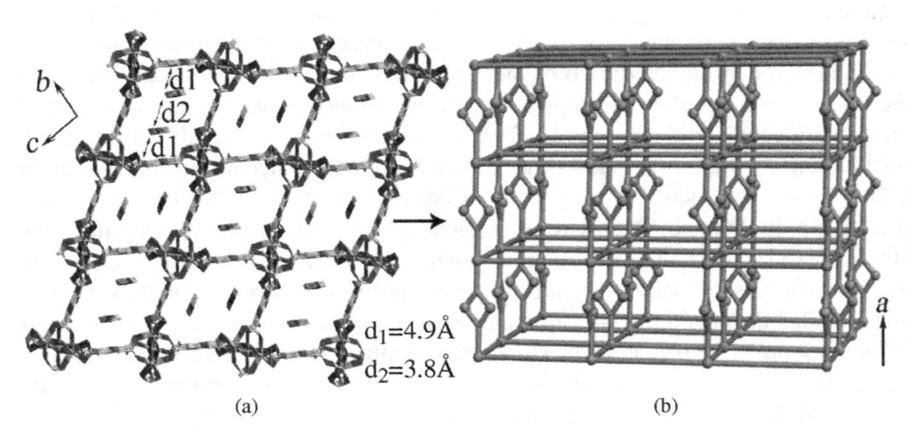

FIGURE 2.34 (a) 3D network in **22** with 1D channels containing Hgpic guests; (b) its conceptual representation (Hgpic guests are omitted for clarity). Please note the difference in view direction.

generating channels along the [0 1 1] and the [0 −1 1] directions. The only channel runs through the crystallographic *a*-axis with diagonal dimensions as long as 13.425×17.544 Å^2. The organic guests, protonated pyridine derivatives, reside in the channels. The pyridine rings of protonated gpic guests are parallel to each other and the centroid-to-centroid distances are longer than 4.5 Å, indicating no π–π contacts between pyridine rings. However, the centroid-to-centroid distances between the pyridine rings and the nearest opposite BTEC benzene rings are 3.790 Å, indicating strong π–π contact. The nitrogen atoms in Hgpic molecules have no hydrogen-bond interactions with the oxygen atoms in the framework. On the basis of Platon calculations, the solvent-accessible volume constitutes 43.3% of the total crystal volume.

Employing base 2,2′-bpy, $[In_2(BTEC)(2,2'\text{-bpy})_2Cl_2]_n$ (**24**) [68] was obtained from the hydrothermal reaction of In(III) and H_4BTEC. In **24**, each indium ion is coordinated by four oxygen atoms from two BTEC groups, two nitrogen atoms from one 2,2′-bpy molecule, and one terminal chlorine atom to form a distorted pentagonal bipyramid. One chlorine atom and one nitrogen atom occupy the axial positions of the bipyramid. Each of the four carboxylate arms of the BTEC group binds to one indium ion via a bidentate chelating mode. Arranged in this way, four indium centers and four BTEC groups give rise to a (4,2) grid (Fig. 2.35a) in which four indium centers are coplanar. The grid possesses C_i molecular symmetry with the inverse center situated at a special position (0,0.5,0), which is also the center of the grid. The opposite BTEC benzene rings are parallel with each other and located at both sides of the plane defined by the four indium centers, with centroid-to-centroid distances of the opposite rings being 16.520(9) and 9.819(6) Å, respectively. The grids are extended through the (110) plane to generate a double-layered sheet (Fig. 2.35b) comprised of two parallel indium-ion layers separated by 3.70 Å. The chelating 2,2′-bpy molecules are attached to both sides of the sheet. The sheets, separated by 6.14 Å, are stacked in ABAB order, which means that two adjacent sheets are related through the *a*-glide plane parallel to the (110) direction (Fig. 2.36). The 2,2′-bpy molecules from the adjacent sheets are interdigitated, resulting in π–π interactions, with centroid-to-centroid distances of 3.715 Å.

Using 3,3′,5,5′-azobenzenetetracarboxylic acid as the organic linker, hydrothermal reaction in a DMF/CH_3CN solution in the presence of piperazine gave orange polyhedral crystals formulated as $[In_3O(C_{16}N_2O_8H_6)_{1.5}(H_2O)_3]_n \cdot 3nH_2O \cdot n(NO_3)$ (**25**). The crystallographic analysis of **25** reveals that its structure contains an indium trimer building block $[In_3O(CO_2)_6(H_2O)_3]$, shown in Figure 2.37a. In the trimer building unit, each indium center is six-coordinated with a slightly distorted octahedral geometry in which the equatorial positions are occupied by four oxygen atoms from two organic ligands, and the axial positions are occupied by one central μ_3-oxo anion and one terminal water molecule. Three $\{InO_5(H_2O)\}$ octahedra sharing a central μ_3-oxo anion constitute the trimer building unit. Each indium trimer unit is linked by six separate organic linkers, which make the trimers act as six-connected nodes. The organic linker 3,3′,5,5′-azobenzenetetracarboxylate ligand can be viewed as a four-connected rectangular planar node, with the four carboxylate groups all adopting bidentate chelating coordination modes. The

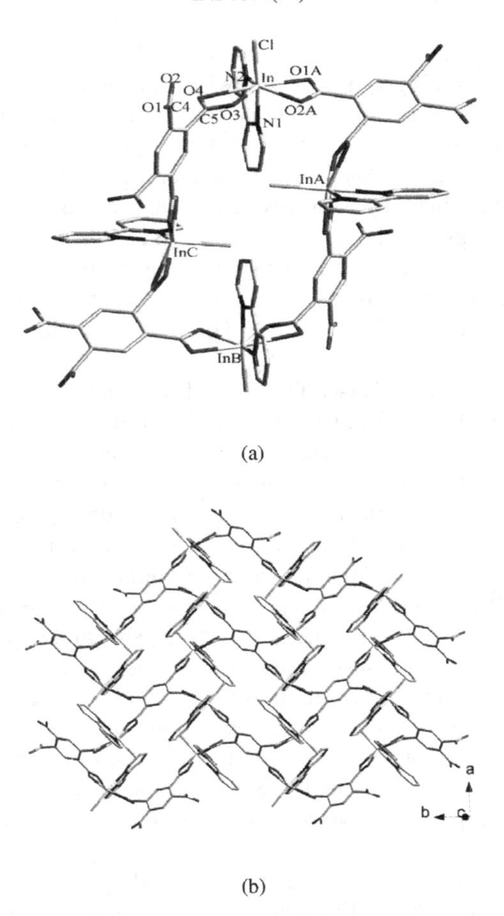

FIGURE 2.35 (a) The (4,2) grid in **24**; (b) double-layer sheet in **24**.

assembly of these two types of nodes constructs the nanosized cuboidal cage described in Figure 2.37a, in which the indium trimer units and organic ligands act as vertices and planes, respectively. By the way of sharing corners, these cubodidal cages aggregate together and result in the generation of a 3D network having the soc (soc = square–octahedron) topology [75] (Fig. 2.37b). It is worthy of note that the 3D framework of **25** exhibits two different types of extra-large infinite channels. In one type of channel, the guest water molecules are located in the free volume and form hydrogen bonds with coordinated water molecules which are pointed inside the channels. Thus, this type of channel can be considered a hydrophilic host. In the other type, the channel is guest-free in an as-synthesized complex approximately 1 nm in diameter [76]. As each indium ion is trivalent, an overall framework possesses one positive charge per formula unit. The cationic framework is balanced by disordered $[NO_3]^-$ ions, which statistically occupy two

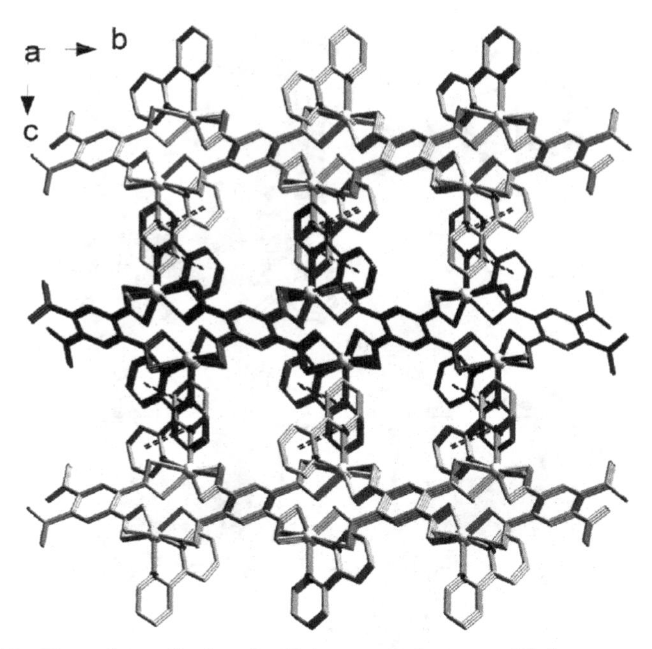

FIGURE 2.36 Three sheets displayed as light gray, dark gray, and light gray, respectively, as well as π–π interactions (dashed lines) in **24**.

positions on the threefold axis with equal probability. The total solvent-accessible volume for **25** determined by Platon software is estimated to be 57.2%.

2.5 LUMINESCENCE, ION EXCHANGE, AND HYDROGEN STORAGE

The architectures constructed by indium(III) and benzenemulticarboxylates have attracted considerable interest due not only to their intriguing topologies, but also to their rich applications in luminescence, ion exchange, and catalysis.

Among the aforementioned In(III)–BMC complexes, **9–10**, **14–16**, **18–20**, and **24** display fluorescent emissions in the solid state at room temperature [26,30,31,67–69]. The luminescent properties of these complexes are summarized in Table 2.1.

Some of the In(III)–BMC architectures possess porous open frameworks with large channels occupied by pyridine or pyridine derivatives. This type of host–guest structure allows In(III)–BMC complexes to show their competence for ion exchange. In typical ion-exchange experiments it was found that py or pyd guests in complexes **11–13**, **19**, and **21–22** can be fully or partially exchanged by K^+, NH^{4+}, Ca^{2+}, Ba^{2+}, or Sr^{2+} ions, with the remaining framework unchanged [66,69,74]. Tables 2.2 to 2.4 show the results of elemental analyses after ion-exchange experiments.

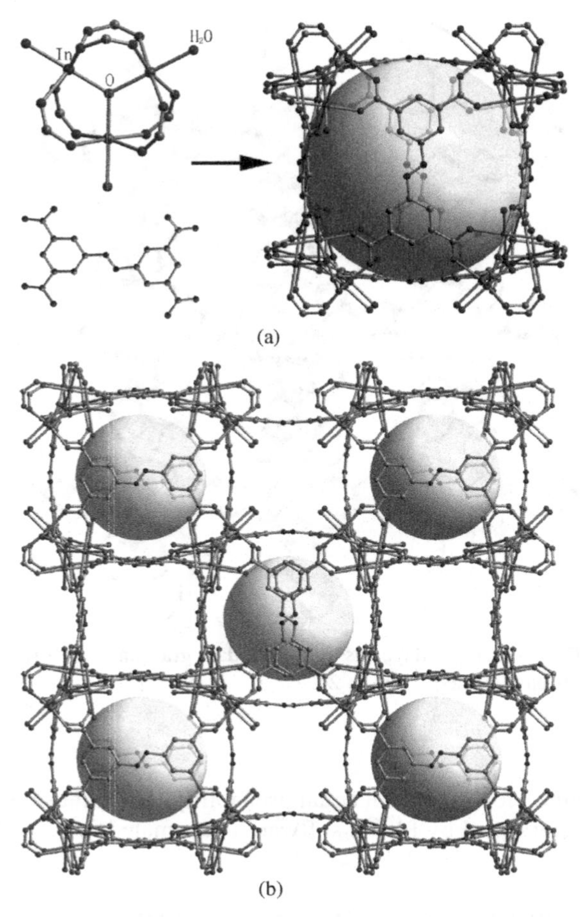

FIGURE 2.37 (a) The assembly of the oxygen-centered indium–carboxylate $[In_3O(CO_2)_6(H_2O)_3]$ and the organic linker 3,3′,5,5′-azobenzenetetracarboxylate generates the nanosized cuboidal cages in **25**; (b) a view of the 3D network exhibiting two types of extra large channels (the half-transparent spheres indicate the large cavities). Hydrogen atoms, water molecules, and $[NO_3]^-$ ions are omitted for clarity.

In addition, according to the literature [28,43], the In(III)–BMC complexes (**2**,**6**,**17**) were found to possess interesting catalytic properties in the hydrogenation of nitroaromatics, oxidation of sulfide, and acetalization of aldehydes.

2.6 CONCLUSIONS

The structures and properties of several recently published or unpublished In(III)–BMC materials have been described briefly. Introducing the main group metal ion indium(III) into the benzenemulticarboxylate system generates a series of

TABLE 2.1 Luminescent Properties of Some In(III)–BMC Complexes

Compound	Excitation Band (nm)	Emission Bands (nm)	Originated from:
9	325	494	LMCT[a]
10	332	436, 504	LMCT
14	330	444, 516	LMCT
15	637	437	LMCT
16	330	366	Energy transfer of BTC ligands
18	396	487	Interligand transition of 4,4′-bpy
19	474	535	LMCT
20	337	489	LMCT
24	340	376	Energy transfer of BTEC ligands

[a] LMCT, ligand-to-metal charge transfer.

TABLE 2.2 Elemental Analysis Results for C, N, H (%) After Ion-Exchange Experiments in 11–13

Compound	**11**	**12**	**13**
Original samples	32.63, 2.06, 2.22	33.68, 2.02, 2.47	33.68, 2.02, 2.47
K^+			
Found	28.65, 0.98, 2.35	29.43, 1.10, 2.19	29.33, 0.97, 2.20
Calc.	29.08, 1.06, 1.84	29.67, 1.05, 1.97	29.67, 1.05, 1.97
Exchange percent	100%	100%	100%
Sr^{2+}			
Found	29.90, 1.44, 2.40	29.66, 1.12, 2.45	32.11, 1.87, 2.76
Calc.	30.45, 1.46, 1.99	29.99, 1.15, 2.01	32.88, 1.83, 2.37
Exchange percent	60%	90%	20%
Ba^{2+}			
Found	29.70, 1.44, 2.56	29.12, 1.14, 2.16	31.21, 1.81, 2.16
Calc.	30.12, 1.45, 1.97	29.50, 1.13, 1.98,	32.76, 1.82, 2.38
Exchange percent	60%	90%	20%

1D, 2D, and 3D M(III)–BMC architectures totally different from those of M(II)–BMC complexes, which is due presumably to the increased valence charge and higher coordination numbers of the indium metal centers.

On the other hand, it has been considered that the hydrothermal–solvothermal method is an efficient way to assemble novel architectures based on indium and benzenemulticarboxylates. The construction of different types of topological structures is influenced primarily by three factors: the mole ratio of the indium ion and benzenemulticarboxylic acids, the pH value of the reaction system, and the utilization of organic templates. By changing the reaction conditions deliberately, systems of indium(III) and benzenemulticarboxylates can be tunable and generate different types of architectures containing diverse topologies. The research not merely enrichs the

TABLE 2.3 Elemental Analysis Results for C, N, H (%) After Ion-Exchange Experiments in 19

Original samples	37.81, 2.39, 4.82	Mg^{2+}	33.65, 1.99, 5.00
Li^{+}		Found	34.75, 1.94, 4.87
Found	33.97, 2.06, 4.96	Calc.	17%
Calc.	34.80, 1.94, 4.88	Exchange percent	
Exchange percent	17%	Ca^{2+}	
Na^{+}		Found	31.57, 1.61, 4.49
Found	34.24, 2.13, 5.14	Calc.	33.31, 1.59, 4.58
Calc.	35.18, 2.07, 4.97	Exchange percent	33%
Exchange percent	10%	Sr^{2+}	
K^{+}		Found	30.92, 1.46, 4.21
Found	32.85, 1.76, 4.56	Calc.	32.22, 1.42, 4.39
Calc.	33.73, 1.75, 4.68	Exchange percent	40%
Exchange percent	50%	Ba^{2+}	
NH_4^{+} (0.5 M)		Found	32.47, 1.85, 4.37
Found	29.83, 2.60, 4.15	Calc.	33.86, 1.84, 4.73
Calc.	29.83, 2.54, 4.39	Exchange percent	20%
Exchange percent	73%	Sr^{2+} (0.3 M)	
K^{+} (0.3 M)		Found	26.84 0.30, 3.26
Found	28.91, 0.81, 3.64	Calc.	27.42, 0.33, 3.44
Calc.	29.70, 0.82, 3.88	Exchange percent	85%
Exchange percent	67%		

TABLE 2.4 Elemental Analysis Results for C, N, H (%) After Ion-Exchange Experiments in 21 and 22

Compound	**21**	**22**
Original samples	36.03, 2.63, 1.99	36.03, 2.63, 1.99
Ca^{2+}		
Found	29.65, 1.21, 2.10	27.69, 1.19, 2.10
Calc.	28.84, 1.30, 2.15	28.84, 1.30, 2.15
Exchange percent	100%	100%
Ba^{2+}		
Found	27.42, 1.30, 1.82	25.82, 1.18, 1.80
Calc.	27.60, 1.24, 2.06	27.60, 1.24, 2.06
Exchange percent	100%	100%

world of metal–benzenemulticarboxylate systems but also discloses an entirely new strategy of design and construction metal–organic coordination polymer architectures.

Acknowledgments

We are thankful for financial support from the National Nature Science Foundation of China and the Nature Science Foundation of Fujian Province.

REFERENCES

1. Li, H. L.; Eddaoudi, M.; O'Keeffe, M.; Yaghi, O. M. *Nature* **1999**, *402*, 276–279.
2. Hagrman, P. J.; Hagrman, D.; Zubieta, J. *Angew. Chem., Int. Ed.* **1999**, *38*, 2638–2684.
3. Batten, S. R.; Robson, R. *Angew. Chem., Int. Ed.* **1998**, *37*, 1460–1494.
4. Sato, O.; Iyoda, T.; Fujishima, A.; Hashimoto, K. *Science* **1996**, *271*, 49–51.
5. Evans, O. R.; Xiong, R.; Wang, Z.; Wong, G. K.; Lin, W. *Angew. Chem., Int. Ed.* **1999**, *38*, 536–538.
6. Eddaoudi, M.; Kim, J.; Rosi, N.; Vodak, D.; Wachter, J.; O'Keeffe, M.; Yaghi, O. M. *Science* **2002**, *295*, 469–499.
7. Harrison, R. G.; Fox, O. D.; Meng, M. O.; Dalley, N. K.; Barbour, L. J. *Inorg. Chem.* **2002**, *41*, 838–843.
8. Cheetham, A. K.; Férey, G.; Loiseau, T. *Angew. Chem., Int. Ed.* **1999**, *38*, 3268–3292.
9. Zaworotko, M. *Nature* **1997**, *386*, 220–226.
10. Miller, J. S. *Adv. Mater.* **2001**, *13*, 525–527.
11. Barton, S. C.; Kim H. H.; Binyamin, G.; Zhang, Y.; Heller, J. *J. Am. Chem. Soc.* **2001**, *123*, 5802–5803.
12. James, S. L. *Chem. Soc. Rev.* **2003**, *32*, 276–288.
13. Plater, M. J.; Foreman, M. R. S. E.; Gómez-García Coronado, C. J.; Slawin, A. M. Z. *J. Chem. Soc., Dalton Trans.* **1999**, *23*, 4209–4216.
14. Yaghi, M.; Li, H. L.; Groy, T. L. *J. Am. Chem. Soc.* **1996**, *118*, 9096–9101.
15. Choi, H. J.; Suh, M. P. *J. Am. Chem. Soc.* **1998**, *120*, 10622–10628.
16. Plater, M. J.; Howie, R. A.; Roberts, A. J. *Chem. Commun.* **1997**, 893–894.
17. Kepert, C. J.; Prior, T. J.; Rosseinsky, M. J. *J. Solid State Chem.* **2000**, *152*, 261–270.
18. Cheng, D. P.; Khan, M. A.; Houser, R. P. *Inorg. Chem.* **2001**, *40*, 6858–6859.
19. Suh, M. P.; Ko, J. W.; Choi, H. J. *J. Am. Chem. Soc.* **2002**, *124*, 10976–10977.
20. Prior, T. J.; Rosseinsky, M. J. *Chem. Commun.* **2001**, 1222–1223.
21. Chen, W.; Wang, J. Y.; Chen, C.; Yue, Q.; Yuan, H. M.; Chen, J. S.; Wang, S. N. *Inorg. Chem.* **2003**, *42*, 944–946.
22. Prior, T. J.; Rosseinsky, M. J. *Chem. Commun.* **2001**, 495–496.
23. Su, J. R.; Yin, K. L.; Xu, D. J. *Chin. J. Struct. Chem.* **2004**, *23*, 399–402.
24. Prior, T. J.; Rosseinsky, M. J. *Inorg. Chem.* **2003**, *42*, 1564–1575.
25. Daiguebonne, C.; Guilloa, O.; Gérault, Y.; Lecerf, A.; Boubekeur, K. *Inorg. Chim. Acta* **1999**, *284*, 139–145.
26. Lin, Z. Z.; Luo, J. H.; Hong, M. C.; Wang, R. H.; Han, L.; Xu, Y.; Cao, R. *J. Solid State Chem.* **2004**, *177*, 2494–2498.
27. Thirumurugan, A.; Natarajan, S. *Dalton Trans.* **2003**, 3387–3391.
28. Gomez-Lor, B.; Gutiérrez-Puebla, E.; Iglesias, M.; Monge, M. A.; Ruiz-Valero, C.; Snejko, N. *Inorg. Chem.* **2002**, *41*, 2429–2432.
29. Chen, Z. X.; Zhao, Y. M.; Weng, L. H.; Zhang, H. Y.; Zhao, D. Y. *J. Solid State Chem.* **2003**, *173*, 435–441.

30. Lin, Z. Z.; Jiang, F. L.; Chen, L.; Yuan, D. Q.; Hong, M. C. *Inorg. Chem.* **2005**, *44*, 73–76.
31. Lin, Z. Z.; Jiang, F. L.; Chen, L.; Yuan, D. Q.; Zhou, Y. F.; Hong, M. C. *Eur. J. Inorg. Chem.* **2005**, 77–81.
32. Sun, J. Y.; Weng, L. H.; Zhou, Y. M.; Chen, J. X.; Chen, Z. X.; Liu, Z. C.; Zhao, D. Y. *Angew. Chem., Int. Ed.* **2002**, *41*, 4471–4473.
33. Hsieh, W. Y.; Liu, S. *Inorg. Chem.* **2004**, *43*, 6006–6014.
34. Luo, B.; Cramer, C. J.; Gladfelter, W. L. *Inorg. Chem.* **2003**, *42*, 3431–3437.
35. Li, H. L.; Kim, J.; O'Keeffe, M.; Yaghi, O. M. *Angew. Chem., Int. Ed.* **2003**, *42*, 1819–1821.
36. Férey, G. *Angew. Chem., Int. Ed.* **2003**, *42*, 2576–2579.
37. Cahill, C. L.; Gugliotta, B.; Parise, J. B. *Chem. Commun.* **1998**, 1715–1716.
38. Holmes, K. E.; Kelly, P. F.; Elsegood, M. R. J. *Dalton Trans.* **2004**, 3488–3494.
39. Li, X. J.; Cao, R.; Sun, D. F.; Sun, Y. Q.; Bi, W. H.; Wang, Y. Q.; Hong, M. C. *Chin. J. Struct. Chem.* **2004**, *23*, 1017–1021.
40. Kepert, C. J.; Rosseinsky, M. J. *Chem. Commun.* **1998**, 31–32.
41. Anokhina, E. V.; Vougo-Zanda, M.; Wang, X.; Jacobson, A. J. *J. Am. Chem. Soc.* **2005**, *127*, 15000–15001.
42. Gao, Q.; Lin, Z. Z.; Hong, M. C.; Unpublished work.
43. Gomez-Lor, B.; Gutiérrez-Puebla, E.; Iglesias, M.; Monge, M. A.; Ruiz-Valero, C.; Snejko, N. *Chem. Mater.* **2005**, *17*, 2568–2573.
44. Wang, Y. L.; Liu, Q. Y.; Zhong, S. L. *Acta Crystallogr.* **2006**, *C62*, m395–m397.
45. Wade, K.; Banster, A. J. *Comprehensive Inorganic Chemistry*, Vol. 1, Dickenson, New York, 1973, pp. 1105–1107.
46. Wells, A. F. *Structure Inorganic Chemistry*, 5th ed., Oxford University Press, Oxford, 1983.
47. Prewitt, C. T.; Shannon, R. D.; Rogers, D. B.; Sleight, W. W. *Inorg. Chem.* **1969**, *8*, 1985–1993.
48. Harris, K. D. M. *Phase Transitions* **2003**, *76*, 205–218.
49. Bourgeois, L.; Toudic, B.; Ecolivet, C.; Ameline, J. C.; Bourges, P.; Guillaume, F.; Breczewski, T. *Phys. Rev. Lett.* **2004**, *92*, 026101-1-4.
50. Hill, T. L. *Biophys. Chem.* **1986**, *25*, 1–15.
51. Li, R.; Petricek, V.; Yang, G.; Coppens, P.; Naughton, M. *Chem. Mater.* **1998**, *10*, 1521–1529.
52. Bulc, N.; Goliè, L. *Acta Crystallogr.* **1983**, *C39*, 174–176.
53. Abram, S.; Maichle-Mössmer, C.; Abram, U. *Polyhedron* **1997**, *16*, 2183–2191.
54. Preut, H.; Huber, F. Z. *Anorg. Allg. Chem.* **1979**, *450*, 120–130.
55. Carlucci, L.; Ciani, G.; Proserpio, D. M. *Coord. Chem. Rev.* **2003**, *246*, 247–289.
56. Barthelet, K.; Riou, D.; Férey, G. *Chem. Commun.* **2002**, 1492–1493.
57. Carlucci, L.; Ciani, G.; Proserpio, D. M.; Sironi, A. *Angew. Chem., Int. Ed.* **1995**, *34*, 1895–1898.
58. Subramanian, S.; Zaworotko, M. J. *Angew. Chem., Int. Ed.* **1995**, *34*, 2127–2129.

59. Prior, T. J.; Bradshaw, D.; Teat, S. J.; Rosseinsky, M. J. *Chem. Commun.* **2003**, 500–501.

60. Dai, J. C.; Wu, X. T.; Fu, Z. Y.; Cui, C. P.; Hu, S. M.; Du, W. X.; Wu, L. M.; Zhang, H. H.; Sun, R. Q. *Inorg. Chem.* **2002**, *41*, 1391–1397.

61. Fan, J.; Zhu, H. F.; Okamura, T. A.; Sun, W. Y.; Tang, W. X.; Ueyama, N. *New J. Chem.* **2003**, *27*, 1409–1411.

62. Kim, J.; Chen, B.; Reineke, T. M.; Li, H.; Eddaoudi, M.; Moler, D. B.; O'Keeffe, M.; Yaghi, O. M. *J. Am. Chem. Soc.* **2001**, *123*, 8239–8247.

63. Guilloa, O.; Boubekeur, K. *J. Alloys Compd.* **2002**, *344*, 179–185.

64. Daiguebonne, C.; Gérault, Y.; Guillou, O.; Lecerf, A.; Boubekeur, K.; Batail, P.; Kahn, M.; Kahn, O. *J. Alloys Compd.* **1998**, *275*, 50–53.

65. Serre, C.; Férey, C. *J. Mater. Chem.* **2002**, *12*, 3053–3057.

66. Lin, Z. Z.; Chen, L.; Yue, C. Y.; Yuan, D. Q.; Jiang, F. L.; Hong, M. C. *J. Solid State Chem.* **2006**, *179*, 1154–1160.

67. Lin, Z. Z.; Chen, L.; Jiang, F. L.; Hong, M. C. *Inorg. Chem. Commun.* **2005**, *8*, 199–201.

68. Lin, Z. Z.; Chen, L.; Yue, C. Y.; Yan, C. F.; Jiang, F. L.; Hong, M. C. *Inorg. Chim. Acta* **2008**, *361*, 2821–2827.

69. Lin, Z. Z.; Jiang, F. L.; Chen, L.; Yue, C. Y.; Yuan, D. Q.; Lan, A. J.; Hong, M. C. *Cryst. Growth Des.* **2007**, *7*, 1712–1715.

70. Spek, A. L. *Acta Crystallogr.* **1990**, *A46*, C34–C37.

71. Plater, M. J.; Foreman, M. R. S.; Howie, R. A.; Skakle, J. M. S.; Coronado, E.; Gómez-García, C. J.; Gelbrich, T.; Hursthouse, M. B. *Inorg. Chim. Acta* **2001**, *319*, 159–175.

72. Wang, Y. Q.; Cao, R.; Sun, D. F.; Bi, W. H.; Li, X.; Li, X. J. *J. Mol. Struct.* **2003**, *657*, 301–309.

73. Barthelet, K.; Riou, D.; Nogues, M.; Férey, G. *Inorg. Chem.* **2003**, *42*, 1739–1743.

74. Lin, Z. Z.; Jiang, F. L.; Yuan, D. Q.; Chen, L.; Zhou, Y. F.; Hong, M. C. *Eur. J. Inorg. Chem.* **2005**, 1927–1931.

75. O'Keeffe, M.; Eddaoudi, M.; Li, H.; Reineke, T.; Yaghi, O. M. *J. Solid State Chem.* **2000**, *152*, 3–20.

76. Yang, G.; Sevov, S. C. *J. Am. Chem. Soc.* **1999**, *121*, 8389–8390.

3

CRYSTAL ENGINEERING OF COORDINATION POLYMERS VIA SOLVOTHERMAL IN SITU METAL–LIGAND REACTIONS

JIE-PENG ZHANG AND XIAO-MING CHEN
MOE Laboratory of Bioinorganic and Synthetic Chemistry, School of Chemistry and Chemical Engineering, Sun Yat-Sen University, Guangzhou, People's Republic of China

3.1 INTRODUCTION

The design and construction of coordination polymers have been studied intensively in recent years, as evidenced by the very rapid growth of publications. It is now well known that crystal engineering of coordination polymers depends not only on judicious design or choice of molecular building blocks but also suitable synthetic conditions. Organic building blocks can be designed to have particular functional groups and coordination sites having specified geometries, which facilitate rational construction of desired structures and functions by the aid of appropriate transition-metal ions [1–4]. Actually, an appropriate organic ligand is usually responsible for the fascinating structure and unique property of a particular coordination polymer. Nevertheless, the syntheses and/or handling of such molecular building blocks are sometimes complicated and time consuming, and the self-assembly processes are sensitive to diversified synthetic conditions. In fact, a variety of superstructures can be obtained for a given set of metal ions and organic ligands by alternation of reaction conditions such as the solvent, pH, counter ion, temperature, and pressure [5–8]. As noted by Moulton and Zaworotko, control over the supramolecular level of coordination polymers or supramolecular isomers lies at the heart of the concept of crystal

Design and Construction of Coordination Polymers, Edited by Mao-Chun Hong and Ling Chen

engineering of coordination polymers [6]. Therefore, seeking new, rational, and effective methods for assembling coordination polymers of specific or desired structures and properties is still an important issue, although crystal engineering of coordination polymers has been explored for almost two decades.

Solvothermal (including hydrothermal) conditions have been used widely in the crystallization of inorganic materials and coordination polymers [7]. Solvothermal reactions are typically carried out in sealed Teflon autoclaves or glass tubes under autogenous pressure at temperatures higher than the normal boiling points of the solvents. Under solvothermal conditions, crystal growth of highly insoluble materials is highly favored, due to the enhanced solubility of the solutes and the reduced solvent viscosity, which significantly enhance the diffusion processes and the recrystallization speed. As the solubility problems are minimized, the solvothermal reaction variables, such as solvent, pH, stoichiometry, template, and additive, can be changed readily to exploit new metastable phases. Actually, compared with direct self-assembly approaches, solvothermal reactions are powerful for the discovery of new crystalline organic–inorganic materials and coordination polymers, leading to the very interesting structural diversity with different, even tunable functionalities of the products.

Moreover, the elevated temperature and pressure, as well as the presence of metal ions, mainly transition-metal ions [9], may promote organic reactions to generate new organic ligands during the solvothermal generation of coordination polymers. Simultaneously, redox-active metal species can also change their valence states under solvothermal conditions. Obviously, the in situ metal–ligand reactions, which produce molecular building blocks not presented in the starting materials, are additional variables in solvothermal reactions. Aside from the advantages of solvothermal conditions, in situ metal–ligand reactions have at least two advantages for crystal engineering of metal–organic complexes or frameworks:

1. The in situ formation of molecular building blocks may provide new routes for crystalline products that are not easily accessible or are even infeasible by direct use of the ligands and metal ions with a particular oxidation state.
2. The in situ formation of molecular building blocks and other species (such as reaction intermediates and by-products) may render new thermodynamic variables and nonequilibrium crystallization environments, leading to the discoveries of new crystalline products that are not accessible by direct methods.

The unique reaction environments of these solvothermal reactions have led to the discovery of many novel organic reactions infeasible for conventional methods. Regarding their unique thermodynamic and kinetic environments, the solvothermal in situ metal–ligand reactions can be considered as an important strategy for crystal engineering of coordination polymers and have been used to synthesize many novel coordination polymers in recent years, some of which are inaccessible by direct self-assembly approaches. More commonly, solvothermal in situ metal–ligand reactions can rationally be used to convert easily available starting materials to new molecular building blocks that are difficult to obtain or handle by conventional methods [10–13].

As reviewed recently with emphasis on the reaction chemistry, various metal–organic reactions can be carried out under solvothermal conditions [12,13]. However, only those that can produce suitable molecular building blocks are of particular interest for the crystal engineering of coordination polymers. The most important reaction types in this context involve the conversion of carboxylic acids, carbon–carbon bond formation, heterocycle formation from small molecules, and transformation of sulfur-containing ligands, as well as the redox conversion of metal species. In this chapter, through a series of selected examples, we summarize in situ metal–ligand reactions as new approaches to the crystal engineering of coordination polymers.

3.2 METAL-REDOX REACTION

Compared to the difficulty of ligand syntheses, most transition-metal ion building blocks (in different oxidation states) are readily available as starting materials. However, some of the compounds targeted are difficult, even impossible, to prepare directly using metal ions with corresponding oxidation states. For example, the homoleptic framework [Eu(tzpy)$_2$] (**1**, Htzpy = 1*H*-1,2,3-triazolo[4,5-*b*]pyridine) was prepared solvothermally by metallic europium, Htzpy ligand, and pyridine, as well as mercury for activation of the metal [14]. Considering the extraordinary oxophilicity of rare earth elements, it would be difficult to get nitrogen donors of ligand to fully occupy the coordination sphere of the Eu(II) ion and to furnish the crystallization of a binary [Eu(tzpy)$_2$] coordination polymer in the presence of multiple components such as anions and oxygen-based solvents. The employment of Eu(0) to Eu(II) in situ oxidation reaction is crucial for assembly of the binary metal–organic framework of **1** without introducing any other coordinative species.

$$n\text{Eu} + 2n\text{Htzpy} \xrightarrow{\Delta T} [\text{Eu(tzpy)}_2]_n + n\text{H}_2$$

More commonly, many transition-metal ions are capable of adopting different oxidation states in solution and solid states. Transition-metal ions with a particular oxidation state may be oxidized or reduced by dissolved oxygen gas and/or organic ligands. For thermodynamic and kinetic reasons, some redox reactions are usually slow or virtually impossible under ambient conditions but are performed readily at elevated temperature and pressure. So far, the most studied in situ metal-redox reaction is Cu(II)-to-Cu(I) reduction. Soluble Cu(I) salts are easily oxidized by air into Cu(II) or spontaneously disproportionate into Cu(0) and Cu(II) in water. In crystal engineering, the use of Cu(I) is limited in nonaqueous solutions (or protected by strong donor ligands) and an inert atmosphere, in contrast to the fact that Cu(II) salts are stable at ambient conditions in both the solution and solid states. Fortunately, in situ reduction of Cu(II) can be a convenient method to produce Cu(I) for organic reactions and crystal growth of coordination polymers. The Cu(II)/Cu(I) couple has a low standard electrode potential ($E^0 = 0.16$ V) and a positive slope against temperature, indicating that the weak tendency of Cu(II)-to-Cu(I) reduction at room temperature can be promoted by increasing the reaction temperature. More important, higher temperatures can significantly increase the reaction rate from a kinetic point of view.

$$Cu^{2+}\,(aq) \xrightarrow{0.16\ V} Cu^{+}\,(aq) \xrightarrow{0.52\ V} Cu$$

Actually, Cu(II) can readily be converted solvothermally into Cu(I) in the presence of different types of heterocycle species [15]. Therefore, Cu(I) and Cu(I,II) coordination polymers can be prepared conveniently from Cu(II) salts in water without inert atmosphere protection. The fact that the reaction speed and equilibrium from Cu(II) to Cu(I) can be tuned continuously by temperature is also important for the crystal growth of some Cu(I) and Cu(I,II) coordination polymers. For example, imidazolate derivatives are very strong exobidentate ligands. Binary univalent metal imidazolates tend strongly to form the kinetic or rapidly deposited products of one-dimensional (1D) chainlike polymers for their very low solubility, arising from the strong coordination bonds and the polymeric nature. Consequently, conventional synthetic approaches usually yield microcrystals not large enough for x-ray single-crystal diffraction [5]. In contrast, solvothermal in situ generation of Cu(I) ions from Cu(II) salts in aqueous ammonia solution can easily lead to crystallization of the rather insoluble 1D chainlike Cu(I) imidazolates into very good quality single crystals for x-ray diffraction [16]. More pronouncedly, the thermodynamically favored zero-dimensional (0D) polygonal isomers become possible under such reaction conditions [17]. Uniform molecular octagons $[Cu_8(mim)_8]$ (mim = 2-methylimidazolate) (**2**) and decagons $[Cu_{10}(mim)_{10}]$ (**3**) have been designed and synthesized by introducing methyl groups as the hydrophobic acceptors and suitable templates such as benzene, toluene, xylene, and mesitylene (Fig. 3.1). During the reactions, ammonia functions as both a base and a coordination-buffering agent to reduce the crystallization speed by stabilizing the Cu(I) ion and soluble intermediate Cu(I)-mim species, thus avoiding rapid crystallization of the chainlike species as the major products, whereas the hydrophobic supramolecular interactions between the methyl groups of imidazolates and templates render additional supramolecular stabilization energy for the formation of crystalline **2** and **3**.

The significance of these tunable in situ Cu(II)-to-Cu(I) reduction reactions is exemplified further by the convenient synthesis of a series of Cu(I), Cu(I,II), and Cu(II) imidazolates from the Cu(II) salt and imidazole starting materials. By simply increasing the reaction temperature in the range 120 to 160°C, the copper ions in the resulting products vary from pure Cu(II), 1 : 1 and 2 : 1 mixtures of Cu(I) and Cu(II), to pure Cu(I), resulting in the diverse structural motifs for a simple copper imidazolate system (Fig. 3.2) [18].

Obviously, the oxidation of the metal to cation involves the formation of molecular hydrogen; hence, the reactions should be carried out in very small amounts. On the other hand, organic ligands are commonly suggested as the reducing agents for Cu(II)-to-Cu(I) conversions, but the oxidized products are

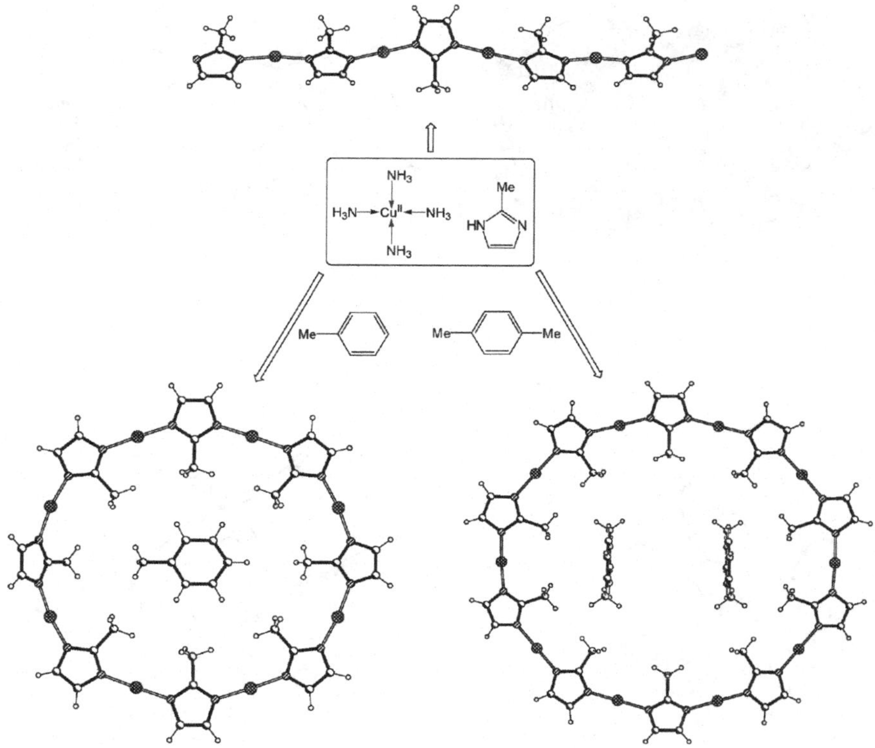

FIGURE 3.1 1D and 0D $[Cu^I(mim)]_n$ isomers prepared by solvothermal reactions of Hmim, Cu(II), and NH_3.

usually unclear. Nevertheless, some of oxidized products are presented in the final products as new ligands. An interesting example worthy of mention is the copper-assisted oxidative hydroxylation of 2,2′-bipyridine-like ligands, which provides not only structural evidence of a covalent hydration mechanism, but also a powerful and convenient synthetic method for novel ligands [19,20]. More oxidative in situ ligand syntheses are discussed in the following sections.

Except for the examples mentioned above, other types of solvothermal in situ metal-valence conversions have not been used widely so far in the crystal engineering of coordination polymers. Nevertheless, this approach is expected to cover other redox-active metals that are capable of adopting different oxidation states in solution and solid states. In particular, this in situ generational approach is critically important for those compounds that are difficult to synthesize by the direct input of metal ions with the targeted oxidation states, either metastable or mixed-valence oxidation states. Sometimes, the in situ metal redox reactions are facile or obligatory for the solvothermal preparation of coordination polymers with the metal ions in a specific oxidation state. For instance, iron(II) can easily be oxidized to iron(III) under ordinary or solvothermal conditions, especially in the presence of air. We have recently

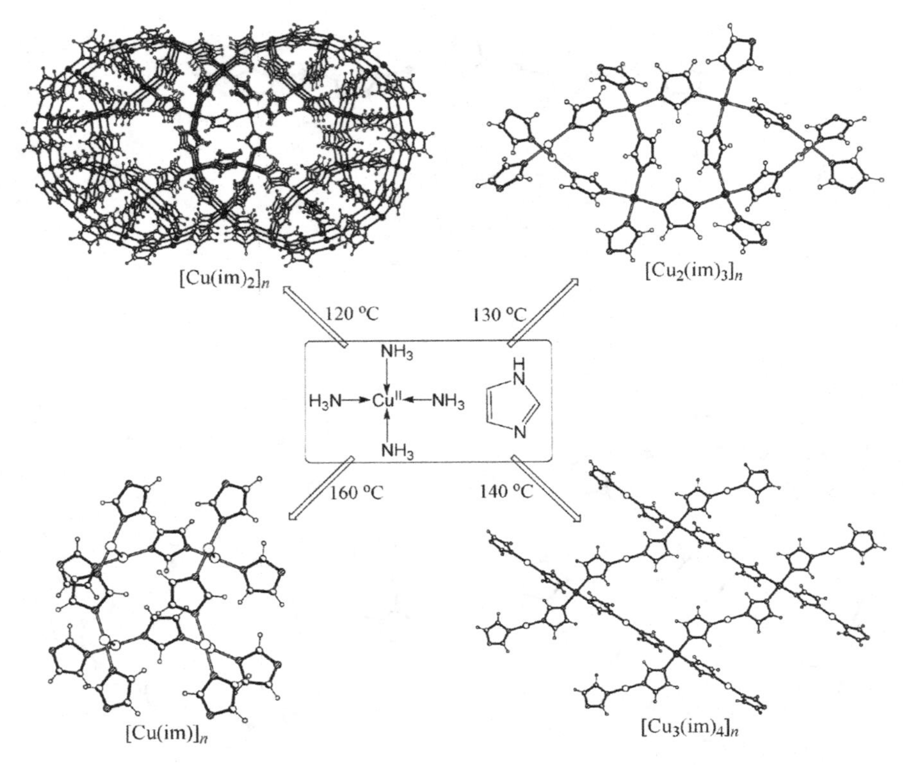

FIGURE 3.2 Metal-valence tuning of the copper imidazolate system.

synthesized a new two-dimensional (2D) iron(II) coordination polymer $[Fe(pyoa)_2]_\infty$ (**4**) [pyoa = 2-(pyridin-3-yloxy)acetate] in very high yield by hydrothermal reaction of a mixture of metallic iron powder and pyoaH in degassed water at 140°C [21]. The unique structure of the carboxylate-bridged iron(II) chain in **4** is crucial for an unusual field-induced magnetic transition from a spin-canted antiferromagnetic state to a single chain magnetic state (Fig. 3.3). Actually, direct hydrothermal reactions between the ligand and iron(II) salts were in vain, resulting in a noncrystalline mixture. The employment of an in situ iron(0) to iron(II) conversion is crucial for the facile synthesis of **4**, in which oxidation of iron(II) to iron(III) during the hydrothermal synthesis was avoided [22].

Although rarely known so far, in situ hydrothermal reaction should be very useful for the generation of functional coordination polymers with mixed-valence metal ions other than Cu(I,II), which could be expected to exhibit interesting magnetic or electronic properties. For example, our preliminary study shows that a three-dimensional (3D) coordination polymer $[Fe^{II}Fe^{III}(\mu_4\text{-}O)(1,4\text{-chdc})_{1.5}]_n$ (1,4-chdc = *trans*-1,4-cyclohexanedicarboxylate) could be prepared hydrothermally by a reaction of metallic iron powder and 1,4-chdcH_2, which features an unprecedented 1D inorganic

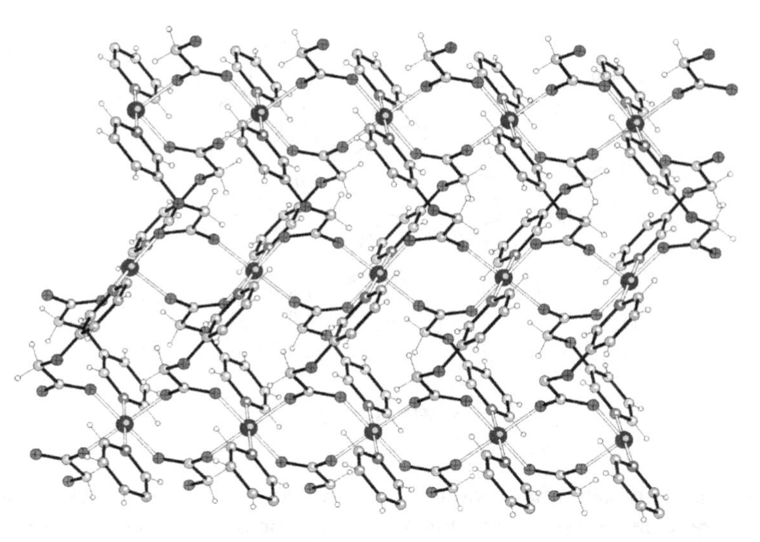

FIGURE 3.3 Carboxylate-bridged Fe(II) chains embedded in the 2D coordination network of **4**.

$[Fe^{II}_2(\mu_4\text{-}O)Fe^{III}_2(\mu_4\text{-}O)]_n^{6n+}$ tetrahedral mixed spin chain and interesting magnetic frustration behavior [23].

3.3 CONVERSION OF CARBOXYLIC ACID

As one of the most studied types of organic ligands, carboxylates can be synthesized by various conventional methods. The solvothermal in situ hydrolysis of organic esters and nitriles has been used rationally by Lin and co-workers to construct acentric metal carboxylate frameworks, or nonlinear optically (NLO) active coordination polymers, that may be inaccessible from the corresponding carboxylic acids; a large number of interesting examples were reviewed rather comprehensively by Evans and Lin in 2002 [10] and will not be duplicated in this chapter.

Carboxylates can also be generated in situ by other reaction types, such as hydrolysis–oxidative cleavage of heterocycles and ethylene carbon–carbon bonds [24–26]. More important, carboxylates can also undergo various solvothermal in situ reactions to form modified carboxylate derivatives [27–32].

Two acentric NLO active coordination polymers, [Zn(na)(ina)] (**5**) and [Cd(na)(ina)(H_2O)]N_2H_4 (**6**), have been synthesized by in situ hydrolysis of an unsymmetric ligand 2-(2-pyridyl)-5-(4-pyridyl)-1,3,4-oxadiazole (see Scheme 3.1) [25]. It is noteworthy that direct reactions of the Zn(II) and Cd(II) salts, respectively, with a 1 : 1 mixture of Hna and Hina under very similar reaction conditions led to white microcrystalline products exhibiting no second harmonic generation effect. This fact implies that in situ ligand generation is crucial for the formation of acentric **5** and **6**, being similar to those reported by Lin's group [10].

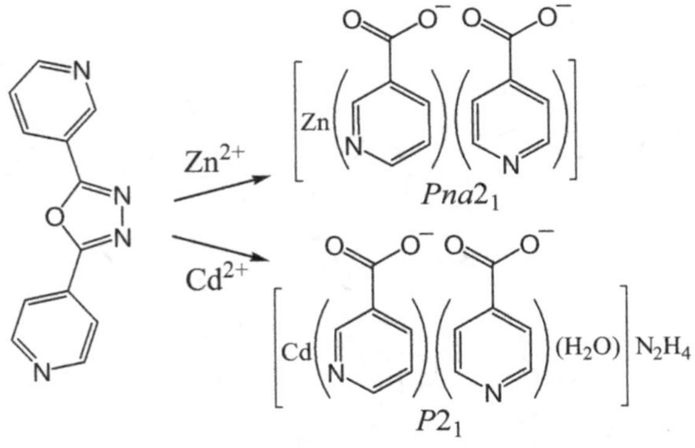

Zn^{2+} → [Zn(nicotinate)(isonicotinate)] *Pna*2_1

Cd^{2+} → [Cd(nicotinate)(isonicotinate)(H_2O)]N_2H_4 *P*2_1

SCHEME 3.1

Copper isonicotinate (ina) coordination polymers possess diverse chemical compositions and extended structures. Interestingly, some novel 3D copper isonicotinate coordination polymers have been obtained via solvothermal in situ metal–ligand reactions. A mixed-valence Cu(I,II) isonicotinate $[Cu_2(ina)_3]$ (**7**) having twofold interpenetrating α-Po networks was synthesized hydrothermally by Cu(II) salt and 4-cyanopyridine (Fig. 3.4) [33]. The ina ligand evidently resulted from the hydrolysis of 4-cyanopyridine, which may also act as a reducing agent to promote the formation of mixed-valence Cu(I,II) units. The hydrothermal reaction of $Cu(NO_3)_2$ with nicotinic acid and *trans*-1,2-bis(4-pyridyl)ethylene produced a novel interpenetrating structure $[Cu_2(ina)_4(H_2O)_3][Cu_2(ina)_4(H_2O)_2]\cdot 3H_2O$ (**8**) with two covalently bonded networks

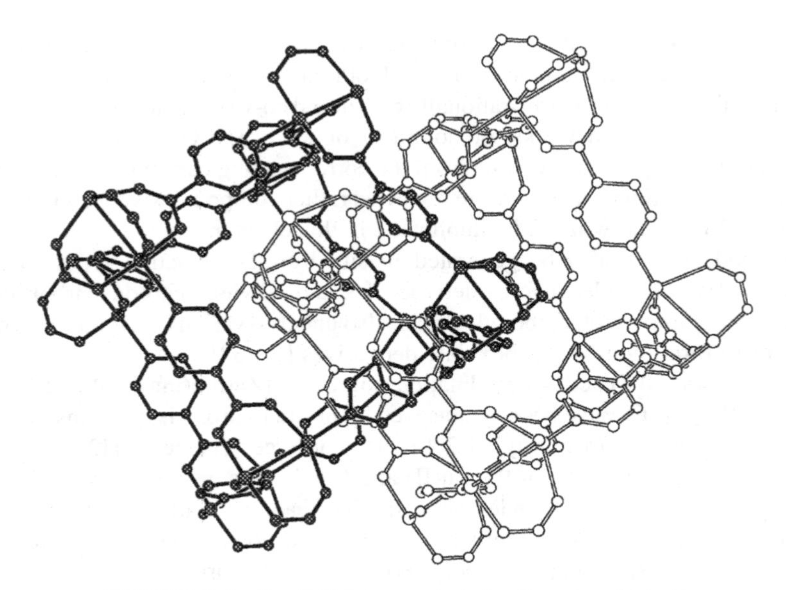

FIGURE 3.4 Twofold interpenetrated α-Po networks in **7**.

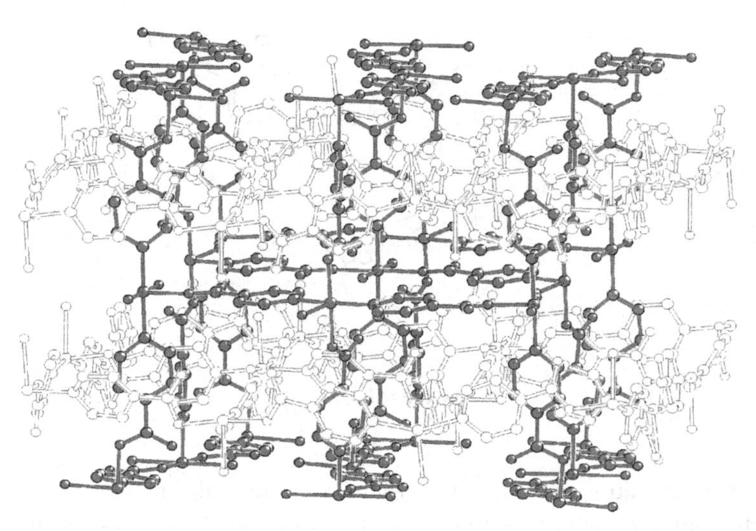

FIGURE 3.5 Interpenetration between a 3D $CdSO_4$ network and multiple 2D (4,4) networks in **8**.

of different dimensionality: a 2D square layer and a 3D $CdSO_4$ net (Fig. 3.5) [26]. A chemical rearrangement of nicotinic acid was proposed to account for formation of the ina ligands, although an oxidative cleavage hydrolysis of *trans*-1,2-bis(4-pyridyl) ethylene appears more likely. Actually, direct hydrothermal reactions of Cu(II) salts and Hina at similar conditions produce structures other than **7** or **8** [34].

Besides the hydroxylation of pyridyl rings, copper-mediated oxidative hydroxylation can also occur at an electron-deficient phenyl ring. An in situ oxidative hydroxylation of aromatic rings was documented for isophthalic acid (H_2ip) [27]. A hydrothermal reaction of H_2ip and 4,4′-bpy with $Cu(NO_3)_2$ at 180°C yielded a mixed-valence Cu(I,II) coordination polymer, [Cu_2(ipO)(4,4′-bpy)] (**9**; ipOH_3 = 2-hydroxyisophthalic acid), which is composed of an in situ–generated ipO ligand and exhibits a 3D structure that has a very strong antiferromagnetic interaction admixture with a weak ferromagnetic interaction. An analogous oxidative hydroxylation of 1,3,5-tricarboxylic acid has also been reported recently [28]. Interestingly, ipO can also be generated hydrothermally by substitution of the 2-carboxylate group of 1,2,3-benzenetricarboxylic acid (1,2,3-btcH_3) with a hydroxyl group upon addition of 4 equivalents (Eq) of NaOH, furnishing **9** in very high yield (95%) (see Scheme 3.2) [29]. When reducing the amount of NaOH to 2 Eq, the 2-carboxylate group was removed without the formation of a hydroxy group, furnishing a mixed-valence Cu(I,II) compound [Cu_2(ip)(Hip)$(4,4'\text{-bpy})_{1.5}$]. However, without addition of NaOH, the 2-carboxylate group was retained, resulting in a 3D porous framework, [Cu_2(1,2,3-btc)(4,4′-bpy)$(H_2O)_2$](NO_3). Since the carboxylate groups of 1,2,3-btc are sterically crowded, the one at the 2-position can potentially undergo a decarboxylation reaction catalyzed by the base and copper ion. The in situ–generated ip ligand further transforms into ipO at a higher pH in the presence of Cu(II).

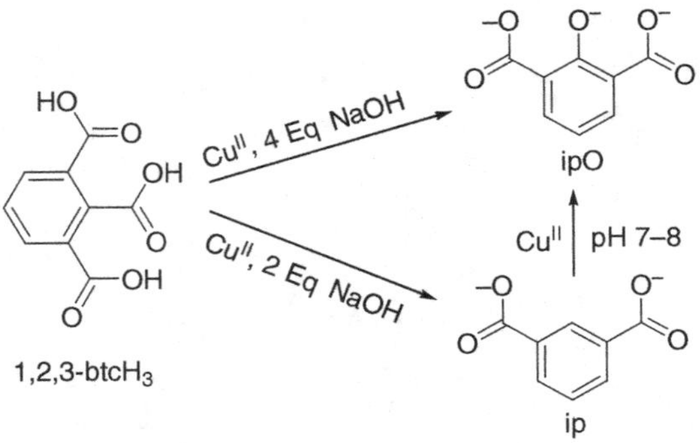

SCHEME 3.2

Covalent hydrations of aromatic groups give unstable, nonaromatic species, which tend to transform back to the original aromatic starting ligands and water by reverse dehydration reactions. Cu(II) ions oxidize the unstable species to stable aromatic phenols and ketones, which serve as good bridging ligands in the crystalline coordination complexes [18,19]. On the other hand, covalent hydration of a C=C double bond may produce stable alcohol derivatives directly, without further oxidation. For example, in the presence of Zn(II), Co(II), or Fe(II), maleic acid and fumaric acid can be covalently hydrated to malic acid under solvothermal conditions at pH 8 (Fig. 3.6a) [30]. Solvothermal reactions of fumaric acid, maleic acid, or maleic anhydride with Co(II), isonicotinic acid (Hina), and NaOH produce a microporous coordination polymer formulated as $[Co_2(ma)(ina)]\cdot H_2O$ (**10**, ma = malate), in which the ma ligand is generated by a covalent hydration of the ethylene dicarboxylic acids [31]. Interestingly, the carboxylate groups and the hydroxyl group bridge the Co(II) ions to a complicated, magnetically frustrated 2D layer (Fig. 3.6b), which is further pillared by ina ligands into a very rigid pillar-layered 3D framework with microporosity. As a result, **9** exhibits interesting magnetic properties sensitive to guest removal and exchange due to the tuning of supramolecular interactions between the magnetic lattice and the guest molecules.

The conversions of carboxylic acids are relatively simple solvothermal in situ reactions. However, such a strategy is very useful in the generation of functional metal–carboxylate frameworks. In addition to the well-documented NLO molecular solids by Lin's and other groups, this strategy can also be applied to other functional materials, such as porous and magnetic materials.

3.4 CARBON–CARBON BOND FORMATION

The carbon–carbon bond formation is an important strategy to extend the length and complexity of simple ligands. Generally, C–C bond formation reactions require

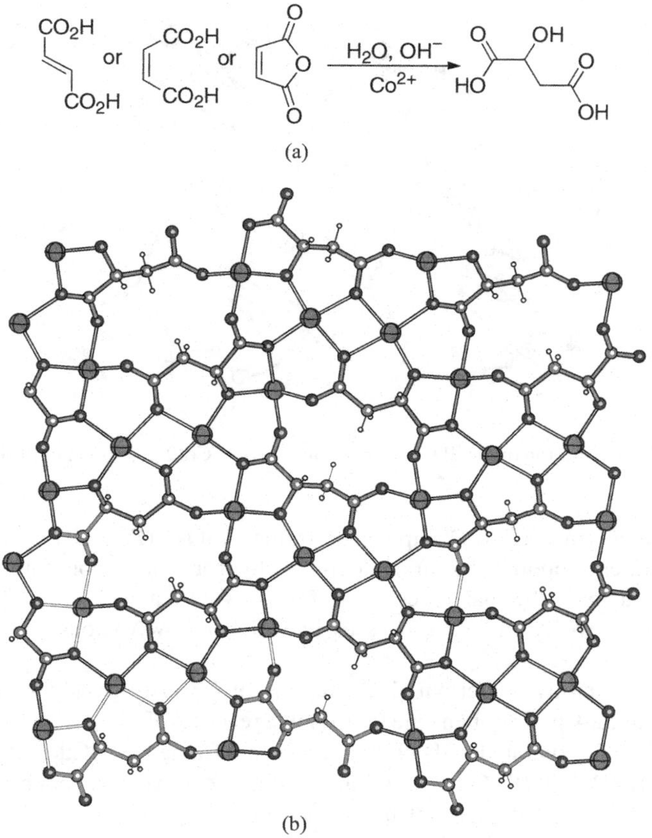

(a)

(b)

FIGURE 3.6 (a) Covalent hydration of fumaric acid, maleic acid, or maleic anhydride to malic acid; (b) frustrated magnetic layer of $[Co_2(ma)]_n^{n+}$ in **10**.

highly reactive, expensive organometallic reagents of late transition metals as catalysts. So far, only a few unconventional C–C bond formation reactions have been obtained under solvothermal preparations of coordination polymers.

An unexpected oxidative coupling of methanol to oxalate (ox) was observed in the reaction of $Zn(NO_3)_2$ and pyridine in methanol at 140 °C, which generated 2D (methylpyridinium)$_2$[$Zn_2(ox)_3$] [35]. It was suggested that the oxidative coupling of methanol into ox was promoted by a nitrate group, and the formation of insoluble [$Zn_2(ox)_3$] polymer is probably the principal driving force for the reaction. Moreover, solvothermal in situ ox formation from the decarboxylation coupling of isonicotinic acid and acetic acid has also been documented [36].

C–C bond formation can produce novel and more complicated ligands uneasily synthesized by conventional methods in a few cases [37]. Liu and co-workers reported a dehydrogenative coupling of phen into 2,2′-biphenanthroline in a hydrothermal reaction of NH_4VO_3, H_3BO_3, $Co(NO_3)_2$, and phen, leading to an interesting layered

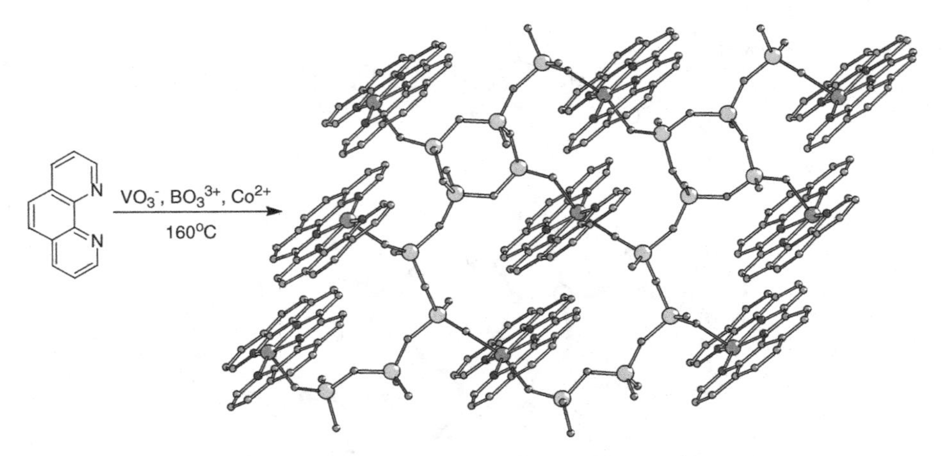

FIGURE 3.7 Formation of the 2D coordination network [Co(2,2′-biphenanthroline)]$V_3O_{8.5}$.

coordination polymer, [Co(2,2′-biphenanthroline)]$V_3O_{8.5}$ (Fig. 3.7) [38]. This reaction was further supported by an analogous dehydrogenative coupling of 2,2′-bpy observed in a hydrothermal reaction of $NiCl_2$, V_2O_5, and 2,2′-bpy [39]. These reactions may involve radical processes in which the V(V) ions may act as the oxidants.

A more unusual hydrothermal C–C coupling of 1,3-bis(4-pyridyl)propane (bpp) into 1,2,4,5-tetra(4-pyridyl)benzene (bztpy) in a reaction of $Cd_{10}S_4(SPh)_{12}$, bpp, and Na_2SO_4 at 190°C generated a 3D photoluminescent framework $[Cd_8(SPh)_{12}(bztpy)_2SO_4](HSO_4)_2{\cdot}4H_2O$ [40]. Obviously, this dehydrogenative C–C coupling reaction requires an oxidant, although not being mentioned in the report. Later, Hu et al. observed a relevant reaction in hydrothermal treatments of fresh $Cu(OH)_2$ with bpp, 1,4-cyclohexanedicarboxylic acid, and water in dilute HCl media at 175 to 190°C [41]. A dihydroxylcyclohexane ligand *a*,*a*-1,4-dihydroxy-*e*,*e*,*e*,*e*-1,2,4,5-tetra (4-pyridyl)cyclohexane (chtpy) was generated from the dehydrogenative coupling and hydroxylation of bpp. By simply tuning the molar ratios of the starting materials, a variety of novel Cu(I)-chtpy coordination polymers have been synthesized. Indeed, the formation of Cu(I) indicates clearly that Cu(II) is the oxidant in the in situ ligand reactions. It should be noted that not only the chtpy ligand, but also a monohydroxylcyclohexane ligand and bztpy ligand could be isolated and characterized structurally, implying that the hydroxylcyclohexane derivatives may be the reaction intermediates from bpp to bztpy (Fig. 3.8a). These in situ–generated tetradentate ligands can be utilized to construct interesting 3D coordination polymers. As shown in Figure 3.8b, the planar chtpy ligands are linked by tetrahedral Cu(I) centers to form a PtS network.

Although only a few C–C bond formation reactions under solvothermal preparations have been documented, the unconventional reactions above imply that solvothermal reactions could possibly serve as a new method for discovery of new C–C bond formation reactions.

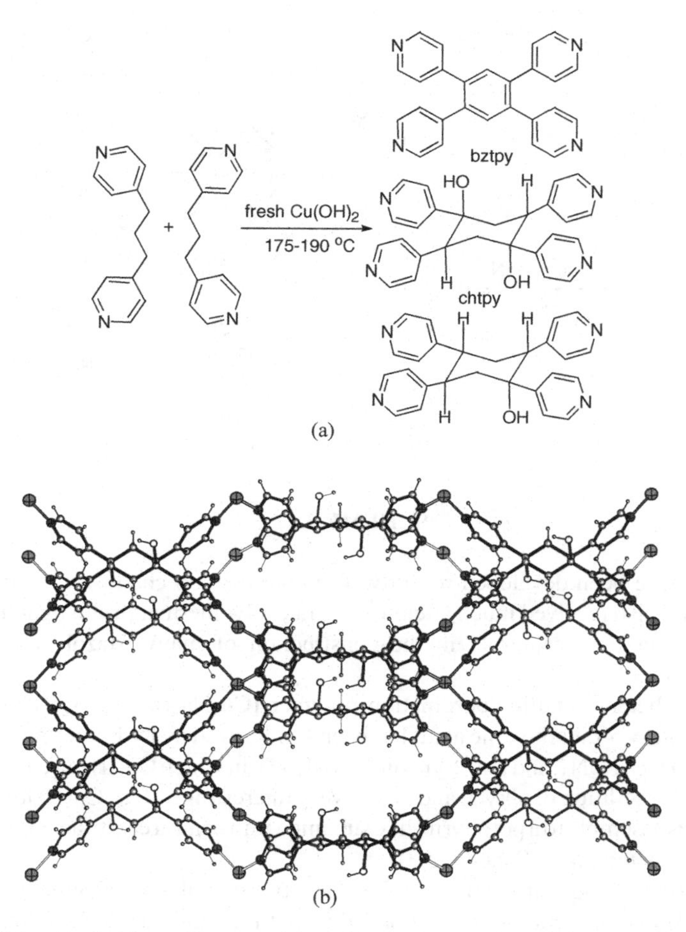

FIGURE 3.8 (a) C–C coupling reactions of bpp; (b) PtS structure of the $[Cu_2(chtpy)_2]$ coordination network in $[Cu_2(chtpy)_2]Cl(NO_3)\cdot\frac{4}{3}(C_6H_6)$.

3.5 HETEROCYCLE FORMATION FROM SMALL MOLECULES

Triazoles and tetrazoles are important heterocyclic organic compounds for their wide-ranging applications. Continuous efforts have been focused on the search for more efficient and simple synthetic procedures for these heterocycles. Thanks to the direct Huisgen 1,3-dipolar cycloaddition route, facile synthetic approaches for 1,2,3-triazole have been established for several decades [42]. On the other hand, syntheses of 1,2,4-triazole and tetrazole are more difficult. 1,2,4-Triazoles are usually synthesized from hydrazine derivatives by multistep reactions. Although tetrazole can also be synthesized by the direct Huisgen 1,3-dipolar cycloaddition between nitriles and azides, this process is neither general nor practical for many unactivated or sterically hindered starting materials. Nevertheless, some important improvements for the facile syntheses of

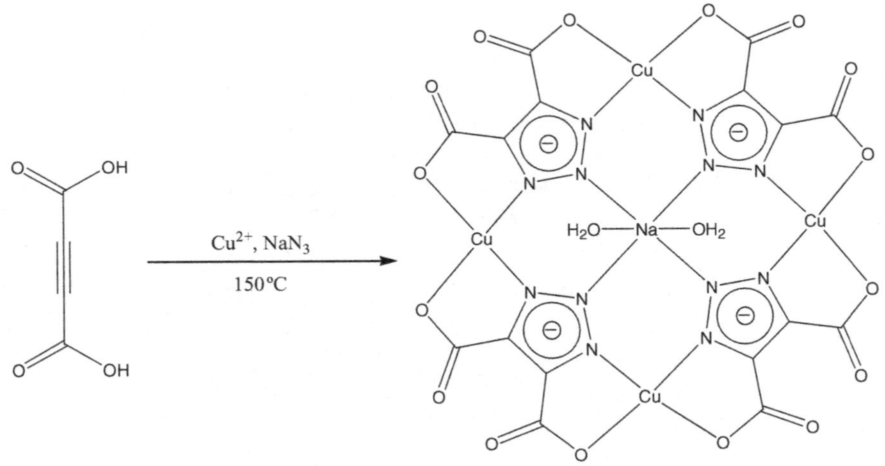

SCHEME 3.3

tetrazoles have been developed recently. The coordination chemistry of triazole and tetrazole has also received much attention. So far, solvothermal in situ ligand synthesis has played an important role in the construction of many triazole and tetrazole coordination polymers.

A novel heterometallic coordination polymer $[Cu_4Na_4(tzdc)_4(H_2O)_7]$ (H_3tzdc = 1,2,3-triazole-4,5-dicarboxylic acid) has been synthesized by hydrothermal reaction of $Cu(NO_3)_2 \cdot H_2O$, NaN_3, and but-2-ynedioic acid [43], in which Cu(II) is a catalyst for the [2 + 3] cycloaddition of alkyne and azide. Very interestingly, four Cu(II) ions and four tzdc ligands assemble to a porphyrin-like subunit with a central cavity accommodating a sodium ion as template (Scheme 3.3).

After establishing that Zn(II) and other transition-metal ions can greatly accelerate the [2 + 3] cycloadditions of organonitriles and inorganic azide in water [44], the 5-substituted 1*H*-tetrazoles become very attractive for the crystal engineering of coordination polymers because of the availability of various organonitriles and azide salts (Scheme 3.4). This is a convenient synthetic procedure and has been utilized in

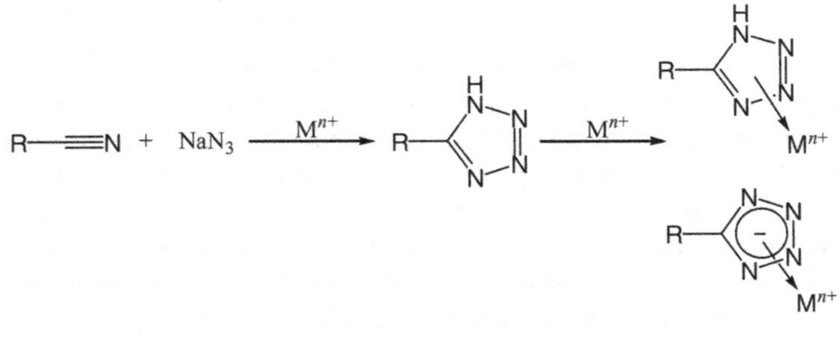

SCHEME 3.4 5-substituted 1*H*-tetrazoles and their transition-metal complexes synthesized by transition metal–catalyzed [2 + 3] cycloadditions of organonitriles and inorganic azide in water.

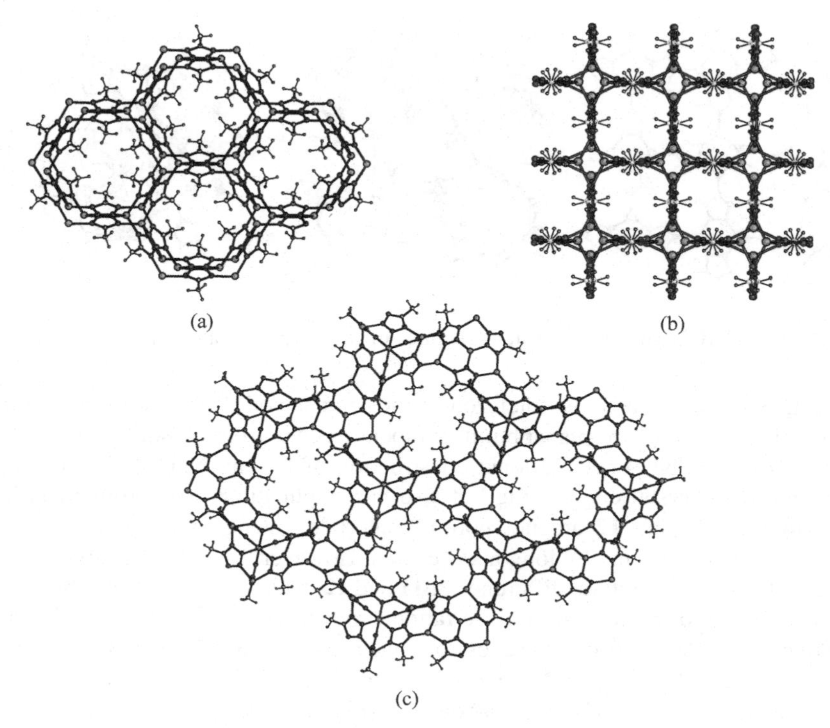

FIGURE 3.9 Crystal structures of (a) α-[Cu^I(mta)]·0.17H_2O, (b) β-[Cu^I(mta)], and (c) [$Cu^I_6Cu^{II}_2(mta)_9$]N_3·$x$$H_2O$ (Hmta = 5-methyltetrazole).

solvothermal reactions to construct functional coordination polymers exhibiting interesting structural diversity and important properties [45–48].

Metal-redox reactions can also be incorporated with this [2 + 3] nitrile–azide cycloaddition. Wu et al. [46] and Zhang et al. [47] have reported several interesting Cu(I) and Cu(I,II) tetrazolate polymers (Fig. 3.9) obtained from the solvothermal reactions of acetonitrile and NaN_3 with various Cu(II) salts and solvents.

The hydrazine-based synthetic approaches of 1,2,4-triazoles can also be used in the solvothermal preparations of metal complexes and coordination polymers [49]. More interestingly, we have found that organonitriles and ammonia can undergo oxidative cycloaddition to give 3,5-disubstituted-1,2,4-triaozle in the presence of Cu(II) [50]. This new and non-hydrazine-based approach has led to the discovery of many novel coordination polymers. For example, using the common solvents acetonitrile and butyronitrile as the starting materials, two Cu(I) triazolates, $[Cu(mtz)]_n$ (**11**; Hmtz = 3,5-dimethyl-1,2,4-triazole) and $[Cu(ptz)]_n$ (**12**; Hptz = 3,5-dipropyl-1,2,4-triazole), have been generated [50]. Despite their simple chemical compositions, **11** and **12** are archetypal coordination polymers of two unique three-connected topologies, namely 4.8.10 and 4.12^2, respectively. These three-connected network topologies are important not only for their low connectivity and high symmetry. Actually, the three-connected 4.8.10 and 4.12^2 nets are equivalent to the four-connected lvt and NbO nets,

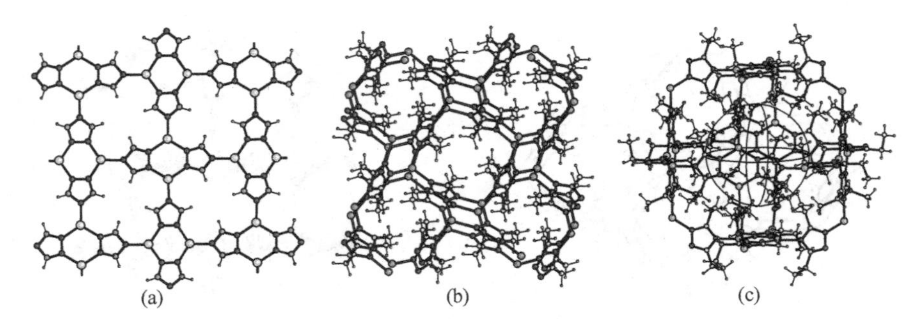

FIGURE 3.10 Crystal structures of (a) **14**, (b) **11**, and (c) **13**.

respectively, which are built solely on square-planar nodes. It is well known that interconnection of square-planar building blocks trend to from 2D square grids rather than a 3D lvt or NbO net. Rational construction of lvt and NbO nets is not only pronounced in crystal engineering but also has implications for porous materials. Nevertheless, **11** and **12** are not porous materials, owing to interpenetration of the networks or clogging by long alkyl chains. However, by careful analysis of the relationship of the structural difference of **11** and **12**, it has been possible to propose a substituent-size directed route for the rational construction of square grid, lvt, and NbO networks, leading to the successful construction of a porous NbO networked Cu(I) triazolate $[Cu(etz)]_n$ (**13**; Hetz = 3,5-diethyl-1,2,4-triazole) and a layered Cu^I triazolate $[Cu(tz)]_n$ (**14**; Htz = 1,2,4-triazole) (Fig. 3.10) [51].

The structural design of these copper(I) triazolates is based on their predictable chemical compositions as required by the strong metal–azolate coordination bonds and the low solubility of the binary salt. These facts indicate that anions and/or some competing ligands may be used as coordination buffering agents for the generation of various supramolecular isomers. Moreover, the enumeration of various meta-phases also takes advantage of solvothermal in situ metal–ligand reactions. The in situ generation speed of molecular building blocks can be controlled by varying the reaction temperature, offering a variety of nonequilibrium crystallization environments. By using 4,4′-bipyridine and nitrate as the coordination buffering agents, a uninodal $8^2.10$-a isomer and a new binodal 6.10^2 isomer of **11** have been discovered (Fig. 3.11) [52]. Moreover, the three supramolecular isomers of **11** exhibit different excitation and emission spectra and luminescent lifetimes, which can be related to the well-defined structures of the isomers.

On the other hand, when a very strong donor is added to the reaction system, it may participate in the coordination polymer. For example, Huang et al. have reported a coordination polymer $[Cu_4I(mtz)_3]_n$ synthesized by the solvothermal reaction of Cu(II) salt, KI, aqueous ammonia, and acetonitrile [53]. While the in situ–generated mtz ligand possesses a coordination mode similar to those in the $[Cu(mtz)]_n$ isomers, the incorporation of iodide produces a novel self-penetrated, binodal, three-connected topology $(12^3)(12^2.14)_3$ (Fig. 3.12).

The more complicated triazolate ligands pyridyl-substituted triazolates 2-pytz, 3-pytz, and 4-pytz (Hpytz = 3,5-dipyridyl-1,2,4-triazole) can be generated using

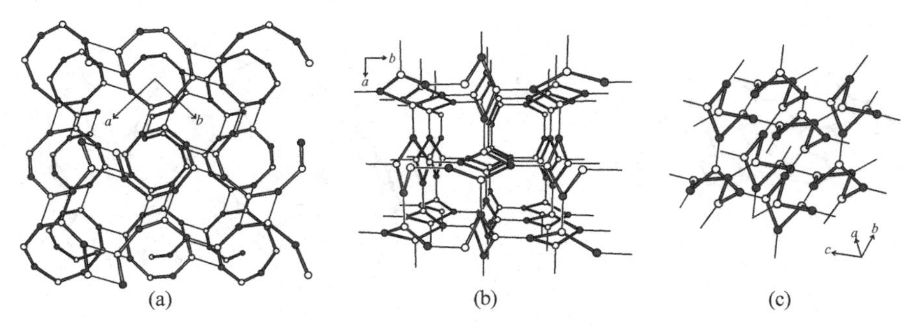

FIGURE 3.11 The (a) 4.8.10, (b) 8^2.10-a, and (c) 6.10^2 topological isomers of **11** (open circles, Cu; filled circles, mtz).

cyanopyridines as the starting materials. It is difficult to fully utilize all five nitrogen donors of these ligands since univalent metal ions such as Cu(I) usually have a coordination number of 2, 3 or 4. Therefore, prediction of the superstructures or the local coordination environments of these systems would be difficult. Actually, several supramolecular isomers with different structural features have been observed for the $[Cu(4\text{-pytz})]_n$ system [54]. More interestingly, 2-pytz is quite different from 3-pytz and 4-pytz, since it contains bidentate chelating sites. We could readily predict two possible bis-bidentate chelating coordination modes for 2-pytz and four-coordinated Cu(I) centers in a $[Cu(2\text{-pytz})]_n$ (**15**) system, as the tridentate mode is probably not possible for Cu(I). Consequently, the Cu(2-pytz) system may be used to synthesize a series of supramolecular isomers, including 0D polygons, 1D zigzag, and helical

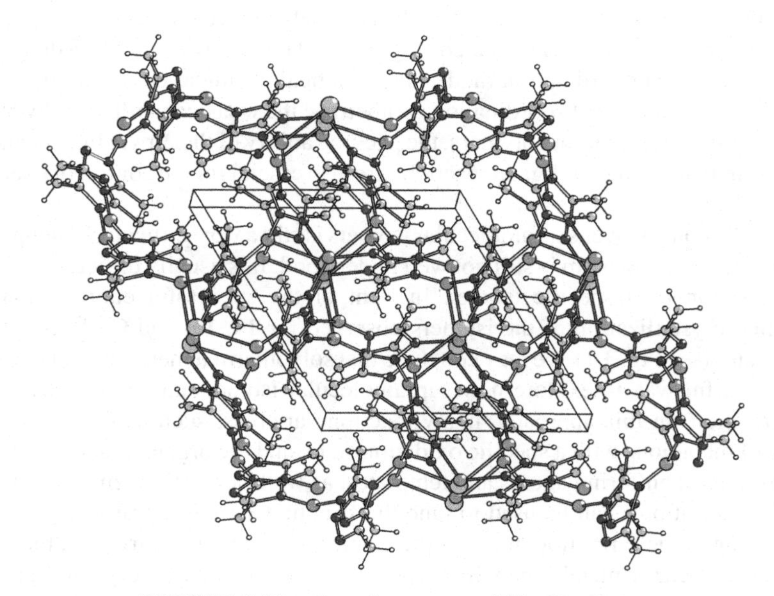

FIGURE 3.12 Crystal structure of $[Cu_4I(mtz)_3]_n$.

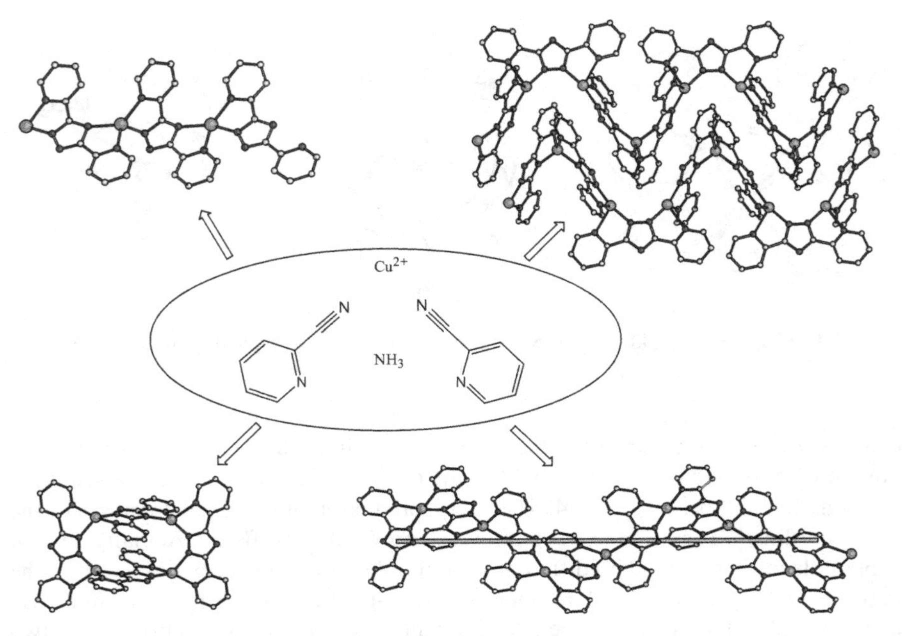

FIGURE 3.13 Four supramolecular isomers of **15** prepared from the solvothermal reactions of 2-cyanopyridine, Cu^{2+}, and ammonia.

chains [55]. By varying the solvothermal conditions, a chairlike tetranuclear metallomacrocycle, a zigzag chain, a homochiral 4_1 helix, and an unexpected zipperlike double-chain compound were isolated (Fig. 3.13). The four isomers of **15** do possess the predicted local coordination modes and identical chemical compositions. Apart from the predictable local coordination geometries, the weak capability for hydrogen bonding of the coordination polymers, the packing ability of the 2-pytz ligand, and the solvothermal in situ metal–ligand reactions are also crucial for the control of chemical composition.

Actually, control over supramolecular isomers is still a very difficult challenge, and supramolecular isomerism is still not well understood. It is usually difficult to control the generation of a specific isomer. In light of the successful enumeration and controlled crystallization of the isomers observed for **11**, **15**, and Cu(I) 2-methylimidazolates (see Fig. 3.1), we may conclude that solvothermal metal–ligand reactions are a powerful way to explore new supramolecular isomers and control over their crystallization. In this approach, many reaction variables, such as the in situ generation of metal ions with a specific oxidation state, and the organic ligands, addition of coordination buffering agents and templates, as well as reaction temperatures and even concentrations, can be used to tune the structures at the supramolecular level, thus leading to the formation of a specific isomer. Among these variables, the in situ generation of either metal ions with a specific oxidation state or organic ligands is quite unique compared with conventional synthetic techniques.

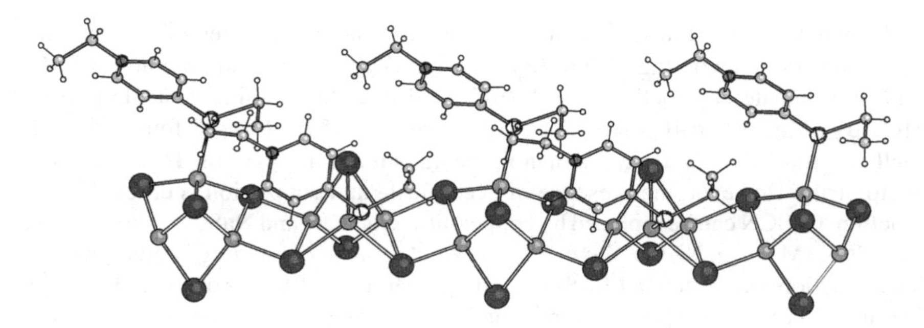

FIGURE 3.14 Crystal structure of **16**.

3.6 TRANSFORMATION OF SULFUR-CONTAINING LIGANDS

Due to the relative weak covalent C–S (276 kJ mol^{-1}) and S–S (317 kJ mol^{-1}) bonds and the rich redox behaviors, sulfur-containing ligands are suitable starting materials for solvothermal in situ ligand reactions. The oxidation of thiol groups to disulfide and sulfonate groups has generated a variety of new coordination polymers [56]. Recently, Cheng et al. reported a simultaneous solvothermal redox alkylation by a reaction of $CuCl_2$, KI, pyridine-4-thiol (HS-4-C_5H_4N), and ethanol, which afforded a photoluminescent chainlike polymer $[(Cu_3I_4)(EtS\text{-}4\text{-}C_5H_4NEt)]_n$ (**16**) [57]. This reaction involves not only double alkylation of the pyridine-4-thiol ligand, but also very complicated redox processes for the copper, sulfur, and iodine species. Compound **16** is the first example of a linear chain consisting of trinuclear Cu_3I_4 subunits. Meanwhile, in situ–generated EtS-4-C_5H_4NEt cations complete the coordination spheres of the Cu(I) ions (Fig. 3.14).

Two unusual in situ transformations of sulfur-containing ligands were reported by Wang and co-workers in the solvothermal reactions of CuI, 4,4′-dithiodipyridine (dtdp), and acetonitrile in different ratios at 120 and 160°C (Scheme 3.5) [58]. In these reactions, dtdp was not converted into a 4-pyridinethiolate (pdt) but converted unexpectedly into 4,4′-thiodipyridine (tdp) and 1-(4-pyridyl)-4-thiopyridine (ptp), leading to a 2D coordination network, $[Cu_4I_4(tdp)_2]$, and two 3D coordination networks, $[Cu_5I_5(ptp)_2]$ and $[Cu_6I_6(ptp)_2]$, respectively.

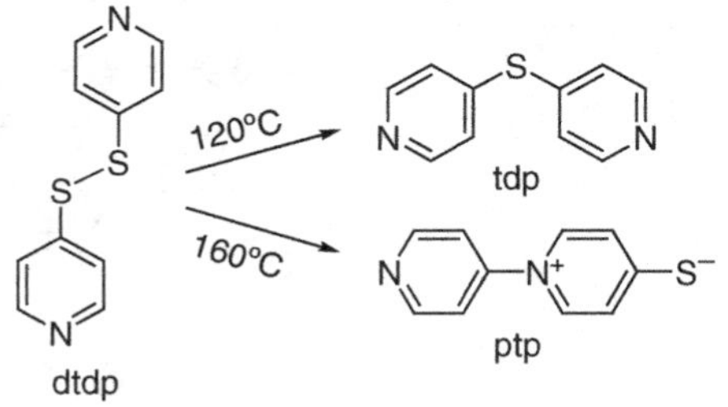

dtdp —120°C→ tdp

dtdp —160°C→ ptp

SCHEME 3.5

Li and Wu have found that thiocyanate can undergo a series of solvothermal in situ reactions involving sulfur atom transfer. A solvothermal reaction of CuSCN with acetonitrile and methanol at 140°C yielded a photoluminescent 3D polymer $\{[Cu(\mu_3\text{-}SMe)_2(CN)]_2[Cu_{10}(\mu_3\text{-}SMe)_4(\mu_4\text{-}SMe)_2]\}_n$ [59]. It was found that the methyl group comes from methanol, being similar to that of **16**, rather than acetonitrile. However, the presence of acetonitrile and copper ion is crucial for this reaction. CuSCN can be replaced by copper nitrate/acetate and NaSCN, but other salts $M(SCN)_n$ ($M = Ag, Ni, Co, Zn$; $n = 1, 2$) are not capable of yielding similar products. This reaction was extended further to the preparation of a cuprous sulfide cluster-based, 12-connected, face-centered cubic network. $[Cu_{12}(\mu_4\text{-}SMe)_6(CN)_6]\cdot 2H_2O$ (**17**) was synthesized from the solvothermal reaction of copper nitrate and NaSCN with methanol–acetonitrile at 160°C [60]. The 12-connected node and the linker of **17** is the $[Cu_{12}(\mu_4\text{-}SMe)_6]$ cluster and the CN ligand, respectively (Fig. 3.15).

Another interesting cuprous sulfide cluster-based, 12-connected, face-centered cubic network was also synthesized by a solvothermal in situ metal–ligand reaction. $[Cu_3(pdt)_2(CN)]$ (**18**; pdt = 4-pyridinethiolate) was prepared by the solvothermal reaction of $Cu(MeCO_2)_2$, (4-pyridylthio)acetic acid, NH_4SCN, and NaOH [61]. The pdt and CN ligands were generated from cleavages of the S–C bonds of (4-pyridylthio)acetic acid and thiocyanate, respectively. In **18**, the Cu_6S_4 clusters are interlinked by 8 pdt and 4 CN ligands to form a 12-connected topology (Fig. 3.16).

Zhou et al. have also demonstrated sulfur atom transfers from a thiocyanate to a phosphine group. The solvothermal reaction of CuSCN and 1,2-bis(diphenylphosphino) ethane (dppe) produced a 3D coordination polymer $[(CuCN)_2(dppeS_2)]_n$ ($dppeS_2$ = 1,2-bis(diphenylthiophosphinyl)ethane) (Scheme 3.6) [62]. It should be noted that the above-mentioned reactions also produce a cyanide anion simultaneously and may serve as a general approach for construction of metal cyanide coordination polymers.

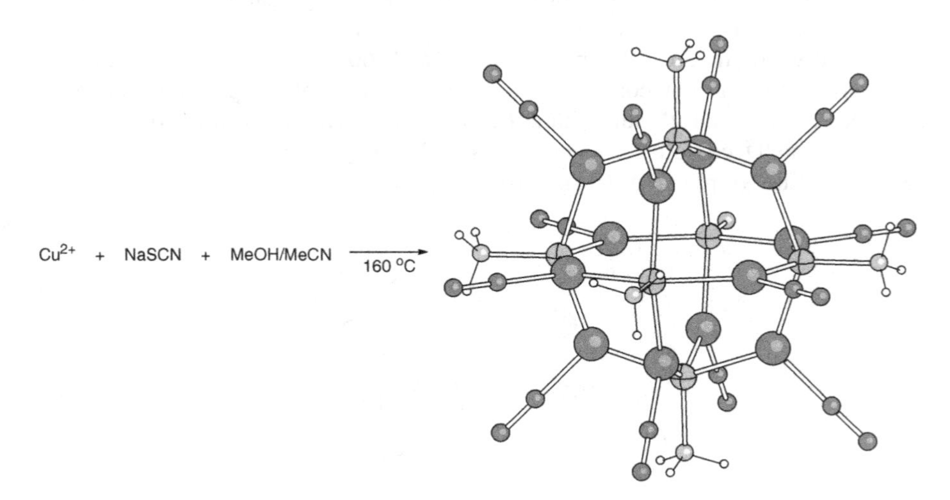

FIGURE 3.15 Synthesis and structure of the 12-connected core of **17**.

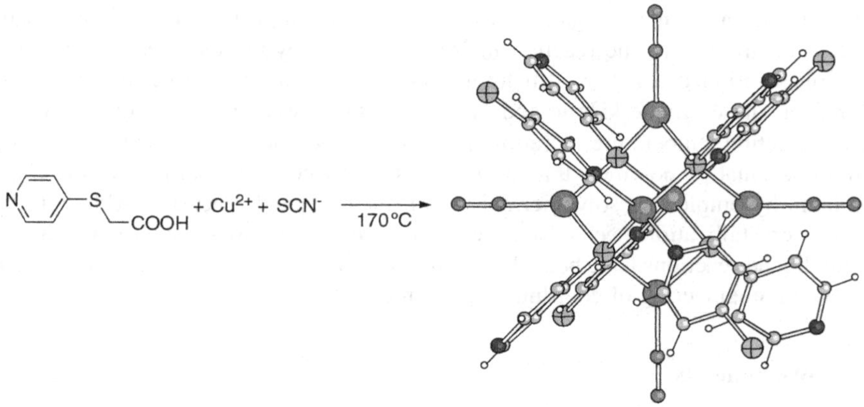

FIGURE 3.16 Synthesis and structure of the 12-connected core of **18**.

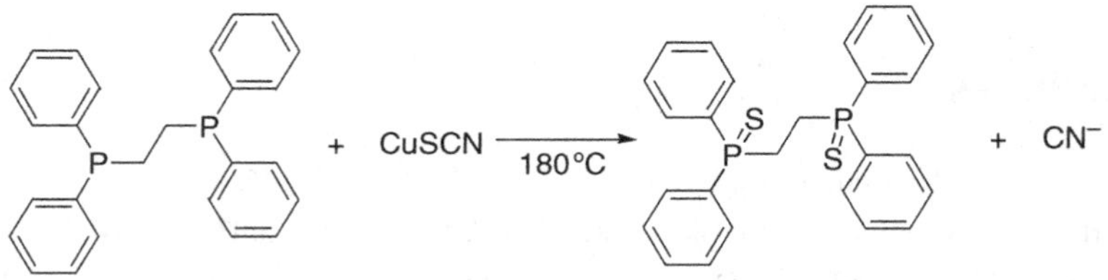

SCHEME 3.6

The observations above clearly demonstrate that sulfur-containing ligands are quite versatile during the solvothermal syntheses of coordination polymers. In addition to other in situ metal–ligand reaction types, the solvothermal in situ reactions of sulfur-containing ligands have also provided a new dimension for the crystal engineering of coordination polymers and discovery of novel metal–organic architectures.

3.7 CONCLUSIONS

As outlined above, solvothermal in situ metal–ligand reactions have been employed widely in the crystal engineering of coordination polymers. These reactions can be recognized as a new bridge between coordination chemistry and synthetic organic chemistry. Although the discoveries of new reactions and novel coordination polymers are sometimes serendipitous, in situ metal–ligand reactions can be utilized rationally in crystal engineering to avoid the difficulty of synthesis and handling of particular molecular building blocks. For example, the in situ reduction of Cu(II) ions has been demonstrated as a convenient, powerful approach to the construction and crystal growth of Cu(I) and Cu(I,II) coordination polymers, especially the highly insoluble polymers. Many important organic ligands, such as carboxylic acids, triazoles, and tetrazoles, can also be in situ–generated from their commercially available precursors. More important, the relatively slow generation of molecular building blocks during solvothermal in situ metal–ligand reaction produces a unique nonequilibrium crystallization environment. The reaction intermediates and by-products presented in the reaction solution can act as structural-directing or coordination-buffering agents. Moreover, these unique kinetic and thermodynamic variables can readily be tuned by the reaction temperature, duration time, and similar factors. Consequently, many novel coordination polymers that are inaccessible by conventional methods become available by employing solvothermal in situ metal–ligand reactions. Although the intrinsic crystallization mechanisms are usually difficult to clarify, solvothermal in situ metal–ligand reactions have been demonstrated as an important practical strategy for the crystal engineering of coordination polymers.

Acknowledgments

The authors thank the 973 Project (2007CB815302) and NSFC (No. 20531070 & 20821001) for financial support.

REFERENCES

1. Batten, S. R.; Robson, R. *Angew. Chem., Int. Ed.* **1998**, *37*, 1461–1494.
2. Carlucci, L.; Ciani, G.; Proserpio, D. M. *Coord. Chem. Rev.* **2003**, *246*, 247–289.
3. Yaghi, O. M.; O'Keeffe, M.; Ockwig, N. W.; Chae, H. K.; Eddaoudi, M.; Kim, J. *Nature* **2003**, *423*, 705–714.
4. Kitagawa, S.; Kitaura, R.; Noro, S. I. *Angew. Chem., Int. Ed.* **2004**, *43*, 2334–2375.
5. Masciocchi, N.; Galli, S.; Sironi, A. *Comments Inorg. Chem.* **2005**, *26*, 1–37.
6. Moulton, B.; Zaworotko, M. J. *Chem. Rev.* **2001**, *101*, 1629–1658.
7. Hagrman, P. J.; Hagrman, D.; Zubieta, J. *Angew. Chem., Int. Ed.* **1999**, *38*, 2638.
8. Zhang, J.-P.; Chen, X.-M. *Chem. Commun.* **2006**, 1689–1699.
9. (a) Transition-metal ions are well known to catalyse organic reactions even at ambient conditions: Iwahori, F.; Inoue, K.; Iwamura, H. *J. Am. Chem. Soc.* **1999**, *121*, 7264–7265. (b) Mokuolu, Q. F.; Kilner, C. A.; Barrett, S. A.; McGowan, P. C.; Halcrow, M. A. *Inorg. Chem.* **2005**, *44*, 4136–4138.
10. Evans, O. R.; Lin, W.-B. *Acc. Chem. Res.* **2002**, *35*, 511–522.
11. Lu, J. Y. *Coord. Chem. Rev.* **2003**, *246*, 327–347.
12. Zhang, X.-M. *Coord. Chem. Rev.* **2005**, *249*, 1201–1219.
13. Chen, X.-M.; Tong, M.-L. *Acc. Chem. Res.* **2007**, *40*, 162–170.
14. Müller-Buschbaum, K.; Mokaddem, Y.; Schappacher, F. M.; Pöttgen, R. *Angew. Chem., Int. Ed.* **2007**, *46*, 4385–4387.
15. (a) Yaghi, O. M.; Li, H. *J. Am. Chem. Soc.* **1995**, *117*, 10401–10402. (b) Lo, S. M.-F.; Chui, S. S.-Y.; Shek, L.-Y.; Lin, Z.-Y.; Zhang, X.-X.; Wen, G.-H.; Williams, I. D. *J. Am. Chem. Soc.* **2000**, *122*, 6293.
16. (a) Huang, X.-C.; Zhang, J.-P.; Lin, Y.-Y.; Chen, X.-M. *Chem. Commun.* **2005**, 2232–2234. (b) Huang, X.-C.; Zhang, J.-P.; Chen, X.-M. *Cryst. Growth Des.* **2006**, *6*, 1194–1198.

17. Huang, X.-C.; Zhang, J.-P.; Chen, X.-M. *J. Am. Chem. Soc.* **2004**, *126*, 13218–13219.
18. Huang, X.-C.; Zhang, J.-P.; Lin, Y.-Y.; Chen, X.-M. *Chem. Commun.* **2004**, 1100–1101.
19. (a) Zhang, X.-M.; Tong, M.-L.; Chen, X.-M. *Angew. Chem., Int. Ed.* **2002**, *41*, 1029–1031. (b) Zhang, X.-M.; Tong, M.-L.; Gong, M.-L.; Lee, H. K.; Luo, L.; Li, J.-H.; Tong, Y.-X.; Chen, X.-M. *Chem. Eur. J.* **2002**, *8*, 3187–3194.
20. (a) Xiao, D.-R.; Hou, Y.; Wang, E.-B.; Wang, S.-T.; Li, Y.-G.; De, G.; Xu, L.; Hu, C.-W. *J. Mol. Struct.* **2003**, *659*, 13–21. (b) Zhang, J.-P.; Lin, Y.-Y.; Weng, Y.-Q.; Chen, X.-M. *Inorg. Chim. Acta* **2006**, *359*, 3666–3670.
21. Zheng, Y.-Z.; Xue, W.; Tong, M.-L.; Zhang, W.-X.; Chen, X.-M.; Grandjean, F.; Long, G. J. *Inorg. Chem.* **2008**, *47*, 4077–4087.
22. Sanselme, M.; Grenèche, J. M.; Riou-Cavellec, M.; Férey, G. *Chem. Commun.* **2002**, 2172–2173.
23. Zheng, Y.-Z.; Xue, W.; Zhang, W.-X.; Tong, M.-L.; Chen, X.-M.; Grandjean, F.; Long, G. J.; Ng, S. W.; Panissod, P.; Drillon, M. *Inorg. Chem.* **2009**, *48*, 2028–2042.
24. (a) Knope, K. E.; Cahill, C. L. *Inorg. Chem.* **2007**, *46*, 6607–6612. (b) Han, Z.; Zhao, Y.; Peng, J.; Gómez-García, C. J. *Inorg. Chem.* **2007**, *46*, 5353–5455. (c) Rodríguez-Diéguez, A.; Cano, J.; Kivekäs, R.; Debdoubi, A.; Colacio, E. *Inorg. Chem.* **2007**, *46*, 2503–2510. (d) Zhao, X.-X.; Ma, J.-P.; Dong, Y.-B.; Huang, R.-Q. *Cryst. Growth Des.* **2007**, *7*, 1058–1068. (e) Wang, R. H.; Hong, M. C.; Luo, J. H.; Cao, R.; Weng, J.-B. *Chem. Commun.* **2003**, 1018–1019.
25. Wang, Y.-T.; Fan, H.-H.; Wang, H.-Z.; Chen, X.-M. *Inorg. Chem.* **2005**, *44*, 4148–4149. The topology of **5** was reported incorrectly to have a short vertex symbol of $6^3.8^2.10$ by an automatic analysis using Olex. Our recent reexamination shows that the topology of **4** should have a short vertex symbol of $6^5.8$.
26. Lu, J. Y.; Babb, A. M. *Chem. Commun.* **2001**, 821–822.
27. Tao, J.; Zhang, Y.; Tong, M.-L.; Chen, X.-M.; Yuen, T.; Lin, C. L.; Huang, X.-Y.; Li, J. *Chem. Commun.* **2002**, 1342–1343.
28. Yucesan, G.; Ouellette, W.; Chuang, Y.-H.; Zubieta, J. *Inorg. Chim. Acta* **2007**, *360*, 1502–1509.
29. (a) Zheng, Y.-Z.; Tong, M.-L.; Chen, X.-M. *New J. Chem.* **2004**, *28*, 1412–1415. (b) Zheng, Y.-Z.; Tong, M.-L.; Chen, X.-M. *J. Mol. Struct.* **2006**, *796*, 9–17.
30. Lu, J.; Chu, D.-Q.; Yu, J.-H.; Zhang, X.; Bi, M.-H.; Xu, J.-Q.; Yu, X.-Y.; Yang, Q.-F. *Inorg. Chim. Acta* **2006**, *359*, 2495–2500.
31. Zeng, M.-H.; Feng, X.-L.; Zhang, W.-X.; Chen, X.-M. *Dalton Trans.* **2006**, 5294–5303.
32. (a) Lu, J. Y.; Babb, A. M. A. *Inorg. Chem.* **2002**, *41*, 1339–1341. (b) Li, X.-J.; Cao, R.; Guo, Z.; Lu, J. *Chem. Commun.* **2006**, 1938–1940. (c) Humphrey, S. M.; Mole, R. A.; Rawson, J. M.; Wood, P. T. *Dalton Trans.* **2004**, 1670–1678.
33. Tong, M.-L.; Li, L.-J.; Mochizuki, K.; Chang, H.-C.; Chen, X.-M.; Li, Y.; Kitagawa, S. *Chem. Commun.* **2003**, 428–429.
34. Lin, C. Z.-J.; Chui, S. S.-Y.; Lo, S. M.-F.; Shek, F. L.-Y.; Wu, M.-M.; Suwinska, K.; Lipkowski, J.; Williams, I. D. *Chem. Commun.* **2002**, 1642–1643.
35. Evans, O. R.; Lin, W.-B. *Cryst. Growth Des.* **2001**, *1*, 9–11.
36. (a) Lu, J. Y.; Macias, J.; Lu, J.-G.; Cmaidalka, J. E. *Cryst. Growth Des.* **2002**, *2*, 485–487. (b) Zhang, J.-P.; Lin, Y.-Y.; Huang, X.-C.; Chen, X.-M. *Eur. J. Inorg. Chem.* **2006**, 3407–3412.

37. Feller, R. K.; Forster, P. M.; Wudl, F.; Cheetham, A. K. *Inorg. Chem.* **2007**, *46*, 8717–8721.
38. Liu, C.-M.; Gao, S.; Kou, H.-Z. *Chem. Commun.* **2001**, 1670–1671.
39. Xiao, D.-R.; Hou, Y.; Wang, E.-B.; Lu, J.; Li, Y.-G.; Xu, L.; Hu, C.-W. *Inorg. Chem. Commun.* **2004**, *7*, 437–439.
40. Zheng, N.; Bu, X.; Feng, P. *J. Am. Chem. Soc.* **2002**, *124*, 9688–9689.
41. Hu, S.; Chen, J.-C.; Tong, M.-L.; Wang, B.; Yan, Y.-X.; Batten, S. R. *Angew. Chem., Int. Ed.* **2005**, *44*, 5471–5475.
42. Kolb, H. C.; Fin, M. G.; Sharpless, K. B. *Angew. Chem., Int. Ed.* **2001**, *40*, 2004–2021.
43. Yue, Y.-F.; Wang, B.-W.; Gao, E.-Q.; Fang, C.-J.; He, C.; Yan, C.-H. *Chem. Commun.* **2007**, 2034–2036.
44. (a) Demko, Z. P.; Sharpless, K. B. *J. Org. Chem.* **2001**, *66*, 7945–7950. (b) Himo, F.; Demko, Z. P.; Noodleman, L.; Sharpless, K. B. *J. Am. Chem. Soc.* **2003**, *125*, 9983–9987.
45. (a) Xiong, R.-G.; Xue, X.; Zhao, H.; You, X.-Z.; Abrahams, B. F.; Xue, Z.-L. *Angew. Chem., Int. Ed.* **2002**, *41*, 3800–3803. (b) Li, J.-R.; Tao, Y.; Yu, Q.; Bu, X.-H. *Chem. Commun.* **2007**, 1527–1529.
46. Wu, T.; Yi, B.-H.; Li, D. *Inorg. Chem.* **2005**, *44*, 4130–4132.
47. Zhang, X.-M.; Zhao, Y.-F.; Wu, H.-S.; Batten, S.-R.; Ng, S. W. *Dalton Trans.* **2006**, 3170–3178.
48. Zhao, H.; Qu, Z.-R.; Ye, H.-Y.; Xiong, R.-G. *Chem. Soc. Rev.* **2008**, *37*, 84–100.
49. Cheng, L.; Zhang, W.-X.; Ye, B.-H.; Lin, J.-B.; Chen, X.-M. *Inorg. Chem.* **2007**, *46*, 1135–1143.
50. Zhang, J.-P.; Zheng, S.-L.; Huang, X.-C.; Chen, X.-M. *Angew. Chem., Int. Ed.* **2004**, *43*, 206–209.
51. Zhang, J.-P.; Lin, Y.-Y.; Huang, X.-C.; Chen, X.-M. *J. Am. Chem. Soc.* **2005**, *127*, 5495–5506.
52. Zhang, J.-P.; Lin, Y.-Y.; Huang, X.-C.; Chen, X.-M. *Dalton Trans.* **2005**, 3681–3685.
53. Huang, X.-H.; Sheng, T.-L.; Xiang, S.-C.; Fu, R.-B.; Hu, S.-M.; Li, Y.-M.; Wu, X.-T. *Inorg. Chem.* **2007**, *46*, 497–500.
54. Zhang, J.-P.; Lin, Y.-Y.; Huang, X.-C.; Chen, X.-M. *Cryst. Growth Des.* **2006**, *6*, 519–523.
55. Zhang, J.-P.; Lin, Y.-Y.; Huang, X.-C.; Chen, X.-M. *Chem. Commun.* **2005**, 1258–1260.
56. (a) Zhu, J.-X.; Zhao, Y.-J.; Hong, M.-C.; Sun, D.-F.; Shi, Q.; Cao, R. *Chem. Lett.* **2002**, 484–485. (b) Humphrey, S. M.; Mole, R. A.; Rawson, J. M.; Wood, P. T. *Dalton Trans.* **2004**, 1670–1678. (c) Li, F.-Y.; Xu, L.; Gao, G.-G.; Qu, X.-S.; Yang, Y.-Y *Dalton Trans.* **2007**, 1661–1664. (d) Xiao, H.-P.; Liu, B.-L.; Liang, X.-Q.; Zuo, J.-L.; You, X.-Z. *Inorg. Chem. Commun.* **2008**, *11*, 39–43. (e) Chen, X.-D.; Wu, H.-F.; Du, M. *Chem. Commun.* **2008**, 1296–1298.
57. Cheng, J.-K.; Yao, Y.-G.; Zhang, J.; Li, Z.-J.; Cai, Z.-W.; Zhang, X.-Y.; Chen, Z.-N.; Chen, Y.-B.; Kang, Y.; Qin, Y.-Y.; Wen, Y.-H. *J. Am. Chem. Soc.* **2004**, *126*, 7796–7797.
58. Wang, J.; Zheng, S.-L.; Tong, M.-L. *Inorg. Chem.* **2007**, *46*, 795–800.
59. Li, D.; Wu, T. *Inorg. Chem.* **2005**, *44*, 1175–1177.
60. Li, D.; Wu, T.; Zhou, X.-P.; Zhou, R.; Huang, X.-C. *Angew. Chem., Int. Ed.* **2005**, *44*, 4175–4178.
61. Zhang, X.-M.; Fang, R.-Q.; Wu, H.-S. *J. Am. Chem. Soc.* **2005**, *127*, 7670–7671.
62. Zhou, X.-P.; Li, D.; Wu, T.; Zhang, X. *Dalton Trans.* **2006**, 2435–2443.

4

CONSTRUCTION OF SOME ORGANIC–INORGANIC HYBRID COMPLEXES BASED ON POLYOXOMETALATES

CAN-ZHONG LU, QUAN-GUO ZHAI, XIAO-YUAN WU, LI-JUAN CHEN, SHU-MEI CHEN, ZHEN-GUO ZHAO, AND XIAO-YU JIANG

State Key Laboratory of Structural Chemistry, Fujian Institute of Research on the Structure of Matter, Chinese Academy of Sciences, Fuzhou, Fujian, People's Republic of China

4.1 INTRODUCTION

The continuous interest in organic–inorganic hybrid frameworks is due to their fascinating structural diversity and promising potential application in chemistry, biology, and material sciences [1–3]. Exploitation of polyoxometalates as building blocks to construct solid-state materials with versatile organic ligands or charge-compensating cations is more intriguing and a major challenge to chemists [4–10]. A contemporary interest in this area is to employ organic components or metal complexes to dictate the formation of novel organic–inorganic hybrid composites, as these show promising applications in the areas of catalysis, magnetism, photochemistry, sensors, and energy storage. A large number of POM-based materials have been self-assembled from aqueous solution in the presence of additional transition-metal ion and organic molecules [11–20]. Previous studies have demonstrated that organic molecules usually serve as charge-compensating units and that metal coordination complexes might adopt a variety of roles: (1) as charge-compensating coordinated units; (2) as covalently bound subunits of the metal oxide framework

Design and Construction of Coordination Polymers, Edited by Mao-Chun Hong and Ling Chen

itself, and (3) as inorganic bridging "ligands" linking polyanion clusters into infinite extended frameworks.

On the other hand, as one type of significant metal oxide cluster with nanosizes and abundant topologies, polyoxometalates have recently been employed as inorganic counterions for constructing inorganic–organic hybrid supramolecular arrays with various organic ligands or metal–organic coordination fragments [21–26]. Supramolecular assemblies based on polyoxometalates have been investigated intensively in many important aspects, such as catalysis, electrical conductivity, and biological chemistry [27–29]. Compared to simple inorganic anions, polyoxometalates are bigger, have more diverse topologies, higher charge, and are more suitable as guest units in the metal–organic host, as they lead to larger pores, channels, and cavities. To date, several unique high-dimensional metal–organic frameworks have been synthesized utilizing polyoxometalates as templates [30–35]. A current development is to explore novel lattice architectures resulting from the association of organometal units and polyoxometalate anions.

In this chapter we discuss some of our recent published or unpublished work concerning organic–inorganic hybrid complexes with extended structures based on polyoxometalates as basic units.

4.2 COMPLEXES BUILT UP BY POMs WITH 1,2,4-TRIAZOLATE AND ITS DERIVATIVES

Polyoxometalates (POMs), an important family of metal oxides, have been investigated extensively for use as building blocks in the synthesis of hybrid materials. To date, a large number of organic–inorganic hybrid solids constructed from these isomers and organic molecules or metal–organic coordination subunits have been synthesized. However, nearly all of the organic components studied in these hybrid frameworks are ligands containing 2-, 3-, or 4-substituted pyridines, such as 2,2-bpy, 2,2-phen, 4,4-bpy, and so on. In fact, pyridine is just one of many readily available heterocyclic ring systems, which differ in their electronic and structural properties [36]. Surprisingly, compared to the 4-substituted pyridines, most other ligands of heterocyclic systems have been ignored by people looking for building blocks to design novel organic–inorganic hybrid solids based on POMs. Recently, 1,2,4-triazole and its derivatives have gained more and more attention as ligands for transition metals, not only because they unite the coordination of pyrazole and imidazole, but also because of their great potential application in many fields (e.g., drugs, agricultural chemicals, anticorrosion coatings, photographic materials, dyes, ferromagnetic materials) [37]. Quite a few metal coordination complexes of 1,2,4-triazole, especially its derivatives, have been reported [38–43]; however, only three organic–inorganic hybrid frameworks containing nonsubstituted 1,2,4-triazole have been obtained to date: $[MoO_3(trz)_{0.5}]$, $[\{Cu_2(trz)_2(H_2O)_2\}Mo_4O_{13}]$, and $[Cu_3(trz)_2V_4O_{12}]$. To the best of our knowledge, no POM-based hybrid solids with substituted 1,2,4-triazole have been reported. In this study, 1,2,4-triazole and six derivatives, which are structurally closely related (Fig. 4.1), have been chosen to investigate their hybrid

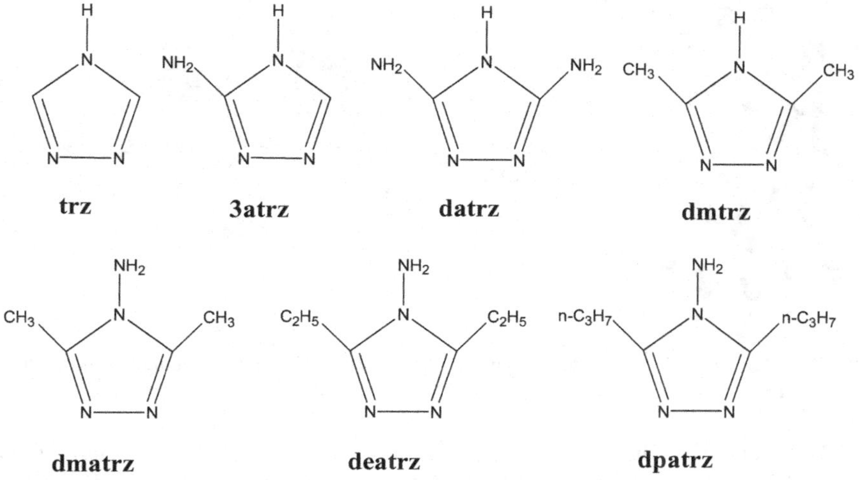

FIGURE 4.1 Schemes of 1,2,4-triazole and its six derivatives.

solid with molybdenum oxides under hydrothermal conditions. Sixteen novel complexes were obtained [44].

4.2.1 Complexes Built Up by Octamolybdate and Tetranuclear Copper(I) 1,2,4-Triazolate

By changing the substituents on a 1,2,4-triazole ring, six novel organic–inorganic hybrid complexes, $[\{Cu_4(L)_x\}Mo_8O_{26}]$ [L = 3,5-diamino-1,2,4-triazole (datrz) and $x = 4$ for **1**; L = 3-amino-1,2,4-triazole (3atrz) and $x = 4$ for **2**; L = 3,5-dimethyl-1,2,4-triazole (dmtrz) and $x = 4$ for **3**; L = 3,5-dimethyl-4-amino-1,2,4-triazole (dmatrz) and $x = 6$ for **4**; L = 3,5-diethyl-4-amino-1,2,4-triazole (deatrz) and $x = 4$ for **5**; L = 3,5-di(*n*-propyl)-4-amino-1,2,4-triazole (dpatrz) and $x = 3$ for **6**] were constructed from tetranuclear copper(I) 1,2,4-triazolate clusters and octamolybdates.

$[\{Cu(datrz)\}_4Mo_8O_{26}]$ **1** $[\{Cu(3atrz)\}_4Mo_8O_{26}]$ **2**
$[\{Cu(dmtrz)\}_4Mo_8O_{26}]$ **3** $[\{Cu_2(dmatrz)_3(H_2O)\}_2Mo_8O_{26}]$ **4**
$[\{Cu(deatrz)\}_4Mo_8O_{26}]$ **5** $[\{Cu_4(dpatrz)_3\}Mo_8O_{26}]$ **6**

Crystal Structure of $[\{Cu(datrz)\}_4Mo_8O_{26}]$ (1) In complex **1**, a structure consisting of tetranuclear copper-coordinated cations $[Cu_4(datrz)_4]^{4+}$ and polyoxoanions $[Mo_8O_{26}]^{4-}$ was obtained. The centrosymmetric $[Mo_8O_{26}]^{4-}$ anion built up from eight edge-shared $\{MoO_6\}$ octahedron was a typical β-octamolybdate. As shown in Fig. 4.2, four Cu(I) atoms were linked by four datrz ligands through the N1,N2 bridging mode, a coordination mode usually found for N4- or 3,5-substituted triazole ligands, to form the coordinated cations $[Cu(datrz)]_4^{4+}$. In such a fashion, a distorted square (3.02 × 2.94 Å) Cu_4 with angles of 116.72(2)°, 64.20(3)°, and 62.36(3)° was generated with π-stacked triazole rings.

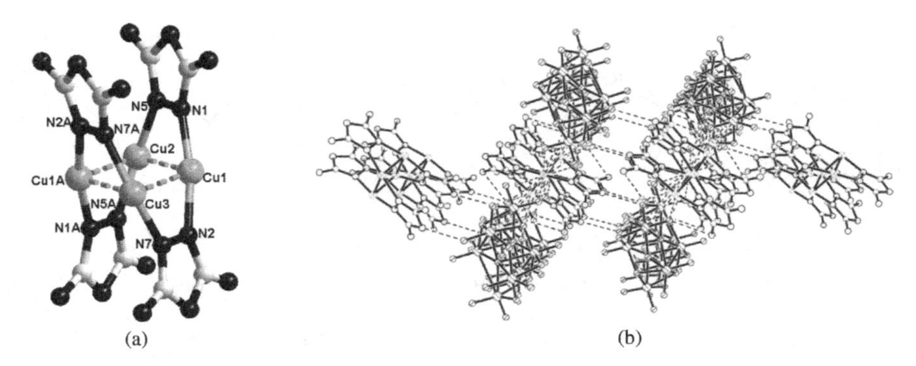

FIGURE 4.2 (a) Tetranuclear copper(I) cluster; (b) 3D supramolecular framework in complex **1**.

The N–N bond of the ligands was markedly longer (average 1.422 Å) than those found in uncoordinated triazole or its derivatives (an approximate value of 1.36 to 1.38 Å). As pointed out by Ehlert et al. [56] this phenomenon could be associated with a Cu(d_π)–datrz (π^*) interaction between metal and triazole orbitals of acidic character. The neutral [{Cu (datrz)}$_4$Mo$_8$O$_{26}$] was further extended into interesting three-dimensional (3D) supramolecular arrays via π–π stacking interactions between triazole rings and complex hydrogen bonds. To our knowledge, complex **1** was the first synthesized tetranuclear complex with N1,N2 bridging triazoles only, which offers a new type of triazole-bridged structure.

Crystal Structure of [{Cu(3atrz)}$_4$Mo$_8$O$_{26}$] (2) When 3atrz (only one amino group on the triazole ring) was used instead of datrz, complex **2** was obtained, which exhibited a one-dimensional (1D) zigzag chain constructed from β-octamolybdate and [Cu(3atrz)]$_4^{4+}$ units (Fig. 4.3). The tetranuclear unit was similar to that in complex **1**. Each β–[Mo$_8$O$_{26}$]$^{4-}$ anion formed covalent interaction with two [Cu (3atrz)]$_4^{4+}$ through the terminal oxo group of the octamolybdate cluster. Cu(1) was coordinated with two nitrogen atoms from different 3atrz [Cu–N = 1.883(8) Å and N–Cu–N = 151.2(5)°]. Cu(2) was linked to two terminal oxygen atoms of different

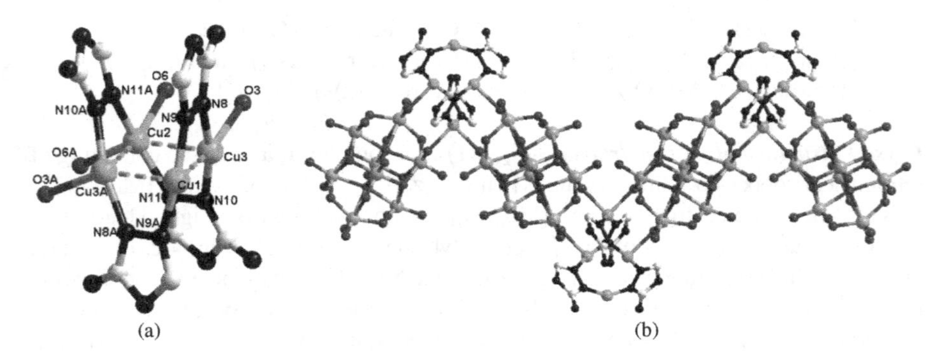

FIGURE 4.3 (a) Tetranuclear copper(I) cluster; (b) 1D chain structure in complex **2**.

β-octamolybdates [Cu–O = 2.370(6) Å] and two nitrogen atoms from 3atrz [Cu–N = 1.905(7) Å] to form an O_2N_2 distorted tetrahedron. Cu(3) was coordinated by two nitrogen atoms [Cu–N = 1.912(7) Å] and one terminal oxo atom [Cu–O = 2.348(6) Å] to generate a T-shaped coordination geometry. It is interesting that three types of distinct Cu(I) coordination geometries were observed in the same structure. The distances between copper atoms (ca. 2.99 and 3.11 Å) were a little longer than those in **1**. Angles for the Cu_4 quadrangle were 63.30°, 118.05°, and 60.59°. The ligand 3atrz is a twofold disorder as described in the complexes [ZnF(3atrz)] solvents [45] (for clarity, only one ligand disorder component has been shown in the corresponding figures).

Crystal Structure of [{Cu(dmtrz)}$_4$Mo$_8$O$_{26}$] (3) When dmtrz was used, complex **3** was obtained with a two-dimensional (2D) layer structure. In the layer (Fig. 4.4), each β-octamolybdate was bonded to four $[Cu_4(dmtrz)_4]^{4+}$ units (Fig. 4.4) through four terminal oxo groups. It could also be seen that each tetranuclear cluster linked four β-octamolybdate units, which was similar to the structure of [{Cu (pyrd)}$_4$($H_4Mo_8O_{26}$)] [46]. All four unique Cu(I)'s were in T-shaped triangle coordination environments, which was defined by two nitrogen atoms of different dmtrz ligands and one oxo atom from octamolybdate [Cu–O, Cu–N bonds and corresponding N–Cu–N angles: 2.387(3), 1.910(3), 1.913(3) Å, and 164.37(14)° for Cu(1); 2.359(3), 1.898(3), 1.903(3) Å, and 152.57(14)° for Cu(2); 2.294(3), 1.898(3), 1.905(3) Å, and 157.28(14)° for Cu(3); 2.370(3), 1.922(3), 1.917(3) Å, and 141.91 (14)° for Cu(4)].

Crystal Structure of [{Cu$_2$*(dmatrz)*$_3$(H$_2$O)}$_2$Mo$_8$O$_{26}$] (4) When dmatrz, a common 3,4,5-substituted-1,2,4-triazole, complex **4**, was used, a new 1D chain was obtained. Complex **4** is the first example of δ-octamolybdates linked by metal coordination units into an infinite extended structure (Fig. 4.5). As shown in Figure 4.5, two $[Cu_2(dmatrz)_2]^{2+}$ rings were bridged by another two dmatrz ligands placed on opposite sides with a N1,N2 bridging mode to form a $[Cu_4(dmatrz)_6]^{4+}$ cluster. This tetranuclear unit had a strictly planar metal core

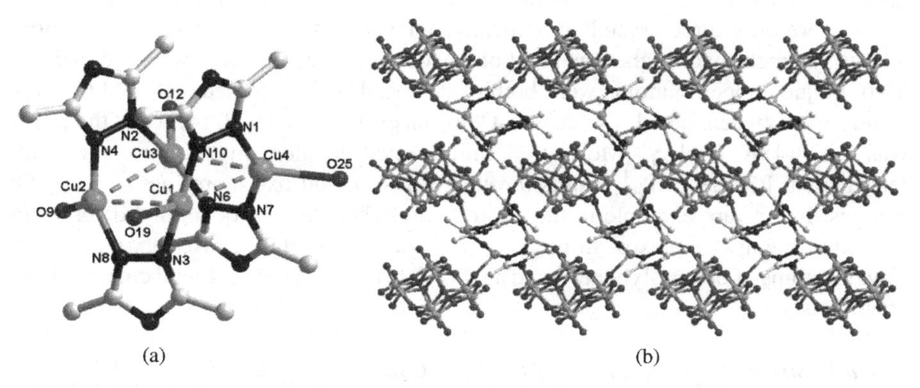

FIGURE 4.4 (a) Tetranuclear copper(I) cluster; (b) 2D layer structure in complex **3**.

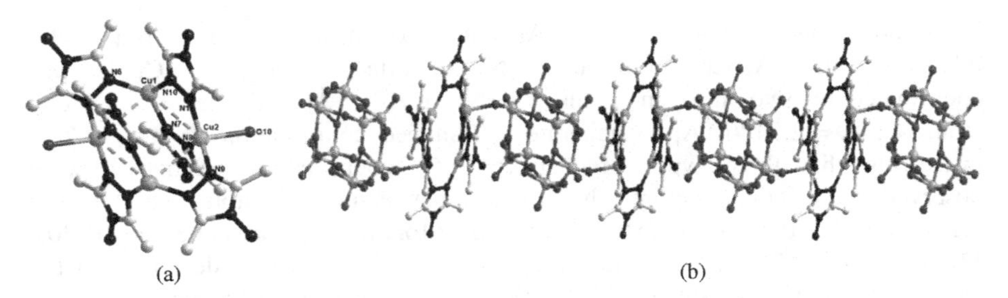

(a) (b)

FIGURE 4.5 (a) Tetranuclear copper(I) cluster; (b) 1D chain structure in complex **4**.

imposed by symmetry. Four copper atoms were arranged at the corner of a parallelogram with side lengths of about 3.08 and 3.52 Å. The anionic complex $[Cu_4(dmnpz)_6]^{2-}$ [47] was structurally akin to this cationic cluster, containing a tetranuclear copper framework bridged by six dmnpz (3,5-dimethy-4-nitropyrazole) ligands. However, this complex exhibited a more irregular metal arrangement [Cu···Cu: 3.324(1) and 3.019 Å, Cu–Cu–Cu: 101.92(2) and 78.08(2)°]. There were two crystallographically independent copper sites in the tetranuclear unit. Cu(1) was coordinated by three nitrogen atoms from dmatrz [Cu–N = 1.953(5) to 2.002(5) Å] to complete a trigonal environment. Cu(2) ion was linked to one terminal oxygen atom of δ-octamolybdate [Cu–O = 2.294(5) Å] and three nitrogen atoms [Cu–N = 1.969(5) to 2.020(5) Å] to form an elongated tetrahedron in the form of ON_3. With the Cu(2)–O bonds, the δ-octamolybdate units were linked by the $[Cu_4(dmatrz)_6]^{4+}$ units into 1D chains.

Crystal Structure of [{Cu(deatrz)}$_4$Mo$_8$O$_{26}$] (5) The use of deatrz, which has larger groups on the 3 and 5 positions, led to the formation of complex **5**. It crystallized in the space group I_{bam}. The octamolybdate anion in **5** was composed of six edge-sharing MoO_6 octahedra and two MoO_5 square pyramids. There existed 14 μ_t-O, six μ_2-O, four μ_3-O, and two μ_4-O, which were in accord with the character of γ-$[Mo_8O_{26}]^{4-}$. However, the structure mode was quite different from that of a typical γ-isomer. For convenient discussion, this novel $[Mo_8O_{26}]^{4-}$ anion was named γ′-octamolybdate. There were only three crystallagraphically independent Mo atoms in the structure, and its symmetry was higher than that of any octamolybdate isomer reported to date. Two unique copper(I) atoms were both of distorted N_2O_2 tetrahedra (Fig. 4.6). The difference between them lay in the Cu–O distances: One was 2.323(4) Å and the other was 2.224(4) Å. Each γ′-$[Mo_8O_{26}]^{4-}$ unit formed covalent interactions with two tetranuclear $[Cu(deatrz)]_4^{4+}$ units through four terminal oxo atoms acting as μ_3-O bridges. It is a rare example in the coordination chemistry of octamolybdates with transition-metal complexes. In our earlier work we reported a case in which μ_2-O oxygen atoms of octamolybdate acted as a μ_3-O bridge linking the metal coordination fragments.

Crystal Structure of [{Cu$_4$*(dpatrz)$_3$}Mo$_8$O$_{26}$] (6) When the length of the substituents on the 3 and 5 positions was elongated further using dpatrz, complex

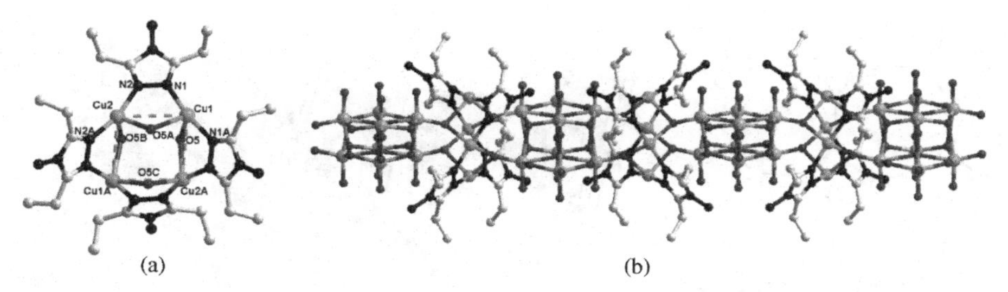

FIGURE 4.6 (a) Tetranuclear copper(I) cluster; (b) 1D chain structure in complex **5**.

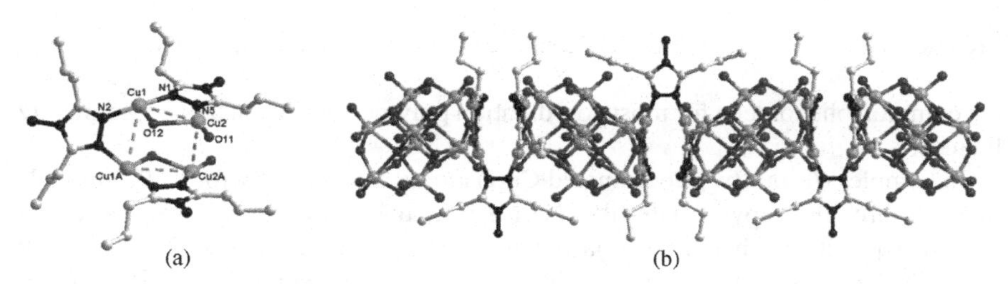

FIGURE 4.7 (a) Tetranuclear copper(I) cluster; (b) 1D chain structure in complex **6**.

6, constructed from β-octamolybdate and $[Cu_4(dpatrz)_3]^{4+}$, was obtained (Fig. 4.7a). Two tetranuclear Cu(I) fragments were attached to one octamolybdate surface by four terminal oxygen atoms to form a 1D zigzag chain. Two of the four terminal oxygen atoms served as μ_2-oxo bridges, and others served as μ_3-oxo bridges. Two T-shaped trigonal Cu(I) atoms existed in the tetranuclear unit (Fig. 4.7b). Cu(1) was coordinated by one terminal oxygen atom [Cu–O = 2.323(4) Å] and two nitrogen atoms [Cu–N = 1.907(5) and 1.914(5) Å]. Cu(2) was defined by one nitrogen atom [Cu–N = 1.889(4) Å] and two terminal oxygen atoms [Cu–O = 1.855(4) and 2.259 (4) Å]. It was different from complexes **1** to **5** in that the four copper atoms in **6** were not planar. The angle between two triangular planes [Cu(1)–Cu(1A)–Cu(2A) and Cu(1)–Cu(2)–Cu(2A)] was 44.04(2)°.

4.2.2 Complexes Built Up by Keggin Polyoxomolybdate and Copper(I)-1,2,4-Triazolate Fragment

Based on our continuous investigations on hybrid inorganic–organic materials containing polyoxometalates, we have succeeded in isolating two novel supramolecular host–guest complexes ($[\alpha\text{-}Cu_{12}(trz)_8][PMo_{12}O_{40}]\cdot H_2O$ (**7**) and $[\beta\text{-}Cu_{12}(trz)_8][PMo_{12}O_{40}]\cdot 2H_2O$ (**8**) (trz = 1,2,4-triazole)) assembled from Cu^I–1,2,4-triazolate and phosphododecamolybdate under hydrothermal conditions [48]. In these two complexes, the metal–organic coordination hosts are two-dimensional (2D) supramolecular isomers which possess 4^4 and 6^3 topologies, respectively. The large Keggin anions occupy the cavities of (4,4) or (6,3) nets. In our opinion, the existence of the POM anions prohibits the Cu^I-trz fragments from forming 3D frameworks, which are

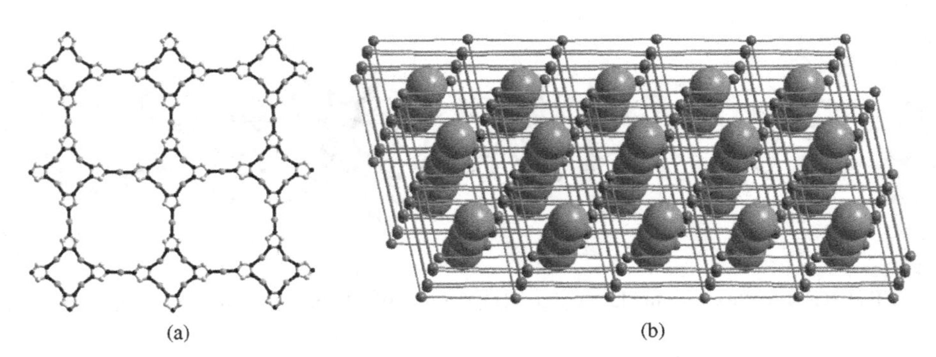

FIGURE 4.8 (a) 2D metal–organic sheet; (b) 3D supramolecular framework in complex **7**.

the common phenomena for most coordination polymers based on $\mu_{1,2,4}$-bridging trz ligands.

In complex **7**, four two-coordinated Cu(I) atoms are linked by four 1,2,4-triazole ligands through a pyrazolate-like N1,N2-bridging mode to generate a cyclic $[Cu_4(trz)_4]$ unit. Further, four adjacent $[Cu_4(trz)_4]$ units are linked by four other linear Cu(I) atoms though the N_4 position to form a 2D host cation. Thus, the entire 2D sheet can be regarded as alternating arrangements of four- and eight-membered Cu-trz rings (Fig. 4.8). From the view of topology, this 2D sheet is a common four-connected 4^4 net, with $[Cu_4(trz)_4]$ units as nodes and the remaining Cu^I ions as linkers. It should be noted that although the 3D porous coordination polymer $[Ag_6Cl(atrz)_4]$ $OH{\cdot}xH_2O$ [49] is constructed by metal-trz cyclic units similar to those in **7**, it has parallel fivefold interpenetration. In our opinion, the existence of large POM ions in **7** prohibits the interpenetration and leads to the formation of a low-dimensional structure.

The metal–organic fragment in **8** is a supramolecular isomer of that in **7**, and the basic unit is not $[Cu_4(trz)_4]$ but trinuclear $[Cu_3(trz)_3]$ (Fig. 4.9). Zhang et al. have pointed out that the trinuclear $[Cu_3(trz)_3]$ units are hypothetically possible from the viewpoint of structural motifs [50], but no structurally characterized example has been documented to date. However, the trinuclear unit in **8** is not a six-connected

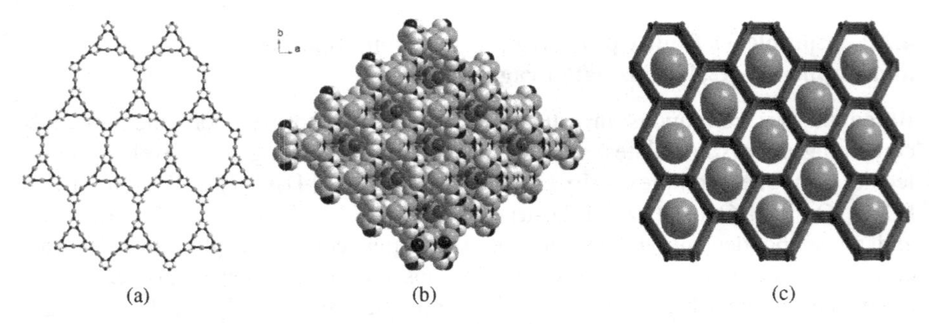

FIGURE 4.9 (a) 2D metal–organic sheet in **8**; (b) space-filling diagram of the structure; (c) schematic presentation of the host–guest assembly.

octahedron as expected, but a three-connected triangle. The [$Cu_3(trz)_3$] units are connected further by other trz ligands and Cu(I) atoms to generate a 2D net. Therefore, the entire structure can be regarded as alternating arrangements of three- and nine-membered Cu-trz rings. This 2D net exhibits three-connected 6^3 topology regarding [$Cu_3(trz)_3$] units and remaining trz ligands as hexagon nodes, and the remaining linearly coordinated Cu^I ions as linkers. To the best of our knowledge, the nine-membered metal-trz circuit in **8** is the largest example reported to date. It is interesting that this sheet presents a wave-shaped single strand viewed from the *b*-axis. Further, short intermetallic $Cu^I \cdots Cu^I$ contacts [2.960(6) Å] without a direct ligand bridge link two sheets together to generate parallel double layers as a rail. It should be noted that two interesting supramolecular isomers containing 1,2,4-triazole derivatives ([Cu^I(mtz)] and [Cu^I(2-pytz)] (mtz = 3,5-dimethyl-1,2,4-triazole, 2-pytz = 3,5-di-2-pyridyl-1,2,4-triazole)) have been reported by Zhang et al. [50]. $[Cu_{12}(trz)_8]^{4+}$ is the first 2D example of supramolecular isomerism in the coordination chemistry of 1,2,4-triazole. In our opinion, construction of these interesting isomer architectures primarily benefits from the variable linking modes of triazole ligands and the structural diversity of Cu-trz secondary building units.

4.2.3 Complexes Built Up by Polyoxomolybdate and Ag(I)-1,2, 4-Triazolate Fragment

Eight members of the Ag/1,2,4-triazole/polyoxometalates (POMs) hybrid supramolecular family—[$Ag_4(dmtrz)_4$][Mo_8O_{26}] (**9**), [$Ag_6(3atrz)_6$][$PMo_{12}O_{40}$]$_2 \cdot H_2O$ (**10**), [$Ag_2(3atrz)_2$]$_2$[$HPMo^{VI}_{10}Mo^{V}_2O_{40}$] (**11**), [$Ag_2(dmtrz)_2$]$_2$[$HPMo^{VI}_{10}Mo^{V}_2O_{40}$] (**12**), [$Ag_2(trz)_2$]$_2$[$Mo_8O_{26}$] (**13**), [$Ag_2(3atrz)_2$][$Ag_2(3atrz)_2(Mo_8O_{26})$] (**14**), [$Ag_4(4atrz)_2Cl$][$Ag(Mo_8O_{26})$] (**15**), and [$Ag_5(trz)_4$]$_2$[$Ag_2(Mo_8O_{26})$]$\cdot 4H_2O$ (**16**)—were synthesized through hydrothermal reactions of 1,2,4-triazole or its derivatives with appropriate silver salts and molybdates [51]. Crystal structure analysis reveals that the POM-dependent Ag–1,2,4-triazolate units in these hybrid complexes are a novel tetranuclear cluster (**9**), a unique double calix[3]arene-shaped hexamer (**10**), a zigzag chain (**13** and **14**), a helix chain (**11**, **12**, and **16**), and an interesting looped chain (**15**). A series of hydrogen bonding–based supramolecular assemblies varying from 0D + 0D (**9** and **10**), 0D + 1D (**11** and **12**), 1D + 1D (**13**, **15**, and **16**), to 1D + 2D (**14**) modes between the organomatic cations and POM anions were observed in these structures.

Structure of 0D + 0D Complex **9**, consisting of tetranuclear silver-coordinated cations $[Ag_4(dmtrz)_4]^{4+}$ and polyoxoanions $[Mo_8O_{26}]^{4-}$, crystallizes in the space group $C_{2/c}$ with four formula units in the unit cell. Four Ag(I) atoms were linked by dmtrz ligands through the N1,N2 bridging mode to form a coordinated $[Ag_4(dmtrz)_4]^{4+}$ cation (Fig. 4.10). In such a fashion, the $Ag \cdots Ag$ distances vary from 3.246(11) to 3.406(3) Å. Neutral [$\{Ag(dmtrz)\}_4Mo_8O_{26}$] was further extended into 3D supramolecular arrays through weak ($N\text{–}H \cdots O$) and ($C\text{–}H \cdots O$) hydrogen bonds, and the overall packing of **1** can be derived from the NaCl structure type.

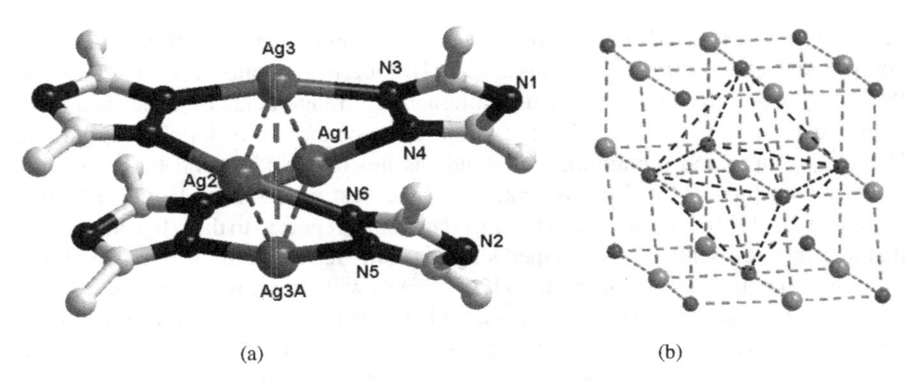

FIGURE 4.10 (a) Structure of the tetranuclear cluster in complex **9**; (b) similarity of the packing of **9** to the NaCl structure type: the silver clusters are packed face-centered cubic, and octamolybdate anions occupy the octahedral voids.

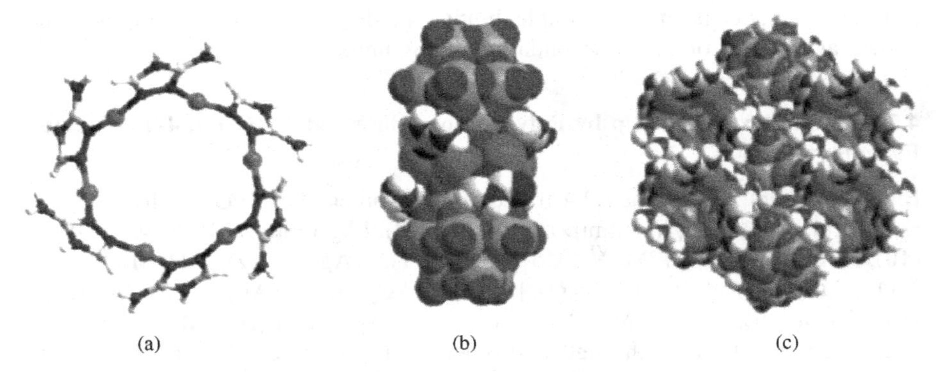

FIGURE 4.11 (a) Structure of the hexanuclear cluster in complex **10**; (b) space-filling diagram of the two polyoxometalate anions in the two bow-shaped cavities of the hexamer; (c) 3D hybrid supramolecular architecture viewed along the *c*-axis.

Complex **10** crystallizes in the space group R_{-3} with three formula units in the unit cell. Six two-coordinated Ag(I) atoms are linked by six 3atrz ligands through the N1, N2 bridging mode (Fig. 4.11) [Ag–N, 2.132(6) and 2.123(5) Å, N–Ag–N, 168.0(2)°] to form a double calix[3]arene-shaped hexamer, where two $[PMo_{12}O_{40}]^{3-}$ are contained as noncoordinated guest anions. Two bottom-shared bowls are open in opposite directions, and due to a limited void, two α-Keggin ions have to protrude out from the cavity, with a distance of about 13 Å between the two phosphorus atoms. The guest anion is a normal α-Keggin structure composed of 12 corner- or edge-sharing MoO_6 octahedra, with the central phosphorus atom ordered and coordinated to four oxygen atoms in a tetrahedral fashion.

Structure of 0D + 1D Complexes **11** and **12**, consisting of 1D Ag–1,2,4-triazolate helix chains and Keggin anions, crystallize isostructurally. The metal–organic helix chain contains two crystallographically unique silver ions linked by two

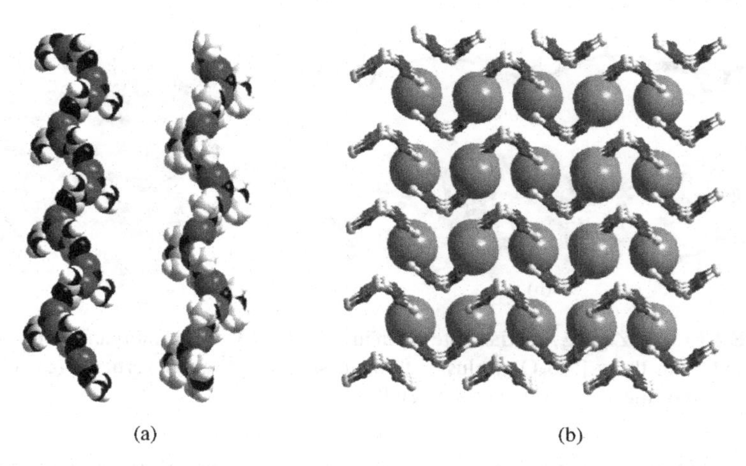

FIGURE 4.12 (a) Space-filling view of the helix chain in complexes **11** (left) and **12** (right); (b) schematic representation of the 0D + 1D supramolecular assembly. The Keggin POM anions are represented as larger gray spheres.

organic ligands through N1,N2 and N1,N4 bridging modes (Fig. 4.12a). The long pitches of the helix chains in **11** and **12** along the *c*-axis are 17.891(31) and 19.950(21) Å, respectively. Both silver atoms are of linear geometry, with Ag–N distances falling in the range 2.103(8) to 2.158(5) Å. The corresponding N–Ag–N angles are of 158.3(3)° and 172.8(3)° for **11** and 153.33(19)° and 151.71(17)° for **12**. It should be pointed out that due to the symmetrical structures of 1,2,4-triazole or its derivatives and the diversities of their bridging modes, 1D helical chain structures with triazole ligands are rare, and to our knowledge, only one example of a single-stranded 4_1 helix has been reported [52]. Moreover, pyrazole- and imidazole-like bridging modes of triazole ligands have never been observed to link metal centers in the same structure. The polyoxoanion $[HPMo^{VI}_{10}Mo^{V}_{2}O_{40}]^{4-}$ in **11** and **12** exhibits a distorted α-Keggin structure with center P atom surrounded by a cube of eight half-occupied oxygen atoms. P–O bonds range from 1.481(6) to 1.592(9) Å. This structural feature often appears in $[XMo_{12}O_{40}]^{n-}$ with the α-Keggin structure, which has been explained by several groups. Three sets of Mo–O distances are Mo–O(t), 1.644(6) to 1.668(4) Å; Mo–O(b), 1.810(7) to 2.000(7) Å; and Mo–O(c), 2.408(10) to 2.522(6) Å. The helices and Keggin anions are linked by complex hydrogen bonds (N–H ··· O or C–H ··· O) to generate a 3D inorganic–organic hybrid supramolecular structure. The distances between N or C atoms of organic ligands and oxygen atoms of the POM anions range from 3.004(11) to 3.222(9) Å. The entire packing schemes of complexes **11** and **12** present layer structures (Fig. 4.12b) following the pattern ··· ABAB ··· (A: POM layer; B: Ag-trz layer) along the *a*-axis. Layers B are parallel to each other along the *a*-axis with a distance of about 12 Å between the planes of two triazole rings, which is slightly longer than the diameter of POM anions (ca. 10 Å).

Structure of 1D + 1D The supramolecular structure of **13** consists of pairs of $[Ag_2(trz)_2]_n^{2n+}$ zigzag chains and one $[Mo_8O_{26}]_n^{4n-}$ chain. Two crystallographically unique silver ions are linked by trz ligands through the N1,N4 bridging mode to

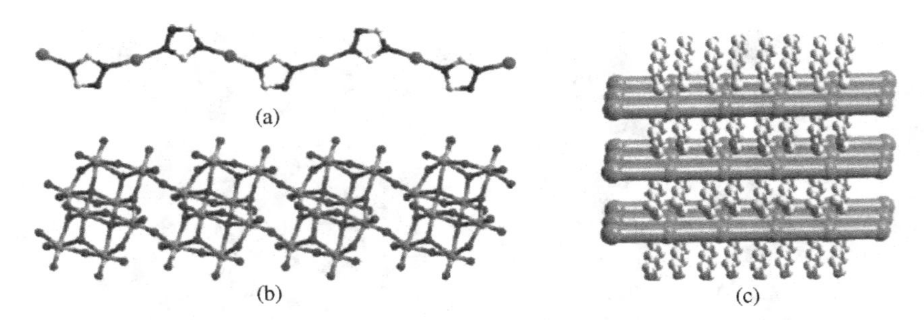

FIGURE 4.13 (a) Zigzag Ag-triazolate chain in complex **13**; (b) 1D inorganic $[Mo_8O_{26}]_n^{4n-}$ constructed from the γ-$[Mo_8O_{26}]$ cluster through sharing common vertices; (c) schematic representation of the 1D + 1D hybrid assembly.

generate a 1D zigzag chain (Fig. 4.13a). Both silver atoms are of linear geometry, with Ag–N distances falling in the range 2.130(4) to 2.200(3) Å and N–Ag–N angles of 170.70(15)° and 175.39(14)°. Each molybdenum atom of the polyoxometalate chain attains a distorted octahedral environment by the coordination of six oxygen atoms. The four asymmetric $[MoO_6]$ octahedra are connected through their edges to form the $[Mo_4O_{13}]^{2-}$ unit, which are stacked together further by edge sharing to give rise to a γ-$[Mo_8O_{26}]^{4-}$ octamolybdate cluster. The γ-octamolybdate units are linked together further by sharing common vertices to form an infinite chain along the *a*-axis (Fig. 4.13b). In the solid state of **13**, two types of 1D chains are held together and extended to a 3D framework via strong N–H···O and weak C–H···O hydrogen bondings. One noncoordinated nitrogen atom and two carbon atoms of the organic ligands all participate in hydrogen bonding with terminal oxo atoms of the POM chain. As shown schematically in Figure 4.13c, the overall 3D framework of **13** also presents layer structures following the pattern ···ABAB··· along the *b*-axis. Two types of chains are basically vertical to each other, and they extend along the *a*- and *c*-axes.

Complex **15** consists of an interesting cationic $[Ag_4(4atrz)_2Cl]_n^{3n+}$ looped chain (Fig. 4.14a) and an unprecedented anionic $[Ag(Mo_8O_{26})]_n^{3n-}$ chain (Fig. 4.14b). Both of the two unique silver atoms in the cationic chain are of linear coordination

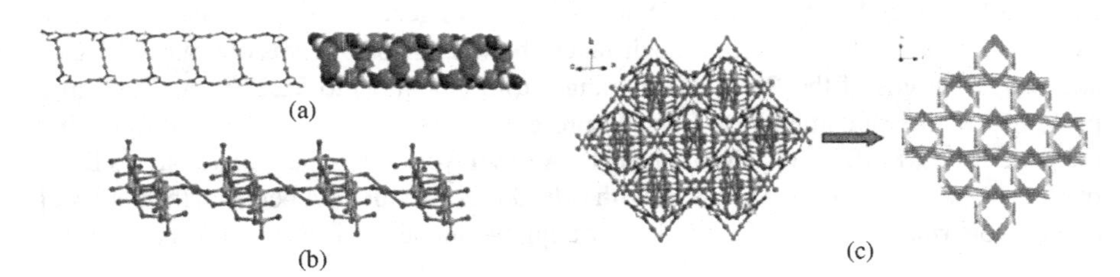

FIGURE 4.14 (a) Ball-and-stick (left) and space-filling (right) representations of the looped chain in complex **15**; (b) inorganic $[AgMo_8O_{26}]_n^{3n-}$ chain in **15**; (c) overall supramolecular assembly between two types of chains (left) and its topological representation (right).

geometry. Each 4atrz ligand links two Ag^+ ions through the N1 (or N2) atom [Ag(2)–N(2): 2.164(6) Å] and amino group [Ag(2)–N(3): 2.225(6) Å] to give a 1D single-chain structure. Then two single chains were connected by $[Ag_2Cl]^+$ units [Ag(1)–Cl(1): 2.435(2) Å] through the uncoordinated N2 (or N1) atoms of the triazole ligands [Ag(1)–N(1): 2.163(6)] to generate an interesting looped chain structure. The $[Ag_2Cl]^+$ units are located alternately at both sides of the chain. Notably, this is the first example in the coordination chemistry of 4-amino-1,2,4-triazole ligands that the amino group participates in bonding. Moreover, the Ag–N bond distance [2.225 (6) Å] is distinctly longer than those between the Ag^+ ion and the nitrogen atom of the triazole ring [2.164(6) and 2.163(6) Å]. On the other hand, this metal–organic chain can also be regarded as hexanuclear $[Ag_6(4atrz)_4Cl_2]$ rings linked together to form an extended structure. Further, the cationic $[Ag_4(4atrz)_2Cl]_n^{3n+}$ are linked through an N–H···Cl hydrogen bond to give a 3D supramolecular porous framework.

The 1D channels formed possess dimensions of about 9.28 × 9.06 Å, which are occupied perfectly by inorganic chains. The anionic $[Ag(Mo_8O_{26})]_n^{3n-}$ chain in **15** is constructed from the β-octamolybdate clusters and single coordinated electrophilic silver ions. The Ag^+ ion is sandwiched between two β-octamolybdate units. Each β-Mo_8O_{26} unit forms covalent interactions with two silver atoms through four terminal oxygen atoms of two $\{Mo_4O_{13}\}$ subunits with Ag–O distances ranging from 2.355(6) to 2.451(6) Å and O–Ag–O angles ranging from 80.24(19) to 163.67 (8)°. Each four-coordinated Ag(I) ion is of a slightly distorted quadrangular plane. The distance between two adjacent silver atoms is 9.198(16) Å. Although a large number of hybrid materials based on octamolybdates and their analog complexes have been reported, such as a 1D infinite chain structure constructed by a $[Mo_8O_{26}]^{4-}$ building block linked only via a single Ag^+ ion. The overall 3D suparmolecular assembly is stabilized further by C–H···O hydrogen bonding between the carbon atoms of 4atrz ligands and the terminal oxygen atoms of the octamolybdate clusters. The entire 1D + 1D supramolecular architecture of complex **15** is expressed simply in Figure 4.14c, which to our knowledge is quite novel.

Complex **16** is constructed from 1D $[Ag_2(Mo_8O_{26})]_n^{2n-}$ anionic chains integrated by pairs of $[Ag_5(trz)_4]_n^{n-}$ chains into a 3D framework. As shown in Figure 4.15a, there exist six crystallographic independent silver(I) centers, which have two types of coordination environment: those with a four-coordinated Ag(2) center and those with a two-coordinated silver center. Linearly coordinated Ag(1), Ag(4), and Ag(5) are linked by three triazole ligands through a pyrazole-like bridging mode to give a $[Ag_3(trz)_3]$ neutral trigonal unit. The Ag–N distances and N–Ag–N angles are 2.085 (12) to 2.117(12) Å and 165.2(5) to 174.5(5)°, respectively. The Ag···Ag distances in this trigonal unit are 3.603(2), 3.609(2), and 3.801(2) Å. The three triazole rings all deviate slightly from the trigonal plane formed by three silver atoms to ease the repulsion of metal ions, and the dihedral angles are about 6.5°, 15.9°, and 19.4°, respectively.

Although the use of $[Cu_3(trz)_3]$ subunits to construct coordination polymers has been reported, $[Ag_3(trz)_3]$ in **16** is the first structurally characterized cyclic trinuclear example with silver centers. This unique $[Ag_3(trz)_3]$ unit is connected further by another two linear silver ions [Ag(3) and Ag(6)] and one triazole ligand

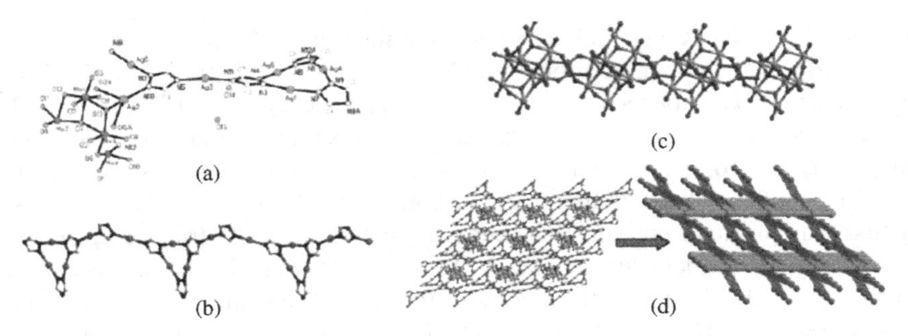

FIGURE 4.15 (a) Ortep diagram of complex **16** showing the local coordination environment in the asymmetric unit and 30% thermal ellipsoids; (b) 1D helix Ag-triazolate chain; (c) 1D inorganic $[Ag_2Mo_8O_{26}]_n^{2n-}$ chain; (d) 3D framework of complex **16** (left) and its topological representation based on two types of rod-shaped building units (right).

(imidazole-like linking mode) to give the 1D chain depicted in Figure 4.15b. In the inorganic chain, the octamolybdate cluster, which contains 14 μ_t-O, six μ_2-O, four μ_3-O, and two μ_4-O, is a γ-isomer like that in complexes **13** and **14**. Each γ-Mo_8O_{26} unit forms covalent interactions with four silver atoms [Ag(2)] through four μ_t-O and two μ_2-O atoms of two $\{Mo_4O_{13}\}$ subunits (Fig. 4.15c). Thus, the Ag^+ ions are sandwiched between two octamolybdate units. Each four-coordinated Ag(I) ion is not of quadrangular plane geometry as in the reported $[Ag_2(Mo_8O_{26})]^{2-}$ examples [53], but in a distorted tetrahedral geometry coordinated with three oxygen atoms and one nitrogen atom [N(10)]. The Ag–O and Ag–N distances are 2.356(10) to 2.510 (9) and 2.165(11) Å, while the N–Ag–O and O–Ag–O angles are 111.4(4) to 139.4(4)° and 80.5(3) to 96.3(3)°, respectively. The distance between the two silver atoms is 4.911(2) Å. Moreover, the Mo(4) atom in the $\{Mo_4O_{13}\}$ subunit is coordinated with the N(12) atom with an Mo–N distance of 2.236(11) Å. Thus, two noncoordinated nitrogen atoms [N(10) and N(12)] of triazole ligands in the metal–organic fragments link the inorganic $[Ag_2Mo_8O_{26}]_n^{2n-}$ chains covalently to give the unique 3D organic–inorganic hybrid framework of complex **16** (Fig. 4.15d). To the best of our knowledge, only limited discrete $[Ag_2Mo_8O_{26}]_n^{2n-}$ chains have been reported by us and other groups to date, and complex **16** is the first multidimensional architecture utilizing $[Ag_2Mo_8O_{26}]_n^{2n-}$ as a building block. The topological representation of the overall network is depicted in Figure 4.15d, which can be described as a 3D framework based on two types of 1D rod-shaped subunits, according to the principles of rod packing.

Structure of 1D + 2D Single-crystal x-ray diffraction analysis reveals that **14** comprises an anionic 2D netlike framework $[Ag_2(3atrz)_2(Mo_8O_{26})]_n^{2n-}$ penetrated by cationic zigzag chains $[Ag_2(3atrz)_2]_n^{2n+}$. Although numerous types of entanglements of multiple motifs in coordination polymeric networks have been reviewed [54], examples of threading 1D chains into 2D sheets are relatively rare. The 1D $[Ag_2(3atrz)_2]_n^{2n+}$ zigzag chain is similar to the $[Ag_2(trz)_2]_n^{2n+}$ chain sin complex **5**. Two unique silver atoms are both of linear geometry [Ag–N: 2.205(4) and 2.120(4) Å] and connected by 3atrz ligands using an imidazole-like bridging mode (Fig. 4.16a). The 2D $[Ag_2(3atrz)_2(Mo_8O_{26})]_n^{2n-}$ framework is constructed from 1D

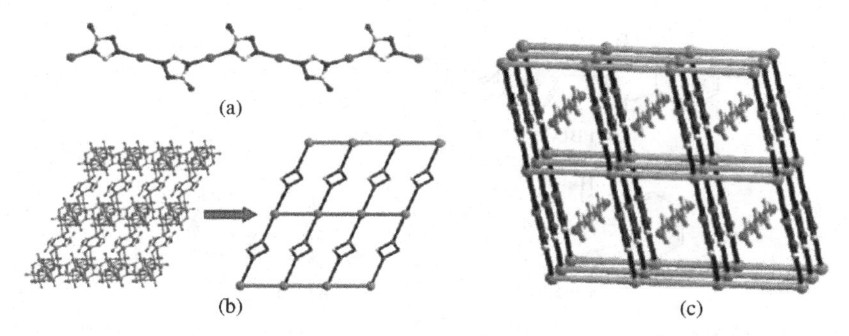

FIGURE 4.16 (a) 1D zigzag metal–organic chain in **14**; (b) 2D $[Ag_2(3atrz)_2(Mo_8O_{26})]_n^{2n-}$ hybrid layer (left) and its topological representation (right); (c) schematic representation of the 1D-in-2D supramolecular assembly of **14**.

$[Mo_8O_{26}]_n^{4n-}$ chains and $[Ag_2(3atrz)_2]^{2+}$ binuclear units. Similar to complex **13**, the basic polyoxometalate building block is also the γ-$[Mo_8O_{26}]^{4-}$ isomer. However, two adjacent subunits are linked not by sharing common vertices as in **13** but by common edges. Therefore, the molybdate ribbon may be regarded as constructed from octamolybdate units joined at two oxo groups or from two groups of *cis*-edge-sharing tetranuclear units united through edge sharing. Similar octamolybdate chains have appeared in $K_2Mo_4O_{13}$: $[(C_5H_7N_2)_2Mo_4O_{13}]$, $[Cu(Hdipyreth)Mo_4O_{13}]$ and $[Cu(bpy)Mo_4O_{13}]$, $[Cu_2(4\text{-}PBIM)_2Mo_4O_{13}]$.

In the binuclear units $[Ag_2(3atrz)_2]^{2+}$, the two silver atoms are linked by two 3atrz ligands through a pyrazole-like bridging mode making up a distance of 3.337(24) Å between the two silver atoms. Each γ-$[Mo_8O_{26}]^{4-}$ cluster in the inorganic chains uses two opposite pairs of oxo groups of adjacent MoO_6 octahedra to coordinate two Ag^+ ions, that is, to connect two binuclear units to form a 2D network. Thus, each silver atom links two nitrogen atoms and two terminal oxygen atoms [Ag–N: 2.181(4) and 2.210(5) Å; Ag–O: 2.411(3) and 2.578(4) Å] to form a distorted tetrahedral environment. As depicted in Figure 4.16b, the overall 2D hybrid layer can be rationalized as a novel 3,4-connected net when silver atoms and octamolybdates are regarded as three- and four-connected nodes, respectively.

If the dinuclear $[Ag_2(3atrz)_2]^{2+}$ subunits are further simplified as linkers, this 2D network has a much simpler (4,4) topology, with parallelogram grids of 8.25×14.49 Å. Further, the $[Ag_2(3atrz)_2]_n^{2n+}$ infinite chains penetrate the $[Ag_2(3atrz)_2(Mo_8O_{26})]_n^{2n-}$ sheets to give the 1D-in-2D supramolecular structure of complex **14**. The spacing of two parallel sheets is 12.75 Å. This supramolecular assembly is stabilized further by nonclassic N $\cdots$ O and C $\cdots$ O hydrogen bondings between nitrogen or carbon atoms of 3atrz ligands in the 1D chains and oxygen atoms of the octamolybdate clusters in the 2D grids. There also exist N $\cdots$ O hydrogen bondings between the nitrogen atoms of 3atrz ligands in the binuclear units and oxygen atoms of the octamolybdate clusters in the 2D grids. The overall supramolecular architecture of rods in grids is shown simply in Figure 4.16c, which is intriguing in the area of chemical topology. To the best of our knowledge, only one 1D-in-2D entangled system incorporating polyoxometalates has been reported to date.

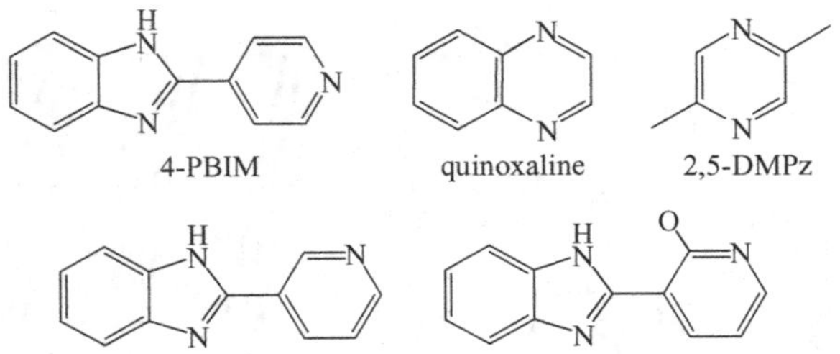

4-PBIM quinoxaline 2,5-DMPz

3-PBIM 3-OPBIM

FIGURE 4.17 Schemes of some organic structure–directing components.

4.3 COMPLEXES BUILT UP BY MOLYBDENUM OXIDE CHAINS WITH PYRIDINE DERIVATIVES

By comparison with a great number of the molybdenum oxide clusters which were modified structurally by incorporating transition-metal complexes, solid materials containing a decorated 1D $[Mo_xO_y]^{n-}$ chain are a fairly small subset of POM-based composites. In the present work, by exploiting Cu^{II}, Ag^{I}, Co^{II}, and Ni^{II} as secondary metal sources and quinoxaline, 2,5-DMPz (2,5-dimethylpyrazine), 3-PBIM [2-(3-pyridyl)benzimidazolium], and 4-PBIM [2-(4-pyridyl)benzimidazole] (Fig. 4.17) as the organic structure–directing components, six inorganic–organic complexes based on $[Mo_xO_y]^{n-}$ chains have been obtained [55]: $[Ag_2(quinoxaline)_2Mo_4O_{13}]$ (**17**), $[Cu_2(DMPz)_2Mo_4O_{13}]$ (**18**) (DMPz = 2,5-dimethylpyrazine), $[Cu_2(4\text{-}PBIM)_2Mo_4O_{13}]$ (**19**) [4-PBIM = 2-(4-pyridyl)benzimidazole], $[Cu(OPBIM)(H_2O)Mo_3O_{10}]\cdot H_2O$ (**20**) [OPBIM = 2-(2-ol-3-pyridinio)benzimidazole], $[Co(3\text{-}HPBIM)_2(H_2O)_2Mo_6O_{20}]$ (**21**) [3-HPBIM = 2-(3-pyridyl)benzimidazolium], and $[Ni(3\text{-}HPBIM)_2(H_2O)_2Mo_6O_{20}]$ (**22**). The molybdate chains in complexes **17** to **19** were based on three isomeric forms of $[Mo_8O_{26}]_n^{4n-}$, respectively: ξ-$[Mo_8O_{26}]^{4-}$ (**17**), ζ-$[Mo_8O_{26}]^{4-}$ (**18**), and γ-$[Mo_8O_{26}]^{4-}$ (**19**). The molybdate chains in complexes **20** to **22** represent two connected models of $[Mo_3O_{10}]_n^{2n-}$ chains: In complex **20** the $[Mo_3O_{10}]_n^{2n-}$ chains showed the ribbons of edge-sharing $[MoO_6]$ octahedra, in complexes **21** and **22** the $[Mo_3O_{10}]_n^{2n-}$ chains were constructed by edge-, corner-, and face-sharing arrangements of $[MoO_6]$ octahedra.

4.3.1 Three Isomeric Forms of $[Mo_8O_{26}]_n^{4n-}$ Chains

Crystal Structure of $[Ag_2(quinoxaline)_2Mo_4O_{13}]$ (17) The structure of **17** is a 3D polymer $[Ag_2(quinoxaline)_2Mo_4O_{13}]$ which is constructed from 1D $[Mo_4O_{13}]^{2-}$ anion chains connected by 1D $[Ag_2(quinoxaline)_2]^{2+}$ belts, forming a 3D framework with channels occupied by the phenyl rings of quinoxaline ligands. As shown in Figure 4.18a, the unsymmetric tetramolybdate subunit in complex **17** consists of three $[MoO_6]$ octahedra and one $[MoO_5]$ square pyramid. These polyhedra

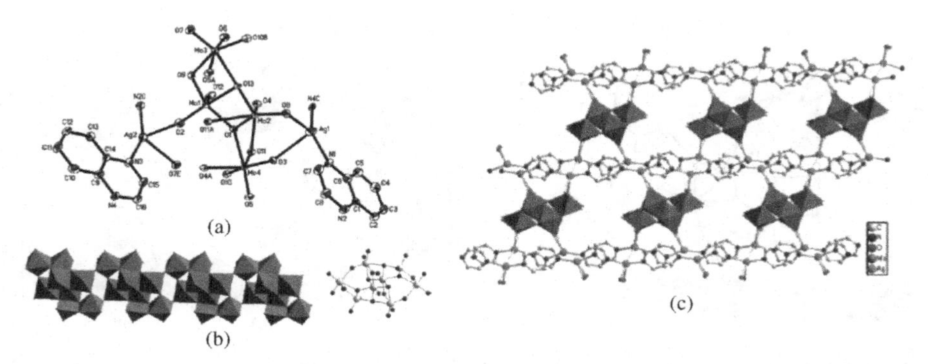

FIGURE 4.18 (a) Ortep drawing of **17** with 50% probability level, showing the coordination environments around Ag and Mo atoms (all hydrogen atoms are omitted for clarity); (b) 1D $[Mo_8O_{26}]_n^{4n-}$ chain in **17** and the ξ-$[Mo_8O_{26}]^{4-}$ cluster embedded in the molybdate chain; (c) 3D structure of **17** along the *a*-axis. The phenyl rings of the quinoxaline and all H atoms are omitted for clarity. The polyhedra represent the $[MoO_x]$ polyhedra.

are further connected through edge or corner sharing to form a 1D tetramolybdate chain running along the *a*-axis (Fig. 4.18b). It is obvious that the tetramolybdate chain may also be viewed as being built of octamolybdate subunits connected by sharing pairs of common vertices. Upon removal of the oxo groups attributed to the adjacent octamolybdate subunits, an ξ-$[Mo_8O_{26}]^{4-}$ isomer (Fig. 4.18b) is found to be the basic building block of the molybdenum oxide chain in complex **17**.

The two silver(I) atoms are both tetrahedrally coordinated by two nitrogen atoms from the quinoxaline ligands [Ag–N = 2.191(7) to 2.227(7) Å] and two terminal oxygen atoms [Ag–O = 2.431(6) to 2.568(5) Å] of the molybdate chain. Along the direction of propagation of the molybdate chain, the face-to-face distances between quinoxaline rings of neighboring Ag-quinoxaline belts are 2.69 to 3.52 Å, indicating the existence of π–π stacking interactions, which may play an important role in stabilization of the 3D framework. It is quite interesting that within the Ag–ligand belt, the oxygen vertices of the neighboring silver sites point away from each other so that the molybdenum oxide chains are linked alternatively to each side of the Ag–ligand belts (Fig. 4.18c).

Crystal Structure of $[Cu_2(DMPz)_2Mo_4O_{13}]$ (18) Single-crystal x-ray structural analysis reveals that the structure of **18** is constructed from 1D $[Mo_4O_{13}]^{2-}$ anion chains, integrated by pairs of $[Cu(DMPz)]^+$ chains into a 3D framework. As shown in Figure 4.19, the 1D tetramolybdate chain is constructed from edge- or corner-sharing $[MoO_5]$ square pyramids and $[MoO_6]$ octahedra and propagates along the *a*-direction. Similar to complex **17**, the tetramolybdate chain in **18** may be described as constructed from octamolybdate units fused at two common vertices. However, upon removal of oxygen atoms donated by adjacent octamolybdate subunits, the basic building block of **18** is revealed to be a ζ-isomer of $[Mo_8O_{26}]^{4-}$ (see Fig. 4.19b), as in the complex $\{[Cu_2(tpyrpyz)]_2Mo_8O_{26}\}\cdot 7H_2O$. This isomer is composed of a Mo_6O_6

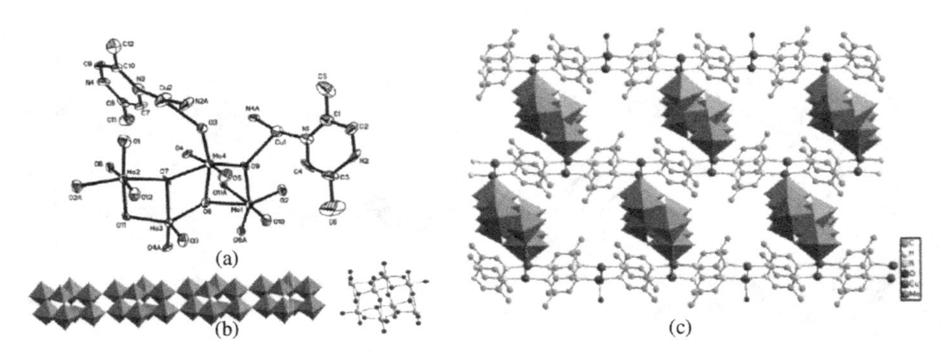

FIGURE 4.19 (a) Ortep drawing of **18** with a 50% probability level, showing the coordination environments around the Cu and Mo atoms (all hydrogen atoms are omitted for clarity); (b) 1D $[Mo_8O_{26}]_n^{4n-}$ chain in **18** and the ζ-$[Mo_8O_{26}]^{4-}$ cluster embedded in the molybdate chain; (c) 3D structure of **18** along the *a*-axis. All H atoms are omitted for clarity. The polyhedra represent the $[MoO_x]$ polyhedra.

ring capped on opposite faces by $\{MoO_5\}$ square pyramids, and the Mo_6O_6 ring contains two pairs of edge-sharing octahedra linked through two edge- and corner-sharing square pyramidal sites.

Crystal Structure of $[Cu_2(4\text{-}PBIM)_2Mo_4O_{13}]$ (19) The structure of $[Cu_2(4\text{-}PBIM)_2Mo_4O_{13}]$ consists of an infinite molybdenum oxide chain decorated with $[Cu_2(4\text{-}PBIM)_2]$ groups on both sides. As shown in Figure 4.20a, the asymmetric unit of the metal oxide chain contains one half of the octamolybdate, denoted as $[Mo_4O_{13}]$; the other half is generated by the inversion center. Each molybdenum atom attains a

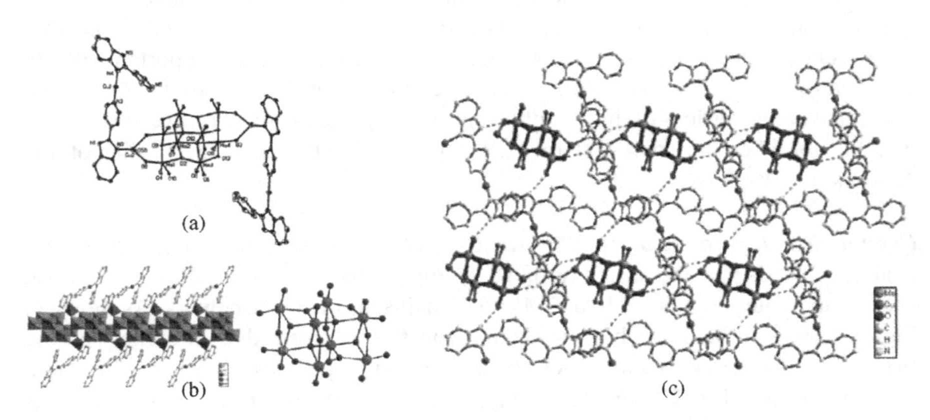

FIGURE 4.20 (a) Ortep drawing of **19** with a 50% probability level, showing the coordination environments around the Cu and Mo atoms (all hydrogen atoms are omitted for clarity); (b) polyhedral representation of the 1D chain of **19** (left) and the γ-$[Mo_8O_{26}]^{4-}$ cluster embedded in **19** (right); (c) 3D structure of **19** along the *a*-axis.

distorted octahedral environment by coordination of six oxygen atoms. The four asymmetric $[MoO_6]$ octahedra are connected through their edges to form an $[Mo_4O_{13}]^{2-}$ unit. Two $[Mo_4O_{13}]^{2-}$ units are stacked together by edge sharing to give rise to γ-$[Mo_8O_{26}]^{4-}$ octamolybdate clusters, which are linked together to form infinite 1D chains along the *a*-axis by sharing common edges. Therefore, the molybdate ribbon may be regarded as being constructed of octamolybdate units joined at two oxo groups or by two groups of *cis*-edge-sharing tetranuclear units united through edge sharing. In the solid state of **19**, the decorated 1D chains are held together and extended to a 3D framework via strong N–H $\cdots$ O hydrogen bonding. As illustrated in Figure 4.20c, both imino groups of the 4-PBIM ligands participate in hydrogen bonding with two terminal oxo atoms of the chain. Each molybdenum oxide chain forms two N–H $\cdots$ O hydrogen bonds with two adjacent chains. Distances of N1 $\cdots$ O13 (symmetry code, $x-1$, $y+1$, z) and N5 $\cdots$ O6 (symmetry code, x, $y+1$, $z-1$) are 2.754 and 2.913 Å, respectively.

4.3.2 Two Isomeric Forms of $[Mo_3O_{10}]_n^{2n-}$ Chains

Crystal Structure of [Cu(OPBIM)(H₂O)Mo₃O₁₀]·H₂O (20) The structure of [Cu(OPBIM)(H_2O)Mo_3O_{10}]·H_2O (**20**) consists of a buckled 1D molybdenum oxide chain linked to peripheral $[CuNO_4]$ square pyramids. As shown in Figure 4.21, three unique Mo atoms are all coordinated with six oxygen atoms in a slightly distorted octahedral geometry. Each molybdenum center has two short Mo–O bonds, with distances ranging from 1.694(6) to 1.763(5) Å; two medium bonds, 1.921(5) to 1.973(5) Å; and two long bonds, 2.096(5) to 2.373(5) Å. The three unsymmetrical $[MoO_6]$ octahedra in the $[Mo_3O_{10}]$ chain are in an edge-sharing arrangement, which is structurally analogous to that observed for the 1D $(NH_4)_2[Mo_3O_{10}]$. A $[Mo_3O_{10}]$ chain of this type has been reported to be extended to a 3D framework by introduction of appropriate transition metals and organic ligands. Nevertheless, no example of 1D or 2D structure based on such a chain has ever been reported to include metal–organic fragment.

It is worthy of note that the 3-PBIM ligand is hydrated to produce 3-PBIM-2-OH during the formation of complex **20**, which may follow a mechanism similar to that of

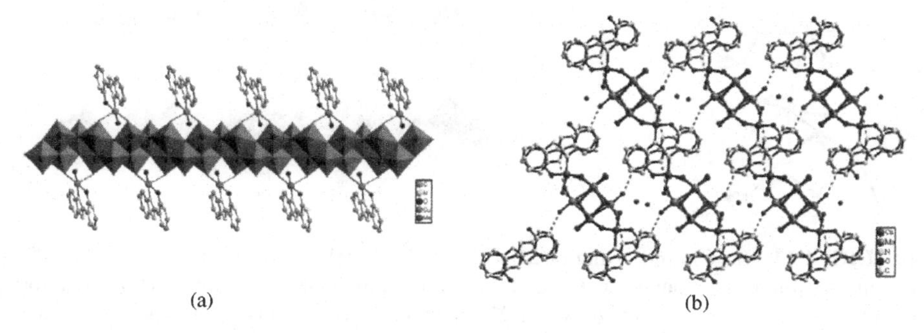

FIGURE 4.21 (a) Polyhedral representation of the 1D chain of **20**; (b) 3D structure of **20** formed by hydrogen bonding.

the long-debated Gillard mechanism. The pyridyl group of the 3-PBIM ligand is protonated to balance the charges of the entire structure according to chemical and structural information. Each Cu site in **20** adopts a distorted square-pyramidal geometry with the apical position defined by one oxo group bridging to a Mo(VI) center, the basal plane generated by one nitrogen atom and the deprotonated hydroxy group from the OPBIM [OPBIM = 2-(2-ol-3-pyridinio)benzimidazole] ligand, and another, terminal oxo group from the molybdenum oxide chain and an aqua group. The overall neutral 1D chain of [Cu(OPBIM)(H_2O)Mo_3O_{10}]·H_2O (**20**), propagating along the a-axis, may also be described as a ribbon of edge-sharing [MoO_6] octahedra, decorated with [Cu(OPBIM)O_2(H_2O)].

There are two types of supramolecular interactions in **20**: N–H···O and O–H···O hydrogen bondings and aromatic π–π stacking interactions. In **20** the N2···O4 (symmetry code, $-x$, $-y$, $-z+1$), N3···O6 (symmetry code, $-x$, $-y$, $-z$), and O2···O12 distances are 2.873, 2.834, and 3.002 Å, respectively, indicating apparent N–H···O and O–H···O hydrogen bonds. The OPBIM ligands between adjacent chains are arranged in an offset fashion with a plane-to-plane separation of 3.75 Å, indicating the existence of weak aromatic π–π stacking interactions. Therefore, the 1D chains of **20** are extended into a 3D structure in the solid (Fig. 4.21b).

Crystal Structure of [Co(3-HPBIM)$_2$(H_2O)$_2$$Mo_6O_{20}$] (21) and [Ni(3-HPBIM)$_2$ (H_2O)$_2$$Mo_6O_{20}$] (22) Complexes **21** and **22** are isomorphous. Only the structure of complex **21** will be described in detail. The structure of **21** is constructed from a 2D bimetallic oxide network $\{Co(H_2O)_2Mo_6O_{20}\}_n^{2n-}$ with the protonated 2-(3-pyridyl) benzimidazole groups projecting into the interlamellar region. As illustrated in Figure 4.22, the title complex features a 1D chain constructed only by {MoO_6}

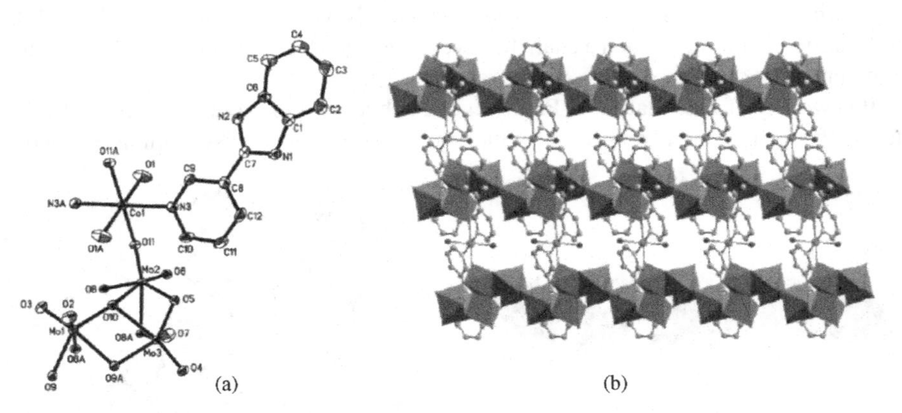

FIGURE 4.22 (a) Ortep drawing of **21** with a 50% probability level, showing the coordination environments around the Co and Mo atoms (all hydrogen atoms are omitted for clarity); (b) view parallel to the b-axis of **21**, showing the covalent connectivity between the chains.

octahedra. The adjacent molybdenum oxide chains are further interconnected by $\{CoN_2O_4\}$ octahedra through corner sharing to form a 2D layer with paddlewheel-shaped holes. Compared to complex **20**, in which the $[MoO_6]$ octahedra are assembled by edge or corner sharing, the three asymmetric $[MoO_6]$ octahedra in complex **21** are in complex patterns of edge-, corner-, and face-sharing arrangements, which is structurally analogous to that observed for the 1D $[NH_3(CH_2)_2NH_3][Mo_3O_{10}]$ chain. The 1D molybdenum oxide chain itself may be assumed to be composed of hexamolybdate subunits linked through edge sharing, and the hexamolybdate subunits may, in turn, be described as two trimolybdate clusters joined at two oxo groups.

The packing diagram of **21** along the *b*-axis is shown in Figure 4.22b. It is observed that there exist two types of hydrogen bonding: one between the O–H from the aqua group and the oxygen atom from the same layer with a $O \cdots O$ distance of 2.86 Å, the other between the N–H groups from ligands and the oxygen atom from a neighboring 2D layer with an average $N \cdots O$ distance of about 2.73 Å. Therefore, the 2D layer of **21** is extended into a 3D structure in the solid.

4.4 CONCLUSIONS

The successful preparation of complexes **1** to **22** provided novel examples of assembling POM-based hybrid materials under hydrothermal conditions to generate new structures with 0D, 1D, 2D, and 3D architectures. The polyoxometalate anions employed in the syntheses may adopt a variety of roles, such as charge-compensating anions (**1** and **9**), space-filling agents (**7** and **8**), bridging ligands (**2** to **6**), or structure-directing agents (**10** to **16**). On the other hand, there exist various polyoxometalate building blocks in one reaction pot, and those blocks can be self-assembled into a related set of complexes with different frameworks in the presence of additional transition-metal ions and organic molecules (**17** to **22**).

Although a great number of inorganic–organic complexes based on the polyoxometalates units have been synthesized to date, control or modification of their structures remains a great challenge. On the other hand, ongoing research toward investigation of the physicochemical properties and even potential applications for these complexes should be carried out in a timely manner. Thus, the chemistry based on polyoxometalate units linked up by organic ligands/transition metals/lanthanum is of tremendous interest in this field, and many intriguing complexes are certainly worthy of exploitation.

Acknowledgments

This work was supported by the 973 Key program of the MOST (2006CB932904, 2007CB815304), the National Natural Science Foundation of China (20425313, 20521101, 50772113), the Chinese Academy of Sciences (KJCX2-YW-M05), and the Natural Science Foundation of Fujian Province (2005HZ01-1, 2006L2005, 2006F3135, 2006F3141).

REFERENCES

1. Muller, A.; Roy, S. *Coord. Chem. Rev.* **2003**, *245*, 153–166.
2. Hagrman, D.; Zubieta, C.; Rose, D. J.; Zubieta, J.; Haushalter, R. C. *Angew. Chem., Int. Ed. Engl.* **1997**, *36*, 873–876.
3. Katsoulis, D. E. *Chem. Rev.* **1998**, *98*, 359–387.
4. Pope, M. T. Heteropoly and Isopoly Oxometalates, Springer-Verlag, Berlin, **1983**.
5. Yuan, M.; Li, Y. G.; Wang, E. B.; Tian, C. G.; Wang, L.; Hu, C. W.; Hu, N. H.; Jia, H. Q. *Inorg. Chem.* **2003**, *42*, 3670–3676.
6. Dolbecq, A.; Mialane, P.; Lisnard, L.; Marrot, J.; Sécheresse, F. *Chem. Eur. J.* **2003**, *9*, 2914–2920.
7. Kim, K. C.; Pope, M. T. *J. Am. Chem. Soc.* **1999**, *121*, 8512–8517.
8. Wu, C. D.; Lu, C. Z.; Zhuang, H. H.; Huang, J. S. *Inorg. Chem.* **2002**, *41*, 5636–5637.
9. An, H. Y.; Lan, Y.; Li, Y. G.; Wang, E.; Hao, N.; Xiao, D. R.; Duan, L. Y.; Xu, L. *Inorg. Chem. Commun.* **2004**, *7*, 356–358.
10. Balula, M. S. S.; Santos, I. C. M. S.; Gamelas, J. A. F.; Cavaleiro, A. M. V.; Binsted, N.; Schlindwein, W. *Eur. J. Inorg. Chem.* **2007**, 1027–1038.
11. Pope, M. T. Polyoxometalate Chemistry: *From Topology via Self-Assembly to Applications*, Kluwer Academic, Dordrecht, The Netherlands, **2001**.
12. Muller, A.; Krickemeyer, E.; Bogge, H.; Schmidtmann, M.; Peters, F. *Angew. Chem., Int. Ed.* **1998**, *37*, 3360–3363.
13. Muller, A.; Shah, S.Q.N.; Bogge, H.; Schmidtmann, M. *Nature 1999*, *397*, 48–50.
14. Liu, C. M.; Hou, Y. L.; Zhang, J.; Gao, S. *Inorg. Chem.* **2002**, *41*, 140–143.
15. Kortz, U.; Savelieff, M. G.; Ghali, F. Y. A.; Khalil, L. M.; Maalouf, S. A.; Sinno, D. I. *Angew. Chem., Int. Ed.* **2002**, *41*, 4070–4073.
16. Mialane, P.; Dolbecq, A.; Lisnard, L.; Mallard, A.; Marrot, J.; Sécheresse, F. *Angew. Chem., Int. Ed.* **2002**, *41*, 2398–2401.
17. du Peloux, C.; Mialane, P.; Dolbecq, A.; Marrot, J.; Sécheresse, F. *Angew. Chem., Int. Ed.* **2002**, *41*, 2808–2810.
18. Wu, C. D.; Lu, C. Z.; Zhuang, H. H.; Huang, J. S. *J. Am. Chem. Soc.* **2002**, *124*, 3836–3837.
19. Xu, L.; Qin, C.; Wang, X. L.; Wei, Y. G.; Wang, E. B. *Inorg. Chem.* **2003**, *42*, 7342–7344.
20. Hagrman, P. J.; LaDuca, R. L.; Koo, H. J.; Rarig, R.; Haushalter, R. C.; Whangbo, M. H.; Zubieta, J. *Inorg. Chem.* **2000**, *39*, 4311–4317.
21. Inman, C.; Knaust, J. M.; Keller, S. W. *Chem. Commun.* **2002**, 156–157.
22. Yang, L.; Naruke, H.; Yamase, T. *Inorg. Chem. Commun.* **2003**, *6*, 1020–1024.
23. Knaust, J. M.; Inman, C.; Keller, S. W. *Chem. Commun.* **2004**, 492–493.
24. Ishii, Y.; Takenaka, Y.; Konishi, K. *Angew. Chem., Int. Ed. Engl. 2004*, *43*, 2702–2705.
25. Han, Z. G.; Zhao, Y. L.; Peng, J.; Tian, A. L.; Li, Q.; Ma, J. F.; Wang, E. B.; Hu, N. H. *CrystEngComm* **2005**, *7*, 380–387.
26. Gamelas, J. A. F.; Santos, F. M.; Felix, V.; Cavaleiro, A. M. V.; Matos Gomes, E. D.; Belsley, M.; Drew, M. G. B. *Dalton Trans.* **2006**, 1197–1203.

27. Hill, C. L.; McCartha, C. M. *Coord. Chem. Rev.* **1995**, *143*, 407–455.
28. Peloux, C. D.; Dolbecq, A.; Barboux, P.; Laurent, G.; Marrot, J.; Sécheresse, F. *Chem. Eur. J.* **2004**, *10*, 3026–3032.
29. Coronado, E.; Galán-Mascarós, J. R.; Giménez-Saiz, C.; Gómez-García, C. J.; Martínez-Ferrero, E.; Almeida, M.; Lopes, E. B. *Adv. Mater.* **2004**, *16*, 324–327.
30. Hagrman, D.; Hagrman, P. J.; Zubieta, J. *Angew. Chem., Int. Ed.* **1999**, *38*, 3165–3168.
31. Zheng, L. M.; Wang, Y. S.; Wang, X. Q.; Korp, J. D.; Jacobson, A. J. *Inorg. Chem.* **2001**, *40*, 1380–1385.
32. Wang, X. L.; Guo, Y. Q.; Li, Y. G.; Wang, E. B.; Hu, C. W.; Hu, N. H. *Inorg. Chem.* **2003**, *42*, 4135–4140.
33. Lisnard, L.; Dolbecq, A.; Mialane, P.; Marrot, J.; Codjovi, E.; Sécheresse, F. *Dalton Trans.* **2005**, 3913–3920.
34. An, H. Y.; Wang, E. B.; Xiao, D. R.; Li, Y. G.; Su, Z. M.; Xu, L. *Angew. Chem., Int. Ed.* **2006**, *45*, 904–908.
35. Kong, X. J.; Ren, Y. P.; Zheng, P. Q.; Long, Y. X.; Long, L. S.; Huang, R. B.; Zheng, L. S. *Inorg. Chem.* **2006**, *46*, 10702–10711.
36. Steel, P. J. *Acc. Chem. Res.* **2005**, *38*, 243–250.
37. Haasnoot, J. G. *Coord. Chem. Rev.* **2000**, *200–202*, 131–185.
38. Klingele, M. H.; Brooker, S. *Coord. Chem. Rev.* **2003**, *241*, 119–132.
39. Ferrer, S.; Lloret, F.; Bertomeu, I.; Alzuet, G.; Borras, J.; Garcia-Granda, S.; Liu-Gonzalez, M.; Haasnoot, J. G. *Inorg. Chem.* **2002**, *41*, 5821–5830.
40. Liu, J. C.; Guo, G. C.; Huang, J. S.; You, X. Z. *Inorg. Chem.* **2003**, *42*, 235–243.
41. Yi, L.; Ding, B.; Zhao, B.; Cheng, P.; Liao, D. Z.; Yan, S. P.; Jiang, Z. H. *Inorg. Chem.* **2004**, *43*, 33–43.
42. Su, C. Y.; Goforth, A. M.; Smith, M. D.; Pellechia, P. J.; zur Loye, H. C. *J. Am. Chem. Soc.* **2004**, *126*, 3576–3586.
43. Meng, X.; Song, Y.; Hou, H.; Han, H.; Xiao, B.; Fan, Y.; Zhu, Y. *Inorg. Chem.* **2005**, *43*, 3528–3536.
44. Zhai, Q. G.; Lu, C. Z.; Zhang, Q. Z.; Wu, X. Y.; Xu, X. J.; Chen, S. M.; Chen, L. J. *Inorg. Chim. Acta* **2006**, *359*, 3875–3887.
45. Goforth, A. M.; Su, C. Y.; Hipp, R.; Macquart, R. B.; Smith, M. D.; zur Loye, H. C. *J. Solid State Chem.* **2005**, *178*, 2511–2518.
46. Hagrman, D.; Sangregorio, C.; O'Connor, C. J.; Zubieta, J. *Dalton Trans.* **1998**, 3707–3709.
47. Ardizzoia, G. A.; Cenini, S.; La Monica, G.; Masciocchi, N.; Maspero, A.; Moret, M. *Inorg. Chem.* **1998**, *37*, 4284–4292.
48. Zhai, Q. G.; Wu, X. Y.; Chen, S. M.; Chen, L. J.; Lu, C. Z. *Inorg. Chim. Acta* **2007**, *360*, 3484–3492.
49. Zhang, J. P.; Lin, Y. Y.; Zhang, W. X.; Chen, X. M. *J. Am. Chem. Soc.* **2005**, *127*, 14162–14163.
50. Zhang, J. P.; Lin, Y. Y.; Huang, X. C.; Chen, X. M. *Dalton Trans.* **2005**, 3681–3685.
51. Zhai, Q. G.; Wu, X. Y.; Chen, S. M.; Zhao, Z. G.; Lu, C. Z. *Inorg. Chem.* **2007**, *46*, 5046–5058.

52. Zhang, J. P.; Lin, Y. Y.; Huang, X. C.; Chen, X. M. *Chem. Commun.* **2005**, 1258–1260.
53. Abbas, H.; Pickering, A. L.; Long, D. L.; Kogerler, P.; Cronin, L. *Chem. Eur. J.* **2005**, *11*, 1071–1078.
54. Carlucci, L.; Ciani, C.; Proserpio, D. M. *Coord. Chem. Rev.* **2003**, *246*, 247–289.
55. Chen, L. J.; He, X.; Xia, C. K.; Zhang, Q. Z.; Chen, J. T.; Yang, W. B, Lu, C. Z. *Cryst. Growth Des.* **2006**, *6*, 2076–2085.

5

SILVER(I) COORDINATION POLYMERS

CHENG-YONG SU, CHUN-LONG CHEN, JIAN-YONG ZHANG, AND BEI-SHENG KANG

MOE Laboratory of Bioinorganic and Synthetic Chemistry, School of Chemistry and Chemical Engineering, Sun Yat-Sen University, Guangzhou, People's Republic of China

5.1 INTRODUCTION

Assembly of silver(I) coordination polymers has received great attention in recent decades as being one of the most exciting research fields in supramolecular coordination chemistry [1]. Due to its flexible coordination sphere, the Ag^+ ion exhibits abundant coordination geometries, varying from linear to trigonal, tetragonal, square pyramidal, and octahedral, corresponding to coordination numbers 2 to 6. Such coordination flexibility contributes greatly to the structural diversity of the silver(I) coordination polymers. So far, a large number of silver (I) coordination polymers with diverse topologies and dimensionalities [one-dimensional (1D), two-dimensional (2D), and three-dimensional (3D)] have been constructed. In addition, the flexibility of the Ag^+ coordination sphere affords a good opportunity to study the mechanism of the self-assembly process, since Ag–ligand interactions are labile and even a slight change of reaction conditions may drastically change the topological structures. The roles of many influencing factors, such as the functionality of the ligand, the ligand-to-metal ratio, the counter anion, the solvent, and the noncovalent interactions, have been investigated widely in the self-assembly process of silver(I) coordination polymers.

Design and Construction of Coordination Polymers, Edited by Mao-Chun Hong and Ling Chen

On the other hand, rational design of organic ligands also contributes to the wide study of silver(I) coordination polymers, because the various organic ligands exploited in the assembly of silver(I) coordination polymers greatly enrich the functional properties of the resulting silver(I) coordination polymers. For example, organic ligands with radical moieties can be used to assemble silver(I) coordination polymers with interesting magnetic properties, despite the fact that the Ag^+ is diamagnetic [2]. On the other hand, the intensive interests in silver(I) coordination polymers are often triggered by their various potential physical and chemical functions. For example, silver(I) coordination polymers have been considered as better therapeutic agents because of their antimicrobial activity against microorganism compared to current silver-based therapeutic agents [3].

Finally, compared with other coordination polymers, silver(I) coordination polymers often have good solubility in many organic solvents, which makes it possible to explore the degraded oligomers and precursors involved in the self-assembly process in solution using techniques such as ESI-MS and VT-NMR [4]. Therefore, a study of silver(I) coordination polymers is beneficial to an insight into the assembly procedure, which is important to in terms of the ambitious goal of rational design and synthesis of functional coordination polymers.

5.2 COORDINATION GEOMETRIES OF Ag^+ IONS

Because of the flexible coordination geometries of Ag^+ ions, silver(I) coordination polymers have exhibited a stream of unusual structures. The number of coordination polymers reported based on Ag^+, in which the coordination number of the Ag^+ ion varies from 2 to 6, with coordination geometries varying from linear to trigonal, tetrahedral, square planar, square pyramidal, and octahedral has increased almost exponentially every year.

5.2.1 Two-Coordinating Ag^+ Ions

Two-coordinating Ag^+ ions with coordination number 2 comprise the most common coordination mode in silver(I) coordination polymers, which usually exhibit linear or bent coordination geometries. Two-coordinating Ag^+ ions can act as either a two-connecting node or a linear connector, thus resulting in a large number of silver(I) coordination polymers, depending on ligands with topologies varying from 1D to 3D, such as a 1D chain or ladder, a 2D (4,4) or (6,3) network, and a 3D diamond or cubic framework, as shown in Figure 5.1. (A detailed structural topology is discussed in Section 5.5.)

1D Structures Based on Two-Coordinating Ag^+ Ions As shown in Figure 5.1a, a linear chain silver(I) coordination polymer can be generated when linearly coordinating Ag^+ ions are connected by a rodlike bis(monodentate) spacer with 1 : 1 stoichiometry. A good example is the complex $\{[Ag(bipy)]\cdot 0\cdot 5H_2btec\cdot H_2O\}_\infty$ (bipy = 4,4′-bipyridine, H_4btec = 1,2,4,5-benzenetetracarboxylic acid) [5], in which the rigid 4,4′-bipyridine connects the Ag^+ ions to afford a 1D linear Ag–bipy–Ag–bipy

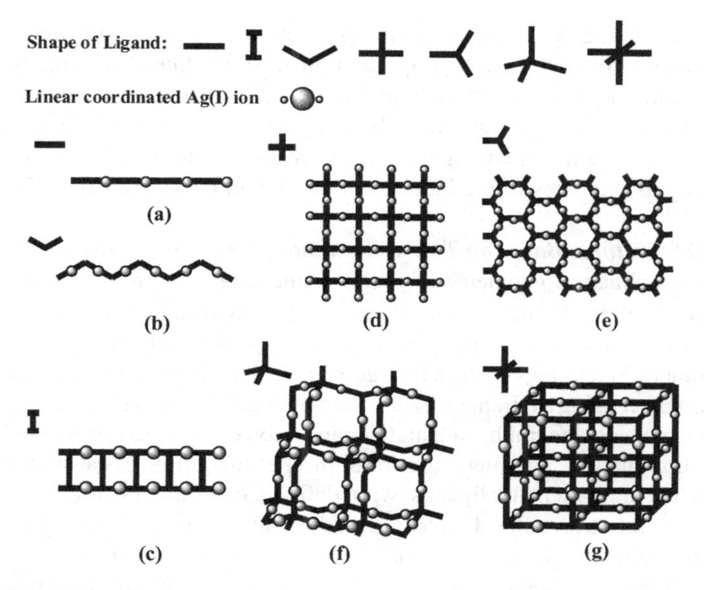

FIGURE 5.1 Silver(I) coordination polymers based on the two-coordinating Ag^+ ion: (a) linear chain; (b) zigzag chain; (c) ladder-like chain; (d) (4,4) net; (e) (6,3) net; (f) diamond net; (g) cubic net.

chain. Some commonly used linear linkers are pyrazine, quinoxaline, diazapyrene, 4,4′-bipyridine, 3,6-bis{4′-pyridyl}tetrazine, diphenazine, and 1,4-bis(4′-pyridyl) butadiyne. On the other hand, if linearly coordinating Ag^+ ions are linked by angular bis(monodenate) organic ligands with 1 : 1 stoichiometry, nonlinear 1D structures such as zigzag or helical chain (Fig. 5.1b) are formed. The complex $\{[AgL1](C_3F_7CO_2)\}_\infty$ (L1 and subsequent L*n*; Fig. 5.2) [6] showed the bent bipyridine type ligand L1 joining Ag^+ ions to give an unfolded zigzag chain.

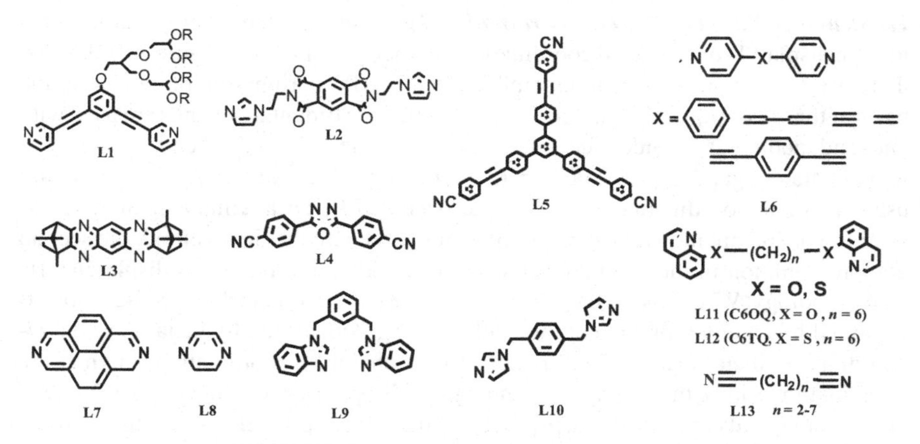

FIGURE 5.2 Some of the ligands mentioned.

Two-coordinating Ag^+ ions can also generate a 1D ladder-like structure when the H-shaped tetradentate ligand is used instead of linear or angular ditopic bis(mondentate) ligands, as shown in Figure 5.1c. Fitchett and Steel obtained a ladder-like coordination polymer when ligand L3 is used to react with silver nitrate, where each tetradentate ligand L3 connects four linear Ag^+ ions, with the Ag atoms acting as the sides of the ladder and the ligand L3 as the rungs [7].

2D and 3D Structures Based on Two-Coordinating Ag^+ Ions Taking advantage of the linear coordination geometry of a two-connecting Ag^+ ion, it is theoretically possible to assemble 2D and 3D coordination polymers as shown in Figure 5.1d–g: (1) the 2D structure of a (4,4) topology using a cross-like tetradentate ligand, (2) the 2D structure of a (6,3) topology using a triangular tridentate ligand, (3) the 3D structure of a diamondoid net using a tetrahedral a tetradentate ligand, and (4) the 3D structure of a cubic net using a octahedral hexadentate ligand. However, these idealized 2D and 3D silver(I) coordination polymers are rare in the literature, because well-shaped multibranched monodentate ligands with definite coordination directions (except probably for the triangular ligands) are difficult to design and synthesize. In addition, to maintain an Ag^+ ion in linear coordination geometry is sometimes not easy. The anions or solvent molecules often participate in coordination, which will make the Ag^+ ion adopt higher coordination numbers.

As shown in Figure 5.1f, one possible strategy for assembly of a 3D silver(I) coordination polymer with diamondoid topology is to design an organic ligand by adopting (pseudo)tetrahedral geometry. Klein et al. have synthesized one 3D fourfold interpenetrating coordination polymer from two-coordinating Ag^+ ions using a ligand containing a [1,1,1,1]metacyclophane backbone and bearing four pyridine units in which the 3D diamond framework is generated through the interconnections of pseudotetrahedral ligands and linear coordinating Ag^+ centers [8].

5.2.2 Three-Coordinating Ag^+ Ions

1D Structure Based on Three-Coordinating Ag^+ Ions A three-coordinating Ag^+ ion can display T- or Y-shaped coordination geometry as shown in Figure 5.3. Usually, there are two possible ways to assemble 1D silver(I) coordination polymers on the basis of three-coordinating Ag^+ ions: (1) Asymmetric tridentate ligands are used with one end acting as a bidentate chelating ligand and the other end acting as a monodentate ligand (Fig. 5.3a); or (2) symmetric bis(monodentate) ligands are used with one coordination Ag^+ ion site occupied by an auxiliary ligand such as an anion or solvent molecule (Fig. 5.3b). Compared with the first method, the second is more common in the assembly of silver(I) coordination polymers displaying 1D chain topology. When one of the coordination sites of three-coordinating Ag^+ ions is occupied by an anion or solvent molecule, the Ag^+ ions actually behaves as a two-coordinating metal center which readily forms chainlike coordination polymers in combination with either a linear or an angular ditopic ligand, showing a 1D linear, zigzag, or helical chain structure [9]. Besides these typical 1D chain structures, three-coordinating Ag^+ ions can generate other interesting 1D structures. For example, the

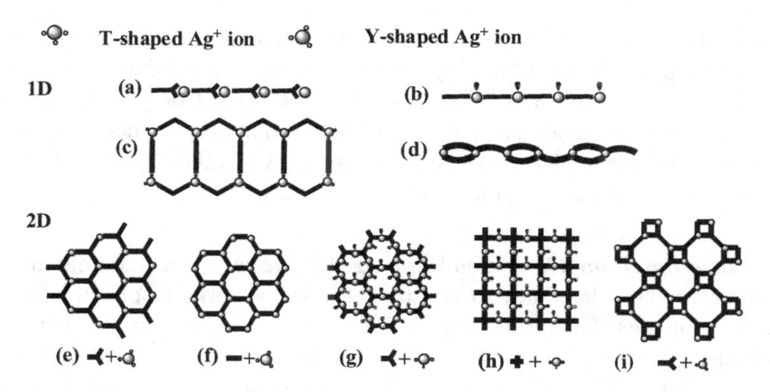

FIGURE 5.3 Silver(I) coordination polymers based on three-coordinating Ag^+ ions: (a,b) 1D chain; (c) stairlike chain; (c) loops-and-rods chain; (e–g) (6,3) net; (h) (4,4) net; (i) (4.8^2) net.

combination of three-coordinating Ag^+ ions with conformational flexible bis (monodenate) ligands in a 2 : 3 ratio gives rise to 1D polymeric complexes showing a stairlike structure [10] (Fig. 5.3c) or a loops-and-rods structure [11] (Fig. 5.3d).

2D Structure Based on Three-Coordinating Ag^+ Ions 2D coordination polymers constructed from three-coordinating Ag^+ ions exhibit various structural topologies. Typical 2D structures include the honeycomb network and the square grid network. As shown in Figure 5.3, the 2D structure of a (6,3) net can be assembled through three different strategies: (1) a Y-shaped Ag^+ ion connects three tripodal tri(monodentate) ligands with a ligand-to-metal ratio of 1 : 1 (Fig. 5.3e); (2) a Y-shaped Ag^+ ion connects three linear bis(monodentate) ligands with a ligand-to-metal ratio of 2 : 3 (Fig. 5.3f); or (3) a T-shaped Ag^+ ion connects two tripodal tri(monodentate) ligands with a ligand-to-metal ratio of 2 : 3, where each T-shaped Ag^+ ion offers only two coordination sites bind tripodal ligands, with the third coordination site occupied by an auxiliary ligand (Fig. 5.3g). By using the first strategy, Lee's group has synthesized a series of porous honeycomb 2D coordination polymers [12]. The second strategy can be exemplified by the polymeric structures of $\{[Ag_2L_3](ClO_4)_2\}_n$ [L = $Me_3CS–(CH_2)_2–SCMe_3$] and $\{[Ag_2L_3](CF_3SO_3)_2\}_n$ [L = bis(4-pyridyl)dimethylsilane] [13].

Since the three-coordinating Ag^+ ion cannot provide a four-connecting node, the only way to fabricate the 2D structure of a (4,4) net is to use a crosslike tetradentate ligand. In this case an Ag^+ ion usually takes a T-shape and leaves one coordination site occupied by an auxiliary ligand, as shown in Figure 5.3h. A good example is the complex $[Ag_2(H_2tpyp)(NO_3)](NO_3)\cdot x$solvent [$H_2$tpyp = 5,10,15,20-tetra(4-pyridyl) porphyrin] [14].

2D structures other than square grid or honeycomb nets can also be assembled from the three-coordinating Ag^+ ion. For example, by using tridentate ligand 2,5-bis

(4-pyridyl)-1,3,4-thiadiazole, Huang et al. synthesized the complex $\{[AgL](PF_6)\}_\infty$ [L = 2,5-bis(4-pyridyl)-1,3,4-thiadiazole] showing three-connected (4.8^2) topology (Fig. 5.3i). This structure may be described as a (4,4) net only if the 4-gons are considered as a single node [15]. A silver(I) coordination polymer with the same topology was also reported by Fan et al. in the complex $\{[AgL](ClO_4)\}_\infty$ [L = 1,3-bis (1-imidazolyl)-5-(imidazol-1-ylmethyl)benzene] [16].

3D Structure Based on Three-Coordinating Ag^+ Ions Three-coordinating Ag^+ ions can offer versatile routes to construct 3D frameworks displaying diversified structural topologies. The most common 3D silver(I) coordination polymers may be the well-known (10,3) frameworks, which exhibit a series of varied subnets. One example is $\{[Ag_2L_3](ClO_4)_2\}_n$ [L = bis(phenylthio)methane], in which the combination of dithioether ligand and three-coordinating Ag^+ ion leads to the formation of a chiral noninterpenetrated (10,3)-a structure (see Fig. 5.6f) [17]. Another interesting 3D framework is the ($4.8.10)(8.10^2$) framework in $\{[Ag(L)](NO_3)\}_\infty$ [L = *cis,trans*-1,3,5-triaminocyclohexane] constructed from the nonplanar *cis,trans*-1,3,5-triaminocyclohexane containing μ_3-bridging groups and the trigonal planar Ag^+ ion (see Fig. 5.6g) [18].

5.2.3 Four-Coordinating Ag^+ Ions

Four-coordinating Ag^+ ions usually adopt distorted square-planar or tetrahedral coordination geometry. Tetrahedral Ag^+ ions are observed much more often than planar ions in four-coordinated silver(I) coordination polymers. The main reason is that the Ag^+ ion has a d^{10} electron configuration, and the tetrahedral arrangement of the donor atoms around a metal center is the most preferable spatial geometry.

1D Structure Based on Four-Coordinating Ag^+ Ions Generally, 1D chain structures can be assembled from bis(bidentate) ligands and the four-coordinating Ag^+ ions in a 1 : 1 ratio, where each Ag^+ ion is chelated by two pairs of coordination donors from two ligands. In such cases, the ligands act as bis-chelators and may arrange in either linear or zigzag fashion, as depicted in Figure 5.4a and b. Known examples are Ag^+ complexes of the ligands C3TQ [1,3-bis(8-thioquinolyl)propane] or C4TQ [1,4-bis(8-thioquinolyl)butane], in which the 8-thioquinolyl group chelates with an Ag^+ ion by providing one S and one N donor [19]. When these bis-chelating ligands have enough conformational flexibility to twist along the Ag^+ centers, 1D helical structures can be constructed [19,20].

A four-coordinating Ag^+ ion can also behave as T-shaped nodes to make an 1D ladder structure when one of its coordination sites is occupied by an auxiliary ligand (including anion or solvent molecules) (Fig. 5.4d). The complex $[Ag_2(bpethy)_5](BF_4)_2$ [bpethy = 1,2-bis(4-pyridyl)ethyne] has a bipyridyl-type ligand, bpethy, displaying both μ_1- and μ_2-coordination modes. The μ_2-ligands connect the four-coordinating Ag^+ ions with flattened tetrahedral coordination geometries to result in a ladder structure, while the μ_1-ligands take the remaining coordination

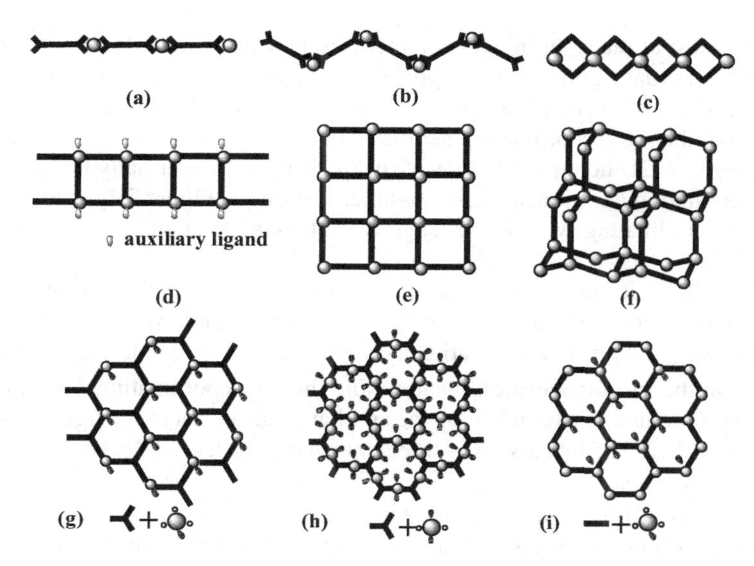

FIGURE 5.4 Silver(I) coordination polymers based on four-coordinating Ag^+ ions: (a) linear chain; (b) zigzag chain; (c) beaded chain; (d) ladder; (e) (4,4) net; (f) diamondoid net; (g–i) (6,3) nets.

sites, which are dangling at the two sides of the ladder [21]. Examples with one coordination site taken by the anion $CF_3SO_3^-$ can be found in the complex $\{[Ag_2(pyrazine)_3(CF_3SO_3)_2]\}_n$ [22], and by the solvent molecule can be seen in the complex $\{[Ag_2(pyrazine)_3](CF_3SO_3)_2{\cdot}2H_2O\}_n$ [23].

As shown in Figure 5.4c, the combination of four-coordinating Ag^+ ion with bis (monodentate) ligand in a 1 : 2 ratio can lead to assembly of 1D polymeric complexes showing beaded chain structural topology. One example is the complex $\{[Ag(L14)_2](PF_6)(C_6H_6)\}_n$, in which the cyclic $Ag_2(L14)_2$ units are extended by sharing the four-coordinating Ag^+ ions [24].

The four-coordinating Ag^+ ion can also be used to assemble 1D coordination nanotubes. By using a flexible tetradentate ligand to react with $AgBF_4$, Dong et al. have obtained a 1D tubular silver(I) coordination polymer. The Ag^+ ion is coordinated by four N donors from different ligands acting as square-planar four-connecting nodes while the ligand adopts *cis*-conformation, acting as a four-connecting linker to offer four N donors [25].

2D Structure Based on Four-Coordinating Ag^+ Ions As shown in Figure 5.4e, the 2D polymeric complex with (4,4) net topology can be generated by a combination of four-coordinating Ag^+ ion with bis(monodentate) ligand in a 1 : 2 ratio. Due to the fact that the bis(monodentate) ligands are very common and the four-coordinating Ag^+ ion is a facile four-connecting node, a large number of (4,4) topological silver(I) coordination polymers have been reported. Because four-coordinating Ag^+ ion usually adopts tetrahedral coordination geometry instead of a square-planar geometry, most of the resulting 2D polymeric complexes with (4,4) topology do not have a perfect square-grid structure, and the Ag_2(Ligand)$_2$ units are not coplanar [26]. Formation of a (4,4) network is normally due to the flexibility of the ligand and the distortion of the Ag^+ coordination geometry [27].

Although the 2D honeycomb network of (6,3) topology is usually typical for the three-coordinating Ag^+ ion, it is also possible, as shown in Figure 5.4g–i, to assemble from four-coordinating Ag^+ ion through the following strategies:

1. Using the tri(monodenate) ligand to connect the four-coordinating Ag^+ ion with one coordination site taken by an auxiliary ligand. In this case the ligand and Ag^+ ion are in a 1 : 1 ratio, and both act as three-connecting nodes [28].
2. Using the tri(monodenate) ligand to join the four-coordinating Ag^+ ion with two coordination sites taken by the auxiliary ligands. Here the ligand-to-metal ratio is 2 : 3. The ligands act as three-connecting nodes, while the Ag^+ ions act as connectors.
3. Using the bis(monodenate) ligand to link the four-coordinating Ag^+ ion with one coordination site taken by an auxiliary ligand with a ligand-to-metal ratio of 3 : 2. The ligands act as linkers, while the Ag^+ ions act as three-connecting nodes [29].

3D Structure Based on Four-Coordinating Ag^+ Ions Due to its tetrahedral (T_d) or pseudo-T_d coordination geometry, the four-coordinating Ag^+ acts as an excellent candidate in the design of 3D diamondoid networks following two strategies: (1) linear bis-monodentate ligands connecting the T_d Ag^+ ions (Fig. 5.4f) [30], and (2) a combination of T_d ligands and T_d Ag^+ ions (see Fig. 5.12) [31].

It has been reported that the polymeric complexes with other 3D framework topologies can also be constructed from the four-coordinating Ag^+ ions. The 3D polymeric complex with (10,3)-α type topology has been assembled using the ligand diquinoxalino[2,3-*a*:2′,3′-*c*]phenazine [32]. In this structure each Ag^+ ion is coordinated with four N donors from two different ligands in tetrahedral fashion, thus simplified as two-connecting linkers while each ligand coordinates to three Ag^+ ions as a tri(bidenate) ligand, acting as three-connecting nodes. By using the tetradentate ligand 2,3,4,5-tetra(4-pyridyl)thiophene and four-coordinating Ag^+ ion, Dolomanov et al. obtained a series of 3D complexes $\{[Ag(L)]X\}_\infty$ (L = 2,3,4,5-tetra(4-pyridyl) thiophene; X = BF_4^-, SbF_6^-, $CF_3SO_3^-$, or PF_6^-), showing a zeolite-like $4^2.8^4$ topological framework [33].

5.2.4 Five-Coordinating Ag^+ Ions

In principle, the five-coordinating Ag^+ ions may possess square-pyramidal or trigonal–bipyramidal coordination geometries. However, they usually exist in coordination polymers such that two bidentate ligands chelate the Ag^+ ion, with the remaining coordination site binding a monodentate ligand. Although there are numerous silver(I) coordination polymers containing the five-coordinating Ag^+ ions, it may not be easy to define a rationally designed strategy to control the specific structural topology because the coordination geometry of the five-coordinating Ag^+ ion is often extremely distorted.

1D Structure Based on Five-Coordinating Ag^+ Ions The 1D coordination polymers formed from the five-coordinating Ag^+ ions are very common, which is evident by witness of a large number of examples, such as $[\{Ag(C3TQ)](ClO_4)\}_n$ [19] in the literature. When five-coordination sites of Ag^+ ion are matched by two bidentate chelators and one monodentate donor, the five-coordinating Ag^+ ion can provide a chance to assemble a 1D coordination polymer with a loops-and-rods structure, similar to the structural topology generated by the trigonal coordination Ag^+ ion shown in Figure 5.3d. For example, the ligands in the complex $\{[Ag(L)_{1.5}]CF_3SO_3\}_n$ [L = 1,4-bis(pyridine-2-yl-methanethio)benzene] display two different coordination modes: one in wich the ligand coordinates with Ag^+ ion through both N and S donors, acting as bis(bidentate) ligand, and one in witch the ligand offers only an N donor to coordinate with the Ag^+ ion, acting as a bis(monodentate) ligand. Therefore, the Ag^+ ion connects two bis(bidentate) ligands and one bis(monodentate) ligand to result in a loops-and-rods structure [34].

Fabrication of a 1D infinite tubelike structure by proper design of the ligand to react with a five-coordinating Ag^+ ion has been achieved. Dong et al. used the ligand 2,5-bis(pyrazinyl)-3,4-diaza-2,4-hexadiene to react with $AgPF_6$ and obtained the tubular coordination polymer $[Ag_4L_4](PF_6)_4{\cdot}CHCl_3$, in which the coordination sites of Ag^+ ion are taken by two bidentate chelators and one monodentate N donor from three different ligands [35].

2D or 3D Structure Based on Five-Coordinating Ag^+ Ions If two of its coordination sites are occupied by nonbridging auxiliary ligands, the five-coordinating Ag^+ ion can generate a 2D network of (6,3) topology with the Ag^+ ion acting as three-connecting center, similar to the approaches shown in Figure 5.4g–i. For instance, in the complex $[Ag_2(L)_3(NO_2)_2]_n$ [L = 2,2′-bis(2-pyridyl)ethane], the nitrate group chelates an Ag^+ ion to occupy two coordination sites. The remaining three sites of Ag^+ ion bind to three bis(monodentate) 2,2′-bpe ligands to form a 2D structure of (6,3) topology [36].

Formation of a 2D (4,4) network may be exemplified by the complex $\{[Ag(bpyz)(MeCN)](BF_4)\}$ (bpyz = 2,2′-bipyrazine) [37]. The Ag^+ center in this complex is coordinated by one chelating and two monodentate bpyz ligands, as well as one MeCN molecule, leading to a distorted trigonal–bipyramidal Ag^+ environment. Since every chelating bpyz ligand is bound further to two other Ag^+ ions, each Ag^+ center is totally linked to four nearest-neighboring Ag^+ junctions. The coordination of the MeCN ligand flattens the extended sheet to give a 2D (4,4) network.

2D coordination polymers of other topologies have also been assembled from five-coordinating Ag^+ ions. Oxtoby et al. [38] reported 2D coordination polymers showing 4.8^2 and $(4^2.8)_4(4^2.8^2)_1(4^3.8)_2$ topologies (Fig. 5.3i) from the ligand 3,6-di-pyrazin-2-yl-[1,2,4,5]tetrazine (dpzta), which exhibits varied coordination modes and therefore direct the structural topology.

3D structures based on five-coordinating Ag^+ ions are relatively fewer than 2D structures. The complex $[Ag(tcm)(pyrazine)]_n$ [tcm = $C(CN)_3^-$] represents a rare type of five-coordinating 3D network [39]. In this complex, Ag(tcm) forms a hexagonal grid and each such sheet is connected on both sides to neighboring sheets by a bridging pyrazine, resulting in a 3,5-connected network of $(6^3)(6^3.8)$ topology.

5.2.5 Six-Coordinating Ag^+ Ions

In contrast to Ag^+ ions of lower coordination numbers, six-coordinating Ag^+ ions are rare, and they usually adopt octahedral or distorted trigonal prismatic coordination geometry. One of the early complexes based on the octahedral coordinating Ag^+ ion is $[Ag(pyriazine)_3](SbF_6)$, in which each Ag^+ ion is coordinated by six bis(monodentate) pyz ligands, forming a six-connected 3D framework of α-polonium topology [40].

When the six-coordinate Ag^+ ion is coordinated by three chelating groups from different ligands, the Ag^+ ion usually adopts a distorted trigonal prismatic geometry instead of an octahedral geometry. In this case, each Ag^+ ion actually acts as a three-connecting node. For example, in the complex $[Ag_2(dpztz)_3](SbF_6)_2$ (dpztz = 3,6-dipyrazin-2-yl-[1,2,4,5]tetrazine) [38], a 2D structure of (6,3) topology is formed because each Ag^+ ion is six-coordinated with three different bis(bidentate) dpztz ligands to provide a three-connecting node, while the bis(bidentate) ligand dpztz acts as a ditopic spacer. The same coordination geometric Ag^+ ion acting as three-connecting node can also be used to assemble a 2D silver(I) polymeric complex with an other topological structure, such as $(4,{}^3{}_6)$ [41].

A few 3D structures have been constructed based on six-coordinating Ag^+ ions [42]. The complex $\{Ag_3[Ag_5(L)_6](BF_4)_{12}{\cdot}12H_2O\}_n$ (L = 3,5-diphenyl-1,2,4-triazolate) displays a 3D porous cationic framework with six-connecting Ag(I) ions in a trigonal–prism adopting uninodal $(4^9,6^6)$ topology [42a], while an example containing cubiclike architecture was found in the complex $[Ag(L)](BF_4)$ [L = bis(1-(pyrazin-2-yl)ethylidene)benzene-1,4-diamine] [42b], which shows two nonequivalent six-coordinating Ag^+ centers, but the overall network is actually the distorted α-Po topology. Other examples include a robust 3D (10,3)-α network containing micropores [43] and a 3D polymeric complex containing channels [44].

From the discussions above we can see that varied coordination geometries of Ag^+ ions have played significant roles in the design and synthesis of silver(I) coordination polymers. The flexible and variable coordination geometries of Ag^+ ions provide versatile approaches to the assemble of diversified silver(I) coordination polymers but also introduce difficulties in the rational design of specific topologic structures. Many factors, such as the length, flexibility, or conformation of ligand, the metal-to-ligand ratio, and the anion, may significantly change the coordination geometry of the Ag^+ ion. Therefore, it is still a big challenge to restrict Ag^+ ions to the desired coordination geometry. A common phenomenon is to find various coordination geometric Ag^+ ions coexisting in the same silver(I) coordination polymer.

5.3 LIGANDS IN SILVER(I) COORDINATION POLYMERS

Besides the versatility of Ag^+ coordination geometry, design and synthesis of organic ligands represent more important factors in the assembly of diversified silver(I) coordination polymers with desired structures and concomitant properties. By judicious selection of specific ligands containing specific coordination donors and displaying adjustable conformations, a large number of silver(I) polymeric complexes have been generated from either a single ligand or from mixed ligands under controlled reaction conditions.

To design organic ligands for assembly of silver(I) coordination polymers, N, P, O, or S atoms are usually chosen as the donors, as they contain one or more lone electron pairs to coordinate readily with an Ag^+ ion. The formation process of Ag^+ coordination complexes based on these ligands is normally reversible because the Ag-donor dative bonds are labile, which makes the Ag^+ polymeric complexes generally crystallizable and ready for single-crystal x-ray diffraction analysis. Although the C atom can also be bonded with Ag^+ ion to form coordination polymers, the number of known silver(I) coordination polymeric structures based on an Ag–C bond are relatively rare compared with those based on an Ag–N, Ag–P, Ag–O, or Ag–S bond.

5.3.1 N-Donor-Containing Ligands

Because of the facile formation of Ag–N bonds, N-donor-containing ligands have been widely exploited in the self-assembly of silver(I) coordination polymers. The commonly used organic ligands containing pyridyl, imidazolyl, benzimidazolyl, pyrazolyl, pyrimidyl, triazolyl, pyrazine, or nitrile terminal groups are good examples. Because the N atom in these terminal groups has only one lone electron pair, each N donor usually coordinates an Ag^+ ion with a μ_1 coordination mode. Therefore, assembly of silver(I) coordination polymers usually demands a ligand containing at least two N donors to bridge and extend the Ag^+ ions in one or more directions. These ligands can be classified conventionally as bidentate, tridentate, tetradentate, hexadentate, and so on, depending on the number of coordination N donors. For example, when the 4,4′-bipyridine coordinates Ag^+ ion through two N donors, it acts as a bidentate ligand functionalizing as a perfect linear linker in many silver(I) coordination polymers [29].

Bidentate N-Donor-Containing Ligands To assemble silver(I) coordination polymers, the bidentate N-donor ligands have to be nonchelating, thus functionalizing as simple "rods" to connect Ag^+ "nodes" in various coordination geometries. The interconnection of these rods and nodes can result in predictable polymeric structures, such as 1D linear or zigzag chain, helical or tubular structure, 2D brick wall, square grid or honeycomb sheet, and 3D diamondoid or cubic framework (Fig. 5.5). Depending on the positions of the N donors and the conformational flexibility, these bidentate ligands can be divided into a linear or nonlinear spacer in the silver(I) coordination polymers.

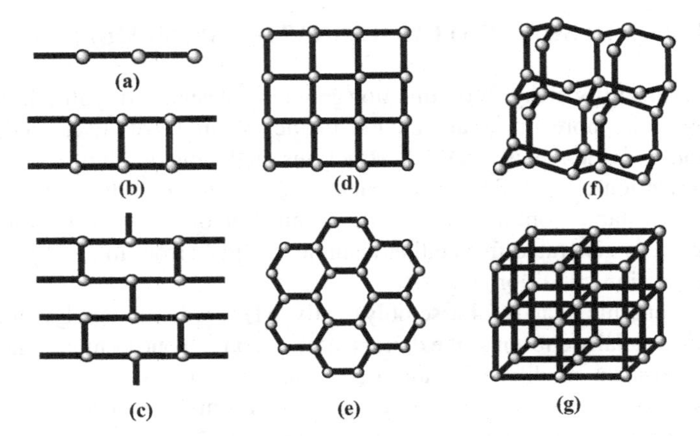

FIGURE 5.5 Some typical structures assembled from bidentate ligands: (a) 1D linear chain; (b) 1D ladder; (c) 2D brick wall; (d) 2D square grid; (e) 2D honeycomb sheet; (f) diamondoid net; (g) cubic net.

In linear bidentate N-donor-containing ligands, exemplified by 4,4′-bipyridine, the two coordination N donors are in linear positions, thus join the Ag^+ ions in opposite directions. In many references, such bidentate ligands, called *rodlike spacers*, offer a straightforward explanation of the conception of crystal engineering. As shown in Figure 5.5, the combinations of linear spacers with nodes having predefined spatial geometries could give rise to direct formation of 1D to 3D structures of well-defined topologies. For example, the 1D linear chain can be formed by combining the linear bidentate ligands with the linear coordination geometric Ag^+ ions in a 1 : 1 ratio, and the 1D ladder-like polymer can be obtained by connecting the linear bidentate ligands with the three-coordinating T-shaped Ag^+ ions in a 3 : 2 ratio. Alternatively, the 2D polymeric silver(I) complexes with (4,4) or (6,3) net topologies and the 3D polymeric silver(I) complexes with diamond or cubic net topologies can be assembled from the linear bidentate ligands when the Ag^+ ions providing planar four-connecting, triangular three-connecting, tetrahedral four-connecting, or octahedral six-connecting nodes.

In nonlinear bidentate N-donor-containing ligands the two coordination N donors are oriented in different directions, which may be due to the conformational isomerism or flexible backbone of the ligand. Many bis(monodentate) ligands, whether symmetric or asymmetric, adopt a nonlinear shape in the silver(I) coordination polymers. For example, the ligands containing 3-pyridyl, 2-pyridyl, imidazolyl, or pyrazole as the terminal coordinating groups usually act as a nonlinear linker in the self-assembly of silver(I) coordination polymers. The nonlinear bidentated ligands have contributed greatly to the structural diversity and supramolecular isomerism in silver(I) coordination polymers [4]. Although such ligands may make the assembly processes less predictable, they afford more opportunities to form fascinating interweaven structures or unusual topologies.

Tridentate N-Donor-Containing Ligands Because of the C_3 or pseudo-C_3 symmetry, the tridentate N-donor-containing ligands are often used as three-connecting junctions in the self-assembly of silver(I) coordination polymers. In combination with Ag^+ ions in different coordination geometries and different metal-to-ligand ratios, many polymeric Ag^+ complexes have been obtained to display diversified structural topologies, as shown in Figure 5.6.

The 1D chain structure can be obtained when tridentate ligands coordinate the three-connecting Ag^+ ions in a 1 : 1 ratio and exhibits both μ_1 and μ_2 coordination modes (Fig. 5.6a). On the contrary, if tridentate ligands exhibit only a μ_1 coordination mode and offer T-shaped building blocks, a 1D ladder structure will be formed (Fig. 5.6b). 2D structures with a (6,3) net topology can be synthesized from tridentate ligands when Ag^+ ions (1) act as three-connecting nodes with a metal-to-ligand ratio of 1 : 1 (Fig. 5.6c), or (2) act as two-connecting nodes with a metal-to-ligand ratio of 3 : 2 (Fig. 5.6d).

In addition to the 2D (6,3) network, a combination of tridentate N-donor-containing ligands with Ag^+ ions which act as three-connecting nodes by 1 : 1 stoichiometry can also generate a 2D polymeric structure of (4.8^2) topology (Fig. 5.6e), and even 3D frameworks of $(4.8.10)(8.10^2)$ (Fig. 5.5f) or (10,3)-a (Fig. 5.5g) net topologies [18]. It is clear that Figure 5.6e, f, and g have exactly the same stoichiometry (metal-to-ligand ratio = 1 : 1) and similar ligand connection and Ag^+ coordination. However, the structural topologies of these structures are completely different. This is a common phenomenon of structural diversity encountered in crystal engineering, demonstrating how the network structures are affected by the ligand design and various influencing factors during self-assembly.

Tetradentate N-Donor-Containing Ligands When tetradentate ligands behave as bis(bidentate) linkers to chelate a four-coordinating Ag^+ ion by a 1 : 1 stoichiometry,

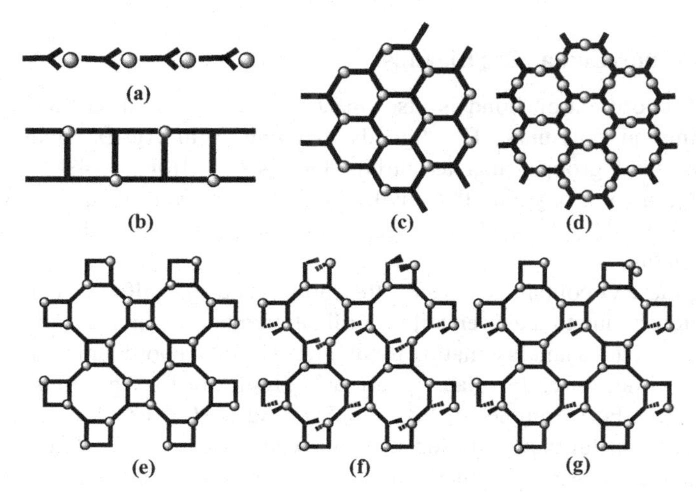

FIGURE 5.6 Representative structures assembled from tridentate ligands: (a) linear chain; (b) ladder; (c,d) (6,3) net; (e) (4.8^2) net; (f) (10,3)-a net; (g) $(4.8.10)(8.10^2)$ net.

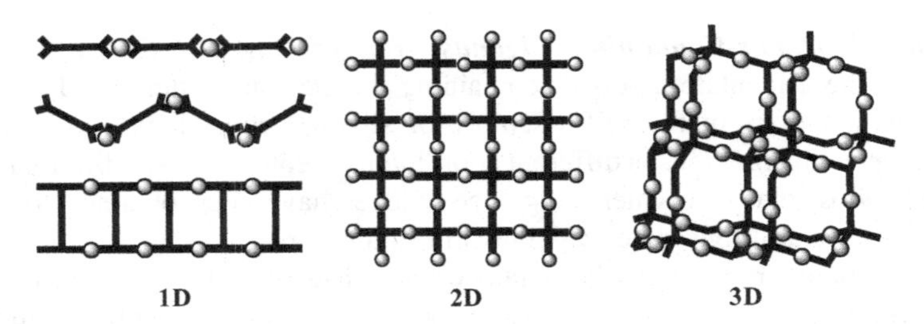

FIGURE 5.7 Representative structures assembled from tetradentate ligands.

it is expected to form a 1D linear chain, as shown in Figure 5.7. However, since the four-connecting Ag^+ ion often adopts a tetrahedral coordination geometry or the tetrahedral ligand itself may not be linear, the resulting 1D coordination polymer usually extends in a nonlinear fashion. For example, the ligand 3,3′-dimethyl-*N,N′*-bis(pyridine-2-ylmethylene)biphenyl-4,4′-diamine reacts with $AgClO_4$ in a 1 : 1 ratio to give the complex $\{[Ag_2L_2](ClO_4)_2(CH_3CN)\}_n$, which displays a 1D zigzag structure. Each Ag^+ ion is coordinated by two ligands via pyridyl and azomethine N atoms, and each ligand coordinates two Ag^+ ions acting as a bis (bidentate) linker [45]. On the other hand, due to the twist of the tetradentate ligand, the combination of a bis(bidentate) ligand with a four-coordinating Ag^+ ion in a 1 : 1 ratio sometimes results in the formation of a helical structure [20a,46].

As shown in Figure 5.7, the tetradentate ligand may also be used to assemble 1D ladders, 2D square grids, or even 3D diamond coordination polymers with the two-connecting Ag^+ ions. An example of a 2D square-grid structure assembled from a tetra(monodentate) ligand shows a crosslike shape, and the Ag^+ ion is two-connected in a 1 : 2 stoichiometry [47].

5.3.2 O-Donor-Containing Ligands

The Ag–O coordination bond is also one of the most common bonds in silver (I) coordination polymers. The ligands that can form Ag–O bonds normally contain terminal groups such as carboxylate [48], sulfonate [49], $-NO_2$ [50], $-PO_3^-$ [51], phenol, oxygen-ether, hydroxy, and so on. Various acid anions often form Ag–O bonds, which play a significant role in adjusting the overall topological structures.

Among these O-containing ligands, the carboxylate and sulfonate derivatives are of great interest, due to their versatile coordination modes, which cause enormously diversified structures and fascinating topologies. Unusual coordination numbers and geometries, either for carboxylate or sulfonate groups, or for Ag^+ ions, have often been observed. For example, in the complex $\{Ag_3[N(CH_2COO)_3]\}_n$, the NTA^{3-} anion (NTA = nitrilotriacetate) adopts an unprecedented 13-coordination mode [50]. On the other hand, Ag^+ polymeric complexes based on carboxylate– or sulfonate group–containing ligands often provide good pathways for the investigation of ligand-supported or ligand-unsupported $Ag \cdots Ag$ interactions [51].

5.3.3 S-Donor-Containing Ligands

The S atom normally participates in coordination with an Ag^+ ion in either thioether or mercapto form, providing two or three lone electron pairs. Therefore, S donors exhibit a strong affinity to Ag^+ ions and provide versatile coordination modes varying from μ_1 to μ_4. As a consequence, S-donor-containing-ligands have been used widely in the assembly of silver(I) coordination polymers. Due to the flexible coordination ability and characteristic coordination behavior, such as folded conformation of the thioether S donors, some unusual polymeric complexes with interesting structures or physical properties can be assembled from S-donor-containing ligands.

S Donors of the μ_1 Coordination Mode When an S atom in a thioether ligand coordinates with an Ag^+ ion in the μ_1 coordination mode, the S-donor-containing ligands show coordination behavior somewhat similar to that of N-donor-containing ligands, except for less directional control. In principle, they can assemble silver(I) polymeric complexes of different topology following the same nodes-and-connectors methods as those followed by N-donor-containing ligands. For example, combination of a tri (monodentate) thioether ligand with a three-connecting Ag^+ ion in a 1 : 1 ratio can afford a 2D (6,3) topological network, as exemplified in the complex $[Ag(L)_2(ClO_4)]_n$, [L = 1,3,5-tri(phenylthiomethyl)benzene]. The Ag^+ ion is tetrahedrally coordinated with three S donors from three different ligands and with one with oxygen donor from the perchlorate anion. The ligands act as a tri(monodentate) juncture by offering three S donors to bind three different Ag^+ ions. As a result, a 2D network of (6,3) topology is formed [28].

The 2D (6,3) net can also be assembled using bidentate thioether ligands and three-connecting Ag^+ ions in 3 : 2 stoichiometry. One example is the complex $[Ag_2L_3(ClO_4)_2]_n$ [L = 1,4-bis(phenylthio)butane], in which each Ag^+ ion joins three S donors from three distinct ligands and each ligand bridges two different Ag^+ ions [52].

S Donors of Bridging Coordination Modes Unlike an N donor, an S donor can also exhibit various bridging coordination modes to connect more than one Ag^+ ion, such as the μ_2 coordination mode [53], the μ_3 coordination mode [54], or even the μ_4 coordination mode [55]. In some cases, S donors may display different coordination modes in the same complex [56]. One example of the μ_2-S coordination mode is found in the complex $\{[Ag_2(StpmH_2)_6](NO_3)_4\}_n$ ($StpmH_2$ = 2-mercapto-3,4,5,6-tetrahydropyrimidine), in which the S donor of $StpmH_2$ bridges two three-coordinating Ag^+ ions, and each Ag^+ ion coordinates with three S donors from different $StpmH_2$ ligands. A 1D infinite ladder-like structure is afforded as shown in Figure 5.8 [53].

5.3.4 P-Donor-Containing Ligands

The phosphorus atom in phosphine has one lone electron pair available to form a coordination bond with a metal ion. The use of phosphine ligands either alone or in

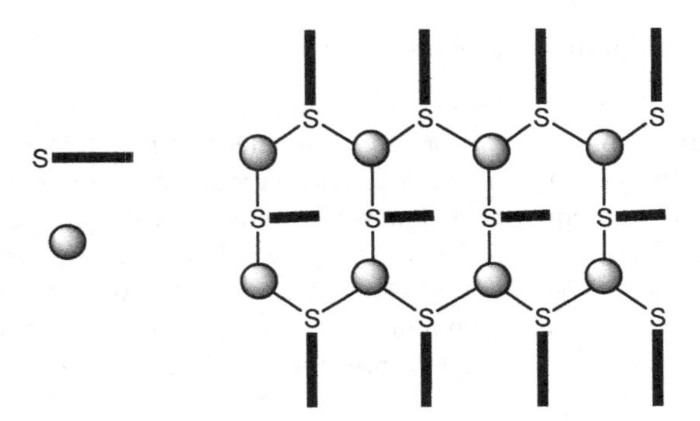

FIGURE 5.8 Ladder-like structure assembled by bridging S donors.

combination with O-/S-/N-donor ligands has much promise in the synthesis of coordination polymers. Various tertiary phosphines have been exploited in the self-assembly of silver(I) coordination polymers. The phosphines that have been used thus far are mainly diphosphines, and only a few triphosphine coordination polymers have been reported. In their coordination polymers, various structures have been generated, such as 1D chains, 2D honeycomb, puckered sheet, 3D diamondoid structure, and so on.

The 1D polymeric chain can be formed via alternating Ag^+ and diphosphine ligands, with extra coordination sites of Ag^+ ion occupied by auxiliary ligands: for example, $\{Ag(\mu\text{-}L)(MeCN)_2\}_n[ClO_4]_n$ [L = 2,6-bis(diphenylphosphino)pyridine] [57]. A 1D chain can also be formed by singly or doubly bridged rings. In $\{AgL_2AgL\}_n$ [L = *trans*-1,2-bis(diphenylphosphino)ethylene (dppe) or bis(diphenylphosphino) acetylene (dppa)], Ag_2L_2 rings are linked into a 1D chain by a third exocyclic bridging diphosphine [58]. In $\{[Ag(dppm)(dmb)]ClO_4\}_n$ (dmb = 1,8-diisocyano-*p*-menthane), a mixed-ligand chain is described as a polymer of $Ag_2(dppm)_2^{2+}$ dimers that are doubly bridged by dmb ligands [59]. A 1D chain has also been obtained from triphosphines [60]. $[Ag_3(L)_2(NCMe)(OTf)_3]_n$ [L = 1,3,5-tris(diphenylphosphino)benzene] and $\{Ag(PCP_2)AgX_2\}_n$ [X = Cl, Br, I; PCP_2 = 1,1,1-tris(diphenylphosphinomethyl) ethane] give rise to 1D infinite chains, in which Ag_2 units are bridged by L or PCP_2. In $\{Ag(PCP_2)AgX_2\}_n$ three- and four-coordinate Ag^+ ions coexist. The three-coordinate Ag^+ environment is almost planar $PAgX_2$, whereas the four-coordinate is distorted tetrahedral P_2AgX_2.

2D silver(I) coordination polymers constructed from P-donor-containing ligands also have diverse structures. In the usual honeycomb layer structure in $\{[Ag_2(dppb)_3(CN\text{-}t\text{-}Bu)_2](BF_4)_2\}_n$, the Ag^+ is four-coordinated with three P donors and one N donor [59]. In honeycomb structures, the network nodes and connectors can be replaced by aggregates. Xu et al. reported that $[Ag_4L_3(OTf)_4]_n$ [L = 1,3,5-tris (diphenylphosphanyl)benzene] is a 2D network of large 72-atom hexagonal rings with trigonal–planar AgP_3 centers as the network nodes and $Ag_2P_2(\eta^1\text{-}OTf)_2$ rings as the connectors [61]. The only puckered sheet structure is reported by Brandys and Puddephatt (Fig. 5.9). In $[Ag(O_2CCF_3)(\mu\text{-}Ph_2P(CH_2)_6PPh_2)_{1.5}]_n$, the Ag^+ ion is four-coordinated with distorted tetrahedral AgP_3O coordination geometry. Each Ag^+

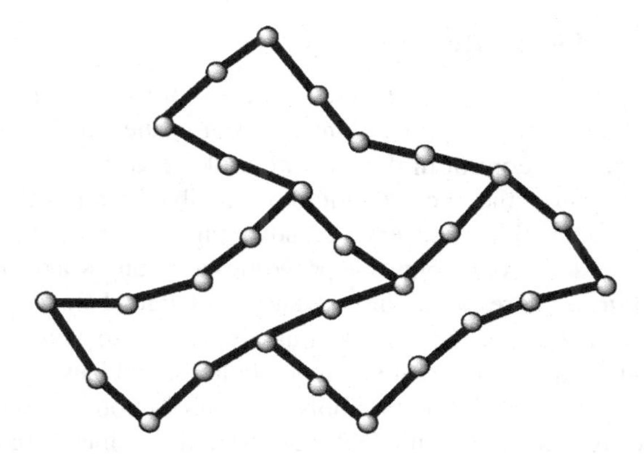

FIGURE 5.9 Representation of the puckered structure assembled from bridging a phosphine ligand.

atom is coordinated with three others by bridging diphosphine ligands, and the trifluoroacetate ligands are terminally coordinated [62].

So far, only a few 3D coordination polymers have been reported based on P-donor-containing ligands. Zhang et al. reported that $[Ag(bix)(dppe)]_n^{n+}$ [bix = 1,4-bis (imidazol-1-ylmethyl)benzene] is a 3D cationic network of diamondoid topology in which the silver centers are 4-coordinated with a AgP_2N_2 environment [63]. $\{Ag[\mu\text{-}Ph_2P(CH_2)_3PPh_2](\mu\text{-bipyen})\}_n(CF_3CO_2)_n$ is another example in which each Ag^+ ion is in a pseudotetrahedral AgP_2N_2 coordination environment and generates a 3D network structure [64].

5.3.5 C-Donor-Containing Ligands

Ag^+ complexes based on an Ag–C bond mainly involve arenas, studied initially over 80 years ago by A. E. Hill [65]. Although this type of complex is essentially organometallic, they can form topological structures similar to those in other Ag(I) coordination polymers. In the past several decades, a large number of aromatic compounds have been used as ligands for Ag^+ complexation, such as corannulene [66], 9,10-diphenylanthracene, rubrene, and oligo(phenylenevinylene) [67]. The renaissance of aromatic Ag^+ chemistry is attributed primarily to the fact that cation–π interactions and aromatic stackings may provide a powerful tool to produce novel organometallic structures. Aromatic hydrocarbons possess versatile coordination modes, such as η^1, η^2, η^3, η^4, and η^6, with transition-metal ions. The silver(I) complexes with conjugated organic π-systems in rings and chains have potential applications in nonlinear optics and molecular electronics. Zhong et al. reported a series of Ag^+ organometallic polymers, such as W-type, pillared brick, helical, double-decker, triple-decker, multidecker, herringbone, and spiral frameworks [68]. Other Ag^+ polymers based on an Ag–C bond include silver(I) acetylides [69] and complexes with carbenes and isocyanide. Fortin et al. showed some interesting examples in which various linear, staircase, and ladder structures were built based on Ag^+ and 1,8-diisocyano-*p*-menthane [70].

5.3.6 The Role of Ligand Rigidity

Rigidity is a key factor to be considered in the rational design of ligands for construction of silver(I) coordination polymers. Usually, the reaction of a rigid ligand is comparatively easy to control in the formation of a coordination polymer with a specific structure, while the incorporation of flexible components may endow molecular architectures with other potential advantages, such as "breathing" ability in the solid state and adaptive recognition properties as a function of coexisting guest or counterions. Furthermore, by taking advantage of ligand flexibility, it becomes possible to investigate the mechanism of self-assembly, thus to gain more information for the directional synthesis of target supramolecular complexes. According to their rigidity, the ligands exploited in the assembly of silver(I) coordination polymers may be classified into rigid ligands, semirigid ligands, and flexible ligands.

Rigid ligands have relatively invariable molecular skeleton and limited conformations, which can decrease remarkably the diversity of structural topologies and isomers. Therefore, conjugated rodlike or planar rigid ligands are frequently used to assemble silver(I) coordination polymers with predictable architectures through the nodes-and-connectors method, especially with symmetric bis(monodentate), tris (monodentate), and tetradentate ligands containing simple terminal groups such as pyridyl, imidazolyl, and –CN (Fig. 5.2).

Semirigid ligands contain one or two flexible joints, such as $-CH_2-$, –O–, or –S–, which can provide limited preferential conformations during self-assembly. For example, ligand L9 or L10 in Figure 5.2 can adopt either *anti* or *syn*conformation, depending on the orientation of their two arms around the C atom [11,63]. Sometimes, different conformations can coexist in the same complex, as in $[Ag_2(bix)_3(NO_3)_2]_\infty$ [bix = 1,4-bis(imidazol-1-ylmethyl)benzene], where the coexistence of *anti* and *syn*conformations of the bix ligand contributes to the formation of an infinite loop-and-chain structure [11].

In contrast to rigid and semirigid ligands, ligands L2, L11, L12, and L13 in Figure 5.2 usually exhibit high conformational flexibility and can display diversified structural conformations subject to the demand of Ag^+ coordination geometry and the fashion of crystal packing. Although the incorporation of flexible ligands may lead to less predictable structures, numerous flexible ligands have been designed and used in the construction of Ag^+ polymeric structures because silver(I) complexes based on flexible ligands often demonstrate advantages not only in studying the roles of supramolecular interactions and other influencing factors (i.e., solvent, counterion, concentration, etc.) in the self-assembly process, but also in the investigation of the assembly mechanism in solution [4,19,71].

5.4 SUPRAMOLECULAR INTERACTIONS AND COUNTER ANIONS IN SILVER(I) COORDINATION POLYMERS

In the self-assembly of silver(I) coordination polymers, the weak supramolecular interactions and counter ions have been shown to play very important roles. These influencing factors not only affect the coordination geometry of Ag^+ ions, but also increase interactions between molecules, thus directing the structural topologies of the resulting silver(I) coordination polymers. In addition, these factors may endow silver(I) coordination polymers with distinctive physical properties. For example, some Ag^+ polymeric complexes show electrical conductivities due to the presence of Ag···Ag or π–π interactions [51a,67b,72].

5.4.1 Supramolecular Interactions

As Ag–ligand interactions are relatively weak and labile (e.g., Ag–N bond energy is comparable with that of a strong hydrogen bond) [1b,4a], a synergistic effect is easily exhibited in combination with other weak intermolecular interactions. The study of silver(I) coordination polymers provides an informative library that abounds with rich supramolecular interactions, such as hydrogen bonding, π···π stacking, weak Ag···X bonding (X = O, S, C, etc.), C–H···X interacting (X = π groups, N, O, Cl, etc.), and Ag···Ag (argentophilic) bonding. Since many excellent reviews and monographs concerning the topic of supramolecular interactions in crystal engineering are available [73], we give only a very brief introduction here.

Hydrogen-bonding interactions, including conventional N/O–H···X (X = N, O, or F) hydrogen bonds and nonclassical weak C–H···X (X = O, N, π, etc.) hydrogen bonds, are most interesting in the field of silver(I) coordination polymers because they can provide comparable connections with the Ag–ligand interactions, and in many cases, direct the overall framework structures. For example, intermolecular hydrogen bonding connects the 1D $[Ag(4\text{-abaH})_2(NO_3)]_\infty$ (4-abaH = 4-aminobenzoic acid) chain into a 2D structure [74].

The second important supramolecular interaction in silver(I) coordination polymers is the π···π aromatic stacking. Many π-electron-rich aromatic ligands have been designed and selected to react with Ag^+ ions to study the role of π···π interaction in the packing of polymeric Ag^+ complexes, although predictable utilization of the aromatic stacking interactions remains a challenge compared to the better-understood hydrogen bonding. A good example is the 8-thioquinolyl-containing ligands 1,3-bis(8-thioquinolyl)propane (C3TQ) and 1,4-bis(8-thioquinolyl) butane (C4TQ) [19]. A series of extended Ag^+ complexes have been assembled in which the 8-thioquinolyl terminal groups in the ligands act as both coordination sites and supramolecular interaction sites for π···π aromatic stacking. Multidimensional frameworks have been achieved via various intra- and intermolecular π···π interactions which can be adjusted by the alkyl linkers between the 8-thioquinolyl terminal ends [19].

A unique supramolecular interaction in the silver(I) coordination polymer is the Ag···Ag argentophilic interaction. As one of d^{10} metal cations, Ag^+ ion can form ligand-supported or ligand-unsupported Ag···Ag interactions. Although Ag···Ag interactions are comparatively weak, they play an important role in the assembly of functional silver(I) coordination polymers, and the existence of d^{10}–d^{10} interactions between two closedshells of silver(I) atoms can not only contribute to the structural diversities of the final structures by increasing the structural dimensionalities, but also possibly endow the related silver(I) coordination polymers with various properties like (e.g., semiconductivity, luminescence). Since the publication of two early papers on Ag···Ag interactions by Robinson and Zaworotko in 1995 [75] and by Yaghi and Li in 1996 [76], argentophilic interactions in either a ligand-supported [77] or ligand-unsupported [56,78] way have received wide attention. Raman spectra [79] and semiconductivity [58] have been used to provide evidence of the existence of Ag···Ag interactions.

5.4.2 Counter Anions

Besides supramolecular interactions, counter anions also play important roles in the self-assembly of silver(I) coordination polymers. Changing anions often alters the structural topologies, sometime even the overall physical properties of polymeric Ag^+ complexes [19,80].

The anions can usually exist in coordinating, weakly coordinating, noncoordinating, or mixed fashion in silver(I) polymeric complexes. Among the commonly used anions, BF_4^-, ClO_4^-, PF_6^-, SbF_6^-, NO_3^-, $CF_3SO_3^-$ and $CF_3CO_2^-$, PF_6^-, and SbF_6^- are usually uncoordinated. BF_4^- and ClO_4^- are also selected as uncoordinated anions in many cases, although they may occasionally exhibit different coordinating modes [67c,81]. Anions such as NO_3^-, $CF_3SO_3^-$, and $CF_3CO_2^-$ usually have relatively complicated situations because they can exist in polymeric silver(I) complexes in either μ_1, μ_2, and μ_3 modes, or in non coordinating fashion. Sometimes, these anions can display mixed modes in the same complex [71a,19,26,82]. The coordination ability of these anions decreases in the approximate order $NO_3^- > CF_3CO_2^- > CF_3SO_3^- > BF_4^-$, ClO_4^-, PF_6^-, SbF_6^- under similar conditions, rendering the investigation of anion exchangeability possible [82c,82d].

It is often observed that the size, coordination ability, and formation of weak supramolecular interactions of these anions exert complicated influences on the resulting silver(I) coordination polymers. The coordinating anions can increase the dimensionality of the polymeric structure by bridging the Ag^+ ions, while the noncoordinating anions may also be able to extend the polymeric structures through weak supramolecular interactions.

5.5 ONE- TO THREE-DIMENSIONAL COORDINATION POLYMERS BASED ON SILVER–LIGAND COORDINATION BONDS

As discussed above the Ag^+ ion has rather flexible coordination geometry and can display versatile coordination modes, depending on the donor set of the ligand. Multidimensional silver(I) coordination polymers can be constructed if a selected ligand can properly coordinate the Ag^+ ions. However, structural diversification often occurs, which is complicated by various supramolecular interactions. Therefore, rational design of silver(I) coordination polymers remains a challenge. Nevertheless, because the Ag–ligand bond represents the basic connecting mode, and topological analysis using the nodes-and-connectors approach can help us to classify the diversified silver(I) coordination polymers, it is useful to give a concise outline of the multidimensional silver(I) coordination polymers based solely on Ag–ligand bonding which may be expected to throw some light on predictable construction of silver(I) coordination polymers.

5.5.1 1D Silver(I) Coordination Polymers

An infinite 1D chain is the simplest and commonest structural topology of silver(I) coordination polymers. However, the simply 1D chain can diversify in two ways, as shown in Figure 5.10. One type (a) is based on the connection mode and the shape of the 1D chains. Such structural diversity is related primarily to Ag^+ ion coordination geometry, ligand dentation, and the metal-to-ligand ratio. The other type (b) is caused by the twist and alignment of 1D chains in crystal packing.

Theoretically, any nonlinear chain can twist in a helical conformation, thus giving rise to a 1D helix. Infinite Ag^+ helices can be constructed from angular ditopic ligands, flexible bis-bidentate chelating ligands, or flexible nonchelating polydentate ligands. Typically, the *P* and *M* infinite chiral chains (Fig. 5.10b) will co-crystallize to result in a racemic crystal [19,20,71a]. However, in a few cases, homochiral Ag^+ helices have been obtained from achiral ligands via spontaneous resolution [83] or from chiral ligands [84]. A rare 1D helical structure is the *meso*-helical chain [85]. In such a structure the twist sense changes every half-turn of the helix, resulting in helical reversal, and forming a *meso*-helix. An early report on Ag^+ *meso*-helices dealt

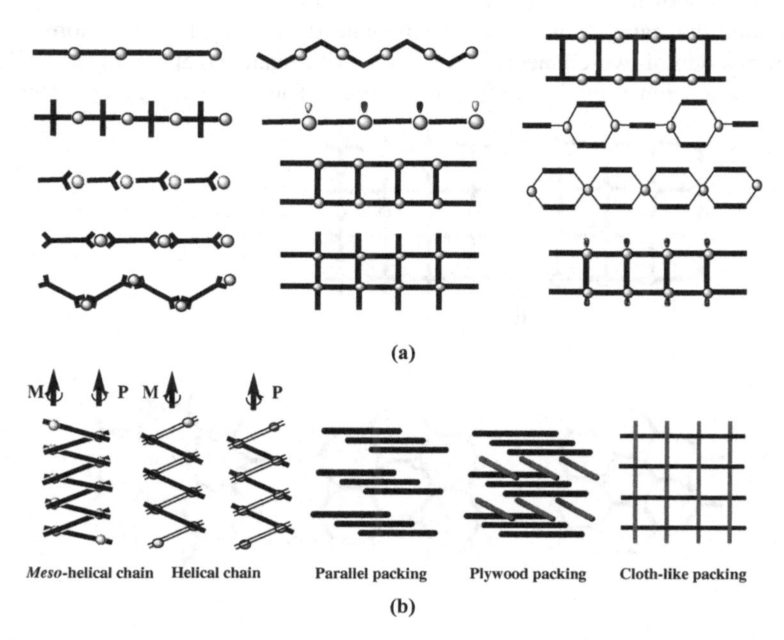

FIGURE 5.10 Structural diversity in 1D silver(I) coordination polymers: (a) different 1D forms; (b) twisting or crossing of 1D chains.

with the complexes $[Ag(C6OQ)]X_n$ [C6OQ = 1,6-bis(8-oxaquinolyl)hexane, $X = ClO_4^-$, BF_4^-, $CF_3SO_3^-$, and $CF_3CO_2^-$] [85b]. In contrast to the classical chiral single-stranded helical chain, the *meso*-helix can be regarded as the alternative propagation of helical conformers of opposite chirality (*M* and *P*) around the inversion center, thus possessing two opposite helical axes, as shown in Figure 5.10b.

The structural diversity of the 1D silver(I) coordination polymers can also be generated by the diverse alignment of 1D chains in crystal packing (Fig. 5.10b). In the usual situation, the 1D chains prefer to run in the same direction in the crystal lattice, to achieve close packing. However, nonparallel alignment with a crosslike arrangement of the chains has also been observed [86]. Such crosslike arrangement of the 1D chain can be described as a 3D "plywood-like array", with two differently oriented neighboring layers stacking in AB fashion.

5.5.2 2D Silver(I) Coordination Polymers

Following the nodes-and-connectors approach, the combination of Ag^+ ions of various coordination geometries with linear, tripodal, tetrapodal, and so on, multifunctional ligands has successfully yielded a number of 2D silver(I) coordination polymers. Two typical 2D net topologies are the square grid (4,4) net and the honeycomb (6.3) net.

As a particularly simple and commonly reported 2D metal–organic polymeric structure, the Ag^+ square grid network can usually be synthesized through two strategies, as shown in Figure 5.11: (1) combination of four-connecting Ag^+ ions with ligands that can functionalize as two-connected rods by 1 : 2 stoichiometry, and (2) combination of two-connecting Ag^+ ions with ligands that can functionalize as four-connected junctures by 2 : 1 stoichiometry. The first strategy has been used

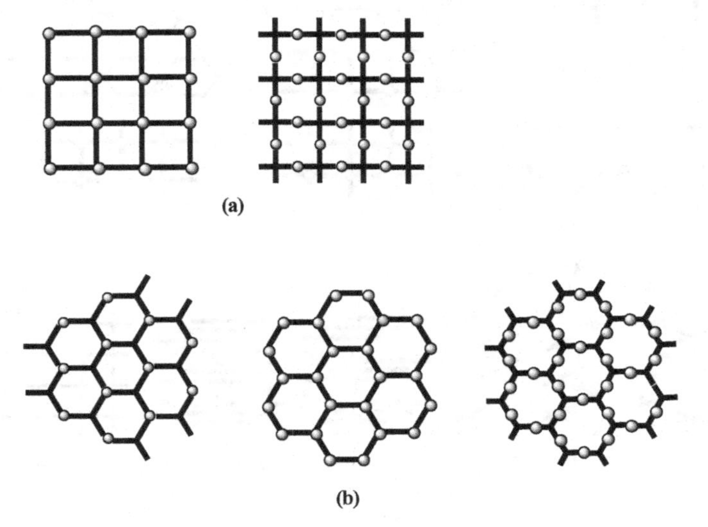

FIGURE 5.11 Strategies to obtain common 2D Ag^+ networks: (a) square grid (4.4) net; (b) honoeycomb (6,3) net.

widely to assemble 2D silver(I) polymeric complexes with a square grid (4.4) net structure, because four-coordinating Ag^+ ions and ditopic organic ligands are readily available. In contrast, ligands used in the second strategy are usually restricted to cross-shaped tetrapodal molecules such as 5,10,15,20-tetra(4-pyridyl)porphyrin or its metal-containing derivatives.

There are three principal routes to the synthesis of 2D polymeric Ag^+ complexes with (6,3) net topology (Fig. 5.11):

1. Combination of ligands and Ag^+ ions that both have C_3 or pseudo-C_3 symmetry by 1 : 1 stoichiometry.
2. The Ag^+ ions provide the required three-fold symmetry for a (6,3) net, while the ligands bridge the Ag^+ ions as linear rods. In this case, the metal-to-ligand ratio has to be 2 : 3, but the Ag^+ ions are allowed to adopt several geometries, such as trigonal geometry, tetrahedral geometry with one coordination site occupied by monodentate anion or solvent, or even six-coordinating Ag^+ ion with tris-chelating environments.
3. The ligands afford the three-fold symmetry, while the Ag^+ ions act as two-connecting nodes. In this case, the metal-to-ligand ratio should be kept at 3 : 2.

5.5.3 3D Silver(I) Coordination Polymers

Compared with 1D and 2D structures, the 3D silver(I) coordination polymers are expected to be more difficult to fabricate in a predictable way because of the highly flexible coordination geometry of Ag^+ ions, which implies formation of more possible supramolecular isomers during the self-assembly process. For instance, the connection of a four-coordinating Ag^+ ion "node" with a bidentate ligand "rod" by 1 : 2 stoichiometry can make either a 2D square grid (4,4) network or a 3D diamondoid framework, as shown in Figure 5.5d and f. In an even more complicated case, the connection of three-connecting Ag^+ ions with tripodal ligands by 1 : 1 stoichiometry can provide 1D ladder, 2D (6,3) or (4.8^2) networks, and 3D (10,3)-a or $(4.8.10)(8.10^2)$ frameworks, as shown in Figure 5.6b, c, e, f and g, respectively. Due to the flexibility of the Ag^+ coordination environment, formation of a specific isomer may be subject to only a subtle influence. Nevertheless, it is possible to engineer some typical 3D frameworks, including diamondoid, (10,3)-a, or cubic frameworks, if reaction conditions are controlled properly.

As shown in Figure 5.12, 3D polymeric Ag^+ complexes of diamondoid topology can be assembled by three strategies: (1) linear bis-monodentate ligands connecting the T_d Ag^+ ions in a 2 : 1 ratio, (2) tetradentate ligands with tetrahedral disposition of the donor atoms (T_d ligand) connecting the linear Ag^+ ions in a 1 : 2 ratio, and (3) combination of T_d ligands and T_d Ag^+ ions. Among them, the first strategy is used most widely in the construction of diamondoid silver(I) coordination polymers because four-coordinating Ag^+ ions with T_d geometry (T_d Ag^+ ion) and bis-monodentate ligands are very common. 3D polymeric Ag^+ complexes of diamondoid topology contain adamantanoid cavities which become larger when the ligand spacers are longer. As a consequence, diamondoid frameworks often interpenetrate each other, generating multifold interpenetrating structures.

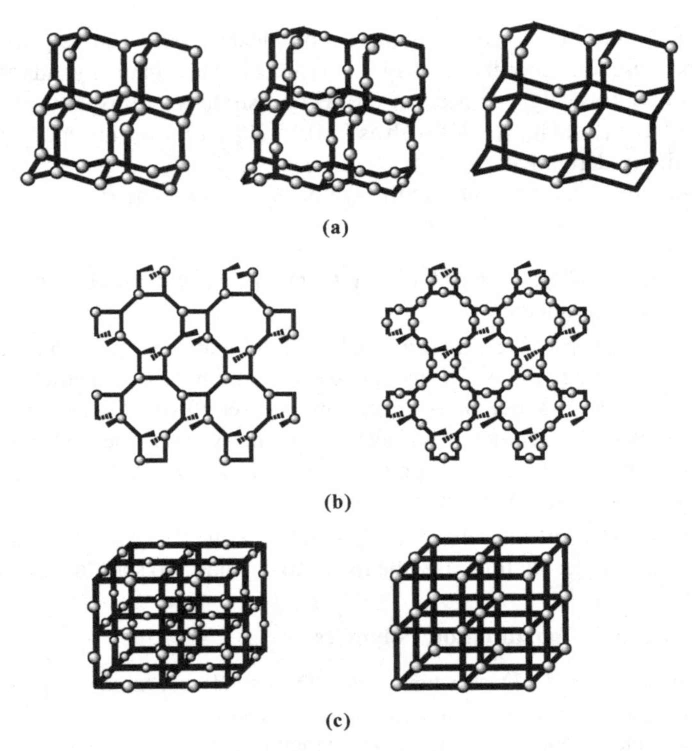

FIGURE 5.12 Strategies to obtain three types of common 3D Ag^+ frameworks: (a) diamond net; (b) (10,3)-a net; (c) cubic net.

Formation of a (10,3)-a topological framework [17,18] is also achieved easily for Ag^+ ions because the trigonal geometry is common for these ions. In principle, there are two ways to construct Ag^+ polymeric structures of (10,3)-a topology, as depicted in figure 5.12: (1) combination of a trigonal Ag^+ ion with a tripodal ligand in a 1 : 1 ratio, and (2) combination of a linear Ag^+ ion with a tripodal ligand in a 3 : 2 ratio. One more complicated assembly method can be found in $[Ag_2(Me_4bpz)]$·guest (H_2Me_4bpz = 3,3′,5,5′-tetramethyl-4,4′-bipyrazole), in which Me_4bpz acts as a tetradentate ligand to coordinate with four linear coordinating Ag^+ ions, and each Ag^+ ion is coordinated with two N atoms from different Me_4bpz ligands. However, the interconnection of Ag^+ ions and Me_4bpz ligands gives $\{Ag_3(pz)_3\}$ trigonal subunits, which behave as three-connected nodes. Therefore, a highly distorted 3D silver(I) coordination framework of (10, 3)-a topology is generated [87].

Assembly of 3D octahedral or cubic frameworks is relatively difficult, although as shown in Figure 5.12, they can theoretically be formed either by using a hexadentate ligand in combination with a linear Ag^+ ion in a 1 : 3 ratio or using a bidentate ligand in combination with an octahedral Ag^+ ion in a 3 : 1 ratio. Therefore, such 3D polymeric Ag^+ complexes of cubic (α-Po) network topology are rare [40].

The reason may be that the six-coordinating Ag^+ ions with octahedral geometry and the hexadentate ligand with octahedral shape are both found sparingly.

5.6 INTERTWINING OR INTERPENETRATING OF SILVER(I) COORDINATION POLYMERS

A well-known and fascinating phenomenon in 1D, 2D, and 3D silver(I) coordination polymers is the occurrence of intertwining or interpenetrating, which involves 1D to 3D structures and usually leads to dimensional increase and structural diversity. A characteristic nature of such intertwining or interpenetrating is that the individual coordination polymers are entangled with each other without forming bonding interactions. Therefore, it is convenient to classify those structural entanglements on the basis of the starting individual coordination polymers.

5.6.1 Entanglements of 1D Silver(I) Coordination Polymers

1D polymeric structures may entangle with each other to result in structural diversification such as 1D-to-1D, 1D-to-2D, and 1D-to-3D variations. The most common observation is that 1D helical chains intertwine to give rise to double- or triple-stranded helices [84,86,88], although triple-stranded Ag^+ helices are still quite rare. Lü et al. have reported an example of single-stranded helices interwoven into a triple-stranded helix (Fig. 5.13a) [86]. The single helical chain is formed in $\{[Ag(3\text{-impmd})]\cdot(NO_3)\cdot MeOH\}_n$ [3-impmd = *N, N′*-bis(3-imidazol-1-yl-propyl)pyromellitic diimide; L2 in Fig. 5.2], in which each Ag^+ ion adopts a linear coordination geometry binding two N donors from different ligand while each ligand acts as a ditopic bis(monodentate) linker connecting two symmetry-related Ag^+ ions. The flexibility of 3-impmd places

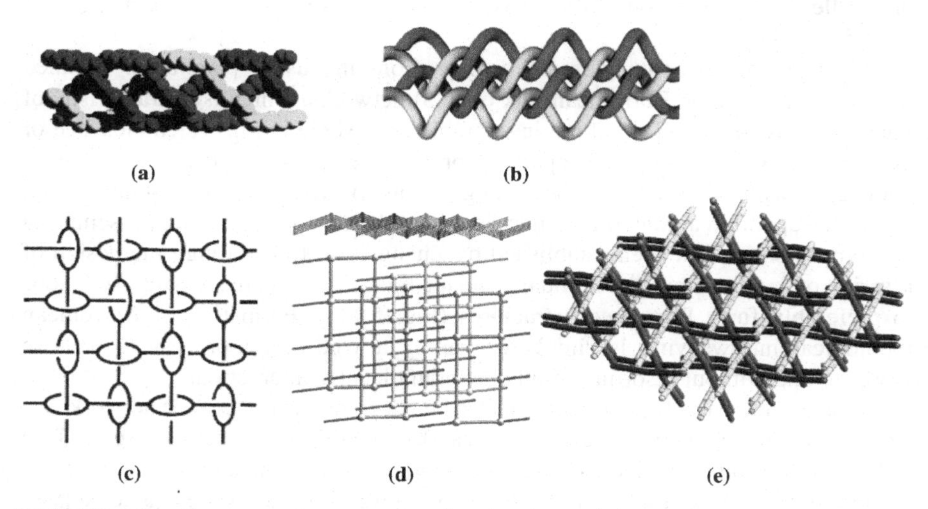

FIGURE 5.13 Representative examples showing intertwining or interpenetrating in 1D silver(I) coordination polymers.

all the ligands in a U shaped conformation, joining Ag^+ ions and forming helical chains. Intertwining of three neighboring helical chains results in a triple-stranded Ag^+ helix encapsulating guest anions and solvent molecules.

The 1D helical chains can also intertwine to from 2D supramolecular structures. Bourlier et al. reported that the entanglement of 1D consecutive helical strands of the same helicity leads to a 2D interwoven chiral architecture (Fig. 5.13b) [10]. Other structural diversifications of 1D polymers have also been observed. Hoskins et al. reported that 1D polymeric chains are knitted together to from a 2D polyrotaxane sheet (Fig. 5.13c) [11]. Carlucci et al. produced a 2D polythreaded layer containing infinite 1D molecular ladders (Fig. 5.13d) [21]. An even more intriguing crossing alignment of 1D chains is the woof-and-weft threading mode [89]. The overall packing of such a clothlike structure actually leads to a 2D sheet (Fig. 5.10b).

3D supramolecular architectures may also be generated based on interpenetrating 1D chains. Carlucci et al. reported that 1D polymeric chains of $[Ag(sebn)](AsF_6)$ (sebn = 1,10-decanedinitrile) can extend in three noncoplanar directions, to give rise to a unique 3D entangled array, as shown in Figure 5.13e [90].

5.6.2 Entanglements of 2D Silver(I) Coordination Polymers

Interpenetration is ubiquitous for some well-known 2D networks, such as (4,4) net, (6,3) net, and (4.8^2) net. These networks usually display two distinctive interpenetration modes: parallel or inclined. The parallel interpenetration will usually not cause dimension increase unless the 2D sheets are highly undulated, whereas inclined penetration will definitely result in 2D-to-3D dimensional increase. Ni and Vittal gave an example of 2D corrugated (4,4) sheets interpenetrating in pairs in the parallel mode, stabilized by π–π interactions (Fig. 5.14a) [91]. Fan et al. provided an example of parallel interpenetration of 4.8^2 nets, producing an infinite 2D layered structure (Fig. 5.14b) [16].

One puzzling 2D entanglement distinct from the usual parallel or inclined interpenetration is the Borromean link of (6,3) networks, which is characteristic of nontrivial three-ring links that are inseparable as a whole but in which cleavage of any ring makes the whole fall apart. Dobrzánska et al. showed that 2D silver(I) complexes with (6,3) nets can be entangled into Borromean sheets stabilized by ligand-unsupported argentophilic interactions [92]. A facile approach to assembling Borromean sheets has been established by Zhang et al. [93] through the design of a series of bulky semirigid tripodal pendant ligands which react with Ag^+ ions to preferably form Borromean structures (Fig. 5.14c). Formation of Borromean topological interweaving is simply driven by a void-filling process to achieve closepacking without resorting to other supramolecular interactions.

3D supramolecular frameworks can also be generated from 2D (6,3) networks via Borromean entanglement. One example can be found in the complex $[Ag_2(H_2L)_3]$ $(NO_3)_2$ [H_2L = bis(salicyidene-1,4-diaminobutane], where the adjacent (6,3) layers are interlinked via Borromean links if the $Ag \cdots Ag$ interactions are neglected [94]. On the other hand, the highly undulated (6,3) nets are also able to polycatenate to generate 3D structures in a parallel fashion (Fig. 5.14d) [95].

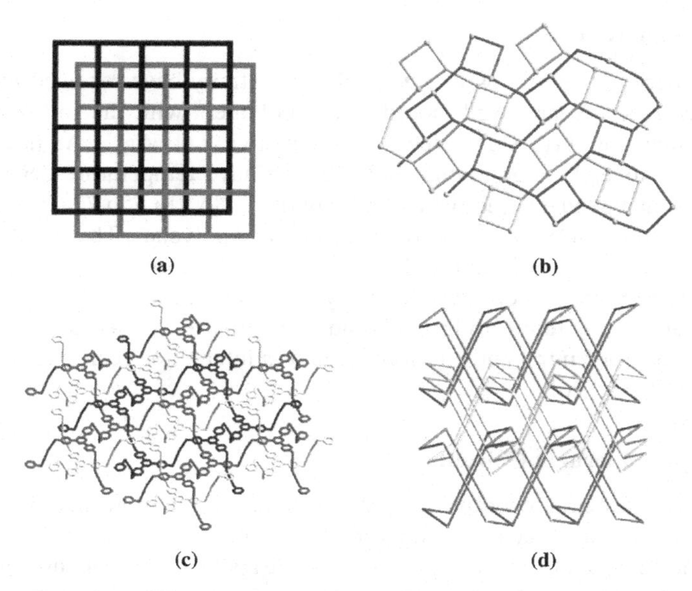

FIGURE 5.14 Representative interpenetrations in 2D silver(I) coordination polymers.

5.6.3 Entanglements of 3D Silver(I) Coordination Polymers

The spacious 3D frameworks themselves can interpenetrate due to the existence of large voids formed within the framework. The diamondoid nets are wellknown to display multifold interpenetration, with the highest level even reaching 18. A series of Ag^+ polymeric coordination networks with the general formula $[Ag(ddn)_2]X$ (ddn= 1,12-dodecanedinitrile; X = NO_3^-, PF_6^-, AsF_6^-, ClO_4^-) have been assembled by Carlucci et al. [30a], in which all structures show diamondoid net topology but the interpenetrating modes are different.

Interpenetration of (10,3)-a nets is also common. It is noticeable that although the (10,3)-a network itself is inherently chiral, interpenetration of two (10,3)-a nets with opposite handedness may give rise to a racemic structure, which may be exemplified by the complex $[Ag_2(2,3\text{-}Me_2pyz)_3(SbF_6)_2]$ [96]. Other interpenetrating 3D frameworks may also exist: for example, interpenetration of two 3,5-connected networks with $(6^3)(6^3.8)$ topology [97].

5.7 PROPERTIES OF SILVER(I) COORDINATION POLYMERS

Silver(I) coordination polymers are interesting not only because of their interesting structural diversity, but also due to their rich physicochemical properties, which might find applications in conductive materials, therapeutic agents, luminescent materials, anion or guest exchange and gas absorption, supramolecular chirality and so on.

5.7.1 Conductivity

Ag^+ polymeric complexes with electrical conductivity have received increasing attention in recent years. Ag^+ coordination polymers with the following characteristics may possibly exhibit electrical conductivity: (1) Ag $\cdots$ Ag interactions, which cause silver(I) coordination polymers to have temperature-dependent or temperature-independent electrical conductivities [48b,51a,55b,72a,72b]; and (2) π-conjugated polycyclic aromatic hydrocarbons and polyenes. The resulting silver (I) coordination polymers exhibit electrical conductivity because the delocalized π-bridges can provide an effective pathway for electron transfer between the two metal centers across these bridging ligands. For this reason, columnar aromatic stacking with strong intra- and intermolecular π–π interactions can give rise to high conductivity [67b,72c].

5.7.2 Supramolecular Chirality

Construction of chiral coordination polymers has been motivated by mimetic synthesis of chiral architectures similar to those molecules in life, on the one hand, while on the other hand, these complexes may have applications in industry such as in asymmetric catalysts, in NLO materials, and in chiral separation. Supramolecular chirality, as proposed by Lehn, can be generated by the following two strategies: (1) using a chiral reagent, and (2) using an achiral ligand, but the interaction between achiral components is dissymmetrizing, thus yielding a chiral association. Based on these two strategies, many chiral Ag^+ polymeric complexes have been synthesized in recent years [83,84a,98]. For example, the chiral silver(I) polymers $[(AgL_2)NO_3]_\infty$ and $[(AgL_2)ClO_4]_\infty$ [L = (*s*)-2,2′-dimethoxy-1,1′-binaphthyl-3,3′-bis(4-vinylpyridine)] were obtained by the first method. The connection of L and Ag^+ ions leads to an interesting 2D lamellar framework. In these two chiral complexes, the chiral ligand L is used as the chiral resource [84a]. One example of generating chiral complexes by using the second method is that of $\{[Ag_2(L)](ClO_4)_2\}_\infty$ [L 2,3-bis(6′-methyl-2′-pyridylmethylsulfanylmethyl)pyrazine], where the formation of intramolecular π–π stacking interactions upon complexation with Ag^+ ions induces planar chirality in the achiral ligand [83].

5.7.3 Luminescence

In many silver(I) coordination polymers there exist Ag $\cdots$ Ag interactions due to strong or weak Ag–Ag bonding which is shorter than the sum of van der Waals radii of two Ag atoms. Such Ag $\cdots$ Ag interactions often contribute to the luminescence of Ag^+ polymeric complexes. On the other hand, coordination of the Ag^+ ion is found to be able to affect the emission wavelength of organic molecules; therefore, judicious design and selection of organic ligands in the syntheses of silver(I) coordination polymers may provide an efficient method of obtaining new types of photo- or electroluminescent materials. In general, the silver(I) polymeric complexes can exhibit emission bands which originate from $n \rightarrow \pi^*$, $\pi \rightarrow \pi^*$ intraligand, or ligand-to-metal charge transfer (LMCT) transitions [1,3,35,80b].

5.7.4 Antimicrobial Activity

Silver complexes have been recognized as antimicrobial agents in curative and preventive health care for decades: for example, in the treatment of burn wounds. Although silver nitrate and silver sulfonamides are the most widely used reagents, new silver therapeutic agents have been discovered [3,100]. The Ag^+ complexes often show different antimicrobial activities against microorganisms from the ligand itself or the hydrated Ag^+ ion. Nomiya's group [3] investigated antimicrobial activities of Ag^+–X (X = P, S, O, and N) bonding complexes in aqueous media, showing that Ag^+ complexes containing easily replaceable Ag–N/O bonds play important roles in exhibiting an effective and wide spectrum of antimicrobial activities, whereas Ag–S complexes often show a narrower spectrum, and Ag–P complexes never have antimicrobial activity.

5.7.5 Guest Exchange and Sorption

It has been reported that guest molecules located in the cavities or channels of porous silver(I) coordination polymers can be removed without causing collapse of the framework, or can be reinserted in some cases. Xu et al. showed that the complex $\{[Ag_4L_3(O_3SCF_3)_4](solvate)\}_\infty$ [L = 1,3,5-tris(diphenylphosphanyl)benzene] was able to be transformed to a solvent/water-free form without destroying the framework [61]. Mäkinen et al. reported that polymeric complexes $[Ag(3\text{-}pySO_3)]_\infty$ and $[Ag(3\text{-}pySO_3)(MeCN)_{0.5}]_\infty$ ($3\text{-}pySO_3$ = 3-pyridinesulfonate) can be interconverted by the reversible sorption and desorption of MeCN guest molecules [100].

Selective intercalation of guests is also reported for silver(I) coordination polymers. The complex $[Ag(4\text{-}pySO_3)]_\infty$ ($4\text{-}pySO_3$ = 4-pyridinesulfonate) was selectively intercalated by amines over other organic guests in the order primary > secondary ≫ tertiary [49d]. The complex $\{Ag_3[Ag_5(L)_6](BF_4)_{12}\cdot 12H_2O\}_n$ [L = 3,5-diphenyl-1,2,4-triazolate] showed anion-exchange behavior. The dehydrated complex has the ability to adsorb nonpolar alkanes such as hexane, pentane, and cyclohexane, but not to adsorb polar solvents such as propyl alcohol, ethyl acetate, and tetrahydrofuran [42a].

5.7.6 Magnetism

Silver(I) coordination polymers may show interesting magnetic properties when the diamagnetic Ag^+ ion is coordinated with organic radical moieties [2,101]. For example, a strong antiferromagnetic interaction was observed in $[Ag(L)(NO_3)_2]_\infty$ [L = m-*N*-methylpyridinium nitronyl nitroxide], in which the Ag^+ ion mediated in the antiferromagnetic spin–spin interactions of the bridging nitronyl nitroxide ligands [2].

REFERENCES

1. (a) Chen, C.-L.; Kang, B.-S.; Su, C.-Y. *Aust. J. Chem.* **2006**, *59*, 3–18. (b) Khlobystov, A. N.; Blake, A. J.; Champness, N. R.; Lemenovskii, D. A.; Majouga, A. G.; Zyk, N. V.;

Schröder, M. *Coord. Chem. Rev.* **2001**, *222*, 155–192. (c) Zheng, S.-L.; Tong, M.-L.; Chen, X.-M. *Coord. Chem. Rev.* **2003**, *246*, 185–202.

2. Zhang, D. Q.; Ding, L.; Xu, W.; Hu, H. M.; Zhu, D. B.; Huang, Y. H.; Fang, D. C. *Chem. Commun.* **2002**, 44–45.
3. Kasuga, N. C.; Sugie, A.; Nomiya, K. *Dalton Trans.* **2004**, 3732–3740, and references therein.
4. (a) Chen, C.-L.; Tan, H.-Y.; Yao, J.-H.; Wan, Y.-Q.; Su, C.-Y. *Inorg. Chem.* **2005**, *44*, 8510–8520. (b) Chen, C.-L.; Yu, Z.-Q.; Zhang, Q.; Pan, M.; Zhang, J.-Y.; Zhao, C.-Y.; Su, C.-Y. *Cryst. Growth Des.* **2008**, *8*, 897–905.
5. Su, D. F.; Cao, R.; Su, Y. Q.; Bi, W. H.; Li, X. J.; Wang, Y. Q.; Shi, Q.; Li, X. *Inorg. Chem.* **2003**, *42*, 7512–7518.
6. Kim, H.-J.; Zin, W.-C.; Lee, M. *J. Am. Chem. Soc.* **2004**, *126*, 7009–7014.
7. Fitchett, C. M.; Steel, P. J. *Dalton Trans.* **2006**, 4886–4888.
8. Klein, C.; Graf, E.; Hosseini, M. W.; Cian, A. D. *New J. Chem.* **2001**, *25*, 207–209.
9. One example of helical structures: Beauchamp, D. A.; Loeb, S. J. *Supramol. Chem.* **2005**, *17*, 617–622.
10. Bourlier, J.; Hosseini, M. W.; Planeix, J.-M.; Kyritsakas, N. *New J. Chem.*, **2007**, *31*, 25–32.
11. Hoskins, B. F.; Robson, R.; Slizys, D. A. *J. Am. Chem. Soc.* **1997**, *119*, 2952–2953.
12. (a) Gardner, G. B.; Venkataraman, D.; Moore, J. S.; Lee, S. *Nature* **1995**, *374*, 792–795. (b) Xu, Z. T.; Lee, S.; Kiang, Y.-H.; Mallik, A. B.; Tsomaia, N.; Mueller, K. T. *Adv. Mater.* **2001**, *13*, 637–641. (c) Gardner, G. B.; Kiang, Y.-H.; Lee, S.; Asgaonkar, A.; Venkataraman, D. *J. Am Chem. Soc.* **1996**, *118*, 6946–6953. (d) Kiang, Y.-H.; Gardner, G. B.; Lee, S.; Xu, Z. T.; Lobkovsky, E. B. *J. Am. Chem. Soc.* **1999**, *121*, 8204–8215.
13. (a) Li, J.-R.; Zhang, R.-H.; Bu, X.-H. *Cryst. Growth Des.* **2003**, *3*, 829–835. (b) Lee, J. W.; Kim, E. A.; Lee, Y.-A.; Pak, Y.; Jung, O.-S. *Inorg. Chem.* **2005**, *44*, 3151–3155.
14. Carlucci, L.; Ciani, G.; Proserpio, D. M.; Porta, F. *Angew. Chem., Int. Ed.* **2003**, *42*, 317–322.
15. Huang, Z.; Du, M.; Song, H.-B.; Bu, X.-H. *Cryst. Growth Des.* **2004**, *4*, 71–78.
16. Fan, J.; Sun, W.-Y.; Okamura, T.-A.; Tang, W.-X.; Ueyama, N. *Inorg. Chem.* **2003**, *42*, 3168–3175.
17. Bu, X.-H.; Chen, W.; Du, M.; Biradha, K.; Wang, W.-Z.; Zhang, R.-H. *Inorg. Chem.* **2002**, *41*, 437–439.
18. Seeber, G.; Pickering, A. L.; Long, D.-L.; Cronin, L. *Chem. Commun.* **2003**, 2002–2003.
19. Chen, C.-L.; Su, C.-Y.; Cai, Y.-P.; Zhang, H.-X.; Xu, A.-W.; Kang, B.-S.; zur Loye, H.-C. *Inorg. Chem.* **2003**, *42*, 3738–3750.
20. (a) Tuna, F.; Hamblin, J.; Clarkson, G.; Errington, W.; Alcock, N. W.; Hannon, M. J. *Chem. Eur. J.* **2002**, *8*, 4957–4964. (b) Liao, S.; Su, C.-Y.; Zhang, H.-X.; Shi, J.-L.; Zhou, Z.-Y.; Liu, H.-Q.; Chan, A. S. C.; Kang, B.-S. *Inorg. Chim. Acta* **2002**, *336*, 151–156.
21. Carlucci, L.; Ciani, G.; Proserpio, D. M. *Chem. Commun.* **1999**, 449–450.
22. Venkataraman, D.; Lee, S.; Moore, J. S.; Zhang, P.; Hirsch, K. A.; Gardner, G. B.; Covey, A. C.; Prentice, C. L. *Chem. Mater.* **1996**, *8*, 2030–2040.
23. You, Z.-L.; Zhu, H.-L.; Liu, W.-S. *Acta Crystallogr.* **2004**, *C60*, m620–m622.
24. Dong, Y.-B.; Ma, J.-P.; Jin, G.-X.; Huang, R.-Q.; Smith, M. D. *Dalton Trans.* **2003**, 4324–4330.

25. Dong, Y.-B.; Jiang, Y.-Y.; Li, J.; Ma, J.-P.; Liu, F.-L.; Tang, B.; Huang, R.-Q.; Battern, S. R. *J. Am. Chem. Soc.* **2007**, *129*, 4520–4521.

26. Carlucci, L.; Ciani, G.; Proserpio, D. M.; Rizzato, S. *CrystEngComm* **2002**, *4*, 121–129.

27. Wu, C.-D.; Ngo, H. L.; Lin, W. B. *Chem. Commun.* **2004**, 1588–1589.

28. Bu, X.-H.; Hou, W.-F.; Du, M.; Chen, W.; Zhang, R.-H. *Cryst. Growth Des.* **2002**, *2*, 303–307.

29. (a) Whang, D.; Kim, K. *J. Am. Chem. Soc.* **1997**, *119*, 451–452. (b) Jiang, J.-J.; Li, X.-P.; Zhang, X.-L.; Kang, B.-S.; Su, C.-Y. *CrystEngComm* **2005**, *7*, 603–607.

30. (a) Carlucci, L.; Ciani, G.; Proserpio, D. M.; Sironi, A. *J. Chem. Soc., Chem. Commun.* **1994**, 2755–2756. (b) Carlucci, L.; Ciani, G.; Proserpio, D. M.; Rizzato, S. *Chem. Eur. J.* **2002**, *8*, 1519–1526.

31. Ferlay, S.; Koenig, S.; Hosseini, M. W.; Pansanel, J.; Cian, A. D.; Kyritsakas, N. *Chem. Commun.* **2002**, 218–219.

32. Bu, X.-H.; Biradha, K.; Yamaguchi, T.; Nishimura, M.; Ito, T.; Tanaka, K.; Shionoya, M. *Chem. Commun.* **2000**, 1953–1954.

33. Dolomanov, O. V.; Cordes, D. B.; Champness, N. R.; Blake, A. J.; Hanton, L. R.; Jameson, G. B.; Schröder, M.; Wilson, C. *Chem. Commun.* **2004**, 642–643.

34. Oh, M.; Stern, C. L.; Mirkin, C. A. *Chem. Commun.* **2004**, 2684–2685.

35. Dong, Y.-B.; Zhao, X.; Huang, R.-Q.; Smith, M. D.; zur Loye, H.-C. *Inorg. Chem.* **2004**, *43*, 5603–5612.

36. Tong, M.-L.; Shi, J.-X.; Chen, X.-M. *New J. Chem.* **2002**, *26*, 814–816.

37. Blake, A. J.; Champness, N. R.; Cooke, P. A.; Nicolson, J. E. B. *Chem. Commun.* **2000**, 665–666.

38. Oxtoby, N. S.; Blake, A. J.; Champness, N. R.; Wilson, C. *Proc. Natl. Acad. Sci. USA* **2002**, *99*, 4905–4910.

39. Batten, S. R.; Hoskins, B. F.; Robson, R. *New J. Chem.* **1998**, *22*, 173–175.

40. Carlucci, L.; Ciani, G.; Proserpio, D. M.; Sironi, A. *Angew. Chem., Int. Ed.* **1995**, *34*, 1895–1898.

41. Kyono, A.; Kimata, M.; Hatta, T. *Inorg. Chim. Acta* **2004**, *357*, 2519–2524.

42. (a) Yang, G.; Raptis, R. G. *Chem. Commun.* **2004**, 2058–2059. (b) Pascu, M.; Tuna, F.; Kolodziejczyk, E.; Pascu, G. I.; Clarkson, G.; Hannon, M. J. *Dalton Trans.* **2004**, 1546–1555.

43. Abrahams, B. F.; Jackson, P. A.; Robson, R. *Angew. Chem., Int. Ed.* **1998**, *37*, 2656–2659.

44. Suenaga, Y.; Yan, S. G.; Wu, L. P.; Ino, I.; Kuroda-Sowa, T.; Maekawa, M.; Munakata, M. *J. Chem. Soc., Dalton Trans.* **1998**, 1121–1126.

45. Wu, H.-C.; Thanasekaran, P.; Tsai, C.-H.; Wu, J.-Y.; Huang, S.-M.; Wen Y.-S.; Lu, K.-L. *Inorg. Chem.* **2006**, *45*, 295–303.

46. (a) Guo, D.; He, C.; Duan, C.-Y.; Qian C.-Q.; Meng, Q.-J. *New J. Chem.* **2002**, *26*, 796–802. (b) Kaes, C.; Hosseini, M. W.; Rickard, C. E. F.; Skelton B. W.; White, A. H. *Angew. Chem., Int. Ed.* **1998**, *37*, 920–922.

47. Carlucci, L.; Ciani, G.; Proserpio, D. M.; Porta, F. *CrystEngComm* **2005**, *7*, 78–86.

48. (a) Chen, C.-L.; Zhang, Q.; Jiang, J.-J.; Wang, Q.; Su, C.-Y. *Aust. J. Chem.* **2005**, *58*, 115-118. (b) Sun, D. F.; Cao, R.; Weng J. B.; Hong, M. C.; Liang, Y. C. *J. Chem. Soc., Dalton Trans.* **2002**, 291–292.

49. (a) Côté, A. P.; Shimizu, G. K. H. *Coord. Chem. Rev.* **2003**, *245*, 49–64. (b) Côté, A. P.; Shimizu, G. K. H. *Inorg. Chem.* **2004**, *43*, 6663–6673. (c) Hoffart, D. J.; Dalrymple, S. A.; Shimizu, G. K. H. *Inorg. Chem.* **2005**, *44*, 8868–8875. (d) May, L. J.; Shimizu, G. K. H. *Chem. Mater.* **2005**, *17*, 217–220.

50. (a) Shimokawa, C.; Itoh, S. *Inorg. Chem.* **2005**, *44*, 3010–3012. (b) Harrowfield, J. M.; Sharma, R. P.; Skelton, B. W.; White, A. H. *Aust. J. Chem.* **1998**, *51*, 735–746.

51. (a) Fu, R.; Xia, S.; Xiang, S.; Hu, S.; Wu, X. *J. Solid State Chem.* **2004**, *177*, 4626–4631. (b) Sagatys, D. S.; Dahlgren, C.; Smith, G.; Bott, R. C.; White, J. M. *J. Chem. Soc., Dalton Trans.* **2000**, 3404–3410.

52. Bu, X.-H.; Chen, W.; Hou, W.-F.; Du, M.; Zhang, R.-H.; Brisse, F. *Inorg. Chem.* **2002**, *41*, 3477–3482.

53. Zachariadis, P. C.; Hadjikakou, S. K.; Haddjiliadis, N.; Skoulika, S.; Michaelides, A.; Balzarini J.; Clercq, E. D. *Eur. J. Inorg. Chem.* **2004**, 1420–1426.

54. (a) Tang, K. L.; Xie, X. J.; Zhao, L.; Zhang, Y. H.; Jin, X. L. *Eur. J. Inorg. Chem.* **2004**, 78–85. (b) Su, W. P.; Cao, R.; Hong, M. C.; Wong, W.-T.; Lu, J. X. *Inorg. Chem. Commun.* **1999**, *2*, 241–243. (c) Chen, J.-X.; Xu, Q.-F.; Zhang, Y.; Chen, Z.-N.; Lang, J.-P. *J. Organomet. Chem.* **2004**, *689*, 1071–1077.

55. (a) Hong, M. C.; Su, W. P.; Cao, R.; Zhang, W. J.; Lu, J. X. *Inorg. Chem.* **1999**, *38*, 600–602. (b) Su, W. P.; Hong, M. C.; Weng, J. B.; Liang, Y. C.; Zhao, Y. J.; Cao, R.; Zhou, Z. Y.; Chan, A. S. C. *Inorg. Chim. Acta* **2002**, *331*, 8–15.

56. Su, W. P.; Hong, M. C.; Weng, J. B.; Cao, R.; Lu, S. F. *Angew. Chem. Int. Ed.* **2000**, *39*, 2911–2914.

57. Kuang, S.-M.; Zhang, Z.-Z.; Wang, Q.-G.; Mak, T. C. W. *Chem. Commun.* **1998**, 581–582.

58. Lozano, E.; Nieuwenhuyzen, M.; James, S. L. *Chem. Eur. J.* **2001**, *7*, 2644–2651.

59. Fournier, É.; Lebrun, F.; Drouin, M.; Decken, A.; Harvey, P. D. *Inorg. Chem.* **2004**, *43*, 3127–3135.

60. (a) Xu, X.; Nieuwenhuyzen, M.; Zhang, J.; James, S. L. *J. Inorg. Organomet. Polym. Mater.* **2005**, *15*, 431–437. (b) Montes, J. A.; Rodriguez, S.; Fernández, D.; Garcia-Seijo, M. I.; Gould, R. O.; García-Fernández, M. E. *J. Chem. Soc., Dalton Trans.* **2002**, 1110–1118.

61. Xu, X.; Nieuwenhuyzen, M.; James, S. L. *Angew. Chem., Int. Ed.* **2002**, *41*, 764–767.

62. Brandys, M.-C.; Puddephatt, R. J. *J. Am. Chem. Soc.* **2002**, *124*, 3946–3950.

63. Zhang, L.; Lü, X.-Q.; Chen, C.-L.; Tan, H.-Y.; Zhang, H.-X.; Kang, B.-S. *Cryst. Growth Des.* **2005**, *5*, 283–287.

64. Brandys, M.-C.; Puddephatt, R. J. *Chem. Commun.* **2001**, 1508–1509.

65. Hill, A. E. *J. Am. Chem. Soc.* **1921**, *43*, 254–268.

66. Elliott, E. L.; Hernández, G. A.; Linden, A.; Siegel, J. S. *Org. Biomol. Chem.* **2005**, *3*, 407–413.

67. (a) Munakata, M.; Wu, L. P.; Kuroda-Sowa, T.; Maekawa, M.; Suenaga, Y.; Ning, G. L.; Kojima, T. *J. Am. Chem. Soc.* **1998**, *120*, 8610–8618. (b) Liu, S. Q.; Kuroda-Sowa, T.; Konaka, H.; Suenaga, Y.; Maekawa, M.; Mizutani, T.; Ning, G. L.; Munakata, M. *Inorg. Chem.* **2005**, *44*, 1031–1036. (c) Munakata, M.; Wu, L. P.; Ning, G. L. *Coord. Chem. Rev.* **2000**, *198*, 171–203.

68. Zhong, J. C.; Munakata, M.; Kuroda-Sowa, T.; Maekawa, M.; Suenaga, Y.; Konaka, H. *Inorg. Chem.* **2000**, *40*, 3191–3199, and references therein.

69. Wang, Q.-M.; Mak, T. C. W. *Angew. Chem., Int. Ed.* **2001**, *40*, 1130–1133. Guo, G.-C.; Wang, Q.-G.; Zhou, G.-D.; Mak, T. C. W. *Chem. Commun.* **1998**, 339–340.

70. Fortin, D.; Drouin, M.; Harvey, P. D. *J. Am. Chem. Soc.* **1998**, *120*, 5351–5352.

71. (a) Chen, C.-L.; Su, C.-Y.; Cai, Y.-P.; Zhang, H.-X.; Xu, A.-W.; Kang, B.-S. *New J. Chem.* **2003**, *27*, 790–792. (b) Muthu, S.; Yip, J. H. K.; Vittal, J. J. *J. Chem. Soc., Dalton Trans.* **2002**, 4561–4568.

72. (a) Lin, P.; Henderson, R. A.; Harrington, R. W.; Clegg, W.; Wu, C.-D.; Wu, X.-T. *Inorg. Chem.* **2004**, *43*, 181–188. (b) Rao, C. N. R.; Ranganathan, A.; Pedireddi, V. R.; Raju, A. R. *Chem. Commun.* **2000**, 39–40. (c) Munakata, M.; Ning, G. L.; Suenaga, Y.; Kuroda-Sowa, T.; Maekawa, M.; Ohta, T. *Angew. Chem., Int. Ed.* **2000**, *39*, 4555–4557.

73. For example, see (a) Roesky, H. W.; Andruh, M. *Coord. Chem. Rev.* **2003**, *236*, 91–119. (b) Beatty, A. M. *Coord. Chem. Rev.* **2003**, *246*, 131–143.(c) Tiekink, E. R. T.; Vittal, J. J., Eds. *Frontiers in Crystal Engineering*, Wiley, Chickester, UK, **2006**.

74. Wang, R. H.; Hong, M. C.; Luo, J. H.; Jiang, F. L.; Han, L.; Lin, Z. Z.; Cao, R. *Inorg. Chim. Acta* **2004**, *357*, 103–114.

75. Robinson, F.; Zaworotko, M. J. *Chem. Commun.* **1995**, 2413–2414.

76. Yaghi, O. M.; Li, H. L. *J. Am. Chem. Soc.* **1996**, *118*, 295–296.

77. Noro, S.-I.; Miyasaka, H.; Kitagawa, S.; Wada, T.; Okubo, T.; Yamashita, M.; Mitani, T. *Inorg. Chem.* **2005**, *44*, 133–146.

78. Maji, T. K.; Konar, S.; Mostafa, G.; Zangrando, E.; Lu, T.-H.; Chaudhuri, N. R. *Dalton Trans.* **2003**, 171–175.

79. Adachi, K.; Kaizaki, S.; Yamada, K.; Kitagawa, S.; Kawata, S. *Chem. Lett.* **2004**, *33*, 648–649.

80. (a) Konaka, H.; Wu, L. P.; Munakata, M.; Kuroda-Sowa, T.; Maekawa, M.; Suenaga, Y. *Inorg. Chem.* **2003**, *42*, 1928–1934. (b) Seward, C.; Jia, W.-L.; Wang, R.-Y.; Enright, G. D.; Wang, S. *Angew. Chem., Int. Ed.* **2004**, *43*, 2933–2936. (c) Carlucci, L.; Ciani, G.; Porta, F.; Proserpio, D. M.; Santagostini, L. *Angew. Chem., Int. Ed.* **2002**, *41*, 1907–1911. (d) Min, K. S.; Suh, M. P. *J. Am. Chem. Soc.* **2000**, *122*, 6834–6840.

81. (a) Haftbaradaran, F.; Draper, N. D.; Leznoff, D. B.; Williams, V. E. *Dalton Trans.* **2003**, 2105–2106. (b) Munakata, M.; Wu, L. P.; Kuroda-Sowa, T.; Maekawa, M.; Suenaga, Y.; Ohta, T.; Konaka, H. *Inorg. Chem.* **2003**, *42*, 2553–2558.

82. (a) Park, K.-M.; Yoon, I.; Lee, Y. H.; Lee, S. S. *Inorg. Chim. Acta* **2003**, *343*, 33–40. (b) Pigge, F. C.; Burgard, M. D.; Rath, N. P. *Cryst. Growth Des.* **2003**, *3*, 331–337. (c) Raehm, L.; Mimassi, L.; Guyard-Duhayon, C.; Amouri, H.; Rager, M. N. *Inorg. Chem.* **2003**, *42*, 5654–5659. (d) Jung, O.-S.; Kim, Y. J.; Lee, Y.-A.; Park, K.-M.; Lee, S. S. *Inorg. Chem.* **2003**, *42*, 844–850.

83. Caradoc-Davies, P. L.; Hanton, L. R. *Chem. Commun.* **2001**, 1098–1099.

84. (a) Wang, R. H.; Xu, L. J.; Li, X. S.; Li, Y. M.; Shi, Q.; Zhou, Z. Y.; Hong, M. C.; Chan, A. S. C. *Eur. J. Inorg. Chem.* **2004**, 1595–1599. (b) Mamula, O.; von Zelewsky, A.; Bark, T.; Bernardinelli, G. *Angew. Chem., Int. Ed.* **1999**, *38*, 2945–2948.

85. (a) Plasseraud, L.; Maid, H.; Hample, F.; Saalfrank, R. W. *Chem. Eur. J.* **2001**, *7*, 4007–4011. (b) Cai, Y.-P.; Zhang, H.-X.; Xu, A.-W.; Su, C.-Y.; Chen, C.-L.; Liu, H.-Q.; Zhang, L.; Kang, B.-S. *J. Chem. Soc., Dalton Trans.* **2001**, 2429–2434.

86. Lü, X.-Q.; Qiao, Y.-Q.; He, J.-R.; Pan, M.; Kang, B.-S.; Su, C.-Y. *Cryst. Growth Des.* **2006**, *6*, 1910–1914.

87. Zhang, J.-P.; Horike, S.; Kitagawa, S. *Angew. Chem., Int. Ed.* **2007**, *46*, 889–892.

88. Sailaja, S.; Rajasekharan, M. V. *Inorg. Chem.* **2000**, *39*, 4586–4590.

89. (a) Li, Y.-H.; Su, C.-Y.; Goforth, A. M.; Shimizu, K. D.; Gray, K. D.; Smith M. D.; zur Loye, H.-C. *Chem. Commun.* **2003**, 1630–1632. (b) Feng, Y.; Guo, Y.; Yang, Y. O.; Liu, Z.; Liao, D.; Cheng, P.; Yan, S.; Jiang, Z. *Chem. Commun.* **2007**, 3643–3645.

90. Carlucci, L.; Ciani, G.; Macchi, P.; Proserpio, D. M.; Rizzato, S. *Chem. Eur. J.* **1999**, *5*, 237–243.

91. Ni, Z.; Vittal, J. J. *Cryst. Growth Des.* **2001**, *1*, 195–197.

92. Dobrzańska, L.; Raubenheimer, H. G.; Barbour, L. J. *Chem. Commun.* **2005**, 5050–5052.

93. (a) Zhang, X.-L.; Guo, C.-P.; Yang, Q.-Y.; Wang, W.; Liu, W.-S.; Kang, B.-S.; Su, C.-Y. *Chem. Commun.* **2007**, 4242–4244. (b) Zhang, X.-L.; Guo, C.-P.; Yang, Q.-Y.; Lu, T.-B.; Tong, Y.-X.; Su, C.-Y. *Chem. Mater.* **2007**, *19*, 4630–4632.

94. Tong, M.-L.; Chen, X.-M.; Ye, B.-H.; Li, L.-N. *Angew. Chem., Int. Ed.* **1999**, 2237–2240.

95. Banfi, S.; Carlucci, L.; Caruso, E.; Ciani, G.; Proserpio, D. M. *Cryst. Growth Des.* **2004**, *4*, 29–32.

96. Carlucci, L.; Ciani, G.; Proserpio, D. M.; Sironi, A. *Chem. Commun.* **1996**, 1393–1394.

97. Abrahams, B. F.; Batten, S. R.; Hoskins, B. F.; Robson, R. *Inorg. Chem.* **2003**, *42*, 2654–2664.

98. (a) Konno, T.; Yoshimura, T.; Aoki, K.; Okamoto, K.-I.; Hirotsu, M. *Angew. Chem., Int. Ed.* **2001**, *40*, 1765–1768. (b) Fitchett, C. M.; Steel, P. J. *New J. Chem.* **2000**, *24*, 945–947.

99. Melaiye, A.; Simons, R. S.; Milsted, A.; Pingitore, F.; Wesdemiotis, C.; Tessier, C. A.; Youngs, W. J. *J. Med. Chem.* **2004**, *47*, 973–977, and references therein.

100. Mäkinen, S. K.; Melcer, N. J.; Parvez, M.; Shimizu, G. K. H. *Chem. Eur. J.* **2001**, *7*, 5176–5182.

101. Yamada, S.; Ishida, T.; Nogami, T. *Dalton Trans.* **2004**, 898–903.

6

TUNING STRUCTURES AND PROPERTIES OF COORDINATION POLYMERS BY THE NONCOORDINATING BACKBONE OF BRIDGING LIGANDS

Miao Du

College of Chemistry and Life Science, Tianjin Normal University, Tianjin, People's Republic of China

Xian-He Bu

Department of Chemistry, Nankai University, Tianjin, People's Republic of China

6.1 INTRODUCTION

Currently, crystal engineering has enabled chemists to rationally design various functional crystalline solids to a large extent, and their well-defined lattice architectures can be prearranged by molecular building blocks with suitable structure-directing groups for intermolecular interactions such as the most familiar coordination and hydrogen-bonding contacts [1–3]. Coordination polymers, known also as metal–organic frameworks (MOFs), represent a rising subgroup of coordination compounds at a well-rounded level, which normally possess polymeric network structures extended by metal–ligand coordination interactions and chemical and physical characteristics desired for solid-state materials [4–6]. At this stage, the rational construction of structurally defined coordination frameworks using crystal engineering strategy seems to be a marvelous success, and a variety of typical examples for such crystalline materials with attractive network structures and

Design and Construction of Coordination Polymers, Edited by Mao-Chun Hong and Ling Chen

potential applications have been reported [7–10]. That is, according to the node-and-spacer concept or method, the arrangement of ligand modules (spacers) can be controlled effectively by the metal centers (nodes) due to the direction of the coordination bonds, and thus the topology and dimensionality of the final coordination networks may also be delicately designed and predicted [11–14]. Nevertheless, in practice, there are still many challenges to perfect projection and regulation of the detailed crystal packing of such materials, because structural control is often readily thwarted by the intricate and noncovalent nature of weak secondary interactions such as hydrogen bonding, aromatic stacking, and van der Waals forces [15–20] as well as synthetic conditions and methods [21–24]. Of further interest, prospective applications of coordination polymers as a new type of inorganic–organic hybrid material have been widely established and demonstrated in several areas, such as gas storage and separation, ion exchange, molecular magnetism, conductivity, luminescence, nonlinear optics, and heterogeneous catalysis [25–37]. Certainly, the useful properties of these crystalline materials are closely related to their well-defined coordination networks and further lattice packing, and some preliminary empirical strategies with regard to their structure–property relationship have been realized and stated [38–40]. It should also be emphasized that even a slight change in the building component in the assembling process may result in failure to achieve the desired supramolecular architecture, and further efforts are required for a full understanding of their subtle roles on the design and preparation of such tailormade molecular-based materials.

6.2 LIGAND DESIGN FOR COORDINATION POLYMERS

The rational construction and synthesis of coordination polymers rely critically on the deliberate design of organic bridging ligands with adjustable connectivity and binding tendency and/or judicious selection of inorganic metal ions with different coordination geometry and ability [41–44]. In this context, multidentate bridging ligands containing two or more nitrogen, sulfur, and/or oxygen donor groups with a distinct ability to bind to metal ions have been explored extensively, and a variety of coordination architectures have been produced using these ligands as building blocks [45–50]. The ligands are normally neutral or anionic in nature and can generally be classified into several distinct types, such as the representative dipyridyl, polycarboxylate, pyridinecarboxylate, and thioether/thiolate, which are considered as effective tectons for the construction of diverse extended coordination networks, due to their intrinsic binding discrepancy. In fact, a majority of coordination polymers have been obtained by utilizing the coordinative ability of pyridyl or carboxylate functional groups with transition-metal ions. For example, 4,4′-dipyridyl is a simple neutral ligand that has been used primarily as a rodlike exobidentate building block for coordination polymer assemblies [51–55]. The coordination fashion for this type of ligand is ordinary and reliable, and the resulting network structures depend heavily not only on the coordination geometry of the metal ions but also on the counter anions that are necessarily involved in the lattice structures for entire charge balance. On the other hand, polycarboxylic acids display completely different connectivity than that of the dipyridyl species, due to the versatile coordination capability of carboxylate and to the sensitivity of such building blocks to acidity [56–64]. As a consequence, they usually display different degrees of deprotonation of the carboxyl groups and variable coordination modes upon metalation under appropriate conditions. Notably, polynuclear metal clusters bridged by carboxylate groups are usually generated, such as the most familiar $[M_2(CO_2R)_4]$ type of paddlewheel of metal–carboxylate, and these secondary building units (SBUs) are regarded as very useful and reliable tools to construct robust and porous metal–organic frameworks. In a word, bridging ligands bearing different functional groups and coordination ability play key roles in constructing the overall coordination networks with specific properties.

As a matter of fact, beyond the coordination groups of ligands, the backbones of the bridging building blocks are also important, and this has been realized in recent years [65–68]. Lately, the flexible counterparts of rigid building blocks have received more and more attention in the construction of extended coordination frameworks [69–76] because such ligands have variable flexibility and can adopt adjustable conformations and coordination modes according to the steric requirements of different metal ions compared with rigid ligands. Such flexible ligands can be obtained by introducing some noncoordinating spacers between the terminal functional groups, which may show distinct coordination features during the self-assembly courses. Generally, the long and flexible building blocks will lead to the formation of large available voids and/or channels and thus produce microporous coordination frameworks or unusual interpenetrating supramolecular architectures [77–80].

One of the most outstanding examples of this is 1,2-bis(4-pyridyl)ethane (bpe), which can be regarded as an analog of 4,4′-dipyridyl by inserting two methylene groups between the pyridyl rings. A variety of coordination networks of ingenious design have been achieved on the basis of this ligand, such as one-dimensional (1D) crooked chains [81–85], two-dimensional (2D) layers [86–91] and three-dimensional (3D) frameworks [92–95], among which some fantastic supramolecular architectures exhibit quite unusual and puzzling structural features. For instance, in the 1D polymeric structure of $[Zn(OAc)_2(bpe)]_n \cdot 2nH_2O$ [81], the carbonyl oxygen atoms from the $[Zn(OAc)_2(bpe)]_n$ chain motifs interconnect with four lattice water molecules to generate 24-membered hydrogen-bonded rings, which are further fused together through O–H $\cdots$ O hydrogen bonds among the four water molecules to form the cyclic water tetramer. As a result, a hydrogen-bonded ribbonlike polymeric array comprising fused alternating 24- and 8-membered hydrogen-bonded rings is produced. Interestingly, one of the bpe ligands passes through the center of the larger ring to form an unexpected single self-penetrating 3D network with polyrotaxane-like association. As for 2D coordination species, the layered supramolecular pattern of the complex $[Co_5(bpe)_9(H_2O)_8(SO_4)_4](SO_4) \cdot 14H_2O$ consists of five homochiral parallel 1D linear coordination polymers connected by penta-metal helical motifs [91]. Furthermore, each layer is entangled with four others to result in a quintuple-catenated array. Another noteworthy example is the 3D nanoporous coordination polymer $[Cu(bpe)_2(SO_4)] \cdot 5H_2O$, which shows a self-catenated architecture with the unprecedented $(6^4.8^2)(6.8^5)$ network topology [94]. The most recent work by Maji and co-workers [95] has presented a unique nickel(II) coordination framework with bpe and dicyanamide, in which the two primitive cubic topological networks are entangled and encompass smaller pores in rectangular and hexagonal shapes within the infinite lattice. Peculiarly, such an interpenetrated framework can expel gases and shift with respect to each other to change its pore dimension, providing highly selective and increased strong binding of ions and gases. This result sufficiently illustrates the "pumping" and "flexing" benefits of entangled metal–organic frameworks based on such flexible building blocks.

In this regard it is notable that the prediction and manipulation of the overall lattice structures of these coordination solids are somewhat difficult, due to the variable configuration and adaptable coordination tendency of the flexible backbones compared with the usual rigid ligands. Even now, the coordination networks can be well regulated by these flexible building blocks, with different spacers in some cases. For example [96], reactions of copper(II) perchlorate with three analog disulfoxide ligands [1,2-bis(phenylsulfinyl)ethane, 1,3-bis(phenylsulfinyl)propane, and 1,4-bis(phenylsulfinyl)butane (bpse, bpsp, and bpsb; see Fig. 6.1a for a schematic view)] produce the corresponding coordination polymers $[Cu(bpse)_2(ClO_4)_2]_n$, $\{[Cu(bpsp)_3](ClO_4)_2\}_n$, and $\{[Cu(bpsb)_3](ClO_4)_2\}_n$. In each case, the disulfoxide ligands bridge the adjacent two copper(II) centers to result in the final macrometallacyclic array. However, it is interesting that the subtle altering of the spacers between the functional groups from $-(CH_2)_2-$, to $-(CH_2)_3-$, and then to $-(CH_2)_4-$, can lead to regular modulation of the coordination networks from a 2D layer with 28-membered rhombic units, to a 2D layer with larger 32-membered square units, to a 3D coordination framework with the largest, 36-membered, macrocycles and available channels (see Fig. 6.1b). Of further importance, the controllable cavity sizes of the coordination networks also allow the accommodation of different numbers of perchlorate counter anions. Comparable cases have been observed for copper(II) perchlorate complexes with other disulfoxide derivative ligands [97].

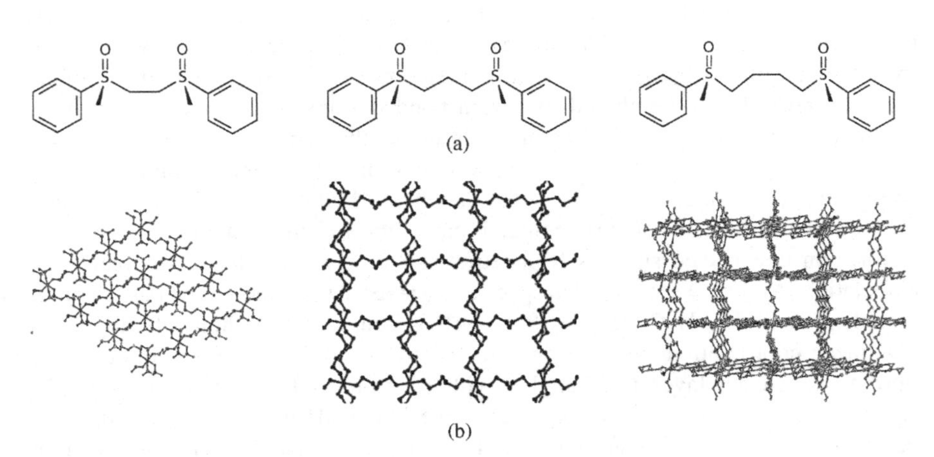

FIGURE 6.1 (a) Schematic view of the ligands bpse, bpsp, and bpsb; (b) crystal structures of their corresponding copper(II) perchlorate complexes.

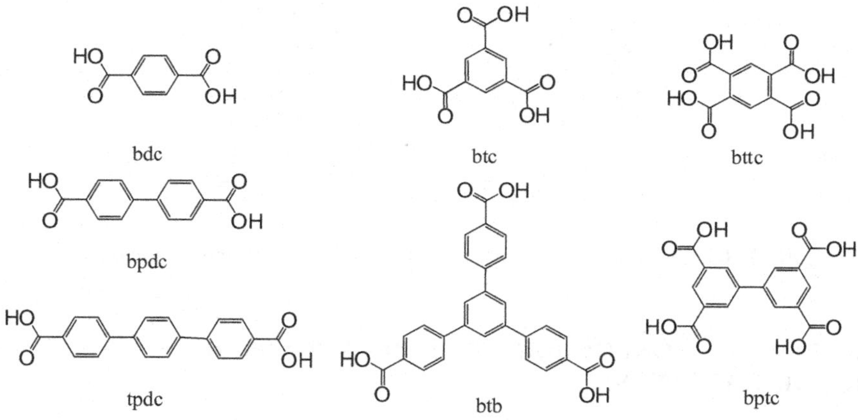

bdc btc bttc

bpdc

tpdc btb bptc

FIGURE 6.2 Chemical structures of representative polycarboxyl ligands with different spacers.

Otherwise, the families of aromatic polycarboxylate compounds, which are based on the backbone of *n*-carboxylbenzene and changed by replacement of the phenyl group with a longer and bulky linkage (see Fig. 6.2), are used purposefully to design and prepare splendidly functional coordination polymers, especially in the field of zeolitelike porous materials for gas adsorption, storage, and separation [7,98–106]. For example, appropriate synthetic strategy toward the target zinc(II) coordination frameworks with linear and rigid dicarboxylate linkers, such as 1,4-benzenedicarboxylate (for IRMOF-5), 4,4′-biphenyldicarboxylate (for IRMOF-10), and 4,4′-terphenyldicarboxylate (for IRMOF-16), has allowed the formation of similar crystalline networks with CaB_6 topology. Notably, the latter two ligands display a tendency to produce doubly interpenetrating networks during self-assembly [7], probably due to their long molecular backbones. On the other hand, owing to the differing nature of the ligand moieties that constitute the pore structures, the free voids of the coordination frameworks vary incrementally from 79.2 to 91.1% of the unit cell volumes. In other words, the pore sizes of this series of metal–organic frameworks can be adjusted effectively by selecting suitable dicarboxylate building blocks with different spacer lengths. Assembly of the trigonal linker 1,3,5-benzenetribenzoate with zinc(II) yields a very effective hydrogen storage material, MOF-177, which can store 7.5 wt% hydrogen with a volumetric capacity of $32\,g\,L^{-1}$ at 77 K and 70 bar [101]. Considering its well-defined network structure and significant H_2 uptake capability, it is predicted that this 3D framework material may serve as an excellent benchmark adsorber.

At this stage it is easily understood that such evident reconstruction of the building blocks will greatly change the ligand backbones and thus their binding features, and further, the resulting network structures and/or potential properties of the corresponding coordination frameworks. The related factors have been noted extensively in the literature and are not covered comprehensively in this chapter.

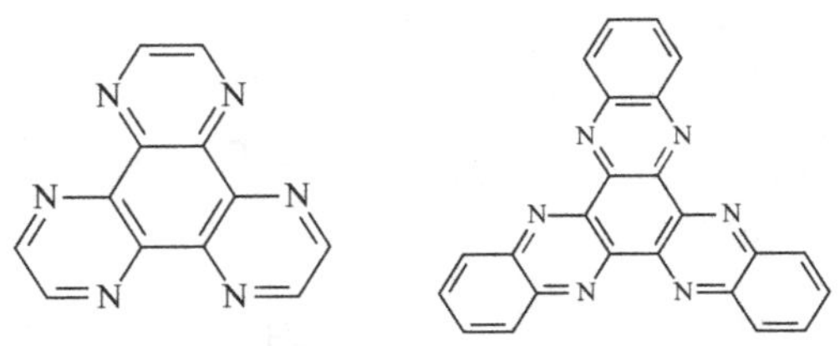

FIGURE 6.3 Chemical structures of the ligands (a) hat and (b) phen-hat.

6.3 ROLE OF NONCOORDINATING BACKBONES OF BRIDGING LIGANDS

In practice, besides the aforementioned strategies for ligand design via the significant alteration of functional groups and ligand backbones, some subtle modification of the ligands by introducing noncoordinating groups into the skeletons can also display an amazing modulating effect on the structures and properties of coordination polymers. For example, self-assemblies of silver(I) salts with 1,4,5,8,9,12-hexaazatriphenylene (hat; see Fig. 6.3a) afford isostructural 3D coordination networks, which are robust and chiral (space group $P4_32_12$). In each case, both the silver ions and ligands serve as three-connected nodes to result in a three-connecting network with well-known (10,3)-a topology (srs) [107]. Also, the ligand diquinoxalino[2,3-*a*:2′,3′-*c*]phenazine (phen-hat; see Fig. 6.3b) shows very similar backbones and coordination sites to hat but has extended aromatic rings (noncoordinating groups). When phen-hat was used, the resulting silver(I) coordination polymers also display the (10,3)-a network structures because this ligand provides similar three-fold bidentate metal-binding sites [108]. Remarkably, such silver complexes crystallize in space group *R*-3, consisting of two enantiomeric and interpenetrating 3D srs networks (class IIa for interpenetration) that are topologically equivalent but differ in the handedness of the threefold helices. The network structures of two types of such coordination frameworks are depicted in Figure 6.4. In the latter case, the larger available channels of each single 3D network may result from the bulky ligand used (a large delocalized system compared to hat), which leads to the formation of an interpenetrating 3D racemate. From this viewpoint, a minor change in the ligand backbone by adding a noncoordinating phenyl group may also play a very important part in structural assemblies of coordination frameworks. Here we present selected examples of this topic to elucidate the importance of noncoordinating groups in the ligand design of coordination polymers, with the aim of providing new insights into the crystal engineering of such crystalline materials.

6.3.1 Terminal Effect of Acyclic Ligands

Dithioether derivatives with acyclic backbones represent an important class of flexible organic ligands with certain binding tendencies in coordination polymer

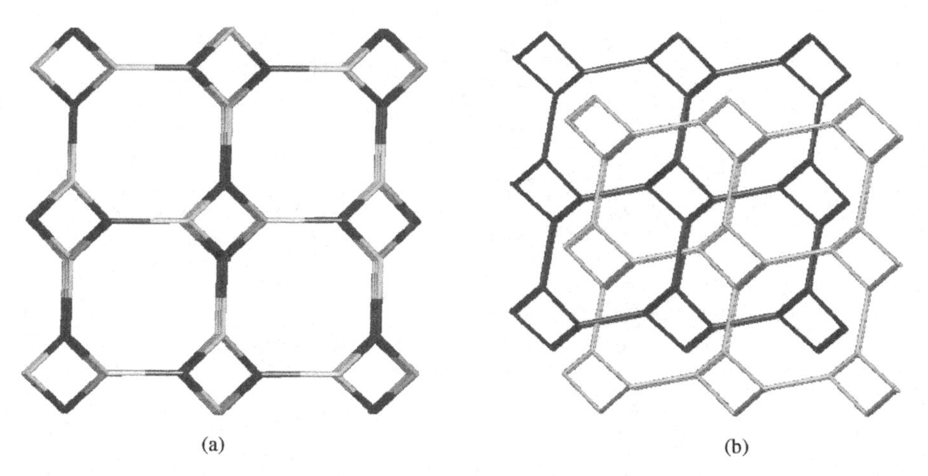

FIGURE 6.4 Schematic view of the network topology of silver(I) coordination polymers with (a) hat and (b) phen-hat ligands.

chemistry. So far, various coordination structures, including discrete molecules and infinite networks with different dimensionalities, have been prepared using dithioether ligands and metal ions. They are considered as soft bases and favor binding to soft acids such as silver(I) and copper(I) ions. These ligands can be synthesized and modified easily by changing the spacers between the sulfur groups or terminals attached to them [109]. In this section, the role of the noncoordinating terminal groups of these ligands in the structural direction and regulation of silver(I) perchlorate coordination polymers are illustrated.

Bis(methylthio)methane (L1a; see Fig. 6.5) is the simplest dithioether ligand, with limited configuration freedom and steric hindrance. Self-assembly of silver(I) perchlorate with this ligand affords a 2D layered coordination polymer $[Ag_2(L1a)_2(ClO_4)_2]_n$ consisting of 14-membered rings as the repeating units [110], in which each silver center serves as a 3-connected node by coordinating to three ligands, and each ligand unusually bridges three silver ions to constitute a (6,3) network topology. When the terminal groups are replaced by $-CH_2CH_3$ [L1b = bis(ethylthio)methane], the corresponding silver complex, $[Ag(L1b)_{1.5}(ClO_4)]_n$, also has a 2D layer structure of (6,3) topology [111]. However, in this structure, each ligand bridges two silver ions through the sulfur atoms, and each 3-coordinated silver ion is coordinated with three neighboring ligands, to result in the centrosymmetric hexagonal 24-membered macrometallacyclic unit. When a *tert*-butyl group is used in this series of ligands (see Fig. 6.5), the resulting coordination polymer, $[Ag(L1c)(ClO_4)]_n$ [L1c = bis(*tert*-butylthio)methane] [112], exhibits a 1D zigzag chain array in which the silver ions are connected by the two-connecting ligands. Furthermore, the bulky benzyl group of the ligand bis(benzylthio)methane (L1e) leads to the formation of a dinuclear silver complex with ligand-supported silver–silver weak interaction [111], whereas the crystal structure determination of complex $[Ag_2(L1d)_3(ClO_4)_2]_n$ [L1d = bis(phenylthio)methane] reveals the

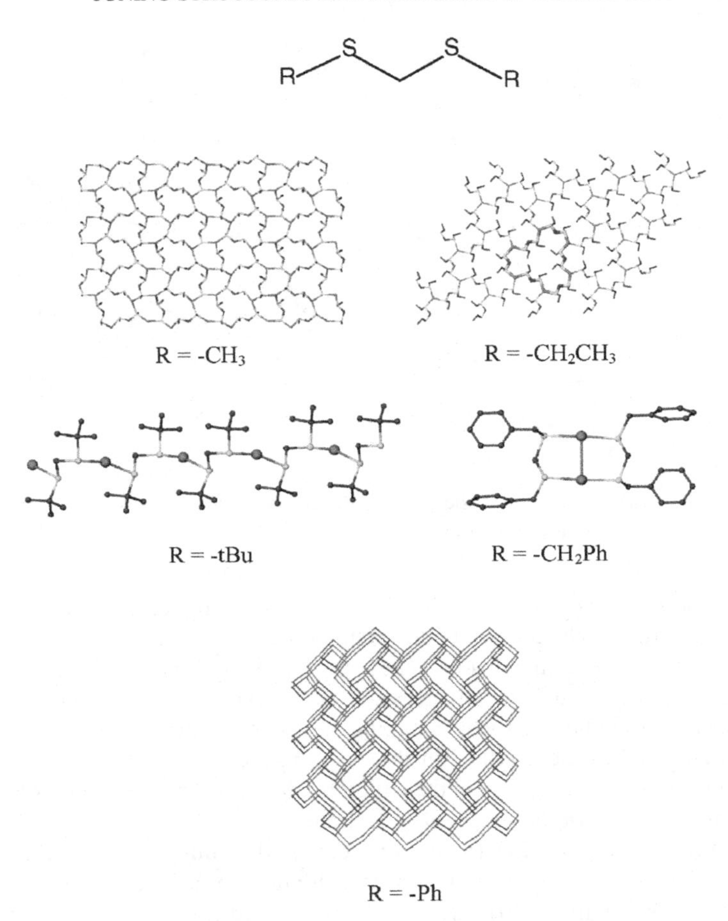

FIGURE 6.5 Schematic view of dithioether methane ligands (L1n) with different terminal groups (R = $-CH_3$, $-CH_2CH_3$, –*t*Bu, $-CH_2Ph$, and –Ph) and crystal structures of their corresponding silver(I) perchlorate complexes.

formation of a 3D noninterpenetrated coordination network with (10,3)-a topology [113]. Thus, from the discussion above it is clear to see that the terminal groups of dithioether ligands also have a strong influence on the coordination architectures of their silver complexes (from discrete dinuclear motif, to infinite 1D, 2D, and 3D networks), as shown in Figure 6.5.

In addition, by employing longer and more flexible spacers between the functional sulfur groups of the dithioether ligands, a variety of coordination structures have been isolated successfully and determined structurally in the solid state, which are also significantly relevant to the noncoordinating terminal groups of the ligands. For example, bis(ethylthio)ethane (L2b) possesses a $-CH_2CH_2-$ spacer between the coordination sulfur groups, which adopts more flexible conformation for structural adjustment during coordination-driven assembly with appropriate metal ions than that

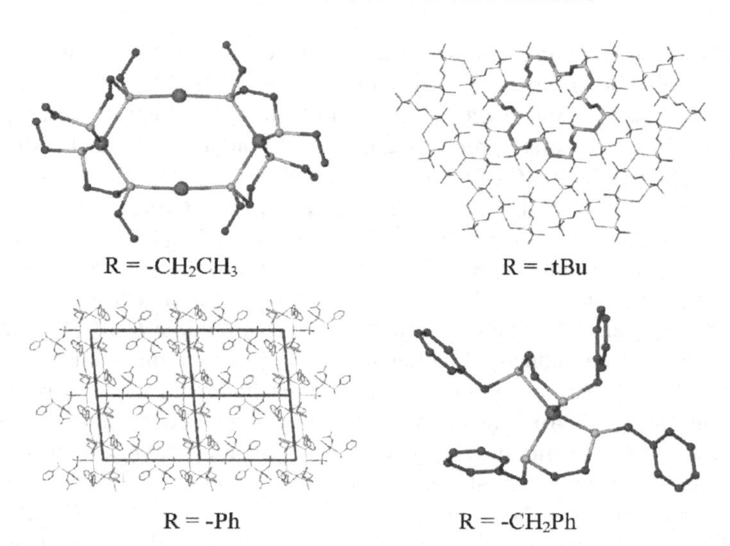

FIGURE 6.6 Crystal structures of the silver(I) perchlorate complexes of dithioether ethane ligands with different terminal groups (L2n, R = $-CH_2CH_3$, –*t*Bu, –Ph, and $-CH_2Ph$).

of bis(ethylthio)methane (L1b). In this case [111], the corresponding silver(I) perchlorate coordination species $[Ag_4(L2b)_8(ClO_4)_4]$ displays an unexpected discrete tetranuclear motif in which the dithioether ligands take both chelating and bridging binding fashions with the tetrahedral and linear silver ions (see Fig. 6.6). When *tert*-butyl, with its larger interspatial hindrance, takes the place of $-CH_2CH_3$ [L2c = bis (*tert*-butylthio)ethane], a 2D layered coordination architecture is produced [112] in which the centrosymmetric snowflake-like 30-membered macrometallacyclic units share their edges and vertexes (see Fig. 6.6). Unexpectedly, a 2D coordination network $[Ag_8(L2d)_5(ClO_4)_8]_n$ is generated on the basis of the chelating/bridging ligand bis(phenylthio)ethane (L2d), μ-perchlorate anion, and 4-connected silver nodes [114], whereas the reaction of ligand bis(benzylthio)ethane (L2e) with silver perchlorate gives a crystalline product $[Ag(L2e)_2(ClO_4)]$ [111] with mononuclear entity in which the tetrahedral silver center is coordinated by two chelated ligands (see Fig. 6.6).

Similar cases have been observed in the silver perchlorate complexes, with dithioether propane ligands bearing different terminals (L3n), which have a very small discrepancy about the spacer with that of the dithioether ethane family. As a result, for bis(ethylthio)propane (L3b) [111], it forms a 1D tapelike coordination array with silver perchlorate in which the trigonal silver centers are interconnected by the ligands as tri-connectors, whereas in the binuclear motif $[Ag(L3c)(ClO_4)]_2$ [L3c = bis (*tert*-butylthio)propane] [112], each silver center is linearly coordinated by sulfur atoms from two symmetrically related dithioether ligands and the intramolecular silver–silver distance is as long as 4.818 Å. This separation indicates the absence of significant argentophilic interaction in this structure, due to the bulky interval spacer between the sulfur atoms compared with the similar dinuclear silver complex with bis (benzylthio)methane (L1e) discussed above. Remarkably, by replacement of the terminal group with a benzyl group, reaction of the ligand bis(benzylthio)propane (L3e) with silver perchlorate produces a novel 3D noninterpenetrated coordination framework $\{[Ag_2(L3e)_3(ClO_4)](ClO_4)\}_n$ [111] which is constructed from cationic 2D (6,3) layers $[Ag_2(L3e)_3]_n$ that are bridged further by perchlorate anions. Finally, in the net structure of $\{[Ag_2(L3d)_4](ClO_4)_2(acetone)_2\}_n$ [L3d = bis(phenylthio) propane] [115], each independent silver ion is bound to the adjacent sulfur atoms in a tetrahedral manner, which leads to the formation of a 2D cationic sheet.

While the spacer is lengthened further, the type of flexible R-$S(CH_2)_4S$-R (L4n) ligands with different noncoordinating terminals is taken into consideration for fabricating the interesting coordination structures. In this case, self-assembly of bis(ethylthio)butane (L4b) with silver(I) [111] results in a corrugated 2D coordination polymer, $\{[Ag(L4b)](ClO_4)\}_n$, in which the layered network consists of left- and right-handed single helical chains. When *tert*-butyl and benzyl groups are utilized in place of ethyl, the resulting silver complexes $[Ag(L4c)_3(ClO_4)]_n$ [L4c = bis(*tert*-butylthio)butane] [112] and $[Ag(L4e)_3(ClO_4)]_n$ [L4e = bis(benzylthio) butane] [111] have similar extended structures in which a pair of silver ions are linked by two bridging ligands to form a binuclear macrometallacyclic unit, and such adjacent subunits are further connected by the ligands in single-bridging fashion to give an alternate single- and double-bridging chain array. Interestingly, reactions of the ligand bis(phenylthio)butane (L4d) with silver perchlorate under different conditions (varying the solvents and metal-to-ligand ratios) lead to the formation of three coordination polymers with different compositions and network structures: $[Ag_2(L4d)_3(ClO_4)_2]_n$, $[Ag_2(L4d)_3(ClO_4)_2(CH_3OH)]_n$, and $\{[Ag(L4d)_2](ClO_4)\}_n$ [116]. In all three complexes, each silver center has tetrahedral coordination geometry, and the final 2D coordination networks consist of large fused macrometallacyclic ring systems. Notably, the "hexagonal" 42-membered rings of $[Ag_6(L4d)_6]$ units are observed in the former two structures, which could be considered as unique examples of self-sustaining noninterpenetrated frameworks formed with flexible ligands. However, the 2D layer in the last complex consists of rectangular $[Ag_4(L4d)_4]$ 28-membered macrometallacycles as the repeating units, in which the perchlorate anions occupy the network voids to stabilize this lattice. This result evidently reveals the structural flexibility of the ligand L4d with $-(CH_2)_4-$ backbone, which allows the ligand to rearrange to minimize the steric interaction upon metal complexation under different conditions, and thus results in the product diversity of its silver coordination polymers.

To examine further the influence of steric hindrance or conformation freedom of such ligands on the resulting coordination architectures, two dithioether pentane compounds with the highest flexibility have been explored as the variable building blocks to construct silver coordination motifs. In this case, one binuclear species is obtained when bis(*tert*-butylthio)pentane (L5c) is used [112], whereas the structure of $\{[Ag(L5d)_2](ClO_4)\}_n$ [L5d = bis(phenylthio)pentane] presents an infinite 2D pattern [117] in which each tetrahedrally coordinated silver ion is ligated via four sulfur donors coming from distinct L5e bridging ligands (as two-connectors).

In conclusion, a variety of coordination architectures can easily be constructed by self-assembly of silver perchlorate and several series of dithioether ligands with different spacers and terminal groups that are closely related as flexible building blocks. In these structures, the silver ions may take linear, trigonal, and tetrahedral coordination geometries; the sulfur donors in the ligands can adopt monodentate or bridging fashion, although the former is probably prior to the latter, in order to reduce the local steric hindrance; and the perchlorate anion is normally noncoordinating (only for charge balance) and may also combine with the silver ion via terminal coordination or even as the bridging ligand. In this connection, the noncoordinating groups of such acyclic ligands display a considerable terminal effect on the structural assemblies of metal–organic coordination architectures.

6.3.2 Substituent Effect of Aromatic Ligands

To date, Yaghi and his co-workers have systematically investigated the influence of a variety of ditopic carboxylate connectors (see Fig. 6.7) derived from the closely related underlying skeleton, formulated as $HO_2C(C_6H_4)_nCO_2H$, on the assemblies of extended network structures of isoreticular metal–organic framework (IRMOF-n). The results reveal that this series of IRMOFs display similar 3D cubic porous networks but with different functionalities and dimensionalities [7,118–124]. Factually, by utilization of each of the linkers R_n-BDC ($n = 1$ to 7) as well as BPDC, HPDC, and PDC (see Fig. 6.7), a series of solid-state crystalline materials have been isolated and display the uniform network topology of CaB_6 adapted by the prototype IRMOF-1. In these framework structures, the oxide-centered Zn_4O tetrahedra are edge-bridged by six carboxylate groups to constitute octahedron-shaped SBUs that reticulate into homogeneous periodic pores. However, these IRMOFs differ in the nature of functional groups decorating the pores and in the metrics of their pore

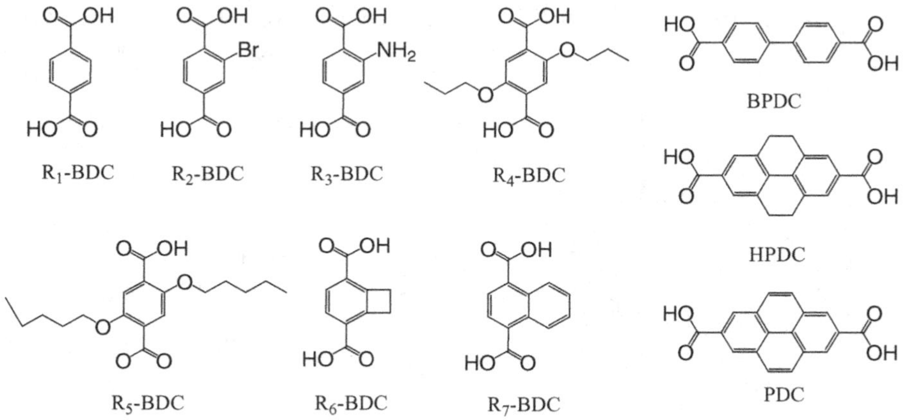

FIGURE 6.7 Chemical structures of two representative types of dicarboxyl ligands with different substituent groups.

structures. In IRMOF-2 through IRMOF-7, BDC linkers with bromo, amino, *n*-propoxy, *n*-pentoxy, cyclobutyl, and fused benzene functional substituent groups reticulate into the desired network structures with porosity wherein these groups point into the voids. On the other hand, pore expansion is also within the scope of this MOF chemistry, and noninterpenetrated structures of IRMOF-10, IRMOF-12, and IRMOF-14 involving BPDC, HPDC, and PDC have been achieved successfully under more dilute conditions. Calculation of the percent of free volume in these 10 crystalline IRMOFs shows that it varies in small increments in the range 55.8% in IRMOF-5 to 87.0% in IRMOF-10 [7]. The crystal densities (in the absence of guests) computed for these materials also vary in small increments between 1.00 g cm^{-3} for IRMOF-5 and 0.33 g cm^{-3} for IRMOF-10. As expected, the porosity of these microporous framework materials will be different due to these discrepancies of crystals, which result from the substituent groups of aromatic dicarboxylate connectors. Remarkably, the gas sorption isotherm measured for IRMOF-6 shows that in the absence of guests, its rigid framework can maintain the porosity, which exhibits a high capacity for methane storage (240 cm^3 at standard temperature and pressure per gram at 36 atm and ambient temperature).

As one type of the most elementary aromatic heterocyclic compounds, pyrazine (pyz) and its derivatives normally play a bridging role during metal-directed coordination assemblies. Nevertheless, a variety of extraordinary framework structures have been constructed from diverse R-substituted pyrazine building blocks (see Fig. 6.8) and metal centers, with the involvement of different counterions, such as dicyanamide [dca, $N(CN)_2^-$], tricyanomethanide [tcm, $C(CN)_3^-$], and $CF_3CO_2^-$.

For the sake of investigating the effect of steric hindrance on R-pyrazine-based coordination architectures, the relevant ligands L6a (pyrazine), L6b (methylpyrazine), and L6f (tetramethylpyrazine), have been explored to assemble with manganese(II) ion in the presence of dca anion [125–127]. Self-assembly of L6a with

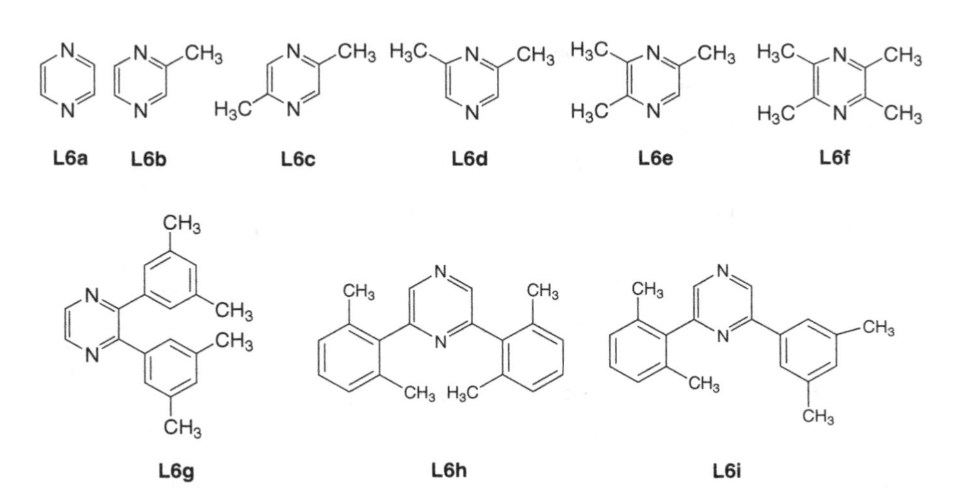

FIGURE 6.8 Chemical structures of R-pyrazine ligands with different substituent groups.

manganese(II) and dca results in a 3D coordination framework [Mn(dca)$_2$(L6a)]$_n$ [125], which has a two-fold interpenetrated lattice with classical α-Po network topology. In this structure, the bidentate dca ligands bridge the metal centers to form square-grid-like Mn(dca)$_2$ sheets, and these sheets are extended further by the pyrazine bridges to give an overall 3D cubiclike network. The large space within a single network allows interpenetration of the second net. It should be worthy of mention here that this compound shows 3D antiferromagnetic order occurring below 2.7 K and displays an interesting phase transformation that induces simultaneous twinning of the crystals. At room temperature the structure is orthorhombic *Pnma* and contains dynamic disorder of the dca ligands. However, on cooling (e.g., to 123 K), this species crystallizes in monoclinic $P2_1/n$ with the dca ligands ordered, and pseudomerohedral twinning is observed since domains within each crystal order differently. In addition, the complexes {[Mn(dca)$_2$(L6b)$_2$][Mn(dca)$_2$(L6b)]$_2$·2MeCN}$_n$ and [Mn$_2$(dca)$_3$(NO$_3$)(L6b)$_2$]$_n$ have been isolated as an intimate mixture from the same reaction in acicular and octahedral shapes, respectively [126]. The molecular framework of the acicular crystals consists of two structural motifs: the 1D polymeric chain, in which the Mn^{II} ions are connected by double $\mu_{1,5}$-dca bridges with *trans*-coordinated monodentate L6b ligands, and the 2D (4,4) sheet, in which similar 1D chains are interconnected by bridging L6b ligands. As for the other crystalline product, it shows the twofold interpenetrating (4,6)-connected 3D network, with $\mu_{1,5}$-bridging dca anions, bridging L6b ligands, and chelating nitrate ions. Notably, when the four hydrogen atoms of the pyrazinyl ring are totally replaced by methyl groups (L6f), only a 1D polymeric chain coordination polymer is generated [127].

When the aromatic building components are kept intact, and the silver ion and tcm anion are considered instead of manganese(II) and dca in the examples above, a series of distinct metal–organic frameworks are obtained [128,129]. The topology of [Ag(tcm)(L6a)]$_n$ is essentially singular, and the Ag(tcm) component in this case forms a planar hexagonal grid. Furthermore, each such sheet is connected on both sides to neighboring sheets by pyrazine bridging ligands whereby the silver center becomes five-coordinate and trigonal bipyramidal, furnishing a twofold interpenetrated (3,5)-connected 3D network [128]. However, the intervention of methyl in the aromatic ring prevents extension of the overall network and affords a 1D ladder-like motif [Ag(tcm)(L6b)$_{3/2}$]$_n$ [129]. Dissimilarly, [Ag(tcm)(L6f)$_{1/2}$]$_n$ has a 2D layer structure composed of 1D Ag(tcm) “tubes” that are linked by the bridging L6f ligands.

More specifically and systemically, the crystal engineering and coordination chemistry of Ag(CF$_3$CO$_2$) salt with R-pyrazine have been studied carefully, producing a considerable variety of typical coordination structures [130–133]. Taking L6a to L6f as building blocks and rodlike linkers, the silver(I) centers are similarly bridged into 2D lamellar structures, except for the subtle discrepancy in the coordination tendencies of the $CF_3CO_2^-$ anion with a substantial contribution to the construction of carboxylate-bridged silver–silver supramolecular synthon [130,131]. As for the remarkable gigantic ligand 2,6-bis(3′,5′-dimethylphenyl)pyrazine (L6g), the coordination chemistry is somewhat complicated and four different crystalline products are produced: two discrete binuclear species, [Ag$_2$(CF$_3$CO$_2$)$_2$(L6g)$_4$] and [Ag$_2$(CF$_3$CO$_2$)$_2$(L6g)$_2$] [132], as well as two 1D chain coordination motifs, {[Ag(CF$_3$CO$_2$)(L6g)]·(CH$_3$COCH$_3$)}$_n$

and $\{[Ag_3(CF_3CO_2)_3(L6g)_3]\cdot(CH_3COCH_3)_2\}_n$ [133]. The minor discrepancy of these crystalline structures lies primarily in the various coordination geometries of the silver centers (square planar vs. trigonal). Correspondingly, the discrepancy between the two 1D polymeric chains comes from whether the CF_3CO_2 anion is involved in metal coordination. By replacement of the aryl group with 2′,6′-dimethylphenyl, a distinct 1D polymeric chain $[Ag_2(\mu_2\text{-}CF_3CO_2)(\mu_3\text{-}CF_3CO_2)(L6h)]_n$ [L6h = 2,6-bis (2′,6′-dimethylphenyl)pyrazine] is afforded, profiting from the stacking forces of the bridging ligands [132]. Significantly, with the existence of unsymmetrical groups toward the pyrazinyl ring, the different binding effect of 2-bis(2′,6′-dimethylphenyl)-6-(3′,5′-dimethyl-phenyl) pyrazine (L6i) and μ_2-/μ_3-CF_3CO_2 is in favor of the formation of a linear coordination array, $\{[Ag_3(\mu_2\text{-}CF_3CO_2)(\mu_3\text{-}CF_3CO_2)_2(L6i)]\cdot(toluene)\}_n$ [132].

Currently, the extension of metal centers by coordinating with two or more types of organic ligands within the given network structures has also been a well-employed strategy to design and synthesize functional metal–organic frameworks. These mixed-ligand coordination polymers are normally built up from pyridyl- and carboxylate-containing ligands that have distinct coordination ability. In this context, very recently, Turner et al. have reported two copper(II) coordination polymers with acetylenedicarboxylate and pyridine or 3,5-dimethylpyridine co-ligands [134]. In each complex, the copper(II) ion shows a square-pyramidal configuration that is provided by two *trans*-coordinated pyridyl ligands and two *trans*-carboxylate groups (monodentate) at the equatorial plane as well as the apical methanol solvent. As a result, the acetylenedicarboxylate ligands act as the dianionic linear connectors, linking the copper(II) centers to form similar 1D polymeric chain arrays, although there are subtle differences in the geometry of the acetylenedicarboxylate ligands and in the relative arrangement of the pyridyl-based co-ligands. Analysis of the crystal packing reveals that the polymeric chains in the former case (with pyridine co-ligand) are extended to a 2D sheet with (4,4) topology via hydrogen-bonding interactions between the methanol ligand and the noncoordinating oxygen atom of carboxylate. However, this topology is not possible in the latter structure (with 3,5-dimethylpyridine co-ligand), due to the extra steric bulk of the methyl groups. As a consequence, a 3D hydrogen-bonded network with $CdSO_4$ (cds) topology (each copper atom acts as a square-planar four-connecting node) is formed in which the lattice methanol molecules are also incorporated into the hydrogen-bonding bridges. This result clearly demonstrates that the steric bulk of the pyridine-based co-ligands may significantly affect the overall structural assemblies of the supramolecular architectures.

Numerous inorganic–organic hybrids on the basis of typical phenyldicarboxylate building blocks incorporating auxiliary co-ligands have also been reported, some of which show intriguing structural motifs and/or potential applications [135–138]. Meanwhile, their analogous compounds possessing coordination-unfavored substituents, such as methyl ($-CH_3$), nitro ($-NO_2$), and hydroxyl ($-OH$) groups, have seldom been explored. To survey the influence of the substituents of isophthalic acid in the direction of coordination crystalline solids, we have prepared three series of cobalt (II), copper(II), and cadmium(II) mixed-ligand complexes with 4-amino-3,5-bis (3-pyridyl)-1,2,4-triazole (3-bpt) and isophthalic acid (H_2ip) or its derivatives,

5-nitroisophthalic acid (–NO_2, H_2nip) and trimesic acid (–COOH, H_3tma) [139]. As a matter of fact, the three polycarboxyl compounds are chemically pertinent and differ only in the 5-position substituents, which may show different geometrical features and chemical functions. For example, the nitro group is seldom engaged in coordination with familiar metal ions; however, as a strong electron-withdrawing group that will impose significantly on the electronic density of the entire ligand, it can not behave only as a reliable hydrogen-bonding acceptor but can also show a spatial effect. On the other hand, it is interesting that the five-site -COOH group of trimesic acid is not involved in metal binding in the coordination polymers presented in this work, but instead, forms head-to-tail hydorgen-bonding dimers to further extend the coordination arrays. This research undoubtedly discovers the substituent effect of R-isophthalate on regulating the mixed-ligand coordination frameworks as well as the overall supramolecular lattices directed by noncovalent contacts, by virtue of the versatility of the dipyridyl co-ligand, which has the potential to show three typical conformations and diverse coordination modes under appropriate surroundings. For example, complex $\{[Cd(3\text{-bpt})(ip)(H_2O)]\cdot(H_2O)_{1.5}\}_n$ displays a 1D molecular ladder coordination array in which a pair of Cd^{II} ions are linked by two 3-bpt ligands with unusual cisoid-II configuration and μ-$N_{pyridyl}$, $N_{triazole}$ binding fashion to generate a dimeric subunit, and such building units are propagated further through ip bridges (see Fig. 6.9a). The 1D arrays are further connected via hydrogen bonds between an amine of 3-bpt and carboxylate oxygen atoms to form a 2D network. $\{[Cd_2(3\text{-bpt})_2(nip)_2(H_2O)_4]\cdot(H_2O)_2\}_n$ represents a distinct 2D coordination layer in which the nip anions connect the Cd^{II} centers to generate crooked 1D chains and the transoid 3-bpt linkers further extend the $[Cd(nip)]_n$ chains to provide a 2D (6,3) layer (see Fig. 6.9b) decorated by monodentate cisoid-I 3-bpt terminals. The layered frameworks are interlinked via interlayer hydrogen bonding to form 3D architectures in which the nitro groups act as the hydrogen-bonding acceptors of amine from 3-bpt. As for the complex $\{[Cd(3\text{-bpt})(Htma)(H_2O)_2]\cdot(H_2O)_2\}_n$, the Cd^{II} centers are connected via cisoid-I 3-bpt and Htma bridges to generate 1D double-stranded arrays. Typically, the coordinated chains in this case are extended further via head-to-tail hydrogen-bonding dimers between the carboxyl groups of Htma, and this results in a 2D porous (6,3) layer (see Fig. 6.9c). Notably, three such 2D motifs are entangled to afford a novel 3D polythreading architecture with the lateral 3-bpt components of each layer penetrating the voids of two adjacent layers.

In a word, it is anticipated that modified aromatic linkers with different substituent groups may afford distinct supramolecular assemblies of coordination architectures, in view of the steric and/or electronic effect of the substituents as well as their particular ability to form secondary interactions such as hydrogen bonding. As a result, product diversity of coordination polymers can also be well regulated by the substituent effect of such aromatic building blocks.

6.3.3 Delocalization Effect of Conjugated Ligands

As discussed at the beginning of this section, the extended ligand phen-hat has molecular backbones and coordination sites very similar to those of Hat but fused

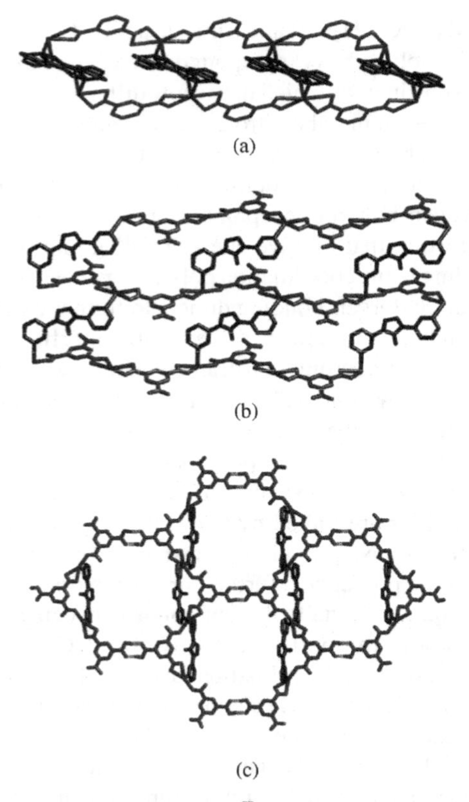

FIGURE 6.9 Network structures of three Cd^{II} coordination polymers with 4-amino-3,5-bis (3-pyridyl)-1,2,4-triazole and R-isophthalate (R = –H, –NO_2, and –COOH) building blocks.

aromatic rings. Thus, it has a larger delocalized π-electron system than Hat, which may allow facile *d*–π interactions between ligands and remote metal centers throughout the extended network. As a result, phen-hat forms different silver(I) supramolecular architectures from those of hat [107,108]. This result indicates that the delocalization effect of such conjugated ligands may also be significant in crystal engineering of unusual coordination polymers.

In this context, considerable attention has been devoted to the coordination network construction based on the small aromatic ligand isonicotinate (L7a) with different functional groups within its backbone, and also recently, its bulky derivatives with one or two fused benzene rings (L7b and L7c; see Fig. 6.10), which represents a type of multifunctional ligand that is potentially able to act as bridging tectons to produce ordered lattice arrangements with various structural topologies [140–144]. On the other hand, work on this topic is also attributed not only to interest in enriching the contents of crystal engineering, but also to targeting particular solid-state materials with tailored functions such as nonlinear optical properties [145,146], guest sorption/desorption [147–149], and molecular magnetism [150].

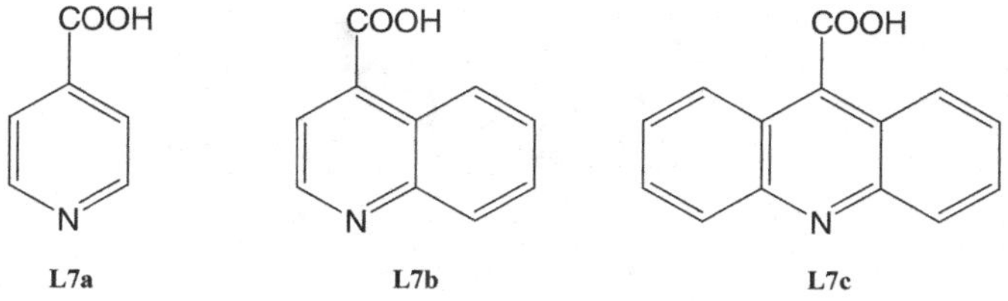

COOH COOH COOH

N N N

L7a L7b L7c

FIGURE 6.10 Schematic view of isonicotinate and its derivative ligands.

Although the coordination sites of the three ligands are very similar, their coordination chemistry is found to be quite different, due to the significant delocalization effect of the latter two building blocks. For example, we have prepared the coordination polymer $[Cu(L7b)_2]_n$ via self-assembly of copper(II) nitrate with 4-quinolinecarboxylate (L7b), in which square-planar Cu^{II} centers are linked by anionic ligands adopting the N,O bridging mode to construct the 2D gridlike polymeric structure with (4,4) network topology [151]. Under similar reaction conditions but with the ligand replaced by L7c, an unexpected trinuclear copper(II) complex $[Cu_3(L7c)_6(CH_3OH)_6]\cdot 3H_2O$ was obtained [151]. In the trinuclear unit, all acridinecarboxylate ligands adopt the O,O-bidentate bridging coordination mode of carboxylate to connect the copper(II) centers, and the uncoordinated acridine nitrogen atoms of each trinuclear unit are hydrogen bonded to the methanol ligands of the adjacent trinuclear units to result in a 3D network. Remarkably, two other distinct copper(II) complexes of L7c could also be isolated under different basicity by adding Et_3N in the reactive solution, and this is not available for other metal ions, such as cobalt(II) and manganese(II), which are not as sensitive to the basicity of the systems (only one type of crystalline product was afforded here). With regard to the dimeric complex $[Cu_2(L7c)_4(CH_3OH)_2]$, the crystal structure consists of a centrosymmetric wheel-shaped neutral molecule in which the ligands adopt the bidentate bridging coordination mode using two oxygen atoms of the carboxylate groups, with the nitrogen atoms remaining uncoordinated. Intermolecular hydrogen bonds occurring between methanol and acridine nitrogen link these dimeric motifs to 1D structures, which are further stabilized by the aromatic stacking interactions between adjacent acridine rings. Of the most interest [152] in the preparation of the complex $[Cu_2(\mu_2\text{-}OMe)_2(L7c)_2(H_2O)_{0.69}]_n$, the basicity seems to be high enough that the methanol molecules are deprotonated and bridge the copper(II) centers to form $Cu_2(\mu_2\text{-}OMe)_2$ dimers (see Fig. 6.11a). These dimeric secondary building units (SBUs) are regarded as 4-connecting nodes to constitute an overall 3D network with NbO topology, which is spacious enough to contain a second interpenetrating counterpart. Notably, this is the first example of an interpenetrating NbO coordination framework. In addition, the structure contains numerous intra- and internetwork $C\text{–}H\cdots\pi$ and $\pi\cdots\pi$ supramolecular interactions that presumably help to stabilize the 3D architecture adopted. A space-filling view of this lattice reveals small channels that run parallel to the z-axis (see Fig. 6.11b), with the solvent-accessible void being 24% of the unit cell volume. Such channels can accommodate different nitrogen gas, methanol, or ethanol content (see Fig. 6.11c), indicating the microporous nature.

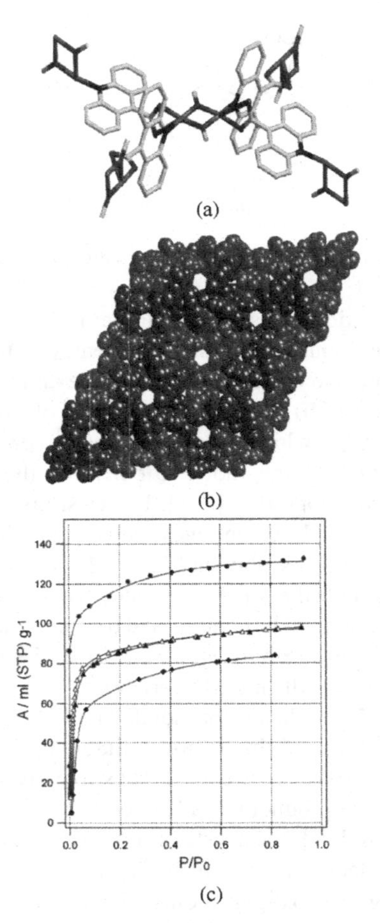

FIGURE 6.11 Crystal structure of $[Cu_2(\mu_2\text{-}OMe)_2(L7c)_2(H_2O)_{0.69}]_n$: (a) dimeric SBU; (b) space-filling diagram of the twofold interpenetrating net showing 1D channels; (c) adsorption isomers for its dehydrate solid of N_2 (circles) at 77 K, EtOH (squares), and MeOH (filled triangles), and desorption of MeOH (open triangles) at 298 K.

On the other hand, secondary intermolecular interactions such as hydrogen bonding and aromatic stacking are also critical in constructing these coordination supramolecular systems. For example [153], in mixed-ligand Mn^{II} complexes with 3-(2-pyridyl)pyrazole and monocarboxylatoanionic building blocks bearing different aromatic backbones (i.e., benzoate, naphthalenecarboxylate, and 9-anthracenecarboxylate), the chelating 3-(2-pyridyl)pyrazole ligand with a noncoordinated pyrazole nitrogen can act as a suitable hydrogen-bonding donor for carboxylate oxygen to facilitate formation of stable coordination species. Interestingly, the binding modes of carboxylate and the final architectures can be adjusted by such intramolecular interactions in accordance with the noncoordinating pendant geometries of these monocarboxylato co-ligands.

To further elucidate the influence of ligand geometries of the carboxylate components and intramolecular/intermolecular weak interactions on the structures and properties of such coordination complexes, a series of pertinent Cd^{II} complexes have been investigated by Liu et al. [154]. This research clearly reveals that both hydrogen bonding and aromatic stacking play important roles in the formation of these coordination architectures, especially in the aspect of extending the discrete polynuclear species or low-dimensional coordination entities into high-dimensional supramolecular networks. In addition, Cu^{II}, Co^{II}, and Ni^{II} complexes with 9-anthracenecarboxylate and other basic co-ligands, such as 2,2′-bipyridine, 1,4-diazabicyclo [2.2.2]octane, 1,10-phenanthroline, and 4,4′-bipyridine, have been studied, which show the effect of the bulky aromatic ring skeleton of the carboxylate ligand on the resulting coordination complexes [155].

Another pair of relevant ligands with different noncoordinating aromatic backbones, 1,4-ditetrazolatebenzene (L8a) and 9,10-ditetrazolateanthracene (L8b), has also been taken into consideration (see Fig. 6.12) and found to display distinct coordination chemistry in the construction of a metal–organic framework [156,157]. With regard to $\{[Zn_3(L8a)_3(DMF)_4(H_2O)_2]\cdot(CH_3OH)_{3.5}\}_n$ [156], its structure contains linear Zn_3 units, which consist of a central Zn^{II} ion that is octahedrally

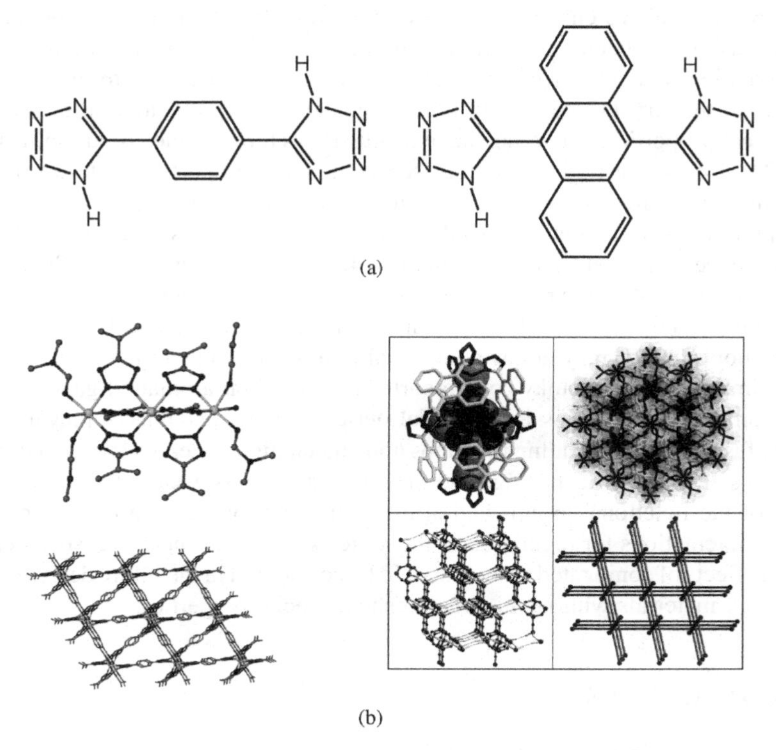

FIGURE 6.12 (a) Schematic view of 1,4-ditetrazolatebenzene (L8a) and 9,10-ditetrazolateanthracene (L8b); (b) crystal structures of their corresponding zinc(II) coordination frameworks.

coordinated by six tetrazolate nitrogen atoms, and two outer octahedral Zn^{II} spheres provided by three tetrazolate nitrogen atoms, two DMF molecules, and one water ligand (see Fig. 6.12). These trinuclear secondary building units (SBUs) are extended via the ligand backbones to form a 3D neutral (4,6)-connected framework in which the trinuclear moieties and linear ligands can be considered as the six-connected nodes and edges, respectively. The Zn_3 subunits within this network (see Fig. 6.12) stack along one direction and define the available 1D channels of the lattice structure, together with the bridging ligands, in which solvent guests such as DMF, water, and methanol are included. It should be noted that this work first demonstrates that such ditetrazolate ligands are indeed capable of forming direct carboxylate-free structural analog of metal–organic frameworks incorporating 1,4-benzenedicarboxylate, and the desolvated form of this microporous MOF can serve as an excellent candidate for hydrogen storage material (a maximum H_2 uptake of 1.46 wt% at 77 K and 880 torr).

In view of the superior constructing capability of 9,10-ditetrazolateanthracene with a bulky skeleton spacer (anthracene ring), a novel 3D coordination polymer $\{[Zn_7(OH)_8(L9b)_3]\cdot H_2O\}_n$ has been synthesized via the in situ solvothermal reaction of the precursor 9,10-dicyanoanthracene and $ZnCl_2/NaN_3$ [157]. In this architecture (see Fig. 6.12), nanosized heptanuclear spindle-shaped cationic SBUs $[Zn_7(OH)_8]^{6+}$ are extended by ligands to furnish a neutral 3D framework with pcu topology (a primitive cubic lattice network). Notably, this unusual heptanuclear cluster has never been observed in coordination chemistry. Furthermore, the corresponding optical diffuse reflectance, absorption, and photoluminescence spectra for this crystalline material have been studied to elucidate the electrical and optical properties. The reflectance result shows the presence of an optical gap ($E_g \approx 2.9$ eV), suggesting a possible semiconductor property. In its absorption spectrum, there are two clear absorption peaks centered at 236 and 372 nm, corresponding to the K- and B-bands. In addition, it exhibits intense blue photoluminescence with an emission maximum at 483 nm upon excitation at 396 nm, and the lifetime of the 483-nm peak was measured as 4.042 ns. These results reveal that this metal–organic framework compound with high thermal stability (decomposition at about 380°C) may act as a potential photoactive material.

To sum up, the more bulky counterparts of the familiar aromatic ligands possess larger conjugated π-systems and potential steric hindrance and thus display different binding fashions and extending functions upon metalation under different experiment conditions. In this regard, the larger conjugated systems may weaken the coordination ability of the functional groups and at the same time, provide important aromatic stacking interactions for regulation of the lattice structures. Therefore, such delocalization effects of conjugated ligands should be considered in the future design of new framework materials with novel structures and special properties.

6.4 CONCLUSIONS

The backbones and compositions of organic ligands have a great influence on the formation of metal complexes and on their structures and properties, so ligand design and selection are the key points for constructing functional coordination architectures. In this regard, subtle variation in the ligands with similar donor groups but different noncoordinating groups may also lead to striking changes in the network structures and available properties of coordination polymers in view of the change in the steric and/or electronic effects of the overall building blocks. In conclusion, a systematic change in the organic ligands through a slight modification of the backbones represents a rational strategy to modulate and control the production of coordination polymers, and further work on this subject may provide profound impetus and new insights into crystal engineering of such tailormade hybrid crystalline materials.

Acknowledgments

We appreciate financial support from the 973 Program of China (2007CB815305) and the National Natural Science Foundation of China.

REFERENCES

1. Zaworotko, M. J. *Chem. Commun.* **2001**, 1–9.
2. Desiraju, G. R. *Angew. Chem., Int. Ed.* **2007**, *46*, 8342–8356.
3. Hosseini, M. W. *Acc. Chem. Res.* **2005**, *38*, 313–323.
4. Kitagawa, S.; Kitaura, R.; Noro, S. *Angew. Chem., Int. Ed.* **2004**, *43*, 2334–2375.
5. Janiak, C. *Dalton Trans.* **2003**, 2781–2804.
6. Evans, O. R.; Lin, W. *Acc. Chem. Res.* **2002**, *35*, 511–522.
7. Eddaoudi, M.; Kim, J.; Rosi, N.; Vodak, D.; Wachter, J.; O'Keeffe, M.; Yaghi, O. M. *Science* **2002**, *295*, 469–472.
8. Blake, A. J.; Champness, N. R.; Hubberstey, P.; Li, W.-S.; Withersby, M. A.; Schröder, M. *Coord. Chem. Rev.* **1999**, *183*, 117–138.
9. Hong, M.-C. *Cryst. Growth Des.* **2007**, *7*, 10–14.
10. Moulton, B.; Zaworotko, M. J. *Chem. Rev.* **2001**, *101*, 1629–1658.
11. Batten, S. R.; Robson, R. *Angew. Chem., Int. Ed.* **1998**, *37*, 1460–1494.
12. Robson, R. *J. Chem. Soc., Dalton Trans.* **2000**, 3735–3744.
13. Brammer, L. *Chem. Soc. Rev.* **2004**, *33*, 476–489.
14. Ockwig, N. W.; Delgado-Friedrichs, O.; O'Keeffe, M.; Yaghi, O. M. *Acc. Chem. Res.* **2005**, *38*, 172–182.
15. Desiraju, G. R. *J. Chem. Soc., Dalton Trans.* **2000**, 3745–3751.
16. Janiak, C. *J. Chem. Soc., Dalton Trans.* **2000**, 3885–3896.
17. Desiraju, G. R. *Acc. Chem. Res.* **2002**, *35*, 565–573.
18. Aakeröy, C. B.; Beatty, A. M.; Helfrich, B. A. *Angew. Chem., Int. Ed.* **2001**, *40*, 3240–3242.
19. Du, M.; Li, C.-P.; Zhao, X.-J.; Yu, Q. *CrystEngComm* **2007**, *9*, 1011–1028.
20. Steiner, T.; Desiraju, G. R. *Chem. Commun.* **1998**, 891–892.
21. Chen, X.-M.; Tong, M.-L. *Acc. Chem. Res.* **2007**, *40*, 162–170.
22. Yaghi, O. M.; Li, H. *J. Am. Chem. Soc.* **1995**, *117*, 10401–11402.

23. Hagrman, P. J.; Hagrman, D.; Zubieta, J. *Angew. Chem., Int. Ed.* **1999**, *38*, 2638–2684.
24. Cheng, L.; Zhang, W.-X.; Ye, B.-H.; Lin, J.-B.; Chen, X.-M. *Inorg. Chem.* **2007**, *46*, 1135–1143.
25. Du, M.; Zhao, X.-J.; Guo, J.-H.; Batten, S. R. *Chem. Commun.* **2005**, 4836–4838.
26. Li, J.-R.; Tao, Y.; Yu, Q.; Bu, X.-H.; Sakamoto, H.; Kitagawa, S. *Chem. Eur. J.* **2008**, *14*, 2771–2776.
27. Beer, P. D.; Gale, P. A. *Angew. Chem., Int. Ed.* **2001**, *40*, 486–516.
28. Zeng, Y.-F.; Zhao, J.-P.; Hu, B.-W.; Hu, X.; Liu, F.-C.; Ribas, J.; Ribas-Arino, J.; Bu, X.-H. *Chem. Eur. J.* **2007**, *13*, 9924–9930.
29. Coronado, E.; Galan-Mascaros, J. R.; Gómez-García, C. J.; Laukhin, V. *Nature* **2000**, *408*, 447–449.
30. Kahn, O. *Acc. Chem. Res.* **2000**, *33*, 647–657.
31. Moulton, B.; Lu, J.; Hajndl, R.; Hariharan, S.; Zaworotko, M. J. *Angew. Chem., Int. Ed.* **2002**, *41*, 2821–2824.
32. Kondo, M.; Shimamura, M.; Noro, S.; Minakoshi, S.; Asami, A.; Seki, K.; Kitagawa, S. *Chem. Mater.* **2000**, *12*, 1288–1299.
33. Seo, J. S.; Whang, D.; Lee, H.; Jun, S. I.; Oh, J.; Jeon, Y. J.; Kim, K. *Nature* **2000**, *404*, 982–986.
34. Evans, O. R.; Ngo, H. L.; Lin, W. *J. Am. Chem. Soc.* **2001**, *123*, 10395–10396.
35. Li, J.-R.; Yu, Q.; Sanudo, E. C.; Tao, Y.; Song, W.-C.; Bu, X.-H. *Chem. Mater.* **2008**, *20*, 1218–1220.
36. Lin, W.; Wang, Z.; Ma, L. *J. Am. Chem. Soc.* **1999**, *121*, 11249–11250.
37. Fujita, M.; Kwon, Y. J.; Washizu, S.; Ogura, K. *J. Am. Chem. Soc.* **1994**, *116*, 1151–1152.
38. Ma, B. Q.; Zhang, D. S.; Gao, S.; Jin, T. Z.; Yan, C. H.; Xu, G. X. *Angew. Chem., Int. Ed.* **2000**, *39*, 3644–3646.
39. Du, M.; Bu, X. H.; Guo, Y. M.; Ribas, J. *Chem. Eur. J.* **2004**, *10*, 1345–1354.
40. Banerjee, R.; Phan, A.; Wang, B.; Knobler, C.; Furukawa, H.; O'Keeffe, M.; Yaghi, O. M. *Science* **2008**, *319*, 939–943.
41. Khlobystov, A. N.; Blake, A. J.; Champness, N. R.; Lemenovskii, D. A.; Majouga, A. G.; Zyk, N. V.; Schröder, M. *Coord. Chem. Rev.* **2001**, *222*, 155–192.
42. Yaghi, O. M.; Li, H.; Davis, C.; Richardson, D.; Groy, T. L. *Acc. Chem. Res.* **1998**, *31*, 474–484.
43. Robin, A. Y.; Fromm, K. M. *Coord. Chem. Rev.* **2006**, *250*, 2127–2157.
44. Steel, P. J. *Acc. Chem. Res.* **2005**, *38*, 243–250.
45. Du, M.; Bu, X. H.; Guo, Y. M.; Liu, H.; Batten, S. R.; Ribas, J.; Mak, T. C. W. *Inorg. Chem.* **2002**, *41*, 4904–4908.
46. Prior, T. J.; Rosseinsky, M. J. *Chem. Commun.* **2001**, 495–496.
47. Zhang, X. M.; Fang, R. Q.; Wu, H. S. *J. Am. Chem. Soc.* **2005**, *127*, 7670–7671.
48. Huang, Z.; Song, H.-B.; Du, M.; Chen, S.-T.; Bu, X.-H.; Ribas, J. *Inorg. Chem.* **2004**, *43*, 931–944.
49. Zhang, X.-L.; Guo, C.-P.; Yang, Q.-Y.; Wang, W.; Liu, W.-S.; Kang, B.-S.; Su, C.-Y. *Chem. Commun.* **2007**, 4242–4244.
50. Feng, Y.; Guo, Y.; OuYang, Y.; Liu, Z.; Liao, D.; Cheng, P.; Yan, S.; Jiang, Z. *Chem. Commun.* **2007**, 3643–3645.

51. Wang, X.-L.; Qin, C.; Wang, E.-B.; Li, Y.-G.; Su, Z.-M.; Xu, L.; Carlucci, L. *Angew. Chem., Int. Ed.* **2005**, *44*, 5824–5827.
52. Fujita, M.; Kwon, Y. J.; Sasaki, O.; Yamaguchi, K.; Ogura, K. *J. Am. Chem. Soc.* **1995**, *117*, 7287–7288.
53. Fujita, M.; Ogura, K. *Bull. Chem. Soc. Jpn.* **1996**, *69*, 1471–1482.
54. MacGillivray, L. R.; Groeneman, R. H.; Atwood, J. L. *J. Am. Chem. Soc.* **1998**, *120*, 2676–2677.
55. Tong, M. L.; Chen, X. M. *CrystEngComm* **2000**, *3*, 1–5.
56. Li, H. L.; Davis, C. E.; Groy, T. L.; Kelley, D. G.; Yaghi, O. M. *J. Am. Chem. Soc.* **1998**, *120*, 2186–2187.
57. Groeneman, R. H.; MacGillivray, L. R.; Atwood, J. L. *Chem. Commun.* **1998**, 2735–2736.
58. Guilera, G.; Steed, J. W. *Chem. Commun.* **1999**, 1563–1564.
59. Groeneman, R. H.; MacGillivray, L. R.; Atwood, J. L. *Inorg. Chem.* **1999**, *38*, 208–209.
60. Deakin, L.; Arif, A. M.; Miller, J. S. *Inorg. Chem.* **1999**, *38*, 5072–5077.
61. Edger, M.; Mitchell, R.; Slawin, A. M. Z.; Lightfoot, P.; Wright, P. A. *Chem. Eur. J.* **2001**, *7*, 5168–5175.
62. Lu, J.; Moulton, B.; Zaworotko, M. J.; Bourne, S. A. *Chem. Commun.* **2001**, 861–862.
63. Vodak, D. T.; Braun, M. E.; Kim, J.; Eddaoudi, M.; Yaghi, O. M. *Chem. Commun.* **2001**, 2534–2535.
64. Du, M.; Bu, X.-H.; Guo, Y.-M.; Ribas, J.; Diaz, C. *Chem. Commun.* **2002**, 2550–2551.
65. Carlucci, L.; Ciani, G.; Proserpio, D. M.; Sironi, A. *Angew. Chem., Int. Ed. Engl.* **1995**, *34*, 1895–1898.
66. Hoskins, B. F.; Robson, R.; Slizys, D. A. *J. Am. Chem. Soc.* **1997**, *119*, 2952–2953.
67. Blake, A. J.; Champness, N. R.; Khlobystov, A.; Lemenovskii, D. A.; Li, W.-S.; Schröder, M. *Chem. Commun.* **1997**, 2027–2028.
68. Losier, P.; Zaworotko, M. J. *Angew. Chem., Int. Ed. Engl.* **1997**, *36*, 1725–1726.
69. Brandys, M.-C.; Puddephatt, R. J. *Chem. Commun.* **2001**, 1508–1509.
70. Schweiger, M.; Seidel, S. R.; Arif, A. M.; Stang, P. J. *Inorg. Chem.* **2002**, *41*, 2556–2559.
71. Wenger, O. S.; Henling, L. M.; Day, M. W.; Winkler, J. R.; Gray, H. B. *Inorg. Chem.* **2004**, *43*, 2043–2048.
72. Ghosh, A. K.; Ghoshal, D.; Zangrando, E.; Ribas, J.; Chaudhuri, N. R. *Inorg. Chem.* **2005**, *44*, 1786–1793.
73. Kitagawa, S.; Matsuyama, S.; Munakata, M.; Emori, T. *J. Chem. Soc., Dalton Trans.* **1991**, 2869–2875.
74. MacGillivray, L. R.; Reid, J. L.; Ripmeester, J. A. *J. Am. Chem. Soc.* **2000**, *122*, 7817–7818.
75. Papaefstathiou, G. S.; Kipp, A. J.; MacGillivray, L. R. *Chem. Commun.* **2001**, 2462–2463.
76. Wu, G.; Wang, X.-F.; Okamura, T.-A.; Sun, W.-Y.; Ueyama, N. *Inorg. Chem.* **2006**, *45*, 8523–8532.
77. Maji, T. K.; Mostafa, G.; Mattsuda, R.; Kitagawa, S. *J. Am. Chem. Soc.* **2005**, *127*, 17152–17153.
78. Carlucci, L.; Giani, G.; Proserpio, D. M.; Rizzato, S. *CrystEngComm* **2002**, *4*, 413–425.
79. Abrahams, B. F.; Hoskins, B. F.; Robson, R.; Slizys, D. A. *CrystEngComm* **2002**, *4*, 478–482.
80. Haasnoot, J. G. *Coord. Chem. Rev.* **2000**, *200–202*, 131–185.

81. Ng, M. T.; Deivaraj, T. C.; Klooster, W. T.; McIntyre, G. J.; Vittal, J. J. *Chem. Eur. J.* **2004**, *10*, 5853–5859.
82. Sharma, C. V. K.; Rogers, R. D. *Chem. Commun.* **1999**, 83–84.
83. Hennigar, T. L.; MacQuarrie, D. C.; Losier, P.; Rogers, R. D.; Zaworotko, M. J. *Angew. Chem., Int. Ed. Engl.* **1997**, *36*, 972–973.
84. Fujita, M.; Kwon, Y. J.; Miyazawa, M.; Ogura, K. *Chem. Commun.* **1994**, 1977–1978.
85. Noro, S.; Horike, S.; Tanaka, D.; Kitagawa, S.; Akutagawa, T.; Nakamura, T. *Inorg. Chem.* **2006**, *45*, 9290–9300.
86. Lu, J. Y.; Babb, A. *Inorg. Chim. Acta* **2001**, *318*, 186–190.
87. Carlucci, L.; Ciani, G.; Proserpio, D. M.; Rizzato, S. *CrystEngComm* **2003**, *5*, 190–199.
88. Power, K. N.; Hennigar, T. L.; Zaworotko, M. J. *Chem. Commun.* **1998**, 595–596.
89. Garcia-Couceiro, U.; Castillo, O.; Luque, A.; Garcia-Teran, J. P.; Beobide, G.; Roman, P. *Cryst. Growth Des.* **2006**, *6*, 1839–1847.
90. Hu, S.; Zhou, A.-J.; Zhang, Y.-H.; Ding, S.; Tong, M.-L. *Cryst. Growth Des.* **2006**, *6*, 2543–2550.
91. Carlucci, L.; Ciani, G.; Proserpio, D. M.; Rizzato, S. *Chem. Commun.* **2000**, 1319–1320.
92. Wang, Q.-M.; Guo, G.-C.; Mak, T. C. W. *Chem. Commun.* **1999**, 1849–1850.
93. Plater, M. J.; Foreman, M. R. S. J.; Skakle, J. M. S. *Cryst. Eng.* **2001**, *4*, 293–308.
94. Carlucci, L.; Ciani, G.; Proserpio, D. M.; Rizzato, S. *J. Chem. Soc., Dalton Trans.* **2000**, 3821–3827.
95. Maji, T. K.; Matsuda, R.; Kitagawa, S. *Nat. Mater.* **2007**, *6*, 142–148.
96. Bu, X. H.; Chen, W.; Lu, S. L.; Zhang, R. H.; Liao, D. Z.; Shionoya, M.; Brisse, F.; Ribas, J. *Angew. Chem., Int. Ed.* **2001**, *40*, 3201–3203.
97. Li, J. R.; Bu, X. H.; Zhang, R. H.; Ribas, J. *Cryst. Growth Des.* **2005**, *5*, 1919–1932.
98. Rowsell, J. L. C.; Yaghi, O. M. *Angew. Chem., Int. Ed.* **2005**, *44*, 2670–4679.
99. Sudik, A. C.; Côté, A. P.; Wong-Foy, A. G.; O'Keeffe, M.; Yaghi, O. M. *Angew. Chem., Int. Ed.* **2006**, *45*, 2528–2533.
100. Férey, G.; Serre, C.; Mellot-Draznieks, C.; Millange, F.; Surblé, S.; Dutour, J.; Margiolaki, I. *Angew. Chem., Int. Ed.* **2004**, *43*, 6296–6301.
101. Furukawa, H.; Miller, M. A.; Yaghi, O. M. *J. Mater. Chem.* **2007**, *17*, 3197–3204.
102. Cao, R.; Sun, D.-F.; Liang, Y.-C.; Hong, M.-C.; Tatsumi, K.; Shi, Q. *Inorg. Chem.* **2002**, *41*, 2087–2094.
103. Fang, Q. R.; Zhu, G. S.; Jin, Z.; Ji, Y.-Y.; Ye, J.-W.; Xue, M.; Yang, H.; Wang, Y.; Qiu, S. L. *Angew. Chem., Int. Ed.* **2007**, *46*, 6638–6642.
104. Chen, B.; Ockwig, N. W.; Millward, A. R.; Contreras, D. S.; Yaghi, O. M. *Angew. Chem., Int. Ed.* **2005**, *44*, 4745–4749.
105. Chen, B.; Ockwig, N. W.; Fronczek, F. R.; Contreras, D. S.; Yaghi, O. M. *Inorg. Chem.* **2005**, *44*, 181–183.
106. Sudik, A. C.; Millward, A. R.; Ockwig, N. W.; Côté, A. P.; Kim, J.; Yaghi, O. M. *J. Am. Chem. Soc.* **2005**, *127*, 7110–7118.
107. Abrahams, B. F.; Jackson, P. A.; Robson, R. *Angew. Chem., Int. Ed.* **1998**, *37*, 2656–2658.
108. Bu, X.-H.; Biradha, K.; Yamaguchi, T.; Nishimura, M.; Ito, T.; Tanaka, K.; Shionoya, M. *Chem. Commun.* **2000**, 1953–1954.

109. Li, J.-R.; Bu, X.-H. *Eur. J. Inorg. Chem.* **2008**, 27–40.
110. Awaleh, M. O.; Badia, A.; Brisse, F. *Cryst. Growth Des.* **2006**, *6*, 2674–2685.
111. Li, J. R.; Bu, X. H.; Jiao, J.; Du, W. P.; Xu, X. H.; Zhang, R. H. *Dalton Trans.* **2005**, 464–474.
112. Li, J. R.; Zhang, R. H.; Bu, X. H. *Cryst. Growth Des.* **2003**, *3*, 829–835.
113. Bu, X. H.; Chen, W.; Du, M.; Biradha, K.; Wang, W. Z.; Zhang, R. H. *Inorg. Chem.* **2002**, *41*, 437–439.
114. Awaleh, M. O.; Badia, A.; Brisse, F.; Bu, X. H. *Inorg. Chem.* **2006**, *45*, 1560–1574.
115. Awaleh, M. O.; Badia, A.; Brisse, F. *Inorg. Chem.* **2005**, *44*, 7833–7845.
116. Bu, X. H.; Chen, W.; Hou, W. F.; Du, M.; Zhang, R. H.; Brisse, F. *Inorg. Chem.* **2002**, *41*, 3477–3482.
117. Bu, X. H.; Hou, W. F.; Du, M.; Chen, W.; Zhang, R. H. *Cryst. Growth Des.* **2002**, *2*, 303–307.
118. Li, H.; Eddaoudi, M.; O'Keeffe, M.; Yaghi, O. M. *Nature* **1999**, *402*, 276–279.
119. Rowsell, J. L. C.; Millward, A. R.; Park, K. S.; Yaghi, O. M. *J. Am. Chem. Soc.* **2004**, *126*, 5666–5667.
120. Rosi, N. L.; Eckert, J.; Eddaoudi, M.; Vodak, D. T.; Kim, J.; O'Keeffe, M.; Yaghi, O. M. *Science* **2003**, *300*, 1127–1129.
121. Kim, J.; Chen, B.; Reineke, T. M.; Li, H.; Eddaoudi, M.; Moler, D. B.; O'Keeffe, M.; Yaghi, O. M. *J. Am. Chem. Soc.* **2001**, *123*, 8239–8247.
122. Braun, M. E.; Steffek, C. D.; Kim, J.; Rasmussen, P. G.; Yaghi, O. M. *Chem. Commun.* **2001**, 2532–2533.
123. Eddaoudi, M.; Kim, J.; Vodak, D.; Sudik, A.; Wachter, J.; O'Keeffe, M.; Yaghi, O. M. *Proc. Natl. Acad. Sci. USA* **2002**, *99*, 4900–4904.
124. Eddaoudi, M.; Li, H.; Yaghi, O. M. *J. Am. Chem. Soc.* **2000**, *122*, 1391–1397.
125. Jensen, P.; Batten, S. R.; Moubaraki, B.; Murray, K. S. *J. Solid State Chem.* **2001**, *159*, 352–361.
126. Kutasi, A. M.; Harris, A. R.; Batten, S. R.; Moubaraki, B.; Murray, K. S. *Cryst. Growth Des.* **2004**, *4*, 605–610.
127. Kutasi, A. M.; Batten, S. R.; Harris, A. R.; Moubaraki, B.; Murray, K. S. *Aust. J. Chem.* **2002**, *55*, 311–313.
128. Batten, S. R.; Hoskins, B. F.; Robson, R. *New J. Chem.* **1998**, 173–175.
129. Abrahams, B. F.; Batten, S. R.; Hoskins, B. F.; Robson, R. *Inorg. Chem.* **2003**, *42*, 2654–2664.
130. Brammer, L.; Burgard, M. D.; Rodger, C. S.; Swearingen, J. K.; Rath, N. P. *Chem. Commun.* **2001**, 2468–2469.
131. Brammer, L.; Burgard, M. D.; Eddleston, M. D.; Rodger, C. S.; Rath, N. P.; Adams, H. *CrystEngComm* **2002**, *4*, 239–248.
132. Schultheiss, N.; Powell, D. R.; Bosch, E. *Inorg. Chem.* **2003**, *42*, 5304–5310.
133. Schultheiss, N.; Powell, D. R.; Bosch, E. *Inorg. Chem.* **2003**, *42*, 8886–8890.
134. Turner, D. R.; Strachan-Hattona, J.; Batten, S. R. *CrystEngComm* **2008**, *10*, 34–38.
135. Shi, X.; Zhu, G. S.; Wang, X.-H.; Li, G.-H.; Fang, Q.-R.; Zhao, X.-J.; Wu, G.; Tian, G.; Xue, M.; Wang, R.-W.; Qiu, S. L. *Cryst. Growth Des.* **2005**, *5*, 341–346.

136. Kesanli, B.; Cui, Y.; Smith, M. R.; Bittner, E. W.; Bockrath, B. C.; Lin, W. *Angew. Chem., Int. Ed.* **2005**, *44*, 72–75.
137. Xiao, D.-R.; Wang, E.-B.; An, H.-Y.; Su, Z.-M.; Li, Y.-G.; Gao, L.; Sun, C.-Y.; Xu, L. *Chem. Eur. J.* **2005**, *11*, 6673–6686.
138. Daiguebonne, C.; Kerbellec, N.; Bernot, K.; Gérault, Y.; Deluzet, A.; Guillou, O. *Inorg. Chem.* **2006**, *45*, 5399–5406.
139. Du, M.; Zhang, Z.-H.; You, Y.-P.; Zhao, X.-J. *CrystEngComm* **2008**, *10*, 306–321.
140. Evans, O. R.; Xiong, R.-G.; Wang, Z.; Wong, G. K.; Lin, W. *Angew. Chem., Int. Ed.* **1999**, *38*, 536–538.
141. Min, K. S.; Suh, M. P. *Eur. J. Inorg. Chem.* **2001**, 449–455.
142. Gu, X.-J.; Xue, D.-F. *Inorg. Chem.* **2007**, *46*, 5349–5353.
143. Lu, J. Y.; Babb, A. M. *Chem. Commun.* **2001**, 821–822.
144. Gheorghe, R.; Andruh, M.; Müller, A.; Schmidtmann, M. *Inorg. Chem.* **2002**, *41*, 5314–5316.
145. Lin, W. B.; Evans, O. R.; Xiong, R.-G.; Wang, Z. *J. Am. Chem. Soc.* **1998**, *120*, 13272–13273.
146. Lin, W. B.; Ma, L.; Evans, O. R. *Chem. Commun.* **2000**, 2263–2364.
147. Aakeröy, C. B.; Beatty, A. M.; Leinen, D. S. *Angew. Chem., Int. Ed.* **1999**, *38*, 1815–1819.
148. Sekiya, R.; Nishikiori, S. *Chem. Eur. J.* **2000**, *8*, 4803–4810.
149. Zhang, J.; Lin, W. B.; Chen, Z. F.; Xiong, R. G.; Abrahams, B. F.; Fun, H. K. *J. Chem. Soc., Dalton Trans.* **2001**, 1806–1808.
150. Chapman, M. E.; Ayyappan, P.; Foxman, B. M.; Yee, G. T.; Lin, W. B. *Cryst. Growth Des.* **2001**, *1*, 159–163.
151. Bu, X.-H.; Tong, M.-L.; Xie, Y.-B.; Li, J.-R.; Chang, H.-C.; Kitagawa, S.; Ribas, J. *Inorg. Chem.* **2005**, *44*, 9837–9846.
152. Bu, X.-H.; Tong, M.-L.; Chang, H.-C.; Kitagawa, S.; Batten, S. R. *Angew. Chem., Int. Ed.* **2004**, *43*, 192–195.
153. Zou, R.-Q.; Liu, C.-S.; Shi, X.-S.; Bu, X.-H.; Ribas, J. *CrystEngComm* **2005**, *7*, 722–727.
154. Liu, C.-S.; Shi, X.-S.; Li, J.-R.; Wang, J.-J.; Bu, X.-H. *Cryst. Growth Des.* **2006**, *6*, 656–663.
155. Liu, C.-S.; Wang, J.-J.; Yan, L.-F.; Chang, Z.; Bu, X.-H.; Sanudo, E. C.; Ribas, J. *Inorg. Chem.* **2007**, *46*, 6299–6310.
156. Dincă, M.; Yu, A. F.; Long, J. R. *J. Am. Chem. Soc.* **2006**, *128*, 8904–8913.
157. Li, J.-R.; Tao, Y.; Yu, Q.; Bu, X.-H. *Chem. Commun.* **2007**, 1527–1529.

7

FERROELECTRIC METAL–ORGANIC COORDINATION COMPOUNDS

HENG-YUN YE, WEN ZHANG, AND REN-GEN XIONG
Ordered Matter Science Research Center, Southeast University, Nanjing, People's Republic of China

7.1 INTRODUCTION

In 1920, ferroelectricity was discovered by Valasek during a study of Rochelle salt ($NaKC_4H_4O_6 \cdot 4H_2O$, potassium sodium tartrate tetrahydrate) [1]. According to an investigation by Solans et al. [2], the ferroelectric phase of Rochelle salt should be confined to the region between 291 and 249 K belonging to monoclinic space group $P2_1$ (polar point group C_2). Fifteen years later, the first type of hydrogen-bonded ferroelectrics was discovered in KH_2PO_4 (KDP) and related materials, including Rochelle salt [3]. In 1945, the first non-hydrogen-bonded ferroelectric $BaTiO_3$ was subsequently discovered by Wul and Goldman in the Soviet Union [4] and by von Hippel's group in the United States [5]. Until this discovery, it was assumed that hydrogen bonds were necessary for ferroelectricity to occur. Later, significant progress in applications was made possible after the discovery of lead zirconate titanate [$Pb(Zr,Ti)O_3$ (PZT)] with a large remanent ferroelectric polarization [6]. Since then, many pure inorganic compounds and organic polymers have been explored and a variety of practical applications have been developed, such as memory devices, ferroelectric random-access memories, infrared sensors, piezoelectric sensors, and actuators [7].

All ferroelectrics are piezoelectric (when a voltage is applied across a material, it can undergo a mechanical distortion in response) and pyroelectric (materials have

Design and Construction of Coordination Polymers, Edited by Mao-Chun Hong and Ling Chen

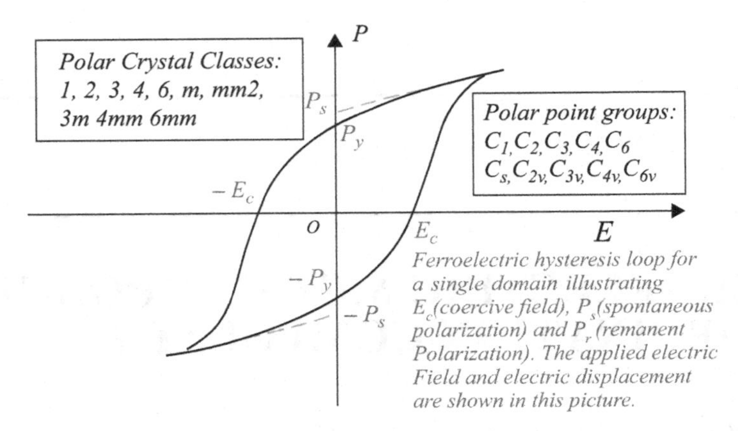

FIGURE 7.1 Ferroelectric hysteresis loop, terminological definitions, and 10 polar point groups as well as polar crystal systems.

a spontaneous polarization whose amplitude changes under the influence of temperature gradients), but they additionally possess a reversible, nonvolatile macroscopic spontaneous electric dipole moment in the absence of an external electric field. In simple words, ferroelectric crystals can be seen as an assembly of batteries with a particular orientation which remain stable unless an external electric field is applied to change their direction. Their polar state is a consequence of the structural transition from a high-temperature, high-symmetry paraelectric phase to a low-temperature, low-symmetry ferroelectric phase. These materials also behave as high-dielectric-constant insulators useful in the development of capacitors and energy storage materials. The ferroelectric definition can be seen clearly in Figure 7.1. Ferroelectric behavior requires the adoption of a space group that is associated with one of 10 polar point groups (C_1, C_s, C_2, C_{2v}, C_3, C_{3v}, C_4, C_{4v}, C_6, C_{6v}). So, generally, a ferroelectric crystal has a mobile atom at its center. Applying an electric field across a face of the crystal causes this atom to move in the direction of the field. Reversing the field causes the atom to move in the opposite direction. Atom positions at the top and bottom of the crystal are stable. Polarization reversal, or ferroelectric hysteresis, may be measured through a Sawyer–Tower circuit [[8]a].

However, much of the attention in this field has been focused on developing ferroelectric and high-dielectric inorganic compounds such as KDP, $BaTiO_3$, $LiNbO_3$, and $CaCu_3Ti_4O_{12}$ [[8]b]. In contrast, studies toward developing ferroelectric and dielectric materials based on metal–organic coordination compounds (MOCCs) have remained relatively sparse despite the fact that the first ferroelectrics is actually *a real MOCC* (Rochelle salt) (see Fig. 7.2a).

Recent success in the synthesis (or self-assembly) and design of novel materials based on metal–organic coordination compounds [or metal–organic coordination polymers (MOCPs)] has prompted us to examine supramolecular crystal engineering of MOCCs (or MOCPs) falling in one of the 10 polar point groups by exploiting the strong and highly directional metal–ligand coordination bonds. Such an effect is

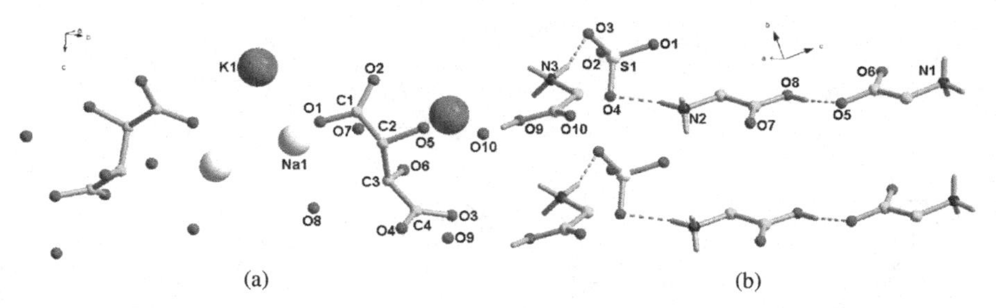

FIGURE 7.2 Molecular structure of (a) Rochelle salt and (b) triglycine sulfate.

expected to result in new complexes (or hybrid materials) bearing the advantages of both the tunalility, π-conjugation, and chirality of organic compounds and the d and f hybrid orbital diversity and electron spin properties of inorganic systems [9]. Noncentrosymmetric (acentric) or homochiral MOCCs (or MOCPs) with desired configuration or topologies can be designed rationally by taking advantage of well-defined metal coordination geometries in combination with carefully chosen bent (or kinked) or homochiral ligands. In this chapter we describe our successful development of several crystal engineering strategies toward the synthesis of acentric or homochiral MOCCs (or MOCPs) that can crystallize in one of the 10 polar point groups. These compounds are attractive targets, as they can potentially display ferroelectric properties with high-dielectric-constant and ferroelectric properties. We cover three categories in detail on the basis of the chirality or acentric origins of MOCCs and MOCPs: homochiral discrete or zero-dimensional (0D) MOCCs, acentric MOCPs produced through supramolecular crystal engineering strategy, and homochiral MOCPs constructed with homochiral organic ligands as building blocks.

7.2 HOMOCHIRAL DISCRETE OR ZERO-DIMENSIONAL MOCCs

To describe a discrete MOCC, it is reasonable to mention the well-known ferroelectric material triglycine sulfate (TGS) (see Fig. 7.2b) [10]. In this salt, the amino acid exists in two different forms, as a zwitterion and as a cation, which are present in a 1 : 2 ratio. Strong hydrogen bonds in the crystal packing of TGS play an important role in its spontaneous polarization.

The first hydrogen-bonded MOCC with a potential ferroelectric property, $[NH_3C_6H_4(CH_2)_2CHNH_3COOH](SnCl_3)_2\cdot(H_2O)_3$ (**1**), was found during the preparation of (*S*)-4-(4′-aminophenyl)-2-aminobutanoic acid by the reduction of (*S*)-2-amino-4-(4′-nitrophenyl)butyric acid (NOHPA) in the presence of $SnCl_2$ and HCl (see Scheme 7.1, Eq. 1) [11]. Compound **1** crystallizes in polar point group C_2 (space group *I*2), a prerequisite for a ferroelectric compound. An x-ray single-crystal diffraction investigation showed that the asymmetric unit contains two crystallographically unique $SnCl_3^-$ anions sharing a similar pseudotrigonal pyramidal Sn(II) coordination environment (see Fig. 7.3a). If longer Sn–Cl interactions (Sn–Cl: 3.666 Å) are taken into account, the Sn atom lies at the center of a distorted octahedron.

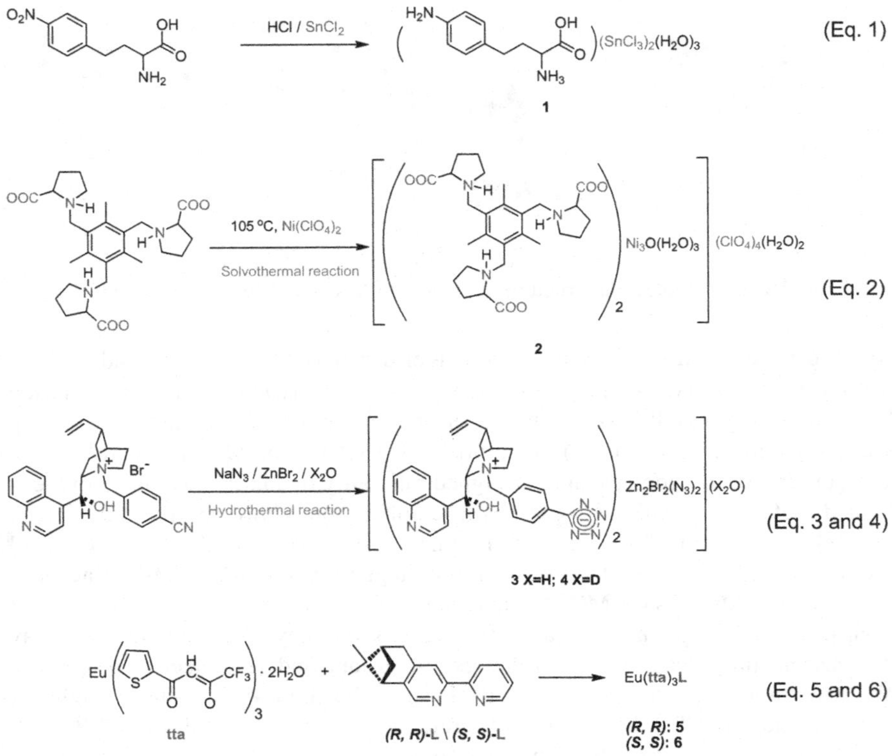

O_2N ... OH ... NH_2 —HCl / $SnCl_2$→ (H_2N ... OH ... NH_3)$(SnCl_3)_2(H_2O)_3$ **1** (Eq. 1)

—105 °C, $Ni(ClO_4)_2$, Solvothermal reaction→ [(...)$_2$ $Ni_3O(H_2O)_3$] $(ClO_4)_4(H_2O)_2$ **2** (Eq. 2)

—NaN_3 / $ZnBr_2$ / X_2O, Hydrothermal reaction→ [(...)$_2$ $Zn_2Br_2(N_3)_2$] (X_2O) **3** X=H; **4** X=D (Eq. 3 and 4)

Eu(tta)$_3$ · $2H_2O$ + *(R, R)*-L \ *(S, S)*-L → Eu(tta)$_3$L *(R, R)*: **5** *(S, S)*: **6** (Eq. 5 and 6)

SCHEME 7.1

Interestingly, the reduction product of NOHPA is a cation, with both amino groups existing as ammonium groups and the carboxylate group protonated. The cation with the NH_3^+ group on the aliphatic chain and the free carboxylic acid in **1** resembles the two non-zwitterionic amino acids in TGS, which play an important role in

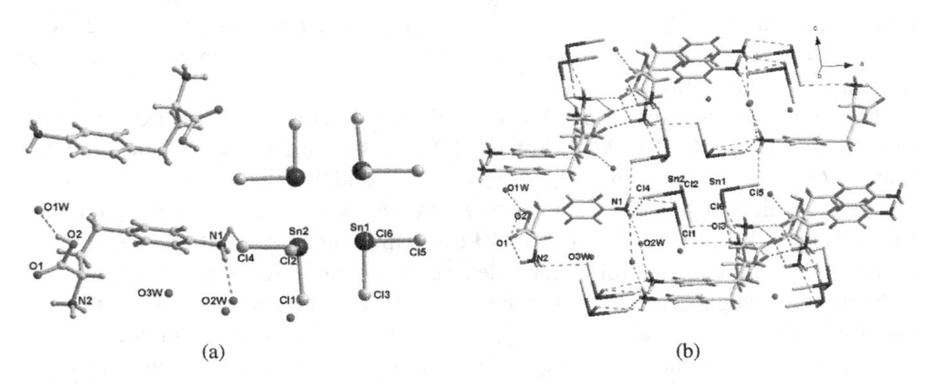

FIGURE 7.3 (a) Asymmetric unit representation of **1**; (b) 3D framework view through hydrogen-bonds in crystal packing state in **1**.

spontaneous polarization. In **1**, both ammonium groups participate in a number of N···HCl hydrogen bonds with $SnCl_3^-$ anions. The N···Cl separations are in the range 3.297(2) to 3.330(2) Å. These hydrogen-bonding interactions result in the formation of a three-dimensional (3D) pillar-layered network. Water molecules participate in two additional, strong hydrogen bonds that extend to the carboxylic acid and the aromatic amino group (see Fig. 7.3b). The O···O and N···O separations are 2.561(2) and 2.694(2) Å, respectively. There is also direct hydrogen bonding between neighboring amino acids (2.894 Å), which results in the formation of chains involving only amino acids that extend in the *b* direction.

Preliminary study of a powdered pellet sample indicates that **1** displays an electric hysteresis loop which can be considered as a typical ferroelectric feature, with a P_r value of about 0.60 μC cm^{-2} and an E_c value of 20 kV cm^{-1}. The possible P_s of **1** is about 3.5 μC cm^{-2}, which is comparable to that of TGS (3.0 μC cm^{-2}) (see Fig. 7.4).

Another hydrogen-bonded discrete homochiral ferroelectric MOCC is $(TBPLA)_2(\mu_3\text{-}O)(ClO_4)_4(H_2O)_5$ (**2**) [TBPLA = (*S*)-1,1′,1″-2,4,6-trimethylbenzene-1,3,5-triyl-tris(methylene)-tris-pyrrolidine-2-carboxylic acid] obtained by the hydrothermal reaction of $Ni(ClO_4)_2{\cdot}6H_2O$ with TBPA (see Scheme 7.1, Eq. 2). Compound **2** crystallizes in a chiral space group ($P2_1$) belonging to polar point group C_{2v}, while the TBPLA ligand is thought to be a zwitterionic neutral molecule, similar to the amino acid in TGS, and acts as a hexadentate chelator with each of the ligand's bidentate carboxylate moieties coordinated to the Ni atoms (see Fig. 7.5a). The molecular charge of **2** is balanced by four free ClO_4^- anions and one μ_3-O atom. This results in the TBPLA ligand taking on a shape resembling that of a parachute with an all-*cis*

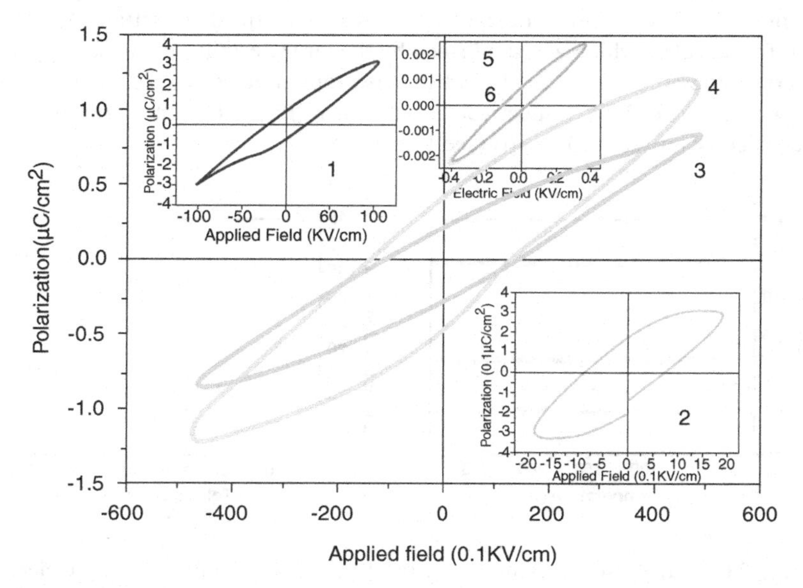

FIGURE 7.4 Hysteresis loops of electric polarization on powdered sample pellets were recorded at room temperature. Ferroelectric magnitude was obtained by referring to the measured magnitude of KDP as a standard value under the same measuring condition.

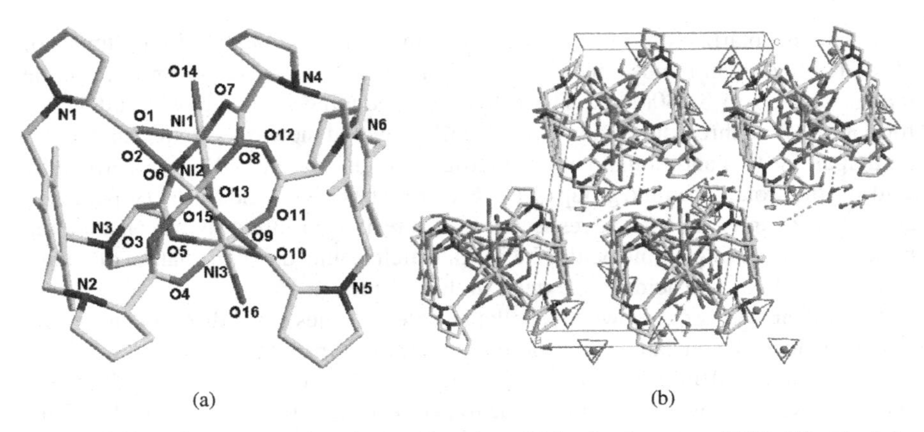

FIGURE 7.5 (a) Asymmetric unit representation of **2** (omitted water and ClO_4^- for clarity); (b) packing view of **2**.

coordination mode. Each Ni center displays a slightly distorted O_6 octahedral coordination geometry in which four of the O atoms are from four different carboxylate groups and two are from the oxo group and H_2O. Similarly, a 3D framework occurs in **2** through strong hydrogen-bonds (see Fig. 7.5b) [12].

As the sample contains water and ClO_4^- anions, the powdered pellet sample was measured on both RT6000 and RT66 ferroelectric testers, showing the occurrence of electronic leakage in RT6000 in a high electric field, while the result on RT66 shows a relatively lower P_r ($\approx 2.1 \times 10^{-3}\,\mu C\,cm^{-2}$) and P_s ($\approx 3.4 \times 10^{-3}\,\mu C\,cm^{-2}$) (see Fig. 7.4). However, the standard hysteresis loop still demonstrates that **2** is a typical ferroelectric which was confirmed by the large permittivity anisotropy along three crystallographic axes (the dielectric anisotropy ratios of $\varepsilon_{r\|c}/\varepsilon_{r\|b}$ and $\varepsilon_{r\|c}/\varepsilon_{r\|a}$ are about 3.47 and 2.22, respectively as shown in Fig. 7.6a) and large powdered dielectric constant ($\varepsilon_r \approx 15$ to 16) (see Fig. 7.6b).

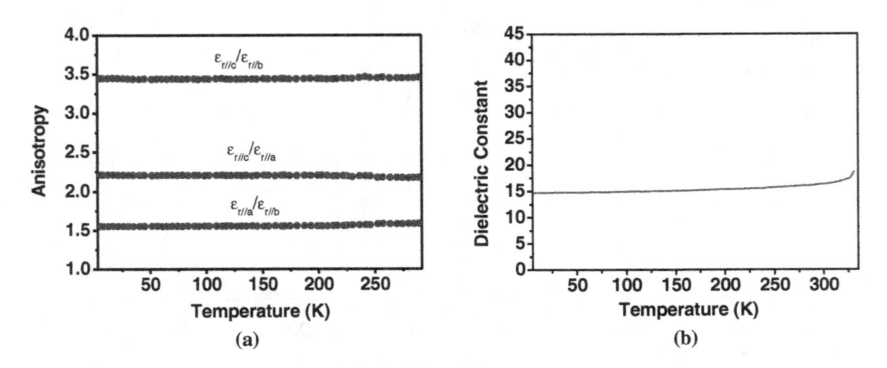

FIGURE 7.6 (a) Anisotropies of **2** as a function of temperature, showing temperature independence; (b) temperature dependence of the dielectric constants of compound **2** in powdered pellet mode.

Taking into account the two samples noted above, it can be seen that the presence of hydrogen-bonds is responsible for ferroelectric behavior. This means that strengthening of hydrogen-bonds is a prerequisite to enhancement of ferroelectric properties. It is worth noting that the deuterium effect (DEF) in exchanging hydrogen with deuterium on the hydrogen bonds plays a very important role not only in the enhancement of many physical properties, such as nonlinear optics (especially second harmonic generation response), ferroelectricity, and permittivity, but also from basic theoretical points of view. Some typical examples are found in pure organic and inorganic compounds, such as Phz–H2ca–Phz–D2ca (Phz = phenazine, H_2ca = bromanilic acid), showing a huge dielectric response with an increase of 50 K in the dielectric constant at 1 MHz as a function of temperature, a significant increase in dielectric DEF in phenylsquaric acid from 8 to 18 [13], and KDP–DKDP (KD_2PO_4) with dielectric DEF increased by 80% and ferroelectric P_s increased by 24% [14].

The following example of homochiral MOCC clearly shows a significant deuterium effect on ferroelectricity in which a novel Zn dimmer, $(CBCN_4)_2Zn_2Br_2(N_3)_2{\cdot}(X_2O)$ (X = H for **3** and D for **4**), was obtained by the reaction of *N*-4′-caynobenzylcinchonidine bromide (CBCBr) with NaN_3 in the presence of $ZnBr_2$ ($CBCN_4$ = *N*-4′-tetrazoylbenzylcinchonidine) (see Scheme 7.1, Eq. 3) [15]. X-ray crystal structural determinations of isostructural **3** and **4** in space group *P*1 revealed that the local coordination geometry around each Zn center can best be described as a slightly distorted tetrahedron in which a tetrazoyl group links two Zn centers to form a dimer composed of three N atoms (two from tetrazoyl group and one from terminal azide) and one terminal Br atom (see Fig. 7.7a and b). The pyridyl group of the quinoline ring does not bind to Zn atoms, while each $CBCN_4$ ligand acts as a bidentate chelator using its 1,2-μ_2-tetrazoyl groups. Due to the formation of many hydrogen-bonds between the O atom of the 9-hydroxy group of $CBCN_4$ and N atom of the azide or Br, the O atom of water or deuterated water, and, the N atom of the azide or tetrazoyl group, a 3D framework is generated formed through hydrogen-bonds

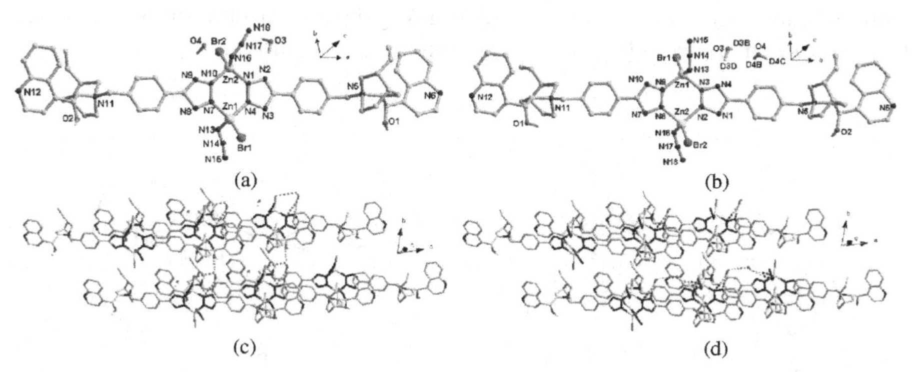

FIGURE 7.7 (a) Dimer representation of **3**; (b) dimer representation of **4**; (c) 3D framework through a hydrogen-bond in **3**; (d) 3D framework through hydrogen-bond in **4**.

(see Fig. 7.7c and d). Interestingly, $CBCN_4$ in **3** and **4** is a zwitterion-like ion with a long charge separation which is essential for strong dipolar moments to result in the possible formation of large P_r or P_s.

Samples **3** and **4** display theoretical ferroelectric behavior. Figure 7.4 shows clearly that there is an electric hysteresis loop in **3** with a P_r of about 0.25 $\mu C\,cm^{-2}$ and an E_c of about 0.12 $kV\,cm^{-1}$. Saturation of P_s in **3** occurs at about 0.84 $\mu C\,cm^{-2}$, which is slightly higher than that for Rochelle salt ($P_s = 0.25\,\mu C\,cm^{-2}$) but significantly lower than that found in KDP (ca. 5.0 $\mu C\,cm^{-2}$). Similarly, the E_c, P_r, and P_s of **4** (see Fig. 7.4) can be estimated to be 0.14 $kV\,cm^{-1}$, 0.46 $\mu C\,cm^{-2}$, and 1.18 $\mu C\,cm^{-2}$, respectively. Thus, a larger ferroelectric DEF (with a P_s increase of 40%) [(1.18 – 0.84)/0.84] on ferroelectric behavior was detected between MOCCs **3** and **4**. Similarly, the large ferroelectric behavior was further confirmed by their large dielectric constants (25.1 for **3** and 69.1 for **4** at low frequency).

Strong geometrical distortion around a central metal ion will result in a large charge separation to induce ferroelectric behavior, as in the non-hydrogen-bonded ferroelectric $BaTiO_3$. A rare earth coordination compound, tris(2-thenoyltrifluoroacetonato)-europium(III) mono(4,5-pinene-bipyridine), $(tta)_3$[(R,-R)-pbpy]Eu (**5**), or $(tta)_3$[(S,-S)-pbpy]Eu (**6**), obtained by the reaction of $[Eu(tta)_3]\cdot 2H_2O$ with chiral 4,5-pinene-bipyridine, may be expected to be a non-hydrogen-bonded ferroelectric (see Scheme 7.1, Eq. 5) [16]. X-ray crystal structure determination reveals that both **5** and **6** crystallize in a chiral space group ($P2_1$) belonging to the polar point group C_2. The central Eu atom is 8-coordinated with a strongly distorted square-antiprism geometry (see Fig. 7.8). A dominant feature is that the ferroelectric value ($P_r \approx 0.022$ $\mu C\,cm^{-2}$), measured on thin film is significantly larger than that in a powdered pellet sample ($P_r \approx 0.006\,\mu C\,cm^{-2}$), with an increase of about 270%. As expected, the ferroelectric behavior of both **5** and **6** is basically identical because they are enantiomeric and their ferroelectric feature is also confirmed by their dielectric behavior.

Compounds **1** to **4** all contain a zwiterion-like ligand leading to the formation of an intramolecular dipolar moment like that in TGS. This gives us a hint that for the design of ferroelectric MOCCs, perhaps an amino acid–like ligand is the best choice, while non-hydrogen-bonded MOCCs (**5** and **6**) generally display strongly geometrical distortion around a high-coordination-number central ion.

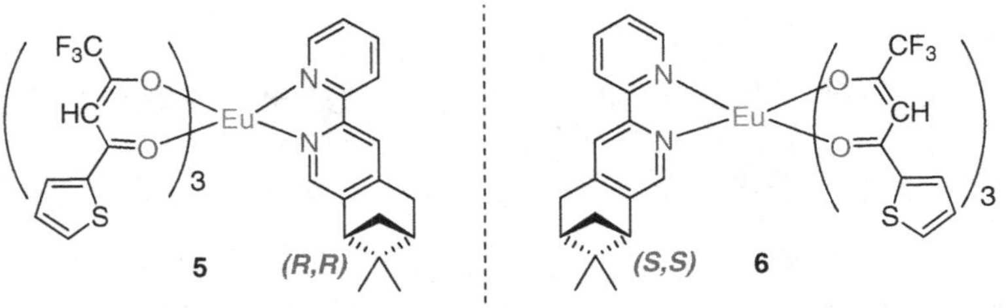

FIGURE 7.8 Molecular structures of **5** and **6**.

7.3 ACENTRIC MOCPs PRODUCED BY SUPRAMOLECULAR CRYSTAL ENGINEERING

In this section we focus primarily on a discussion of potential ferroelectric MOCPs constructed through nonhomochiral organic ligands by helical packing in crystals. The acentric complex $[Cd(L)_2(H_2O)_2]_n$ (**7**) nicely illustrated our supramolecular crystal engineering strategy to self-assemble ferroelectric MOCPs. Compourd **7** was obtained by the hydrothermal reaction of $CdCl_2$ with L (L = *trans*-2,3-dihydro-2-(4′-pyridyl)-3-(3′-cyanophenyl)benzo[*e*]indole) in the presence of NaN_3 (see Scheme 7.2, Eq. 7). The precursor L ligand was chosen for the construction of acentric MOCPs because it contains two chiral centers (highly kinked) and a cyano group that can be converted into a tetrazole functionality via the in situ Sharpless [2 + 3] cycloaddition reaction [15]. X-ray crystallography shows that the local coordination geometry around the Cd(II) center is a slightly distorted octahedral (see Fig. 7.9a). The Cd(II) metal center is bonded to the four N atoms of the L ligands (two from the pyridyl rings and two from the tetrazole rings) and to two water molecules. The four bonded N atoms are located in the equatorial plane, while the two bonded O atoms are situated in axial positions. Each Cd center is bridged by four L ligands to two adjacent metal centers, resulting in the formation of an

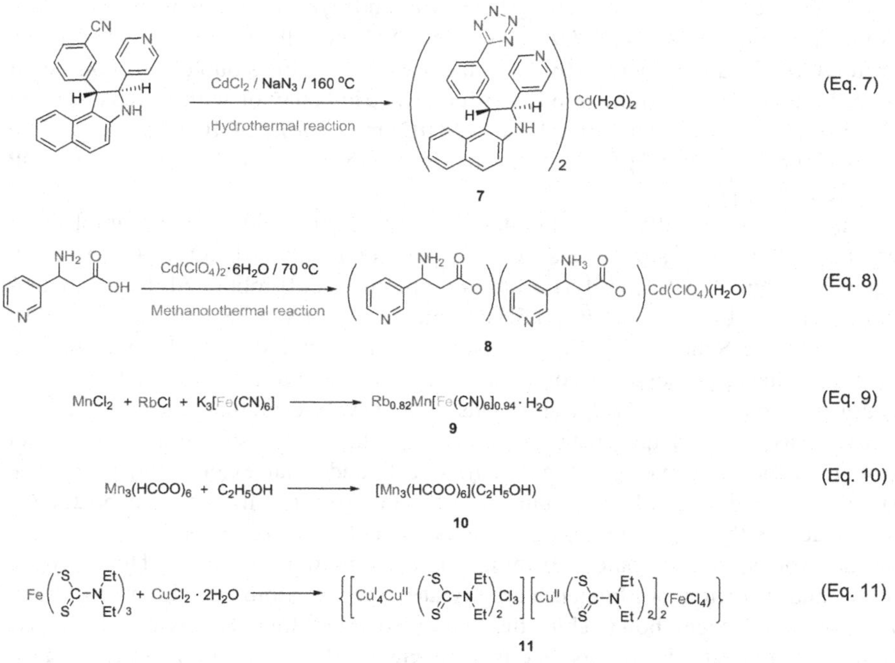

$CdCl_2$ / NaN_3 / 160 °C, Hydrothermal reaction → $(L)_2Cd(H_2O)_2$ **7** (Eq. 7)

$Cd(ClO_4)_2 \cdot 6H_2O$ / 70 °C, Methanolothermal reaction → $Cd(ClO_4)(H_2O)$ complex **8** (Eq. 8)

$MnCl_2 + RbCl + K_3[Fe(CN)_6] \longrightarrow Rb_{0.82}Mn[Fe(CN)_6]_{0.94} \cdot H_2O$ **9** (Eq. 9)

$Mn_3(HCOO)_6 + C_2H_5OH \longrightarrow [Mn_3(HCOO)_6](C_2H_5OH)$ **10** (Eq. 10)

$Fe(S_2CNEt_2)_3 + CuCl_2 \cdot 2H_2O \longrightarrow \{[Cu^I_4Cu^{II}(S_2CNEt_2)_2Cl_3][Cu^{II}(S_2CNEt_2)_2]_2(FeCl_4)\}$ **11** (Eq. 11)

SCHEME 7.2

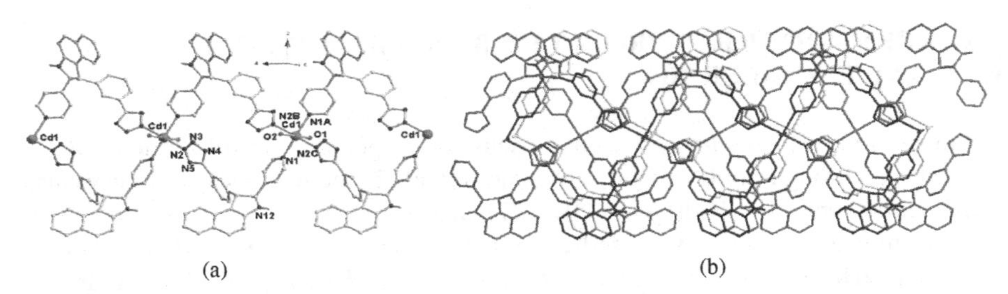

FIGURE 7.9 (a) Chain representation in **7** with M_2L_2 box; (b) AB-stacking packing view in **7**.

infinite Cd–Cd molecular box along the crystallographic *a*-axis. Thus, each L acts as a bidentate linker that bridges two Cd centers. Two adjacent Cd centers in this infinite molecular box form a 28-membered ring $[Cd(NC9N3)_2Cd]$ with a four-bridging-ligand compound. Complex **7** displays a 1D M_2L_2 rhombohedral molecular square with approximate dimensions of $10.37 \times 6.64\,\text{Å}^2$ with no interpenetration. Furthermore, no open channel is found in the crystal structure, and adjacent layers are stacked in a staggered ABAB arrangement. The alignment of collinear adjacent chains to each other prevents cancellation of the dipolar moment, probably due to the occurrence of many cooperative weak supramolecular interactions (hydrogen-bonds, π–π stacking, hydrophobic and hydrophilic as well as van der Waals interactions) (see Fig. 7.9b). As complex **7** crystallizes in the acentric space group *Aba*2, its assignment to the crystal class *mm*2 (point group C_{2v}) is consistent with the fact that it exhibits moderate ferroelectric behavior with a weak electric hysteresis loop, a P_r value of 0.12 to 0.28 $\mu C\,cm^{-2}$, an E_c of 10 $kV\,cm^{-1}$, and an approximate P_s of 4.5 $\mu C\,cm^{-2}$, probably comparable to that found in KDP (5.0 $\mu C\,cm^{-2}$) [17].

Another successfully self-assembled acentric MOCP through supramolecular crystal engineering strategy is $[Cd(rac\text{-}papa)(rac\text{-}Hpapa)]ClO_4{\cdot}H_2O$ (**8**) [*rac*-Hpapa = racemic 3-(3-pyridyl)-3-aminopropionic acid] obtained by a methanothermal reaction of $Cd(ClO_4)_2{\cdot}6H_2O$ with *rac*-Hpapa at 78°C over a period of one week (see Scheme 7.2, Eq. 8). X-ray single-crystal structural determination discloses that **8** possesses a complex 3D anionic network constructed from one cadmium atom and two bridging ligands in the asymmetric unit. The connectivity of the network is perhaps most easily understood in terms of square-grid sheets that extend in the *ab* plane (see Fig. 7.10). The ligands that extend down the page (vertical ligands) are identical but are different from the ligands that bridge Cd atoms across the page (horizontal ligands). For the vertical ligands, the primary amino groups are protonated, forming primary ammonium groups. This nitrogen atom and the noncoordinated carboxylate oxygen atom are involved in an intraligand hydrogen bond. Thus, the two independent ligands play different roles in the net connectivity; that is, one is connected to three cadmium centers and the other to only two. This topology can be described as a twofold interpentrating

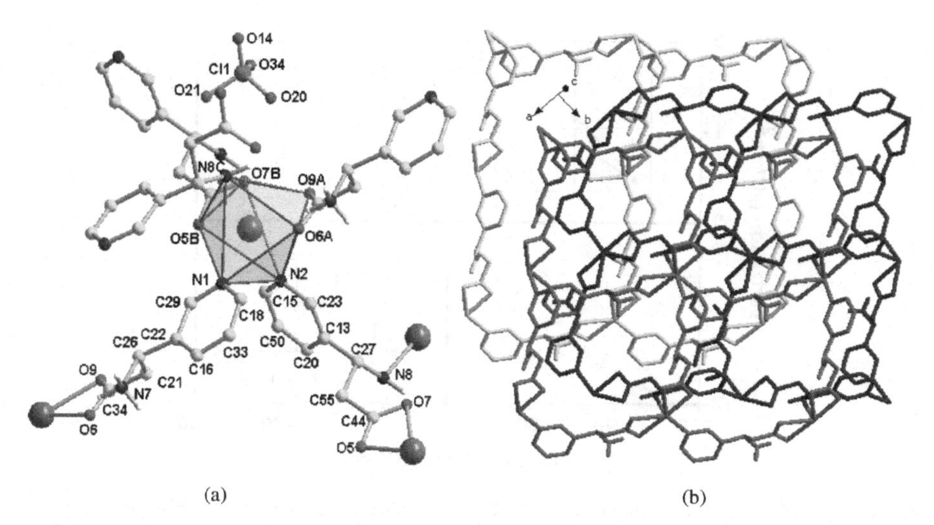

FIGURE 7.10 (a) Octahedral Cd center composed of 4 rac-Hpapa ligands in **8**; (b) twofold interpenetrating 3D framework with a (10,3b) topology.

(10,3b) net. Interestingly, there is extensive hydrogen bonding among protonated amino groups, perchlorate anions, and water, as is the case in TGS.

Compound **8** crystallizes in the acentric space group *Cc*, corresponding to the polar point group C_s. Experimental results indicate that **8** displays ferroelectric behavior showing a clear electric hysteresis loop with a P_r of 0.18 to 0.28 μC cm^{-2} and an E_c of 12 kV cm^{-1}. The possible P_s of **8** is about 1.2 to 1.8 μC cm^{-2} smaller than those found in KDP and TGS, respectively (see Fig. 7.11). It is interesting to note that both **8** and TGS contain protonated amino groups participating in extensive hydrogen bonding, which is responsible for the large value of P_s [18].

Magnetism and ferroelectricity are essential to many forms of current technology, and the quest for multiferroic materials, where these two phenomena are intimately coupled, is of great technological and fundamental importance [19]. Despite their usefulness, magnetic ferroelectrics are rare in nature. Most of them are antiferromagnets with small responses to an external magnetic field, because ferroelectricity and magnetism tend to be mutually exclusive and interact weakly with each other when they coexist. An MOCP successfully incorporating the coexistence of ferroelectricity and ferromagnetism is $Rb^{I}_{0.82}Mn^{II}_{0.20}Mn^{III}_{0.80}[Fe^{II}(CN)_6]_{0.80}[Fe^{III}(CN)_6]_{0.14}\cdot H_2O$ (**9**), which was reported by Ohkoshi et al. at low temperature (see Scheme 7.2, Eq. 9) [20]. Variable-temperature x-ray diffraction measurements have revealed that **9** displays a phase transition between high-temperature (276 K, in centrosymmetric space group *F*43*m*) and low-temperature (184 K, in acentric space group *F*222) forms. The crystal structure may resemble that of Prussian blue (see Fig. 7.12a). Magnetization versus temperature plots of the LT phase show ferromagnetism with a Curie temperature of 11 K, while the polarization (*P*) versus electric field (*E*) plot for the LT phase at 77 K, when applying a field up to 100 kV cm^{-1},

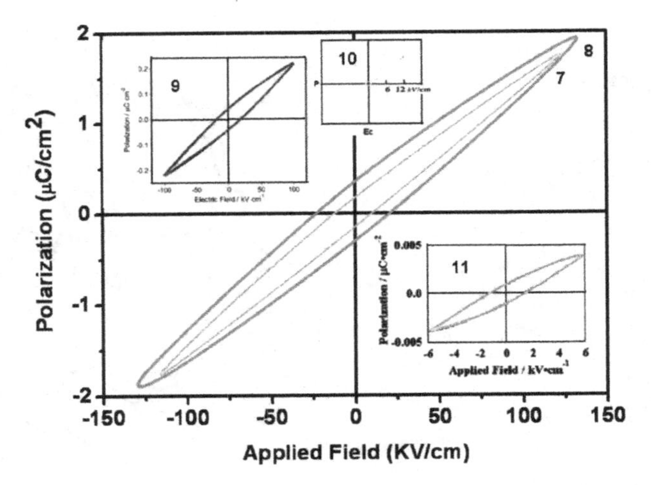

FIGURE 7.11 Hysteresis loops of electric polarization on powdered sample pellets were recorded at room temperature. Ferroelectric magnitude was obtained by referring to the measured magnitude of KDP as a standard value under the same measuring condition.

shows an electric hysteresis loop with a P_r of 0.041 μC cm^{-2} and an E_c of 17.5 kV cm^{-1}. The ferroelectricity may be related to mixing of Fe^{II}, Fe^{III}, Fe vacancy, Mn^{II}, and Jahn–Teller-distorted Mn^{III}. One possible mechanism of ferroelectricity may be explained as follows. In **9**, a multimetallic Prussian blue analog, a local electric dipole moment is created because of an Fe vacancy. In addition, the difference in ionic radii among four metal ions and Mn^{III} Jahn–Teller distortion enhance local structural distortion such as deviation of the M–CN–M′ linkage from a 180° configuration. In such a distorted structure, polarization will probably be induced by the applied electric field, and the polarization can be held by the structural flexibility of the cyano-bridged 3D network.

Another possible ferroelectric and ferrimagnetic MOCP with structural phase transition is the porous molecular crystal $[Mn_3(HCOO)_6](CH_3OH)$ (**10**), where CH_3OH functions as a guest molecule (see Scheme 7.2, Eq. 10; see Fig. 7.12b). Compound **10** is a porous molecular ferrimagnet ($T_c = 8.1$ K) which displays a structural phase transition at LT [21]. LT x-ray structural determination shows that its space group may be an acentric one belonging to a ferroelectricity-active point group. A ferroelectric test shows (see Fig. 7.11) that **10** has a ferroelectric order below 8.5 K, as confirmed by its dielectric behavior, with no distinct anomaly at low temperature.

A mixed-valence coordination polymer, $[Cu^{I}Cu^{II}(Et_2dtc)_2][Cu(Et_2dtc)_2]_2(FeCl_4)_n$ ($Et_2dtc^- =$ diethyldithiocarbamate) (**11**), displaying a 2D ferroelectric structural order with a weak antiferromagnetic interaction ($\theta = -1.03$ K), was obtained by the reaction of a $CHCl_3$ solution of $Fe(Et_2dtc)_3$ and an acetone solution of $CuCl_2{\cdot}2H_2O$ (see Scheme 7.2, Eq. 11; see Fig. 7.13) [22]. Ferroelectric measurements show clearly that there is a hysteresis loop in the temperature range 260 to

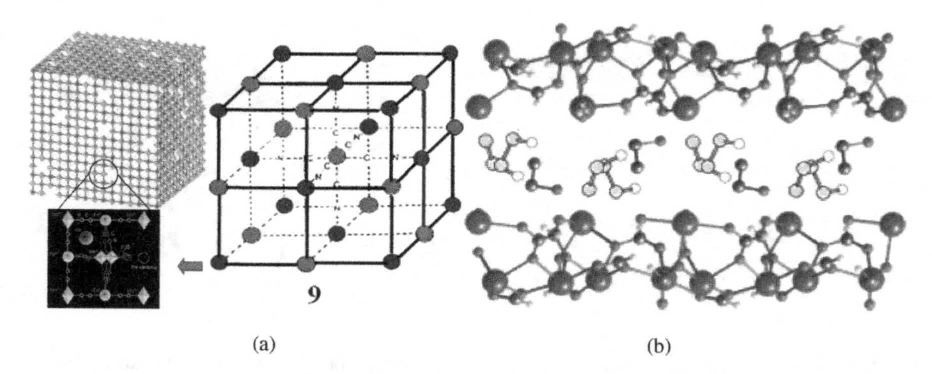

FIGURE 7.12 (a) Molecular structures of **9**; (b) molecular structure of **10**.

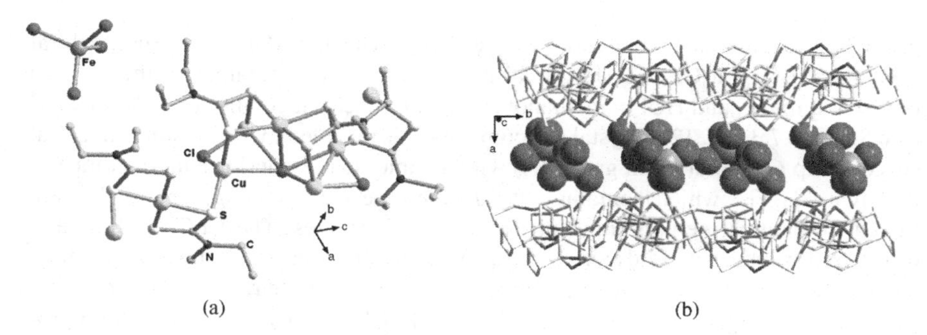

(a) (b)

FIGURE 7.13 (a) Asymmetric unit representation of **11**; (b) 2D laminar view in **11**.

300 K with a low P_s (0.001 and 0.0015 μC cm^{-2}) and possible P_s (0.0035 and 0.0075 μC cm^{-2}), as confirmed by a relatively large dielectric constant ($\varepsilon_r \approx 6.1$) observed for **11**. However, crystal structural determination at room temperature showed that **11** belongs to centrosymmetric space group *Pnma,* which does not agree with its ferroelectric behavior, probably due to the presence of the Fe^{III} ion [20].

Although phase transition (cases **9** to **11**) will change the solid-state structure from high-symmetric to low-symmetric space groups where ferroelectric behavior occurs, it is worth noting that the effect is still weak. However, cases **7** and **8** tell us that acentric MOCPs with magnetism may be constructed using ligands with multichiral centers (highly kinked) as building blocks, in combination with transition metals.

7.4 HOMOCHIRAL MOCPs CONSTRUCTED WITH OPTICAL ORGANIC LIGANDS

KDP is one well-known ferroelectric compound that crystallizes in space group *Fdd*2 (belonging to polar point group C_{2v}), in which extensive hydrogen-bonding

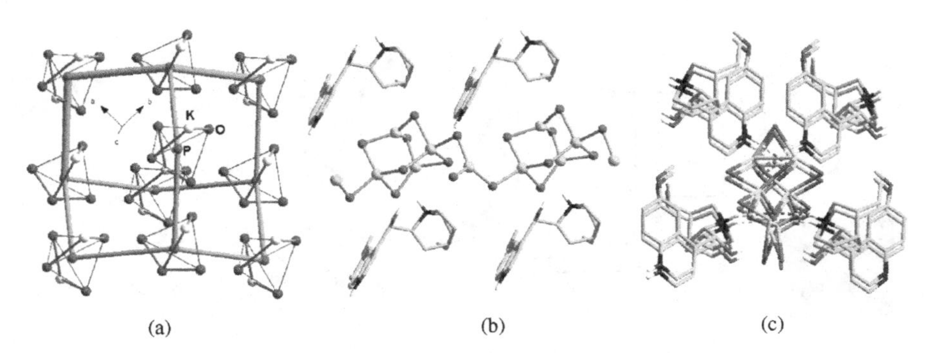

FIGURE 7.14 (a) Diamondoid-like net in KDP (hydrogen bonds composed of diamond net); (b) 1D polymeric diamondlike structure in **12**; (c) 3D framework through hydrogen bonds in **12**.

gives rise to a diamondoid network (see Fig. 7.14a). The first KDP-like 1D homochiral MOCP is $(H_2Q)_2Cu_5Cl_9$ (**12**) (H_2Q = bi-protonated quinine) obtained by the reaction of CuCl, quinine, and HCl in methanol under solvothermal conditions at 60 to 70°C (see Scheme 7.3, Eq. 12). Crystal structure analysis revealed that **12** belongs to chiral space group *C*2 (polar point group C_2) with an anionic polymeric chain of composition $[Cu_5Cl_9^{4-}]_n$, which contains adamantane-type $Cu_4Cl_6^{2-}$ aggregates that are linked to identical neighboring units by $CuCl_3^{2-}$ bridges. The $CuCl_3^{2-}$ bridge is disordered around a twofold axis. There are three independent Cu atoms, two of which are in trigonal coordination geometries, while the third is in a tetrahedral environment. The anionic charge of this highly unusual Cu(I)–Cl polymer is balanced by diprotonated quinine cations. The protons that are bound to the nitrogen atoms of the quinoline and quinuclidine rings participate in hydrogen bonding with chlorine atoms belonging to two Cu_4Cl_6 aggregates that are each part of separate polymeric chains. As a consequence of extensive coordinate and hydrogen bonding, a homochiral 3D network is produced (see Fig. 7.14b and c). Ferroelectric measurement confirmed that **12** is a typical ferroelectric with a clear hysteresis loop, a P_r of 0.28, and a P_s of 2.0 $\mu C\,cm^{-2}$, respectively. The similarity in the values found in **12** and KDP is interesting given that both form 3D networks that involve hydrogen bonding. Furthermore, although **12** does not form a diamond-type net like that formed by KDP, it does possess an adamantane-type unit (in the anionic polymer), which is a characteristic structural feature of diamondoid nets [23].

To investigate the influence of anions on structure motif, a similar reaction of CuBr with quinine(Q) in the presence of HCl was carried out to obtain $Cu_3Br_7(H_2Q)_2(H_2O)$ (**13**), in which diprotonated quinines play a similar role, serving as counter cations for anionic $[Cu_3Br_7^{4-}]_n$ chains (see Scheme 7.3, Eq. 13). The Cu^I centers in this structure have a distorted tetrahedral geometry composed of bridging and terminal bromide ions. As was the case with **12**, there are significant interactions between the H atoms bound to the N atoms of the quinoline and quinuclidine rings and Br centers of the $[Cu_3Br_7^{4-}]_n$ polymer, resulting in the formation of a 3D homochiral network (see Fig. 7.15). Similarly, **13** belongs to chiral space group *C*2, and ferroelectric

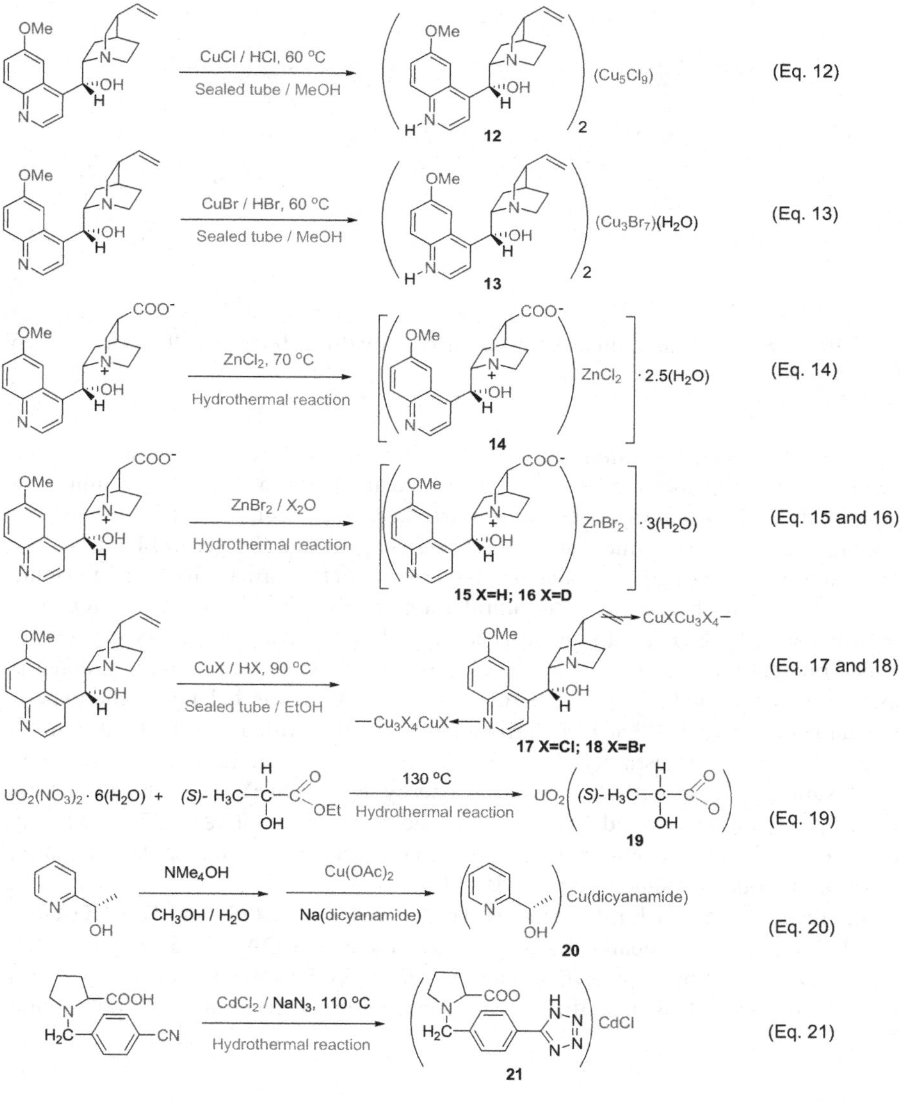

SCHEME 7.3

measurement showed that it is a ferroelectricity-active compound with a P_r of $0.14\,\mu C\,cm^{-2}$ and a P_s of $1.6\,\mu C\,cm^{-2}$, respectively [23].

Utilizing the oxidation of Q by $KMnO_4$ at LT to yield quitenine [6-methoxyl-(8*S*,9*R*)-cinchonan-9-ol-3-carboxylic acid HQA], a chainlike homochiral complex $\{(HQA)(ZnCl_2)(2.5H_2O)\}_n$ (**14**) containing quitenine as a building block was obtained by the reaction of $ZnCl_2$ with HQA (see Scheme 7.3, Eq. 14). X-ray crystal structure analysis revealed two crystallographically independent Zn centers, each

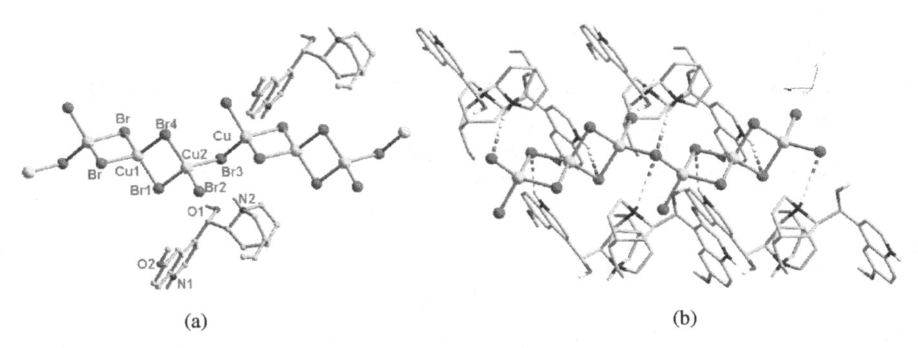

FIGURE 7.15 (a) Anionic dimeric Cu_2Br_2 chain in **13**; (b) 3D framework through hydrogen bonds in **13**.

surrounded by two Cl anions, a quiniline N atom, and an oxygen atom of the carboxylate group to form a slightly distorted tetrahedron. Thus, each quitenine ligand acts as a bidentate spacer that links two Zn centers, resulting in the formation of a coordination chain polymer (see Fig. 7.16). To balance the charges in **14**, the N atom of the quinicludine ring is protonated, which results in the formation of a zwitterionic moiety similar to that found in α-natural amino acids, while in this case, HQA is a β-amino acid. In the crystal structure each wavelike chain adopts AA-type packing, which prevents cancellation of the dipolar moment. The space group $P1$ adopted is associated with point group C_1, as required for ferroelectric behavior. The experimental results revealed that **14** is ferroelectricity-active with a P_r (of 0.04 $\mu C\,cm^{-2}$) (see Fig. 7.17a). Interestingly, the measurement of dielectric loss clearly indicated a relaxation process with an activation energy $E_a = 0.94$ eV and relaxation time $\tau = 1.6 \times 10^{-5}$ s. Combined with its large dielectric constant ($\varepsilon_0 = 37.3$), **14** may execute a chain microvibration, resulting in an overall chain–chain dipolar relaxation process to induce ferroelectric behavior [24].

Interestingly, homochiral chain $[Zn(HQA)Br_2(H_2O)_3]_n$ (**15**) and $[Zn(HQA)Br_2(D_2O)_3]_n$ (**16**) were prepared by thermal treatment of HQA with $ZnBr_2$ in either H_2O or D_2O and 2-butanol at 70°C for 1 to 2 days (see Scheme 7.3, Eqs. 15 and 16). Isostructural **15** and **16** crystallize in chiral space group $C2$, belonging to polar point

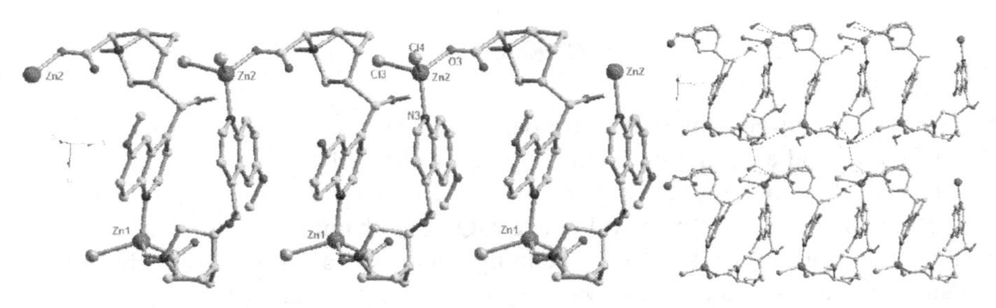

FIGURE 7.16 (a) Wavelike chain in **14**; (b) 3D framework through hydrogen bonds in **14**.

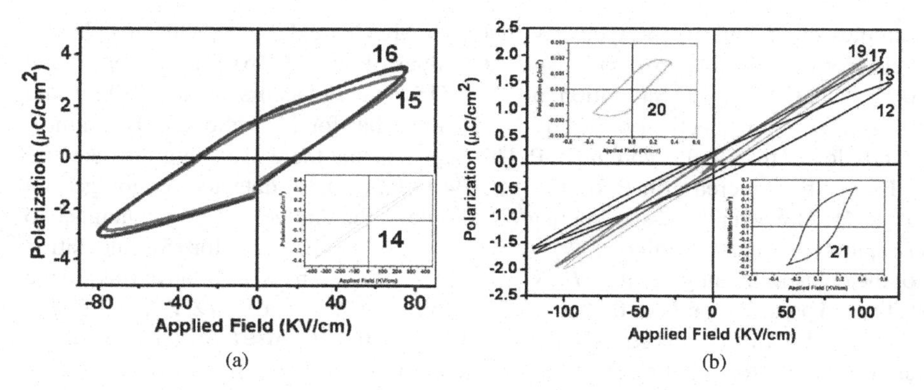

FIGURE 7.17 Hysteresis loops of electric polarization on powdered sample pellets were recorded at room temperature. Ferroelectric magnitude was obtained by referring to the measured magnitude of KDP as a standard value under the same measuring conditions.

group C_2. The N atom of the quinuclidine ring is protonated, implying that the HQA ligand exists in its zwitterionic form. X-ray single-crystal structure determination of **15** revealed that the coordination geometry around the Zn center is slightly distorted tetrahedral, being composed of two terminal Br anions, an oxygen atom from the carboxylate group and a N atom from the quinoline ring (see Fig. 7.18). Each HQA acts as a bidentate spacer that links two Zn centers together to furnish a wavelike infinite chain. The dipolar moment (μ_1) along each chain can be considered with the negative and positive charges sitting on the oxygen atom of the carboxylate group and the N atom of the quinuclidine ring, or Cl and Zn atoms, respectively, and a small dipolar moment (short arrow) between these atoms. This results in the dipolar moment (μ_1 and μ_2) always pointing in the same direction, strengthening the total dipolar moment (μ) within each chain. The presence of numerous hydrogen-bonding interactions in **15** results in the formation of a 3D framework. Pellets of powdered

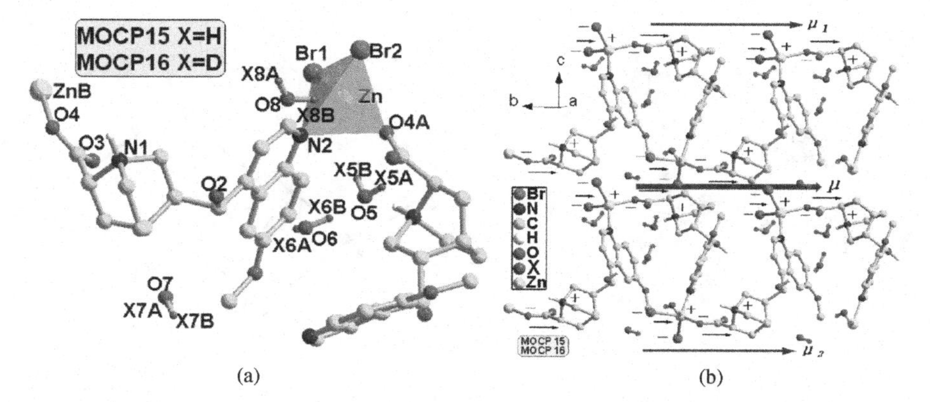

FIGURE 7.18 (a) Asymmetric unit representation in **15** and **16**; (b) 3D framework with same direction alignment in **15** and **16**.

samples of **15** and **16** show clearly that both MOCPs exhibit typical ferroelectric behavior (hysteresis loop), and display P_r values of about $0.16\,\mu C\,cm^{-2}$ for **15** and $0.17\,\mu C\,cm^{-2}$ for **16**. Saturation of P_s in **15** and **16** occurs at about 0.30 and $0.34\,\mu C\,cm^{-2}$, respectively, slightly higher than that for Rochelle salt, but significantly lower than that found in KDP. This suggests that there is a slight ferroelectric DEF (with an increase in P_s of 13%) [(0.34 to 0.30)/0.30] detected on going from **15** to **16**. However, by assuming that one dipole in the unit cell ($Z = 2$) contains two *DA* pairs (the density of dipoles, $N_1 = Z/V_{cell} = 1.5657 \times 10^{27}\,m^{-3}$ for **15**), calculation of the cubic moment gives $\mu_s = P_s/N_1 = 1.916 \times 10^{-30}\,C \cdot m \approx 0.57$ debye (taking the saturated polarization P_s as $0.30\,\mu C\,cm^{-2}$). Similarly, N_2 is equal to $Z/V_{cell} = 1.568 \times 10^{27}\,m^{-3}$ for **16** and μ_s is equal to $P_s/N_2 = 2.168 \times 10^{-30}\,C \cdot m \approx 0.65$ debye, which is slightly larger than that of **15**. These ferroelectric features are further confirmed by their large dielectric constant (362 and 485) [25].

The first organometallic ferroelectric MOCP is $Cu_8Cl_{10}(H\text{-}Q)_2$ (**17**), prepared by solvothermal reactions of Q with Cu^ICl in the presence of HCl (see Scheme 7.3, Eq. 17). Crystal structure analysis of **17** revealed that Q is only monoprotonated and acts as a bridging ligand. In this structure a twofold axis passes through an irregular discrete $[Cu_8Cl_{10}]^{2-}$ aggregate (see Fig. 7.19). The terminal copper centers (Cu1) are coordinated by a quinoline nitrogen of one H–Q, a C=C moiety of another H–Q, and a bridging chloride ion. The remaining copper centers, Cu2, Cu3, and Cu4, adopt approximately trigonal, pyramidal, and linear coordination geometries with chloride ions as ligands, respectively. An unusual feature of this anion is the sharp kink at Cl6 with a Cu2–Cl6–Cu4 angle of 74.20(8)°. This brings Cu2 and Cu4 into close contact [2.700(2) Å] and allows Cl5 to form a weak interaction with Cu3 [2.727(3) Å]. Each $[Cu_8Cl_{10}]^{2-}$ aggregate is linked to four equivalent aggregates through the bridging H–Qs, resulting in a layered network. The packing arrangement of adjacent layers of the 2D network is of the AA type. Compound **17** crystallizes in space group *C*2, and its ferroelectric property is expected. The electric hysteresis loop was observed with a P_r of about $0.12\,\mu C\,cm^{-2}$, an E_c of $5.0\,kV\,cm^{-1}$,

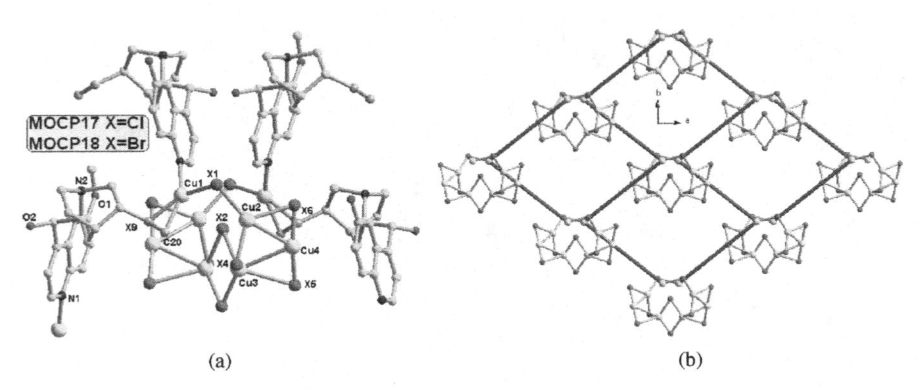

FIGURE 7.19 (a) Asymmetric unit representation with Cu_8X_{10} cluster as supporter in **17** and **18**; (b) 2D laminar framework view while long line is H–Q in **17** and **18**.

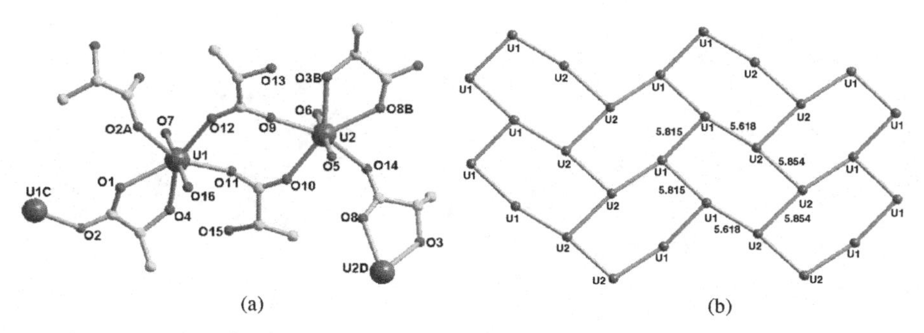

FIGURE 7.20 (a) Asymmetric unit representation in **19**; (b) 2D wall-brick net in **19**.

and a P_s of $2.0\,\mu C\,cm^{-2}$ (see Fig. 7.17b). For the isostructural analog $[Cu_8Br_{10}(H\text{–}Q)_2$ (**18**)] of **17**, its hysteresis loop could not be measured reliably, probably due to current leakage [26].

A 2D laminar ferroelectric MOCP is uranyl-bis[(*S*)-lactate] (**19**), which was obtained in situ from the hydrothermal reaction of $UO_2(NO_3)_2$ with ethyl (*S*)-lactate (see Scheme 7.3, Eq. 19). X-ray crystal structure determination reveals that the local coordination geometry around each U atom center conforms to a distorted pentagonal bipyramid (see Fig. 7.20). However, there are two crystallographically different U centers and three types of lactate ligands: One serves as a bidentate ligand to chelate one U atom using two O atoms from one carboxylate and one hydroxy group; the second lactate ligand acts as a bidentate bridge to link two U atoms using one carboxylate group; the last lactate ligand is a monodentate terminal group with coordination of its carboxylate to one U atom, resulting in the formation of a 2D brick-wall type of grid network. The relationship between two layers is also an ABAB-type stack with no interpenetration. Compound **19** crystallizes in chiral space group $P2_1$, belonging to crystal class 2, and optical activity can occur as a specific physical effect such as ferroelectricity. The experimental results showed a weak electric hysteresis loop with a P_r of $0.12\,\mu C\,cm^{-2}$ and an E_c of $5\,kV\,cm^{-1}$. The P_s value was not observed, probably due to current leakage (see Fig. 7.17b). Possibly the formation of hydrogen bonds and the layered structure are responsible for the ferroelectric behavior of **19** [27].

To explore possible multiferroic MOCPs, homochiral $[Cu(R)\text{-mpm})(dca)]_n$ (**20**) [(*R*)-mpm = (*R*)-α-methyl-2-pyridinemethanol, dca = dicyanamide] was prepared by the reaction of (*R*)-mpm, $Cu(OAc)_2{\cdot}H_2O$, and Na(dca) in a 1 : 1 : 1 molar ratio in a methanol–water mixture at room temperature (see Scheme 7.3, Eq. 20) [28]. Compound **20** crystallizes in chiral space group $P2_1$, and the local coordination polyhedron around the Cu^{II} ion is a distorted square pyramid in which the two apical positions are occupied by N atoms of two dca bridges and the equatorial plane is composed of two O atoms and one pyridyl N atom (see Fig. 7.21a). Magnetic measurement showed that **20** displays strong antiferromagnetic coupling between spin centers. However, ferroelectric testing indicated that a typical hysteresis loop with a P_r of 0.001 and a P_s of $0.002\,\mu C\,cm^{-2}$ was found to confirm the presence of ferroelectricity.

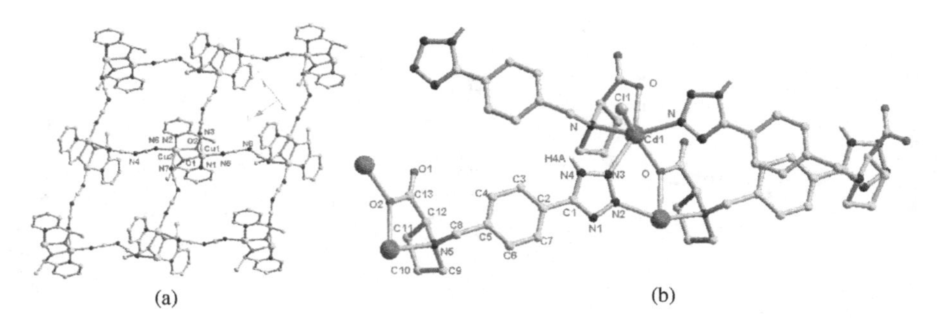

FIGURE 7.21 (a) 2D laminar view in MOCP **20**; (b) asymmetric unit view in **21**.

CdCl(H-TBP) (**21**) [H-TBP = *N*-(4-(1*H*-tetrazol-5-yl)benzyl)proline] was prepared by in situ hydrothermal [2 + 3] cycloaddition reaction of *N*-(4-cyanobenzyl)-(*S*)-proline with NaN_3 in the presence of $CdCl_2$ as a Lewis acid catalyst (see Scheme 7.3, Eq. 21). Due to HT, the racemization occurred during the synthesis of H-TBP. Compound **21** crystallizes in the acentric space group *Cc*, in which the Cd center sits inside a slightly distorted octahedron that consists of three N atoms from two tetrazoyl groups and a pyrrolidinyl group, a terminal Cl atom, and two O atoms from two carboxylate groups (see Fig. 7.21b). Each H-TBP ligand acts as a pentadentate bridging linker that connects five Cd atoms, resulting in the formation of a 3D framework. Furthermore, a carboxylate O atom of the H-TBP ligand binds to a Cd center that is also linked to a pyrrolidinyl N atom to give a stable five-membered ring. A second five-membered ring is found along the Cd–N–N–Cd–O bonding sequence, with the μ-O atom from the H-TBP carboxylate group connecting two Cd ions that are also linked together via two N atoms from the tetrazoyl group (see Fig. 7.22) [29].

The ferroelectric behavior of **21** was also examined, finding that there is an electric hysteresis loop with P_r of about 0.38 μC cm^{-2} and E_c of about 2.10 kV cm^{-1} as well as P_s of about 0.50 μC cm^{-2}, which is significantly higher than that of the prototypical Rochelle salt (see Fig. 7.17b). To evaluate the possible mechanism of ferroelectric behavior, dielectric loss measurement indicates a relaxation process with an activation energy of 1.96 eV and a relaxation time of $1.60 \times 10^{-5.5}$ s, suggesting that **21** is probably a relaxation-type ferroelectric whereas relaxation

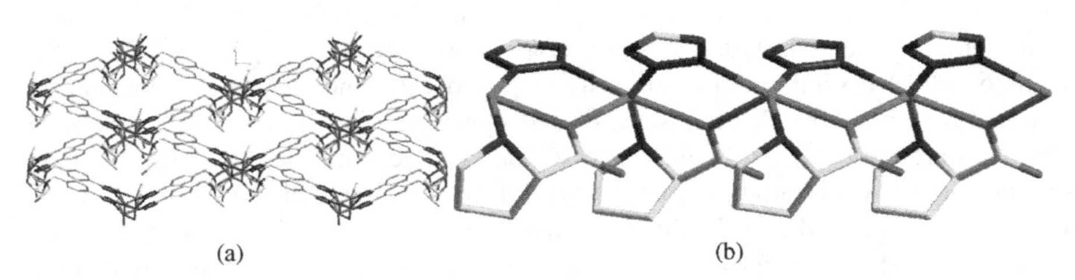

FIGURE 7.22 (a) 3D square-like net framework in **21**; (b) two types of five-membered rings in **21**.

TABLE 7.1 Ferroelectric Properties of Acentric or Homochiral MOCCs or MOCPs

Compound Code	Structure Motif	Crystal System	Point Group	Space Group	P_r (μC cm^{-2})	P_s (μC cm^{-2})
1	0D	2	C_2	$I2$	0.6	3.5
2	0D	2	C_2	$P2_1$	0.0021	0.0034
3	0D	1	C_1	$P1$	0.25	0.84
4	0D	1	C_1	$P1$	0.46	1.18
5	0D	2	C_2	$P2_1$	0.0055	0.02
6	0D	2	C_2	$P2_1$	0.006	0.02
7	1D	$mm2$	C_{2v}	$Aba2$	0.28	4.5
8	3D	m	C_s	Cc	0.18–0.28	1.2–1.8
9	3D	222	D_2	$F222$	0.041	0.22
10	3D	—	Acentric[a]	—	—	—
11	2D	mmm	$D2h$[b]	$Pnma$	0.001	0.0035
12	1D	2	C_2	$C2$	0.28	2
13	1D	2	C_2	$C2$	0.25	1.7
14	1D	1	C_1	$P1$	0.03	0.4
15	1D	2	C_2	$P2_1$	0.16	0.3
16	1D	2	C_2	$P2_1$	0.17	0.34
17	2D	2	C_2	$C2$	0.12	5
18	2D	2	C_2	$C2$	—	—
19	2D	2	C_2	$P2_1$	0.12	2
20	2D	2	C_2	$P2_1$	0.001	0.002
21	3D	m	C_s	Cc	0.38	0.5

[a] The compound displays phase transfer, and there are no ferroelectric data in detail.
[b] Nonpolar point group.

originates from dipolar Cd–Cl bond vibration or proton displacement on the tetrazoyl group.

All homochiral or acentric MOCPs (cases **12** to **21**) definitely contain an optically active organic ligand as the building block. This suggests that the design and synthesis of potential ferroelectric MOCPs depend primarily on the choice of suitable homochiral organic ligands, and for the construction of multiferroelectric MOCPs, transition-metal ions should be used. Finally, the relationships among ferroelectric properties, crystal system, and point group as well as space group are summarized in Table 7.1.

7.5 CONCLUSIONS

Although ferroelectric properties such as P_r and P_s in MOCCs or MOCPs in powder pellet mode are much smaller (by as much as two orders of magnitude) than those found in pure inorganic crystalline salts such as $BaTiO_3$ and $PbTiO_3$, the use of thin films should improve the ferroelectric behavior significantly by reaching an increase of two orders of magnitude, as in **5** and **6**. On the other hand, the search for potential ferroelectric MOCCs or MOCPs theoretically prompted us to focus on the choice of homochiral or optically active organic ligands. Some suggestions on how to construct technologically important multiferroelectric MOCCs or MOCPs are given based on the results of our studies.

Acknowledgments

We thank the 973 Project (2006CB80614) and the National Natural Science Foundation of China. Former Ph.D. and M.S. degree students are gratefully acknowledged for their distinct contributions. We also thank Dr. Philip Wai Hong Chan for his work on the English revision.

REFERENCES

1. Valasek, J. *Phys. Rev.* **1921**, *17*, 475–481.
2. Solans, X.; Gonzalez-Silgo, C.; Ruiz-Perez, C. *J. Solid State Chem.* **1997**, *131*, 350–357.
3. Busch, G. *Ferroelectrics* **1987**, *71*, 43–47.
4. (a) Wul, B. M.; Goldman, I. M. *Dokl. Akad. Nauk SSSR* **1945**, *46*, 154–157. (b) Wul, B. M.; Goldman, I. M. *C. R. Acad. Sci. URSS* **1945**, *49*, 139–142.
5. von Hippel, A.; Breckenridge, R. G.; Chesley, F. G.; Tisza, L. *Ind. Eng. Chem.* **1946**, *38*, 1097–1109.
6. (a) Shirane, G.; Suzuki, K.; Takeda, A. *J. Phys. Soc. Jpn.* **1952**, *7*, 12–18. (b) Shirane, G.; Suzuki, K. *J. Phys. Soc. Jpn.* **1952**, *7*, 333–333.
7. Haertling, G. H. *J. Am. Ceram. Soc.* **1999**, *82*, 797–818.
8. (a) Ok, K. M.; Chi, E. O.; Halasyamani, P. S. *Chem. Soc. Rev.* **2006**, *35*, 710–717. (b) Homes, C. C.; Vogt, T.; Shapiro, S. M.; Wakimoto, S.; Ramirez, A. P. *Science* **2001**, *293*, 673–676.
9. Zhang, H.; Wang, X. M.; Zhang, K. C.; Teo, B. K. *Coord. Chem. Rev.* **1999**, *183*, 157–195.
10. Hoshino, S.; Okaya, Y.; Pepinsky, R. *Phys. Rev.* **1959**, *115*, 323–330.
11. Li, Y.-H.; Qu, Z.-R.; Zhao, H.; Ye, Q.; Xing, L.-X.; Wang, X.-S.; Xiong, R.-G.; You, X.-Z. *Inorg. Chem.* **2004**, *43*, 3768–3770.
12. (a) Fu, D.-W.; Song, Y.-M.; Wang, G.-X.; Ye, Q.; Xiong, R.-G.; Akutagawa, T.; Nakamura, T.; Chan, P. W.-H.; Huang, D. S.-P. *J. Am. Chem. Soc.* **2007**, *129*, 5346–5467. (b) Zhao, H.; Qu, Z.-R.; Ye, H.-Y.; Xiong, R.-G. *Chem. Soc. Rev.* **2008**, *37*, 84–100.
13. (a) Pastor, A. C.; Pastor, R. C. *Ferroelectrics* **1987**, *71*, 61–75. (b) Koval, S.; Kohanoff, J.; Lasave, J.; Colizzi, G.; Migoni, R. L. *Phys. Rev. B* **2005**, *71*, 184102.
14. (a) Horiuchi, S.; Kumai, R.; Tokura, Y. *J. Am. Chem. Soc.* **2005**, *127*, 5010–5011. (b) Takasu, I.; Izuoka, A.; Sugawara, T.; Mochida, T. *J. Phys. Chem. B* **2005**, *108*, 5527–5531.
15. Ye, Q.; Song, Y.-M.; Fu, D.-W.; Wang, G.-X.; Xiong, R.-G.; Chan, P. W.-H.; Huang, D. S.-P. *Cryst. Growth Des.* **2007**, *7*, 1568–1570.

16. Li, X.-L.; Chen, K.; Liu, Y.; Wang, Z.-X.; Wang, T.-W.; Zuo, J.-L.; Li, Y.-Z.; Wang, Y.; Zhu, J.-S.; Liu, J.-M.; Song, Y.; You, X.-Z. *Angew. Chem., Int. Ed.* **2007**, *46*, 6820–6823.
17. Ye, Q.; Tang, Y.-Z.; Wang, X.-S.; Xiong, R.-G. *Dalton Trans.* **2005**, 1570–1573.
18. (a) Qu, Z.-R.; Zhao H.; Wang, Y.-P; Wang, X.-S.; Ye, Q.; Li, Y.-H.; Xiong, R.-G.; Abrahams, B. F.; Liu, Z.-G.; Xue, Z.-L.; You, X.-Z. *Chem. Eur. J.* **2004**, *10*, 54–60. (b) Zhao, H; Li, Y.-H.; Wang, X.-S.; Qu, Z.-R.; Wang, L.-Z.; Xiong, R.-G.; Abrahams, B. F.; Xue, Z.-L. *Chem. Eur. J.* **2004**, *10*, 2386–2390.
19. Eerenstein, W; Mathur, N. D.; Scott, J. F. *Nature*, **2006**, *442*, 759–765.
20. Ohkoshi, S.; Tokoro, H.; Matsuda, T.; Takahashi, H.; Irie, H.; Hashimoto, K. *Angew. Chem., Int. Ed.* **2007**, *46*, 3238–3241.
21. Cui, H. B.; Wang, Z. M.; Takahashi, K.; Okano, Y.; Kobayashi, H.; Kobayashi, A. *J. Am. Chem. Soc.* **2006**, *128*, 15074–15075.
22. Okubo, T.; Kawajiri, R.; Mitani, T.; Shimoda, T. *J. Am. Chem. Soc.* **2005**, *127*, 17598–17599.
23. Zhao, H.; Qu, Z.-R.; Ye, Q.; Abrahams, B. F.; Wang, Y.-P.; Liu, Z.-G.; Xue, Z.-L.; Xiong, R.-G.; You, X.-Z. *Chem. Mater.* **2003**, *15*, 4166–4168.
24. Tang, Y.-Z.; Huang, X.-F.; Song, Y.-M.; Chan, P. W.-H.; Xiong, R.-G. *Inorg. Chem.* **2006**, *45*, 4868–4870.
25. Zhao H.; Ye, Q.; Qu, Z.-R.; Fu, D.-W.; Xiong, R.-G.; Huang, S. D.; Chan W. P. H. *Chem. Eur. J.* **2008**, *14*, 1164–1168.
26. (a) Qu, Z.-R.; Chen, Z.-F.; Zhang, J.; Xiong, R.-G.; Abrahams, B. F.; Xue, Z. *Oganometallics* **2003**, *22*, 2814–2816. (b) Ye, Q.; Wang, S.-X.; Zhao, H.; Xiong, R.-G. *Chem. Soc. Rev.* **2005**, *34*, 208–325.
27. Xie, Y.-R; Zhao, H.; Wang, X.-S; Qu, Z.-R.; Xiong, R.-G.; Xue, X.; Xue, Z.; You, X.-Z. *Eur. J. Inorg. Chem.* **2003**, 3712–3715.
28. Gu, Z.-G.; Zhou, X.-H.; Jin, Y.-B.; Xiong, R.-G.; Zuo, J.-L.; You, X.-Z. *Inorg. Chem.* **2007**, *46*, 5462–5464.
29. Ye, Q.; Song, Y.-M.; Wang, G.-X.; Chen, K.; Fu, D.-W.; Chan, P. W.-H.; Zhu, J.-S.; Huang, D. S.-P.; Xiong, R.-G. *J. Am. Chem. Soc.* **2006**, *128*, 6554–6555.

8

CONSTRUCTING MAGNETIC MOLECULAR SOLIDS BY EMPLOYING THREE-ATOM LIGANDS AS BRIDGES

XIN-YI WANG, ZHE-MING WANG, AND SONG GAO
State Key Laboratory of Rare Earth Materials and Applications, College of Chemistry and Molecular Engineering, Peking University, Beijing, People's Republic of China

8.1 INTRODUCTION

After the early work of pioneers such as Richard L Carlin and Oliver Kahn, the last two decades have witnessed great success in research on molecule-based magnetic materials [1]. Although full control is still difficult, the rational design of molecular magnetic materials seems to be more and more realistic, based on experimental and theoretical accumulation of knowledge of this fascinating field. Pure organic magnets [2], high critical temperatures [3], and bistable spin crossover compounds [4] have been achieved. Furthermore, the discovery and development of single-molecule magnets (SMMs) [5] and single-chain magnets (SCMs) [6] have crossed paramagnetism and long-range-ordered magnetism and provide a wide range of spin dynamics. Parallel to these developments and achievements, there is increasing interest in creating materials that combine magnetism and other properties, such as conductivity, optical properties, porosity [7], and most recently, ferroelectricity [8].

The rational design of new molecular magnets relies mainly on the building block approach, which takes a step-by-step synthesis of materials with desired structures and magnetic properties as the final goal. Before we can consider the building block approach further in practice, two basic aspects need to be considered: spin carriers and bridges. For the former, there are not many choices except pure organic radicals and transition-metal ions. Attention is focused on the magnetic 3d

Design and Construction of Coordination Polymers, Edited by Mao-Chun Hong and Ling Chen

transition-metal ions such as Mn, Fe, Co, Ni, and Cu. These ions, including the isotropic spins (Mn^{2+} and Cu^{2+}), the highly anisotropic spins (Mn^{3+} and Co^{2+}), and one with a large residual orbital contribution (Co^{2+}), provide good opportunities for magnetic investigation, both theoretically and experimentally.

For bridges, on the other hand, the short ligands of one to three atoms, such as O^{2-}, OH^-, CN^-, N_3^-, $HCOO^-$, and $C_2O_4^{2-}$, play a key role in magnetic exchange between magnetic centers. Generally, the shorter and more conjugated the bridges are, the more efficient the magnetic coupling transmitted will be. Thus, metal oxides with the oxygen atom as a single-atom bridge and cyanide-bridged complexes, including some room-temperature molecular magnets, present two of the most widely explored and broadly used magnetic systems [9]. Although they are efficient for magnetic coupling, short bridges are somewhat lacking in the diversity of bridging modes. In contrast, magnetic coupling between two spins through a long pathway, especially longer than four single bonds, is thought to be very weak. Long bridges for magnetic systems are either conjugated in some form, such as oxalate [10] and pyrazine [11], or act as spin carriers themselves, such as the radicals like TCNE and TCNQ [12]. Three-atom bridges, such as azido (N_3^-), formato ($HCOO^-$), nitrito (NO_2^-), thiocyanato (SCN^-), and hydrogen cyanamido ($NCNH^-$), lie in the middle and justify themselves as good candidates to construct molecular magnets. These three-atom bridges are particularly well suited in several aspects. First, they are not too long for magnetic coupling and can always transfer moderate or strong magnetic interaction, ferromagnetic (FO) or antiferromagnetic (AF). Second, as shown below, they show very rich and varied coordination characteristics and bridging modes to transition-metal centers, which makes them able to construct various interesting structures with efficient magnetic interactions. This will dramatically benefit the discovery of new molecular magnetic materials. Third, these three-atom bridges have an important feature more or less in common: that they will usually preclude the inversion center between the two bridged metal ions. This aspect has proved to be quite important in many systems because it will symmetrically allow antisymmetric interaction [Dzyaloshinsky–Moriya (DM) interaction], which has the Hamiltonian $H = -\boldsymbol{D}_{ij}[S_i \times S_j]$, with $\boldsymbol{D}_{ij}$ being the DM vector [1a–c,13]. This interaction acts to cant the spins because the coupling energy is minimized when two adjacent spins are perpendicular to each other. The nonlinear antiparallel alignment of the spins from two sublattices of the antiferromagnets cannot be canceled and will lead to a net, although small in most cases, magnetization below the Néel temperature. The system is referred to as a canted antiferromagnet or weak ferromagnet (WF) since it will show some characteristics as a ferromagnet, such as spontaneous magnetization and a hysteresis effect below T_c. DM interaction requires the lack of an inversion center between the two related spins, and its strength is proportional to the magnetic anisotropy of the system. Thus, weak ferromagnetism can reasonably be anticipated for many three-atom bridged compounds; and this is what we actually observed.

In this chapter, to demonstrate design strategy, the various complexes, and their rich magnetic properties, we concentrate on three-atom bridged magnetic systems that we have explored. The characteristics of the three-atom bridges in coordination and magnetic transmission and the role of the co-ligands and template are first discussed briefly, followed by the main results obtained in this laboratory, from zero- to three-dimensional (0D to 3D) materials.

8.2 COORDINATION CHARACTERISTICS OF THREE-ATOM BRIDGES AND THEIR ROLE IN MEDIATING MAGNETIC INTERACTION

The coordination characteristics of some three-atom bridges (N_3^-, $HCOO^-$, SCN^-, NO_2^-, and $NCNH^-$) and their roles in mediating magnetic interaction are summarized briefly here. Considering the bridging characters, some ligands, such as 1,3-dca (dicyanamide), im (imidazole), carboxylate, and phosphonate, can also sometimes be viewed as three-atom bridges. Several examples of them will also be discussed.

8.2.1 Azido (N_3^-) Ion [14]

As one of the most versatile ligands, azido ion was widely used for the construction of magnetic systems. Different types of bridging modes of azido, such as the 2.20 (end-on, EO), 2.11 (end-to-end, EE), 3.21, 3.30, 4.40, or even 6.33, as depicted in Figure 8.1, can be found in the literature (for the Harris notation 2.20, 2.21, etc., see ref. 15). Depending on these bridging modes, the azido ion can efficiently transmit various magnetic interactions. Generally, the EE azido is responsible for the AF coupling and EO azido for the FO coupling. Of course, the magnetic coupling also depends on the detailed coordination geometries, such as the M–N_{azido}–M

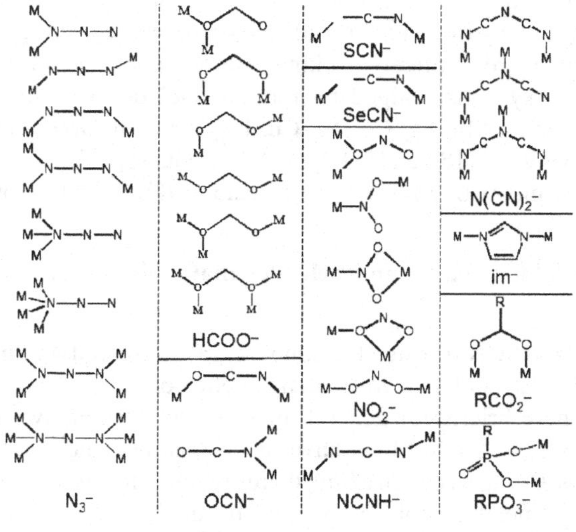

FIGURE 8.1 Three-atom and three-atom-like bridges (N_3^-, $HCOO^-$, OCN^-, SCN^-, $SeCN^-$, NO_2^-, $NCNH^-$, $N(CN)_2^-$, imidazolate, carboxylate, and phosphonate) and their main bridging modes.

angles and the dihedral angle between the mean planes M–N–N–N and N–N–N–M′. The rich bridging modes and the ability to transfer different magnetic couplings lead to abundant magnetic behaviors, such as ferromagnetism, antiferromagnetism, ferrimagnetism, weak ferromagnetism, spin flop, single-molecule magnets, and single-chain magnets, observed in the azido-bridged systems. The investigation of azido-bridged systems has been the focus of many researchers [14]. Numerous interesting compounds have been reported and the structures and magnetism of Mn^{2+} and Ni^{2+} coordination complexes have been well reviewed [14a]. Surprisingly, investigations for Co^{2+} were still lacking when we began our work, which provided us with a good opportunity to explore the Co^{2+}–N_3^- system in addition to other Mn^{2+} and Ni^{2+} systems.

8.2.2 Formato ($HCOO^-$) [16–19] Ion

Since the hydrogen atom is not coordinated or involved in magnetic coupling, the formate anion is actually a three-atom bridge. It has been observed to display multiple bridging modes—2.20, 2.11, 3.21, and 4.22—to link two or more transition-metal ions forming various complexes (Fig. 8.1). Depending on the geometries of the formato bridge (syn-syn, anti-anti, or syn-anti), it can mediate FO or AF coupling in different situations. Interestingly, all of its bridging modes can find their corresponding analog in azido-bridged systems and transfer similar magnetic coupling; for example, with the 2.11 mode both usually mediate AF coupling. These considerable similarities may lead to compounds with similar structures and magnetic properties. Meanwhile, due to the differences in coordinating atoms (O or N), electronic structures (degree of conjugation), and the geometric characteristics (bent or linear) of azido and formato, the resulting structures and the magnetic properties can be very different. Although many formato-bridged compounds have been reported, magnetic studies in this area are strangely lacking. Before our systematic investigation, long-range-ordered systems bridged by formato include the following examples: $[M(HCOO)_2(L)_2]$ [M = Mn, Fe, Co, Ni, Cu, L = H_2O, $HCONH_2$, $(NH_2)_2CO$] [17] dehydrated $[M(HCOO)_2]$ [18], and $[Mn(HCOO)_3] \cdot$ Guest [19]. As we demonstrate in this chapter, many interesting results can be achieved in a metal–formate system.

8.2.3 Cyanato, Thiocyanato, and Selenocyanato (XCN^-, X = O, S, and Se) Ions [20–24]

Although only two SCN^--bridged examples are presented in this chapter, the coordination chemistry and their roles on magnetic coupling of these pseudohalides are summarized here briefly for general interest. Due to their two different donor atoms (O/S/Se and N), XCN^- ions have less versatility and are less efficient as magnetic couplers. Compared with azido, there are much fewer reports on the divalent first-row transition-metal complexes bridged by them; and most are paramagnets above 2 K. Fully studied complexes are mainly compounds of Ni^{2+} and Cu^{2+} [20,21]. For OCN^- bridged Co^{2+} compounds, the first structurally characterized example [22] was reported in 2001. A similar situation pertains for SCN^-

and $SeCN^-$. The 2D compounds $[M(SCN)_2(ROH)_2]$ ($M = Mn^{2+}, Co^{2+}$; $R = CH_3, C_2H_5, C_3H_7$), the 1D chain proposed, $[M(SCN)_2$(2,2-bipyridine)] ($M = Mn^{2+}, Co^{2+}$) [23], and the 1D compound $[M(NCS)_2(HIm)_2]$ ($M = Co^{2+}, Ni^{2+}$) [24] are rare examples investigated thoroughly for SCN^-. As for $SeCN^-$, no example of Mn^{2+} and Co^{2+} can be found in the Cambridge Structural Database (November 2007 version). According to the existing structures, all of them can act as EE bridges and transmit weak magnetic interaction. Actually, the SCN^- and $SeCN^-$ ions adopt the EE mode in most cases. The M–N–C and M–X–C angles and the M–XCN–M dihedral angle influence the sign of the magnetic coupling greatly [21a]. The existing data suggest that EE OCN^- and SCN^- can transmit both AF and F coupling, whereas the $SeCN^-$ transmits mainly ferromagnetic coupling. As for the EO mode, the existing examples are mainly for the OCN^- (using N atoms exclusively) with Ni^{2+} and Cu^{2+} ions. The magnetic coupling transmitted by EO OCN^- is generally weak, with the signs depending on the M–N–M angle.

8.2.4 Nitrito (NO_2^-) Ion [25–28]

The nitrito (NO_2^-) is very unique because unlike the other three-atom bridges, all three of its atoms can coordinate with the metals and it can act as a terminal, bi-, and tri-dentate ligand. As a bridge it can have several different bridging modes, such as 2.200, 2.110, and so on (Fig. 8.1) [25–28]. Interestingly, although it was proposed long ago, the 2.101 mode has not been observed before. The magnetic exchanges through nitrito are mainly antiferromagnetic [27a, 26d, 26f], although ferromagnetic coupling was also observed [26d, 26f, 27h]. Many polynuclear compounds bridged by nitrito, such as 0D clusters [26] and 1D polymers [27], have been synthesized and characterized. Interestingly, 1D compounds with Ni^{2+} bridged by nitrito were investigated thoroughly because the antiferromagnetically coupled 1D chain of $S = 1$ presents the famous Haldane gap phenomenon [27a,27b,28]. Despite these reports, no 2D or 3D polymeric complexes containing nitrito bridges have been documented so far, and this promoted further investigations of this unique bridge.

8.2.5 Hydrogencyanamido ($NCNH^-$) Ion [29]

The $NCNH^-$ ion is one of the two basic forms of cyanamide ($NCNH_2$). Although it is believed that coordination polymers bridged by $NCNH^-$ should exist with regard to well-characterized polymers bridged by its isoelectronic species azido, the coordination chemistry and ability of $NCNH^-$ to mediate magnetic coupling remained almost unexplored. Only three $NCNH^-$ bridged coordination complexes have been reported before we started to explore this interesting ligand. Although there are not enough examples to draw a conclusion on the magnetic interactions transferred by $NCNH^-$, there is a widely investigated analog of $NCNH^-$: the dca ($N(CN)_2^-$) anion, especially in μ-1,3 bridging mode, which was found to provide strong coupling between metal ions. These aspects inspired us to explore $NCNH^-$ as a bridging ligand to construct molecular magnets.

8.2.6 Three-Atom-Like Bridges: Dicyanamido ($N(CN)_2^-$) [30], Imidazolato ($C_3H_4N_2^-$) [31, 32], and Others

Many bridges that are not three-atom ligands themselves can act as three-atom bridges. They use part of the ligand (three atoms involved) to bridge the metal centers, and the bridging part dominates the magnetic properties. In this regard, the dicyanamido-(dca), imidazolato-(im), carboxylato-, and phosphonato-bridged complexes can actually be classified as such and represent a huge category of compounds. We summarize here briefly the normal characteristics of these bridges and their roles in magnetic exchange; one or two examples for each are presented to demonstrate their ability to construct magnetic polymers. For comprehensive views, readers are directed to relevant papers. For dca we focus our attention on the μ-1,3 bridging mode, which can effectively mediate the magnetic coupling, mostly AF, and justifies itself as a good candidate to construct molecular magnets. As for imidazolate, it was found to be very unique in constructing various MOFs, and a good review can be found in ref. 31. Magnetically, efficient magnetic coupling, mostly antiferromagnetic, can be transferred through the ring and lead to weak ferromagnetism in some 3D compounds [32]. Finally, for the carboxylate and phosphonate, the magnetic exchanges through them depend on the structure details, although antiferromagnetic coupling seems more common in the examples reported. On the other hand, other heterocyclic rings, such as triazole, tetrazole, benzeneimidazole, 2,2′-biimidazole, and so on, can also use the imidazolate part to bridge transition-metal ions. These are outside the scope of this chapter.

8.3 CO-LIGANDS, TEMPLATING CATIONS, AND OTHER SHORT BRIDGES

Although magnetic coordination polymers could be constructed by the metal ions and the bridging ligands only, many other components could be involved. Depending on their roles, these components can be divided into two categories: co-ligands and templates. The *co-ligand* is an ancillary ligand, usually an organic molecule coordinating with the metal; while the *template*, being neutral or charged and of specific size and shape, does not coordinate but requires specific space and interactions in the resulting structure. Both factors are very important to the formation of particular architecture and to structure modification, whether or not they are favorable to magnetic interaction.

8.3.1 Co-Ligands [33,34]

The co-ligands can be either terminal (*endo*) or bridging (*exo*). The terminal co-ligands will always block the metal's available coordination sites and decrease the possibility of forming higher-dimensional structures. Magnetically, they are normally adverse to the purpose of achieving high-T_c magnets. However, they make it easier to obtain low-dimensional magnetic materials. Isolated clusters such as SMM,

1D magnetic chains such as SCM, and 2D magnetically anisotropic layers are common products when some terminal co-ligands are included structurally. Due to recent ongoing interests in low-dimensional magnetic systems, the terminal co-ligands have been reinvestigated extensively. In three-atom bridged systems, these terminal co-ligands also play very important roles.

The bridging co-ligands are very important in our study. Considering their ability to transfer magnetic interactions, they can be divided into two groups: magnetically active and inactive. The former co-ligands, such as pyrazine and N-oxidized pyrazine, can not only influence the structures of the complexes but can also modify the magnetic properties directly. On the other hand, magnetically inactive bridges, which are actually the majority of cases, are usually long ditopic organic compounds. Their roles lie mainly in the structural modulation, and they can subtly adjust the magnetic property of the materials by modifying the weaker magnetic interaction, such as interchain or interlayer exchange. Furthermore, the bridging co-ligands can be either rigid or flexible. For rigid co-ligands, prediction of the resulting structures is sometimes possible, although it is very difficult in most cases. On the contrary, the conformational flexibility of the flexible co-ligands adds more freedom and may induce a variety of structures, such as the formation of supramolecular isomers.

8.3.2 Templating Cations

Templated synthesis is of great importance in modern chemistry. It has been employed naturally in the syntheses of many magnetic systems, such as oxalates, azides, and dicyanamides. While co-ligands can have some templating effect, cations (or anions) can act as templates to guide the formation of anionic (cationic) metal–organic frameworks with special structures and magnetic properties. Ammonium cations are perhaps the most popular templates in synthesizing systems such as artificial zeolites [35], metal phosphates [36], polyoxometalates [33], mesoporous MCM-41 materials [37], and metal oxalates [38]. The template effects of ammonium cation lie in (1) the formation of hydrogen bonds between the cation and the templated components, (2) imprinting the transfer of structural information (size and shape) regarding the cation onto the templated architecture, and (3) charge compensation. These effects guide and bring the building blocks (metal ions and ligands) together to form templated assemblies. Recently, we have employed this approach to construct magnetic frameworks of metal formates, metal azides, and so on, and interesting results have been obtained both structurally and magnetically.

Listed in Figure 8.2 are the main co-ligands and template cations incorporated in the three-atom bridged systems. Whether they are terminal or bridging, magnetically active or inactive, rigid or flexible, big or small, they resulted in many interesting magnetic materials, especially some low-dimensional magnetic systems.

8.3.3 Mixed Short Ligands

The short bridges noted above, being effective magnetic mediators, have formed several of the most important families of molecular magnets. However, combining

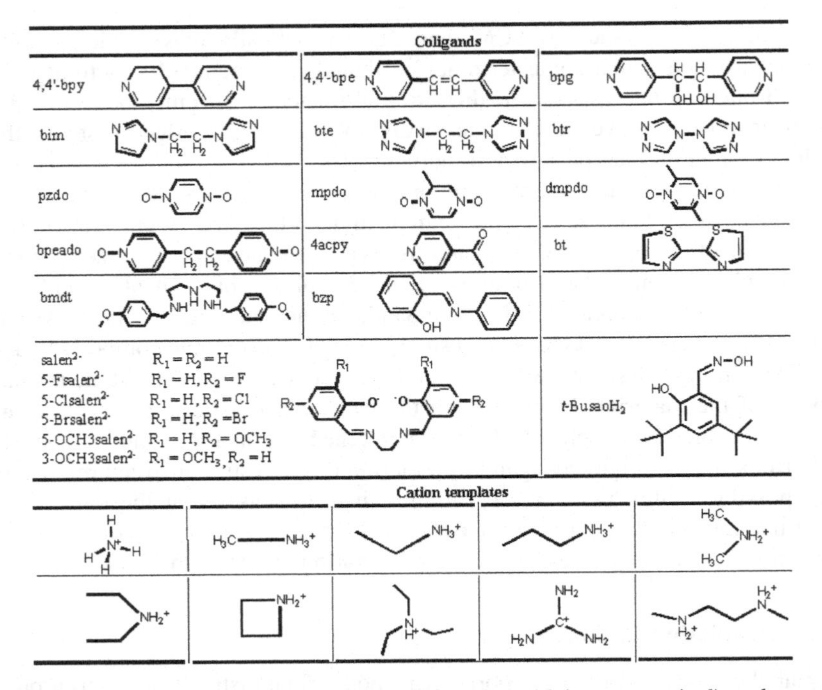

FIGURE 8.2 Schematic structures of the coligands (bridging or terminal) and template cations used in our study.

two or more of these short bridges in one material is still a challenge and of great interest for the rational design and construction of new molecular magnets with special structures and interesting properties with regarded to exploiting their individual advantages of both coordination and magnetism. Several attempts have been made to combine oxalate and azide bridges, or cyanide and oxalate bridges. We demonstrate that, by selecting suitable building blocks carefully, it is possible to prepare new hybrid magnetic materials containing cyanide and azide, azide and hydrazine, azide and carboxylates, and so on.

8.4 MAGNETIC MOLECULAR SOLIDS BASED ON THREE-ATOM BRIDGES

With the three-atom bridges, co-ligands, and templates noted above, we have synthesized many interesting magnetic molecular solids successfully. Some of them will be described briefly, from low to high dimensionality. As for the dimensionality, we refer only to the part bridged by the three-atom bridges, so the dimensionality is specifically that of strong magnetic interactions via these short bridges. The entire framework can be of higher structural dimensionality when connection via the bridging co-ligand is considered. No attempt is made to describe exhaustively the complexes bridged by these bridges. Attention is focused on our own results together with those from some closely related systems.

8.4.1 Single-Molecule Magnets and Single-Chain Magnets

These two amazing systems behave like real magnets but with slow magnetic relaxation and quantum effects. For this reason, SMMs and SCMs have attracted much attention ever since their discovery [5,6]. This prompted us to synthesize such compounds using three-atom bridges and synthetic strategies. The conditions to get a SMM or SCM are very critical. First, an overall easy axis (Ising) type of magnetic anisotropy is definitely required for both. Although no causality exists between the anisotropy of an individual spin and the entire cluster or chain, it is still preferred to try Ising-type metal ions, such as Mn^{3+}, Co^{2+}, and Ni^{2+}. Second, a ground state with large spin and a strong intramolecular magnetic interaction are needed to increase the energy barrier or blocking temperature. These depend strongly on the bridges between spins. Furthermore, to avoid 3D long-range ordering, the intercluster and interchain interactions need to be very weak. For this purpose, terminal co-ligands and long bridging co-ligands can be used to isolate the clusters and the chains. Considering these aspects, several SMMs and SCMs were obtained.

With the terminal chelating co-ligand bzp, a disklike heptanuclear cluster $[Co_7(bzp)_6(N_3)_9(CH_3O)_3](ClO_4)_2(H_2O)_2$ (**1**) can be synthesized (Fig. 8.3) [39]. The azido ligands adopt the EO mode to bridge Co^{2+} centers and transmit efficient FO interaction. The structure is represented by a closest-packing arrangement of donor N/O and Co atoms with local S_6 symmetry. As expected, the bulky bzp capping every Co^{2+} center sufficiently isolates the Co_7 disks from each other. The FO interactions cause the ground state to be $S_T = \frac{7}{2}$ (efficient $S = \frac{1}{2}$ for Co^{2+} at low temperature), and slow relaxation at both zero and nonzero dc fields was observed. However, micro-SQUID measurements on the single crystal revealed a thin hysteresis and suggested

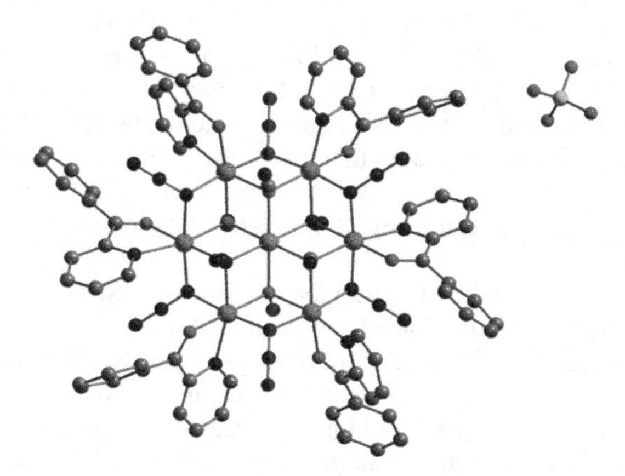

FIGURE 8.3 Azido-bridged heptanuclear Co7 structure of **1**.

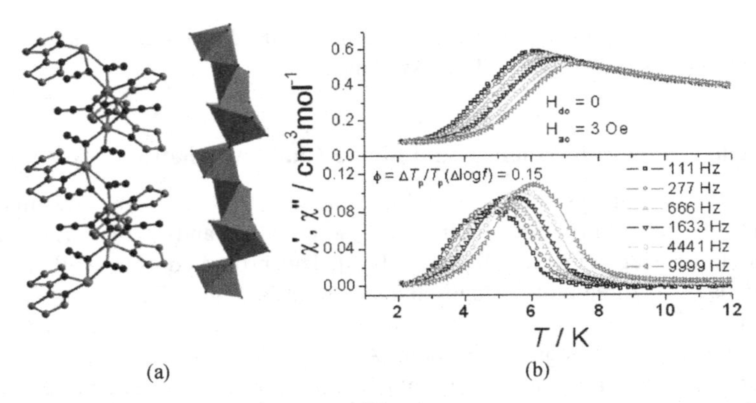

FIGURE 8.4 (a) Helix chain of **2** with Co^{2+} bridged by EO azido; (b) ac susceptibility of **2** under a zero dc field, showing strong frequency dependence.

that **1** is not a real SMM. The main reasons are likely to be the significant spin–orbit interaction and rhombic ZFS parameter E.

The use of another terminal co-ligand, bt, also chelating, gave an infinite chain compound $[Co(N_3)_2(bt)]$ (**2**) with double EO azido bridges (Fig. 8.4) [40]. The bulky co-ligand keeps the helix chains isolated. As expected, the EO azido transmits strong FO interactions along the chain, and the interchain coupling is small enough to prevent **2** from long-range ordering. Slow magnetization relaxation and hysteresis effects confirmed the SCM behavior of **2** with a blocking temperature of 5 K and an energy barrier of 90 K. It is worth noting that compound **2** is the first SCM with homospins [6].

Being confident that the EO azido-bridged Co^{2+} chains are SCMs, provided that there is sufficiently limited interchain interaction, we turned to another approach to generate an isolated 1D Co^{2+} chain by using a long co-ligand. This proved to be successful for compound **3** with the formula $[Co(N_3)_2(H_2O)_2]$(bpeado), where the 1D Co^{2+}–N_3^- chains are isolated by bpeado (Fig. 8.5) [41]. The connections between the Co^{2+} ions in **3** are also double EO azido linkages as in **2**. However, **3** has only one unique Co^{2+} center and the chain is perfectly straight. The bpeado, not acting as a ligand, is hydrogen-bonded to the coordinated water and separates the chains well, with an interchain distance greater than 10 Å. Magnetic measurements on both the powder and a single crystal of **3** show very interesting properties, including strong Ising-type magnetic anisotropy, slow magnetization relaxation, and very large hysteresis loops.

All examples above use the EO azido ion to construct SMMs and SCMs because of its contribution of FO interaction. However, the AF-favored EE azido ion can also serve as a good bridge for SCM and SMM purposes. Specifically, spin canting due to this three-atom bridge can lead to a weak-ferromagnetic ground state under certain conditions, which will also generate high magnetization values in an infinite chain and lead to SCM behavior if other conditions are fulfilled [6,42]. This is the case for the compound $[Ni(\mu\text{-}N_3)(bmdt)(N_3)](DMF)$ (**4**) (Fig. 8.6) [43]. The Ni^{2+} ions were

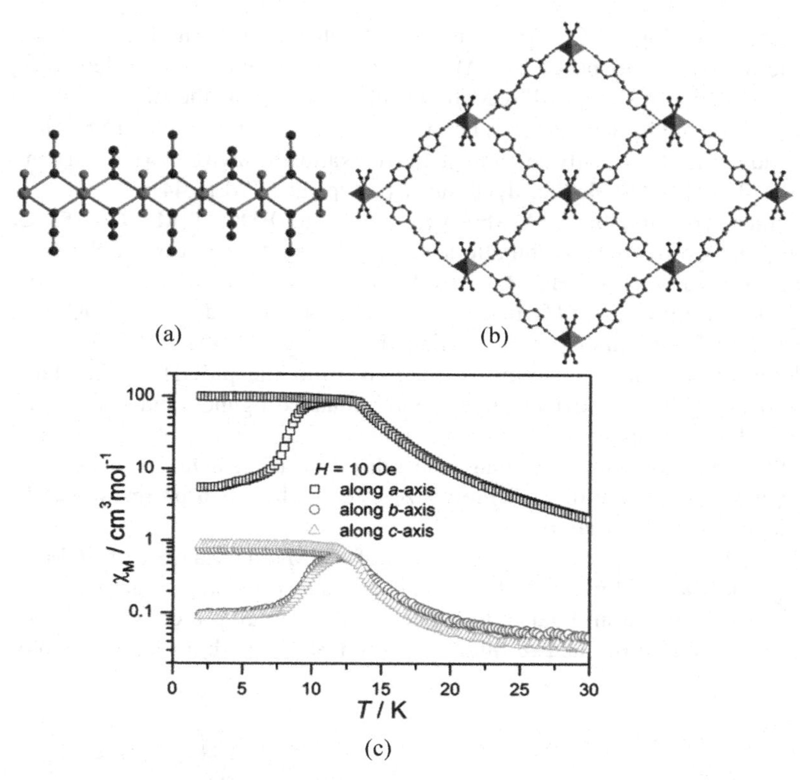

FIGURE 8.5 (a) 1D cobalt-azido chain of **3**; (b) interchain relationship with bpeado hydrogen bonding to the coordination water of Co^{2+} and chain isolation; (c) ZFC/FC curves for **3** under 10 Oe along three different crystallographic axes, showing Ising-type anisotropy.

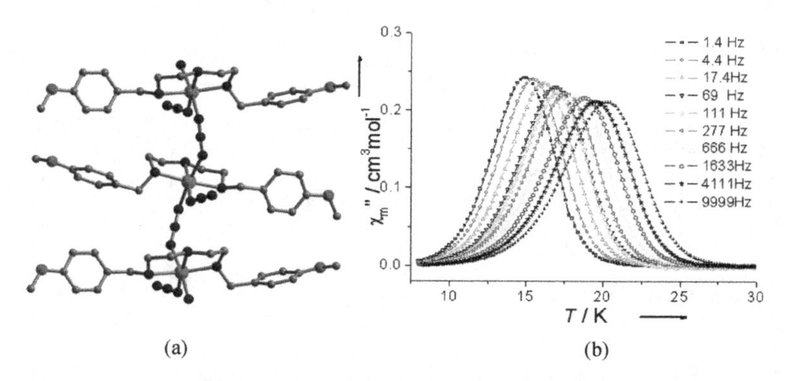

FIGURE 8.6 (a) 1D Ni^{2+} chain bridged by an EE azido of **4**, where the bmdt ligands separate the chains from each other; (b) out-of-phase part of ac susceptibility under a zero dc field.

connected by a single EE azido to form a 1D chain, while the big bmdt helps to separate the chains in space (>9.5 Å). Magnetic measurements revealed strong AF interaction between Ni^{2+} and SCM-like behaviors, such as the slow magnetization relaxation and big hysteresis loops below a blocking temperature of 15 K. However, more study is needed to fully understand the relaxation behavior of **4** since it cannot be interpreted simply by Glauber dynamics as a typical SCM [6,44].

Another two compounds, $[M_3(bim)_2(\mu_3\text{-}OH)_2(HO\text{-}BDC)_2]$ [M = Co, (**5**); Cu (**6**), HO-H_2BDC = 5-hydroxyisophthalic acid], can be synthesized using the co-ligand bim and the carboxylate HO-BDC [45]. Both **5** and **6** exhibit similar structural frameworks resulting from 1D metal–oxygen chains extended by HO-BDC (Fig. 8.7a and b). The HO-BDC and *gauchi* bim ligand help to isolate the 1D chains. Although the dominant magnetic coupling in them is from the μ_3-OH group, these two compounds fit the whole idea of using terminal co-ligands and carboxylates to construct low-dimensional structures. Magnetic measurements reveal that **5** shows SCM-like behavior with slow magnetic relaxation (Fig. 8.7c), a large hysteresis value (Fig. 8.5d), and a distinct finite-size effect, whereas **6** presents overall antiferromagnetic chain behavior.

Although there are now many examples of SMMs, the number of SCMs is still limited. The final example in this section shows a step-by-step strategy to construct SCMs: using three-atom ligands to connect SMM building blocks into 1D chains. It has been used before and resulted in several SCMs with an improved blocking

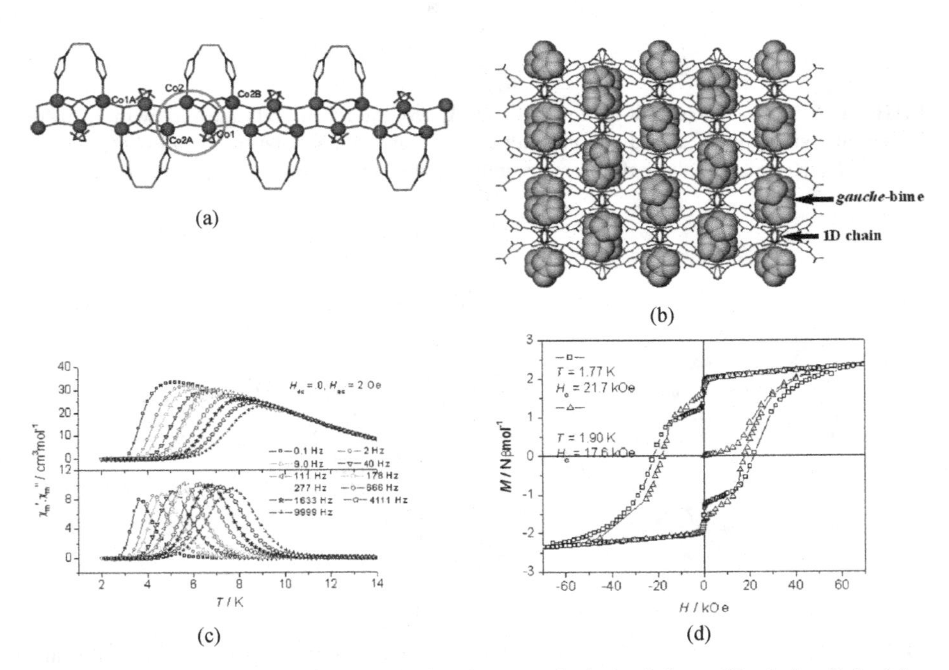

FIGURE 8.7 (a) 1D chain in **5**; (b) 3D framework derived from 1D chains linked by HO-BDC; (c) ac susceptibility of **5** under a zero dc field; (d) hystereses loop of **5**.

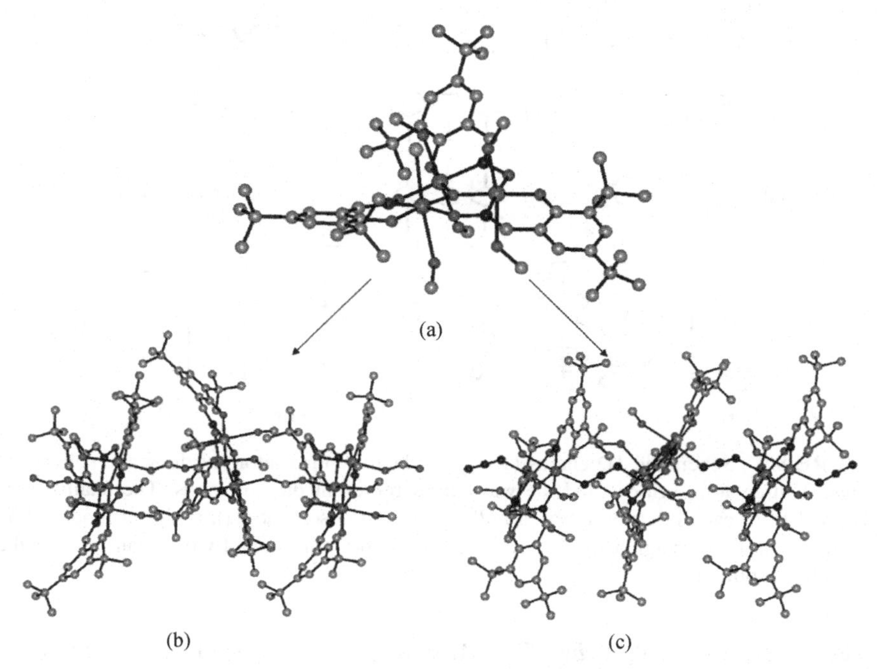

FIGURE 8.8 (a) Molecular structure of **7**. Chain of $[Mn^{3+}{}_3O]$ units bridged (b) by $HCOO^-$ ions in compound **8** and (c) by N_3^- ions in compound **9**.

temperature and a much higher energy barrier than those of the original SMMs [46]. By stringing the oxo-centered trinuclear $[Mn^{3+}{}_3O]$ units [as in the compound $[Mn_3O(t\text{-Busao})_3Cl(CH_3OH)_5]\cdot H_2O$ (**7**)] along the easy axes with formato or azido bridges, we successfully synthesized two 1D compounds, $[Mn_3O(t\text{-Busao})_3(HCOO)(CH_3OH)_4]\cdot CH_3OH\cdot 0.5H_2O$ (**8**) and $[Mn_3O(t\text{-Busao})_3(N_3)(CH_3OH)_4]\cdot 0.5CH_3OH$ (**9**) (Fig. 8.8) [47]. The $[Mn^{3+}{}_3O]$ units remained similar to those in **7**. The 1D chains in **8** and **9** are well isolated from each other by the bulk alkyl groups of *t*-Busao. Magnetically, weak antiferro- and ferromagnetic exchanges were transferred through formato and azido, respectively. Both **8** and **9** show SCM behavior at low temperature, with obvious hysteresis loops and frequency-dependent ac susceptibilities. The blocking temperatures and the energy barriers are indeed much higher than those of **7**, as might arise from enhanced uniaxial anisotropy and intrachain coupling.

8.4.2 Low-Dimensional Molecular Magnetic Systems

As demonstrated above, by using terminal or long bridging co-ligands, discrete clusters and chains that behave as SMM and SCM materials can be achieved successfully. Due to the rigorous conditions necessary for the SCMs, most of the 1D chains obtained do not behave as SCMs. Because they are easier to analyze theoretically, these chains offered good model systems to study the magnetostructural correlations and other magnetic phenomena.

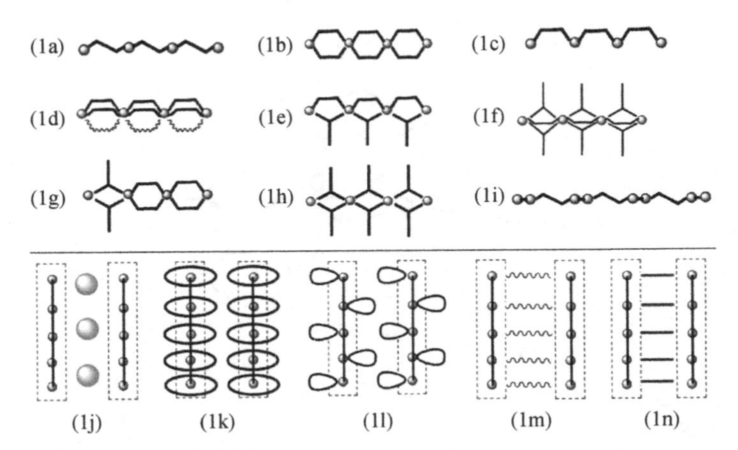

FIGURE 8.9 Schematic 1D chains (1a–1j) bridged by three-atom bridges; the solid lines represent three-atom bridges and the wavy lines represent the coligands. The chains were separated from each other either by a bulky counter ion/solvent (1j), a big terminal ligand (1k and 1l), a long bridging ligand (1m), or a short bridging ligand which can transfer the magnetic coupling (1n).

Magnetic Chains Bridged by Three-Atom Bridges Depicted in Figure 8.9 are some of the chain topologies and interchain relations that we observed (for compound **2**: 1h + 1l; **3**: 1h + 1m, **4**: 1a + 1k, **5** and **6**: 1m, **8** and **9**: 1a + 1l). We are not trying to cover all the possible topologies observed so far. For other 1D topologies, the readers can refer to other reviews for details [9b,48].

Since azido-bridged chain compounds are quite versatile [49], we will focus on only some of our results. When the co-ligand bzp was involved in the Cu^{2+}–N_3^- system, two structurally related exotic 1D tapes, $[Cu_4(N_3)_8(CH_3CN)_3(bzp)_2]$ (**10**) and $[Cu_5(N_3)_{10}(bzp)_2]$ (**11**) (Fig. 8.10), with serial and parallel cyclic eight-membered copper rings were obtained successfully in different solvent systems [50]. All azido are of EO modes (2.20 and 3.30) to bridge the Cu^{2+}. Magnetically, both of them

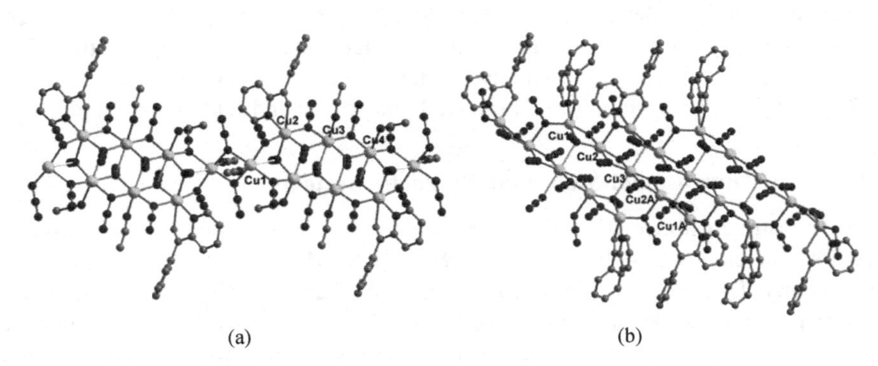

FIGURE 8.10 1D tapes of (a) **5** and (b) **6** with Cu^{2+} bridged by azido.

remain paramagnetic down to 2 K. By analyzing the susceptibility data and broken-symmetry DFT calculation, the EO azido ligands are found to transmit FO coupling in **10** and both AF and FO coupling in **11**. The nature of the magnetic interaction coupled by EO azido was found to be dependent on the Cu–N–Cu angle, the dihedral angle between two basal planes of copper, and the distortion of the coordination geometry of Cu^{2+}.

Besides the SCM compounds **2** and **3**, a series of azido-bridged 1D compounds, $[Mn(N_3)_2(H_2O)_2](bpeado)]$ (**12**) [41] and $[Mn(N_3)_2(bim)]$ (**13**) [51], can be constructed with long bridging co-ligands bpeado and bim. They have the same chain type (1 h). Compound **12** is isostructural to **3** with bpeado hydrogen bonding to the coordinated water, while bim in **13** coordinates to Mn^{2+} and further connects the $[Mn(N_3)_2]_n$ chains into a distorted 3D diamondoid structure. Interestingly, although **12** and **13** have the same type of $[Mn(N_3)_2]_n$ chain, their magnetic properties are totally different. Compound **12** remains paramagnetic above 2 K with the AF interaction, whereas in **13** the magnetic exchange is FO and it has a field-induced spin flop and a metamagnetic transition below $T_C = 3.0$ K. This difference should arise from the Mn–N_{azido}–Mn angle: 99.9° in **7** and 103/106° in **13** [14]. These two examples show us the importance of the co-ligands in the modification of three-atom bridged magnets.

Using formato and 4,4′-bpy, two isostructural compounds, $[M(CHOO)_2(bpy)](H_2O)_5$ [M = Co (**14**), Ni (**15**)], with isolated chains of type 1c + 1m bridged by 2.11-*anti-anti* formato were constructed [52]. They have a 3D uninterpenetrated $CdSO_4$ framework with channels filled with water. Moderate AF coupling was observed and **14** is an antiferromagnet below 3.0 K, while **15** is a weak ferromagnet below 20 K.

Since both azido and formato show excellent performances in constructing new magnetic materials, their combination in one system should also be of interest. In this respect, two new compounds, $[(CH_3)_2NH_2][M(N_3)_2(HCOO)]$ [M = Fe (**16**), Co (**17**)], were synthesized, consisting of chains of $[M(N_3)_2(HCOO)^-]_n$ (type 1f + 1j) isolated by the $[(CH_3)_2NH_2]^+$ cation [53]. They are good examples of obtaining isolated 1D chains by using counter ions. Below about 10 K, metamagnetism arising from the AF-coupled ferromagnetic chain was observed for both of them. Replacement of $[(CH_3)_2NH_2]^+$ by other bulky diamagnetic cations might decrease the interchain interaction further and result in better isolation of the ferromagnetic chains, which is required for the construction of SCMs.

For SCN^- and some flexible bridging co-ligands, we also get two SCN^- bridged cobalt chains: $[Co(SCN)_2(bim)]$ (**18**) and $[Co(SCN)_2(bte)]$ (**19**) [51]. Compound **18** is a triple-bridging chain of type 1d + 1k with double EE SCN^- and one bim as a bridge; and **19** contains 1D chains with double EE SCN^-, and these chains are extended further to (4,4) layers by the *anti*-bte (type 1b + 1m). The EE SCN^- was found to transmit weak FO interaction between the cobalt ions.

Similar to compounds **14** and **15**, another two isostructural compounds $[M(pyrazine)_2NO_2]ClO_4$ [M = Co (**20**), Cu (**21**)] of type 1a + 1n can be constructed by using the nitrito ligand [54]. They were the first examples containing a 2.101-nitrito bridge. The metal centers were bridged by nitrito to form 1D chains, which were

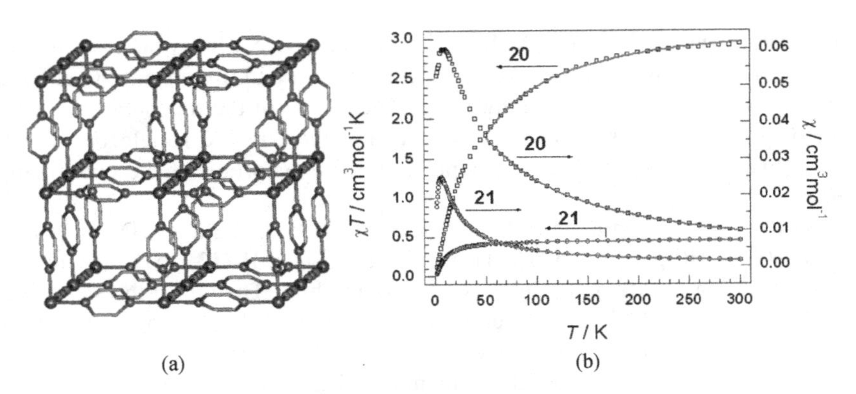

FIGURE 8.11 (a) Three-dimensional structure bridged by nitrite and pyrazine; (b) temperature-dependent susceptibilities for compounds **20** and **21**.

further connected by pyrazine to form a 3D NaCl-type structure (Fig. 8.11a). The magnetic interaction through $\mu_{1,3}$-nitrito bridges were found to be antiferromagnetic in **20** and expected to be rather weak in **21**. Different from the 4,4′-bipyridine in **14** and **15**, pyrazine ligands here were magnetic active and can transfer efficient exchange, leading to the antiferromagnetism in both **20** and **21** at low temperature (Fig. 8.11b).

By introducing various tetradentate Schiff base co-ligands (salen and its derivatives) into the $NCNH^-$ system, a series of chain compounds with Mn^{3+} bridged by $NCNH^-$ were obtained [55]. In these complexes, the resonance structure of $N{\equiv}C{-}NH^-$ dominates the bonding mode of the $NCNH^-$ ligand. Adopting the EE mode, the $NCNH^-$ ligand bridges the mononuclear [Mn(L)] units or the dinuclear $[Mn_2(L)_2]$ units to form 1D chains of types 1a and 1i (Fig. 8.12). The asymmetric bridge of $NCNH^-$ was found to transmit AF interactions between Mn^{3+} ions (J is in the range -0.6 to $-1.2\ cm^{-1}$). The asymmetric character of $NCNH^-$ induces the establishment of spin canting in these chains. Depending on the sign and strength of the interchain coupling, paramagnetism, weak ferromagnetism, or metamagnetism were observed for these chains at low temperatures. Interestingly, after we synthesized the 1D chains for the Mn^{3+}–$NCNH^-$ system, we tried to replace the $NCNH^-$ ligand with the azido ion; and two new azido-bridged Mn^{3+} compounds were obtained successfully [56]. They are of the very similar 1D chains, with the mononuclear Mn(L) units being bridged by EE-N_3^-. Similar magnetic properties, such as weak ferromagnetism at low temperature, were also observed. Although this approach is opposite the original idea of expanding the research from the azido system to the very similar hydrogen cyanamido system, it proves that three-atom bridges are very good for the construction of molecular magnetic materials. We can actually explore more and more interesting systems by referring one bridge to another.

2D Layers Bridged by Three-Atom Bridges The introduction of co-ligands to three-atom bridged magnetic systems can also generate 2D layers. Depicted in

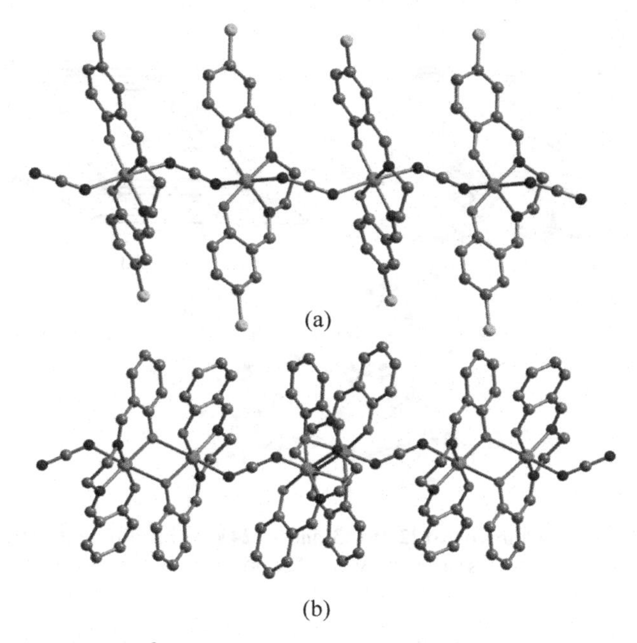

FIGURE 8.12 Typical Mn^{3+} chains bridged by $NCNH^-$ with the repeating unit to be (a) a mononuclear [Mn(L)] or (b) a binuclear $[Mn_2(L)_2]$.

Figure 8.13 are the main topologies of the three-atom bridged 2D layers we obtained, together with the interlayer relationships. Most of these 2D compounds show long-range magnetic ordering, provided that the layers are not too far from each other.

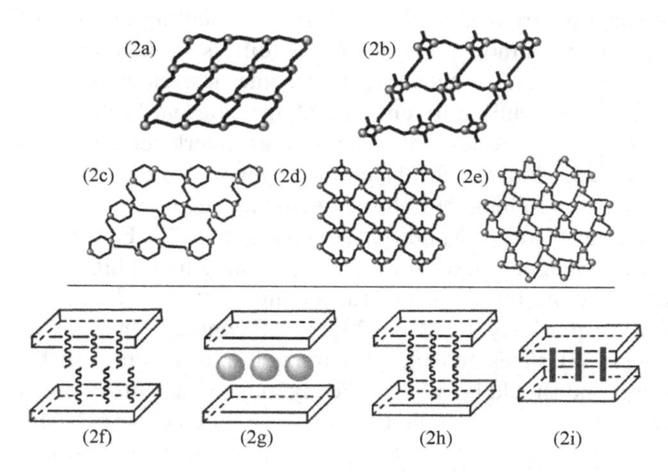

FIGURE 8.13 Schematic 2D layers (2a–2e) and the interlayer relationship.

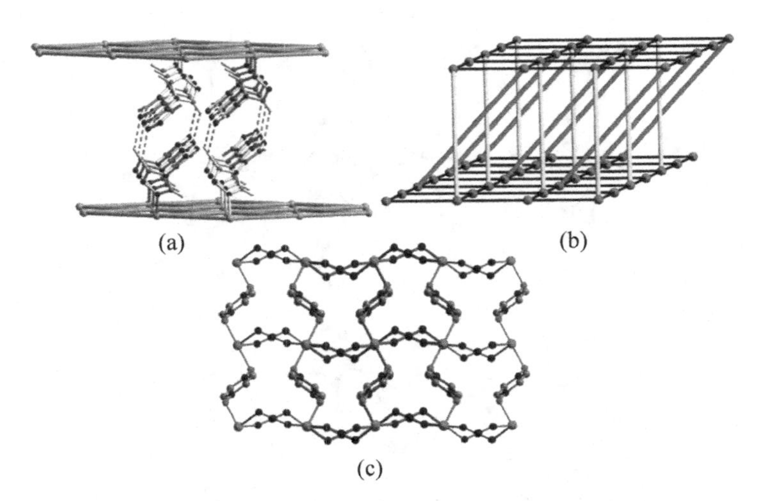

FIGURE 8.14 3D structures of (a) **22**, (b) **23**, and (c) **24** showing different interlayer relations with similar Mn^{2+} (4,4) layers bridged by an EE azido.

For azido-bridged systems, layers of types 2a to 2d had been reported before [14,57]. However, most of the examples concerned only Mn^{2+} and Ni^{2+} ions. For Co^{2+}, the existing examples were relatively few. Therefore, we focused mainly on cobalt systems, together with some Mn and Ni systems along with some new co-ligands. For the Mn–azido system, a series of compounds with a type 2a layer with EE azido can be obtained with different co-ligands: the terminal ligand btr ($[Mn(N_3)_2(btr)_2]$, **22**) [58], the magnetic inactive bridging ligand bpg ($[Mn(N_3)_2(bpg)]$, **23**) [59], and the magnetic active bridging ligand pzdo ($[Mn(N_3)_2(pzdo)]$, **24**) [60] (Fig. 8.14). The basic structural characteristics of the (4,4) layer are similar to each other and to some reported examples [14,61]. AF coupling through EE azido was found for all of them, although the magnitude differs slightly, due to some subtle differences in the details of the bridging geometries. However, their interlayer connections are significantly different. For **22**, the adjacent layers are connected by weak CH $\cdots$ N hydrogen bonds, where the shortest interlayer Mn $\cdots$ Mn distance is 12.4 Å (2f). For **23**, although the layers are connected covalently, the interlayer Mn–Mn distance is even longer (13.8 Å), due to the long ligand bpg (2 h). As for **24**, the shortest interlayer Mn $\cdots$ Mn distance is about 6 Å (2i). Furthermore, pzdo can transmit moderate magnetic exchange [60,62], leading to the higher T_c value of **24**. Magnetic investigations revealed the critical temperatures for **22**, **23**, and **24** as 23.7, 10.8, and 62 K, respectively. In **22** and **23**, the AF-coupled Mn^{2+} spins cant to each other, resulting in WF states. In fact, spin canting is quite common in these 2D azido-bridged systems and should be related directly to the noncentrosymmetric character of the EE azido. Interestingly, detailed magnetic measurements on the single crystal of **22** revealed the coexistence of hidden spin canting, metamagnetism, and spin flop. Magnetic phase diagrams in both the T–H^{a^*} and the T–H^b planes were determined

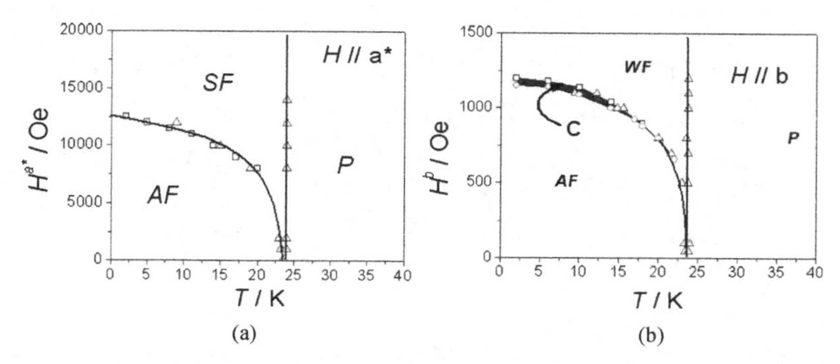

FIGURE 8.15 Magnetic phase diagrams of **22** in (a) the T–H^{a^*} and (b) the T–H^b planes.

(Fig. 8.15) and possible spin configurations before and after the phase transitions were proposed. The interlayer coupling, despite its weakness compared to intralayer coupling through azido, plays a very important role in its magnetic behavior.

For **24** with layers pillared by pzdo, the efficient coupling between the layers increases the T_N remarkably up to 62 K, which is proved by the magnetic measurements and powder neutron diffraction on the fully deuterated sample. In fact, **24** is a 3D system regarding the magnetic interactions along the three dimensionalities. This entire series of results demonstrates clearly the importance of the co-ligands in the three-atom bridged magnetic materials. Besides the ability to modify the structures, they can influence the magnetic properties greatly, not only by means of generating new structures and adjusting the interlayer distances to modify the magnetic interactions indirectly, but also by participating directly in the magnetic coupling.

Similar and even more interesting is the less investigated 2D Co^{2+}–azido system. For Co^{2+}, another important issue, the strong anisotropy, deserves much more attention because it will generally lead to the WF state with remarkable spin canting, together with DM interaction. With bpg as a co-ligand, a series of Co^{2+}-azido layers of types 2a, 2b, and 2e with the formula $[Co(N_3)_2(bpg)](S)_n$ (**25**, S = nothing; **26**, S = DMSO, $n = 1$; **27**, S = DMF, $n = 4/3$) were obtained successfully simply by using different solvents during the syntheses (Fig. 8.16) [63]. The bpg links the 2D layer to form pillared layer architectures (2 h). DMSO and DMF are crucial for the generation of **26** and **27** since various hydrogen bonds exist between the OH groups in bpg and the guest molecules. For **27**, the topology can be considered to be the same as **25** if the EO-N_3^- bridged $[Co_2]$ dimer is considered as one node. The layer of **27**, a Kagomé lattice, was constructed by corner-sharing triangles and was interesting due to its possible geometrical frustration [64]. Significant AF exchange is transmitted through the EE azido for **25** to **27**. As mentioned above, no inversion centers exist in the middle of the adjacent AF-coupled Co^{2+} centers ($[Co_2]$ for **26**), although these compounds all belong to centrosymmetric space groups. As expected, all of them are weak ferromagnets below T_c. The rare molecule-based Kagomé compound **27** shows strong spin frustration, as revealed by the large f ($f = |\theta|/T_c = 10$) parameter [64] and reenters a spin glass state at low temperature.

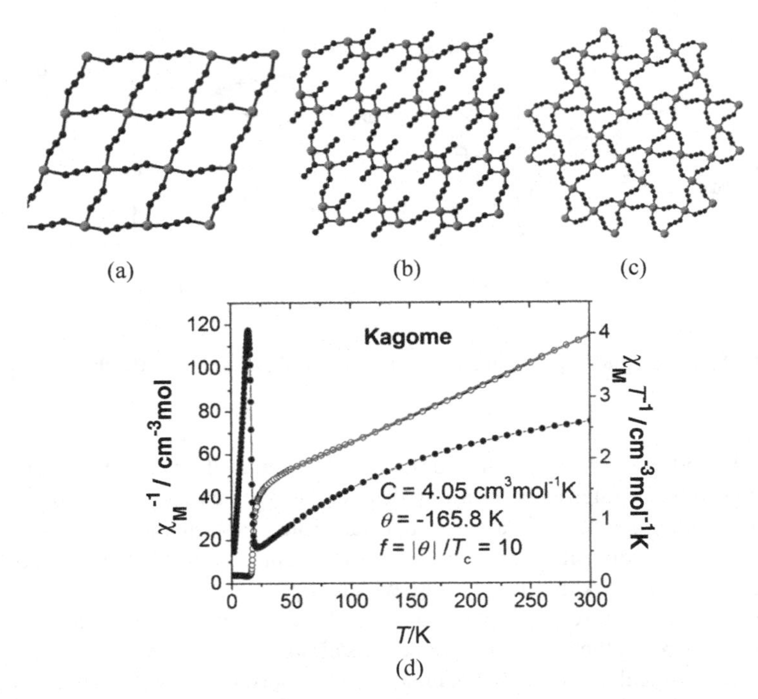

FIGURE 8.16 Azido bridged Co^{2+} layers of (a) **25,** (b) **26,** and (c) **27** with the topology of square, honeycomb, and Kagomé; (d) temperature-dependent inverse susceptibility of **27**, showing the strong spin frustration.

As for future applications, molecular magnets with large and permanent spontaneous magnetization below T_c continue to be pursued. However, because FO coupling is not as common and strong as AF coupling, designed ferromagnets are limited. Other approaches to this target should be explored. As a substitute, ferrimagnets with high magnetization arising from the uncompensated magnetization of two magnetic lattices have attracted much attention [1c,65]. Most synthesized high-T_c Prussian Blue magnets are actually ferrimagnets [3b,3c,9a,9e]. On the other hand, compound **26** shows a hint of another efficient approach to obtain strong magnetization magnets using AF interaction. We call it a weak ferromagnetic approach. Although numerous molecular weak ferromagnets were reported, these investigations remained mainly on the theoretical level since most of them have small canting angles and thus small spontaneous magnetization. However, as long as the canting is efficient, large magnetization can also be generated [30c,34c,66,67]. When the canting angle is 20°, the resulting net moment could be 34.2% (sin 20°) of the total. This is even more efficient than the ferrimagnetic alignment of two heterospins with S and $2S$ (the net moment is S, 33.3% of the total $3S$). The canting angle is 5.2° in **26**, resulting in about a 10% residue of the spin. As a matter of fact, in another compound, $Co(N_3)_2(4acpy)_2$ (**28**), which also has the (4,4) layer of Co^{2+}

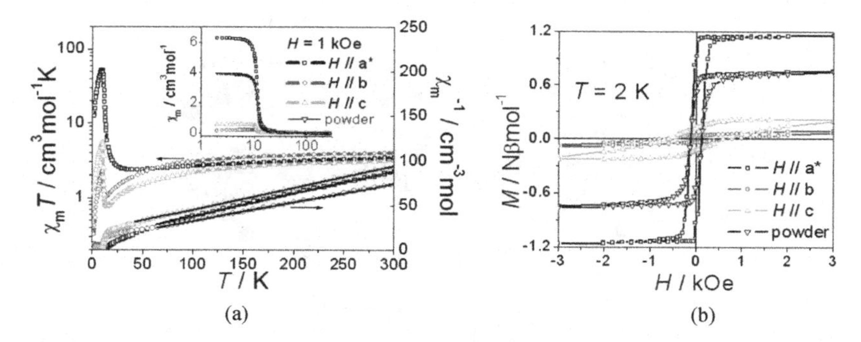

FIGURE 8.17 Anisotropic magnetic measurements for compound **28**: (a) susceptibilities; (b) hysteresis loops.

bridged by EE azido, the canting angle can reach 15° and the residual magnetization is up to 25% of the total spin [67]. As can be seen from detailed anisotropic magnetic measurement of the susceptibilities (Fig. 8.17a) and hysteresis loops (Fig. 8.17b) along different magnetic axes, compound **28** shows very high spontaneous magnetization, similar to that of a typical ferromagnet. Among the molecular weak ferromagnets reported, those with large canting angles are quite rare [30c,34c,66]. Interestingly, all these examples have two important common aspects: strong anisotropic spins and three-atom bridges. This will lead to some rational design of molecular weak ferromagnets with high spontaneous magnetization.

In azido compound **22**, the spin-flop transition occurred under external field parallel to the spin. However, due to the strong AF coupling between Mn^{2+} transmitted by N_3^-, the critical field H_c (defined as the field transferring the spin-flop state to a totally saturated state) [1a] is too high to achieve experimentally. Considering the similarity and difference of azido and formato, 2D layers of Mn^{2+} bridged by $HCOO^-$ may have a similar spin-flop transition and the H_c might be lowered remarkably, due to the weaker exchange through formato. This is the case in $[Mn_2(HCOO)_3(4,4'\text{-bpe})_3(H_2O)_2](H_2O)(ClO_4)$ (**29**) [68], which is also a layer structure with Mn–HCOO herringbone (6,3) layers pillared by 4,4′-bpe, with a ClO_4^- anion and lattice 4,4′-bpe residing in between (Fig. 8.18a and b). Investigation of a single crystal revealed an antiferromagnet with spins all perpendicular to the layer below $T_c = 2.5$ K, and a spin-flop transition occurred under an external field parallel to the spins (Fig. 8.18c). Due to the weaker interaction through $HCOO^-$, its T_c and H_c values are both lower than those of **22**, falling in the measurement range of the SQUID system. Analysis of the data gives a detailed phase diagram and further intrinsic parameters. In addition, its magnetic properties, especially the critical field H_c, can be finely tuned by anions of different sizes, such as NO_3^-, Br^-, I^-, and BF_4^-, which can be included by choosing different starting materials. Furthermore, the stack of the 4,4′-bpe in the lattice satisfies the Schmidt's geometric criteria for [2 + 2] photodimerization [69]. Upon exposure to ultraviolet light, the lattice bpe molecules become photoactive to dimerize to tetrakis(4-pyridyl)cyclobutane. Although it is now difficult to study the influence of the photoreaction on its magnetic property,

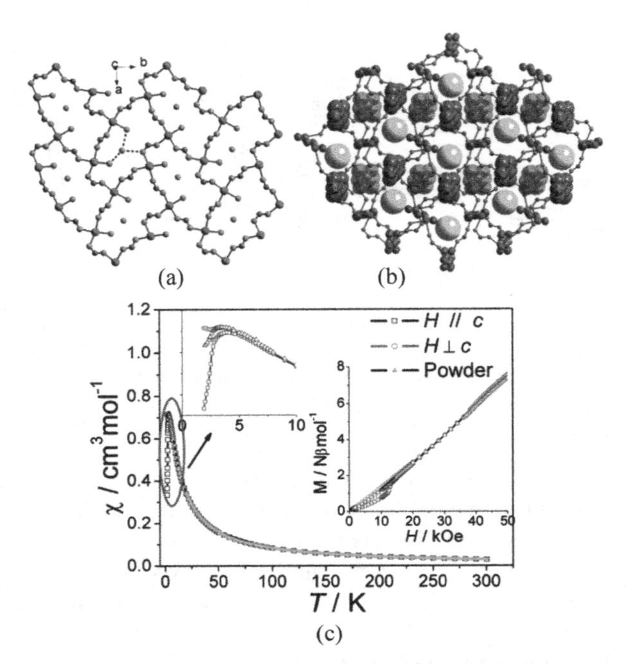

FIGURE 8.18 (a) 2D herringbone layer of **29** with Mn^{2+} bridged by $HCOO^-$; (b) pillared-layer structure where the big balls are for the anion and the lattice bpe molecules form 1D columns along b with distances smaller than 4.2 Å; (c) temperature-dependent χ and field-dependent magnetizations at 2 K for a single crystal ($H \parallel c$ and $H \perp c$) and a powder sample.

29 provides a good example of the design of the multifunctional materials using functional co-ligands.

As for the dicyanamide (dca) system, the employment of the co-ligands pzdo, mpdo, and 2,5-dmpdo also works to give layer compounds $[Co[N(CN)_2]_2(L)]$ [L = pzdo (**30**), mpdo (**31**), and 2,5-dmpdo (**32**)] [70]. Both **30** and **31** contain similar 2D triangular layers with Co^{2+} bridged by mixed 1,5-μ_2- and μ_3-dca (Fig. 8.19a). The pzdo and mpdo use one of the two N-oxido atoms to coordinate to cobalt and act as terminal ligands to separate the adjacent layers (type 2f). Owing to the μ_3-dca, long-range ferromagnetic ordering was observed for both **30** and **31** below about 2.5 K. Actually, this 2D network was also observed in the compounds $[M(dca)_2(H_2O)]$(phenazine) (M = Co, Ni), where long-range order was observed in the Ni^{2+} complex [70c]. Similarly, the dca ligands in **32** use 1,5-μ_2 and 1,3-μ_2 modes to link Co^{2+} into a (4,4) layer (Fig. 8.19b). Interestingly, the co-ligand 2,5-dmpdo also acts as a bridge to pillar the 2D layers and as an intermediate for the magnetic exchange (type 2i). Thus, T_c of **32** is increased to 10.8 K. Here, we can see the sensitivity of these systems because even a small change in the co-ligands can change the final structures and magnetic properties dramatically.

Similar to the idea of combining N_3^- and $HCOO^-$ into one system, the N_3^- can cooperate with other short bridges, the hydrazine N_2H_4 and cyanide CN^-, and gave

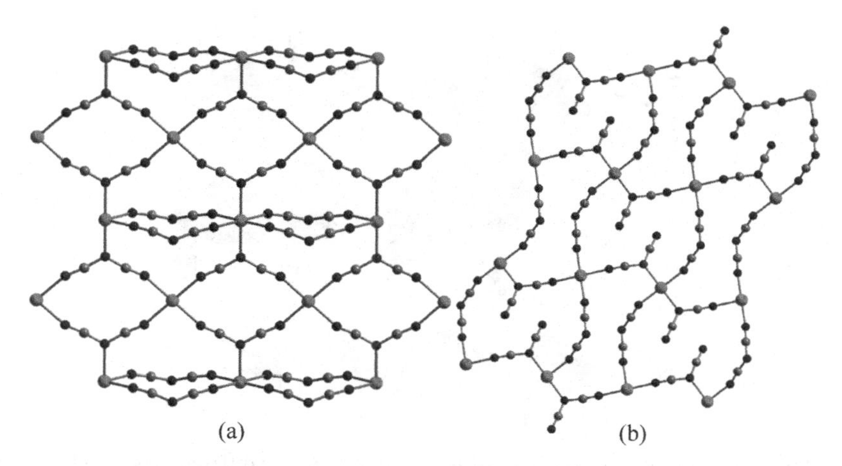

FIGURE 8.19 2D layers bridged by dca of (a) **30–31** and (b) **32**.

some other interesting results. The first example here is the simple 2D compound with the formula $Co(N_2H_4)(N_3)_2$ (**33**) [71]. It is a honeycomb layer with N_2H_4 and N_3^- (of both EE and EO modes) bridging at the same time (Fig. 8.20). Magnetic measurement reveals that compound **33** exhibits spin-canted weak ferromagnetism with a T_c value of 13.5 K. On the other hand, including N_3^- in the CN^- bridged compound $[Cr(phen)(CN)_4]_2[Mn(H_2O)_2](H_2O)_4$ (**34**) extends the structural dimension and gives a 2D compound, $Mn(N_3)(CH_3OH)[Cr(phen)(CN)_4](CH_3OH)$ (**35**) [72]. All the N_3^- bridges are in the EO mode and responsible for FO coupling. With the structure increasing from 1D to 2D, the T_c also increases dramatically, from 3.4 K for **34** to 21.8 K for **35**. This mixed-bridge approach is currently attractive for the cyanide system. Combination of the same system with other bridges, such as oxalate and dca, also leads to other new compounds with higher structural dimensionalities and other interesting aspects [73,74]. This lies outside the scope of this chapter and is not discussed in detail.

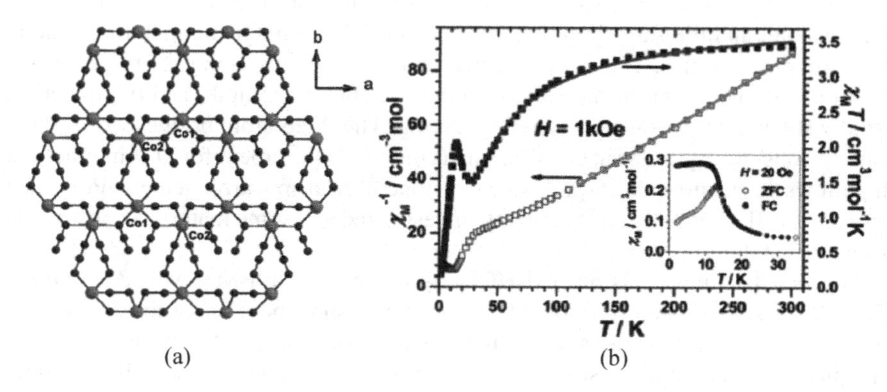

FIGURE 8.20 Honeycomb layer of compound **33** bridged by EO/EE azido and hydrazine.

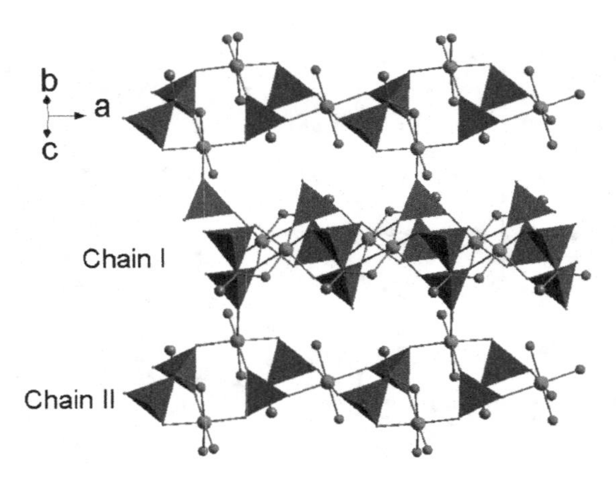

FIGURE 8.21 2D layer of compound **36** constructed from two different types of chains. The phosphonates shown in polyhedra act as three-atom bridges.

The final example given in this section for the 2D system is a phosphonate bridged compound, $[Na_6Co_7(hedp)_2(hedpH)_4(H_2O)_4](H_2O)_8$ (**36**; hedp = 1-hydroxyethylidenediphosphonate) [75]. The Co^{2+} ions in the structure were all bridged by the O–P–O and single oxygen bridges to form a 2D layer, which can be viewed as made up of two types of chains (Fig. 8.21). The 2D layers are then connected to each other further by sodium ions. The phosphonate groups in **36** transmit efficient AF coupling, leading to antiferromagnetic ordering below $T_c = 4.0$ K. Interestingly, two-step field-induced transition, probably from an AF state to a ferri- and to a final ferromagnetic state, was observed below T_c. We hope to use this example as a representative for three-atom-like phosphonate ligands.

8.4.3 3D Systems, Weak Ferromagnets, and Porous Magnets

As mentioned before, the 3D systems discussed here are also magnetically three-dimensional, and the short bridging ligands usually link magnetic metal ions along three dimensionalities. Therefore, the magnetic interactions along the three dimensionalities in the materials are similar or comparable in strength, and the materials usually show 3D long-range magnetic ordering. The three-atom bridges we used are usually noncentrosymmetric bridges, satisfying the requirement for DM interaction; this allows the occurrence of spin canting or noncollinear spin arrangement. Presented here are the 3D metal–formates recently investigated and some materials constructed from imidazolate.

The first two materials are $[M(HCOO)_2(4,4'\text{-bpy})]$ with M = Mn (**37**) and Co (**38**) [52]. They are interesting mainly because they show both many similarities and differences between azido and formate. The two compounds are isostructural, showing a 3D diamondoid structure if the formates are considered as linkers only (Fig. 8.22). Interestingly, the compound $[Mn(N_3)_2(4,4'\text{-bpy})]$ has a very similar

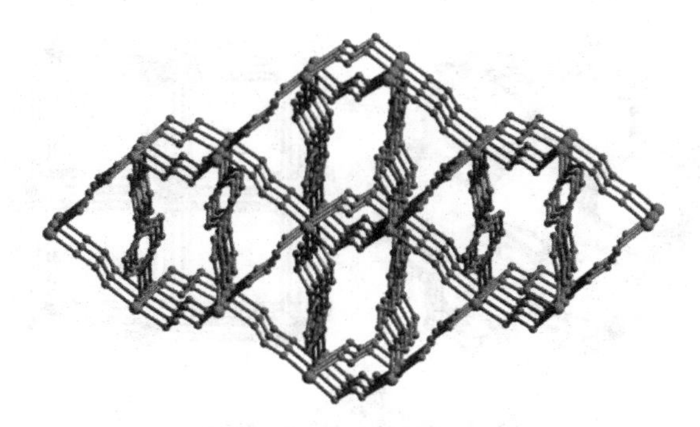

FIGURE 8.22 Formate bridged diamondoid structure of **37** and **38**.

structure except that the linkers are azido [34a,b]. Furthermore, due to the similar AF coupling transmitted through formato and azido and their noncentrosymmetric characters, **37**, **38**, and their azido analog all show weak ferromagnetism below T_c. Of course, the critical temperature of **37** is significantly lower than that of $[Mn(N_3)_2(4,4'\text{-bpy})]$ because of the weaker coupling through $HCOO^-$. This similarity and difference of $HCOO^-$ and N_3^-, as also revealed in the amine-templated metal–formate frameworks and their azido analog $[(CH_3)_4N][Mn(N_3)_3]$ [76], provided us with a unique synthetic approach to these compounds.

The second series of materials described here are the metal–formate frameworks templated by cations of protonated amines [77–80]. Systematic investigations demonstrated that these materials can be obtained by employing protonated amine cations of different sizes in noncoordinating solvent, and the outcome is dependent on the size and shape of the cations. The smallest ammonium NH_4^+ resulted in the chiral magnetic salt $[NH_4][M(HCOO)_3]$ [77], although the starting materials are achiral. The structure is a framework with the rarely observed $4^9{\cdot}6^6$ topology (Fig. 8.23a), consisting of octahedral metal centers connected by 2.11-*anti-anti* formato ions and the ammonium cation arrays located in the channels. For the midsized mono-ammonium cations (AmineH$^+$) $CH_3NH_3^+$, $(CH_3)_2NH_2^+$, $CH_3CH_2NH_3^+$, and $(CH_2)_3NH_2^+$, the outcome is a series of perovskite compounds [AmineH$^+$] $[M(HCOO)_3^-]$ [78] with a NaCl type of framework (Fig. 8.23b). The link between metal sites are also 2.11-*anti-anti* formates, and the cations are in the cubic cavities of the framework. We have recently found that guanidinium ion can also be incorporated in this type of metal–formate framework [79]. However, employment of bulky AmineH$^+$ cations such as $(CH_3CH_2)_3NH^+$ $(CH_3CH_2)_2NH_2^+$ and $CH_3CH_2CH_2NH_3^+$ resulted in the porous $[M_3(HCOO)_6]$ family discussed below. Very recently, this study has been expanded into investigating diamines and polyamines, and the outcomes appear to be very interesting. For example, the employment of protonated *N,N'*-dimethylethylenediamine (dmenH$_2^{2+}$) has templated the formation of a novel binodal metal–formate framework of $(4^{12}{\cdot}6^3)(4^9{\cdot}6^6)$ topology (Fig. 8.23c), closely related to the minerals niccolite (NiAs) and colquiriite ($LiCaAlF_6$) [80]. These materials, in most cases, show 3D long-range antiferromagnetic ordering with weak ferromagnetism arising from the antisymmetric exchange via the formato bridges,

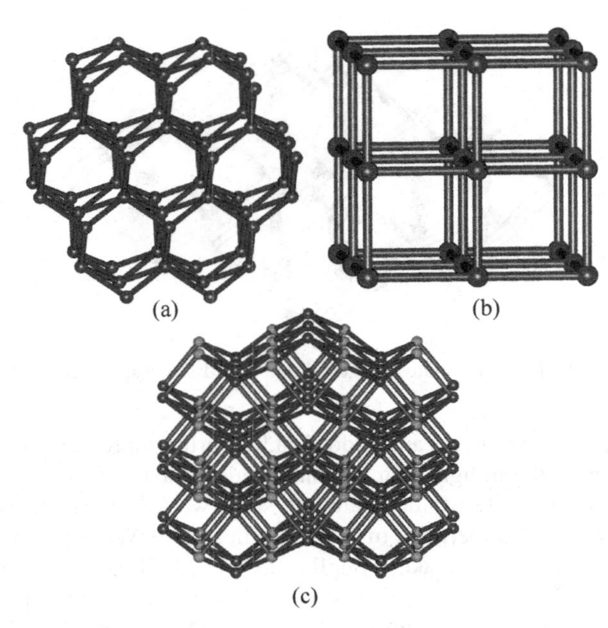

FIGURE 8.23 Topological plots of 3D anionic metal–formate frameworks templated by (a) NH_4^+, (b) $CH_3NH_3^+$, and (c) $dmenH_2^{2+}$. The spheres are metal ions and the sticks are 2.11-*anti-anti* formato bridges. In (c) one cavity is highlighted in red, green spheres are $(4^{12}{\cdot}6^3)$ node, blue spheres $(4^9{\cdot}6^6)$ node.

and in some cases, spin reorientation occurs. The critical temperatures can be as high as 30 K (for Ni compounds). We believe that the metal–formate frameworks display significant malleability and adaptability to protonated amine templates, and we expect that new metal–formate compounds with novel structures and topologies and interesting magnetic properties will probably result when different di- or polyammonium templates are employed.

The final series of formato-bridged compounds are a novel family of porous magnets $[M_3(HCOO)_6]$ (M = Mg, Zn, Mn, Fe, Co, and Ni), which has recently been discovered independently by several groups and by us [81–85]. Although other groups synthesized the compounds through solvothermal routes [82–84], we first demonstrated that the compounds could be prepared easily by a simple solution chemistry method at ambient temperature, in which bulky protonated amines such as triethylamine, diethylamine, or propylamine were employed [81,85]. We believe that the bulky ammonium ions inhibit the aforementioned anionic frameworks as what was observed for the amine cations of smaller size. The isostructural members of this family possess porous diamondoid frameworks consisting of the apex-sharing M-centered MM_4 tetrahedral nodes (Fig. 8.24a and b), in which all metal atoms are octahedral. The MM_4 node has one central M^{2+} ion, four apical M^{2+} ions, and six edges of $HCOO^-$ groups, which link the metal ions in 3.12-*syn-syn/anti* mode. The framework has high stability and flexibility to allow easy removal of solvents

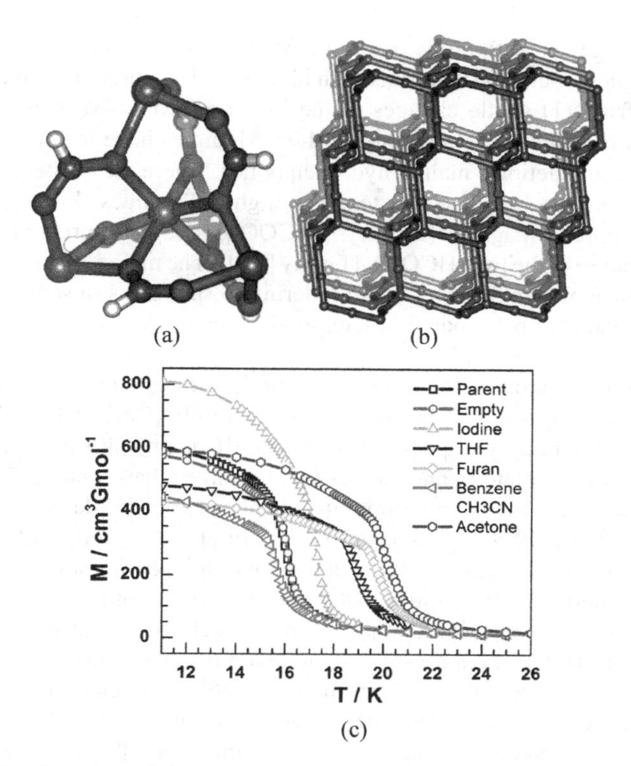

FIGURE 8.24 Structure of the porous diamond framework of $[M_3(HCOO)_6]$ formed by M-centered MM_4 tetrahedra as nodes sharing apices and showing open channels: (a) MM4 tetrahedron unit; (b) diamond framework of $[M_3(HCOO)_6]$; (c) FC measurements of $[Fe_3(HCOO)_6]$ and its guest-inclusion compound series with the guest names shown under an applied field of 10 Oe. "Parent" is the parent compound with guests of methanol and water and "Empty" is the framework without a guest.

and further inclusion of many types of guest without destroying the crystallinity of the crystals. This provides the opportunity to study the guest-inclusion behavior by single-crystal x-ray diffraction. The guest molecules usually form zigzag molecular arrays in the channels of host framework, and host–guest hydrogen bonding provides the primary host–guest interaction. With respect to magnetism, Mn and Fe members are considered as ferrimagnets, with critical temperatures of 8.0 and 16.1 K, respectively; the Co analog is probably an antiferromagnet with a spin canting below 2 K; and the Ni compound might show 3D long-range ferrimagnetic ordering at 2.7 K [81,82]. The Mn and Ni compounds are soft magnets, whereas the Fe compound is quite a hard magnet. Guest-dependent magnetism has been observed and investigated for Mn and Fe frameworks, which display guest-modulated critical temperature. For the Mn frameworks, T_c values were modulated by guests in the range 4.8 to 9.7 K, compared with 8.0 K for $[Mn_3(HCOO)_6]$ without guests, whereas for Fe, T_c changes from 15.6 to 20.7 K compared to 16.1 K for the guest-free $[Fe_3(HCOO)_6]$

(Fig. 8.24c), together with higher coercive fields (H_c) and remnant magnetizations (R_M) for the guest-inclusion materials with higher T_c. The guest-sensitive magnetism might arise from (1) subtle changes in the important M–O–M angles of the host framework, where materials with larger M–O–M angles have lower T_c values, and (2) host–guest interactions, mainly hydrogen bonds, where the existence of stronger host–guest hydrogen bonds seems to favor higher T_c values. Finally, a series of mixed-metal porous magnets $[Fe_xZn_{3-x}(HCOO)_6]$ could be prepared due to the isomorphic nature of the $[M_3(HCOO)_6]$ family [81d]. The mixed-metal series exhibits systematic change from 3D long-range ordering to spin glass to superparamagnets, and finally, to paramagnets upon the increase of diamagnetic component of zinc in the series.

Finally, as an example of the imidazolato-bridged magnetic system, we present a series of cobalt compounds: $[Co(im)_2](pyridine)_{0.5}$ (**39**), $[Co(im)_2](cyclohexanol)_{0.5}$ (**40**), $[Co(im)_2]$ (space group $I4_1cd$, **41**), and $[Co_5(im)_{10}](3\text{-methyl-1-butanol})_2$ (**42**) [86]. All imidazolate ligands act as three-atom bridges; and all Co^{2+} centers show tetrahedral geometry, resulting in silicalike frameworks for all of them. Compounds **39** and **40** are porous and isostructural with a uninodal (6,4)-net, **41** shows a network topology of banalsite, and **42** has an open framework with zeolitic topology. Magnetically, the imidazolates transmit AF coupling between the Co^{2+} ions in all cases. Except for compound **39**, which is an antiferromagnet with $T_N = 13.1$ K, the rest of them all show spin-canted magnetic structure. As discussed above, spin canting should arise from the three-atom characteristic of imidazolate as observed in another two imidazolate-bridged iron compounds [32d,32f]. However, spin canting is sensitive to the structures: Compound **40** shows a very weak ferromagnetism below 15 K, **41** displays a strong ferromagnetism below 15.5 K with a coercive field of about 7 kOe at 1.8 K, and compound **42** behaves as a hidden canted antiferromagnet with a magnetic ordering temperature of 10.6 K.

8.5 CONCLUSIONS

In this chapter we have demonstrated that the versatility of both coordination mode and magnetic transmission of several three-atom bridges allowed us to prepare various molecular magnetic systems, varying from SMMs, SCMs, to higher-dimensional magnets, and even to multifunctional materials such as photoactive and porous magnets. These effective magnetic couplers, whether having been widely employed (such as azido) or not (such as formato and hydrogencyanamide), still provide new opportunities to design and create novel molecular-based magnetic materials. The incorporation of various co-ligands and/or templates provides a route for systematic design of these materials. Furthermore, these organic components in inorganic–organic hybrid materials can be carefully tailored, with specific functions and properties, which will result in multifunctionality. Other transition-metal centers, such as the air-sensitive V^{2+} and Cr^{2+}, the 4d/5d elements, and other bridging co-ligands/templates with different connectivities and additional functions, such as chirality and photoreactivity, will be featured in future investigations.

Acknowledgments

We thank our graduate students and academic collaborators for their valuable contributions, and acknowledge the support of the National Natural Science Foundation of China (20221101, 20490210, and 20571005), the National Basic Research Program of China (2006CB601102), and the Research Fund for the Doctoral Program of Higher Education (20050001002).

REFERENCES

1. (a) Carlin, R. L.; Van Duyneveldt, A. J. *Magnetic Properties of Transition Metal Compounds*, Springer-Verlag, New York, **1977**. (b) Carlin, R. L. *Magnetochemistry*, Springer-Verlag, Berlin, **1986**. (c) Kahn, O. *Molecular Magnetism*, VCH, New York, **1993**. (d) Coronado, E.; Delhaes, P.; Gatteschi, D.; Miller, J. S., Eds., *Molecular Magnetism: From Molecular Assemblies to the Device*, NATO ASI Series, Washington, DC, **1995**. (e) Miller, J. S., Drillon, M., Eds., *Magnetism: Molecules to Materials*, vols. I to V, Wiley-VCH, Weinheim, **2002–2005**. (f) Thompson, L. K., Ed. Special issue on Magnetism: Molecular and Supramolecular Perspectives, *Coord. Chem. Rev.* **2005**, *249*, 2549–2730.
2. (a) Veciana, J. Ed. *Struct. Bonding* **2001**, *100*, 1. (b) Bushby, R. J. In *Magnetism: Molecules to Materials II*, Miller J. S., Drillon, M., Eds. Wiley-VCH, New York, **2002**, p. 149.
3. For example: (a) Manriquez, J. M.; Yee, G. T.; McLean, R. S.; Epstein, A. J.; Miller, J. S. *Science* **1991**, *252*, 1415. (b) Ferlay, S.; Mallah, T.; Ouahès, R.; Veillet, P.; Verdaguer, M. *Nature* **1995**, *378*, 701. (c) Holmes, S. M.; Girolami, G. S. *J. Am. Chem. Soc.* **1999**, *121*, 5593.
4. (a) Gütlich, P.; Carcia, Y.; Goodwin, H. A. *Chem. Soc. Rev.* **2000**, *29*, 419. (b) Gütlich, P.; Goodwin, H. A., Eds. *Top. Curr. Chem.* **2004**, 233–255. (c) Gaspar, A. B.; Ksenofontov, V.; Seredyuk, M.; Gütlich, P. *Coord. Chem. Rev.* **2005**, *249*, 2661.
5. (a) Gatteschi, D.; Sessoli, R. *Angew. Chem., Int. Ed.* **2003**, *42*, 268. (b) Beltran, L. M. C.; Long, J. R. *Acc. Chem. Res.* **2005**, *38*, 325. (c) Berchin, E. K. *Chem. Commun.* **2005**, 5141. (d) Aromí, G.; Brechin, E. K. *Struct. Bonding* **2006**, *122*.
6. (a) Coulon, C.; Miyasaka, H.; Clérac, R. *Struct. Bonding* **2006**, *122*, 163. (b) Lescouëzec, R.; Toma, L. M.; Vaissermann, J.; Verdaguer, M.; Delgado, F. S.; Ruiz-Pérez, C.; Lloret, F.; Julve, M. *Coord. Chem. Rev.* **2005**, *249*, 2691. (c) Miyasaka, H.; Clérac, R. *Bull. Chem. Soc. Jpn.* **2005**, *78*, 1725.
7. (a) Sato, O.; Tao, J.; Zhang, Y. Z. *Angew. Chem., Int. Ed.*, **2007**, *46*, 2152. (b) Sato, O. *Acc. Chem. Res.* **2003**, *36*, 692. (c) Decurtins, S.; Pellaux, R.; Antorrena, G.; Palacio, F. *Coord. Chem. Rev.* **1999**, *190–192*, 841. (d) Coronado, E.; Day, P. *Chem. Rev.* **2004**, *104*, 5419. (e) Gütlich, P.; Garcia, Y.; Woike, T. *Coord. Chem. Rev.* **2001**, *219–221*, 839. (f) Talham, D. R. *Chem. Rev.* **2004**, *104*, 5479.
8. (a) Eerenstein, W.; Mathur, N. D.; Scott, J. F. *Nature* **2006**, *442*, 759. (b) Cheong, S. W.; Mostovoy, M. *Nat. Matter.* **2007**, *6*, 13. (c) Khomskii, D. I. *J. Magn. Magn. Mater.* **2006**, *306*, 1. (d) Ohkoshi, S.; Tokoro, H.; Matsuda, T.; Takahashi, H.; Irie, H.; Hashimoto, K. *Angew. Chem., Int. Ed.* **2007**, *46*, 3238. (e) Bai, Y. L.; Tao, J.; Wernsdorfer, W.; Sato, O.; Huang, R. B.; Zheng, L. S. *J. Am. Chem. Soc.* **2006**, *128*, 16428, and references therein.
9. (a) Verdaguer, M.; Bleuzen, A.; Marvaud, V.; Vaissermann, J.; Seuleiman, M.; Desplanches, C.; Scuiller, A.; Train, C.; Garde, R.; Gelly, G.; Lomenech, C.; Rosenman, I.; Veillet, P.; Cartier, C.; Villain, F. *Coord. Chem. Rev.* **1999**, *190–192*, 1023. (b) Černák, J.;

Orendáč, M.; Potočňák, I.; Chomič, J.; Orendáčová, A.; Skoršepa, J.; Feher, A. *Coord. Chem. Rev.* **2002**, *224*, 51. (c) Sieklucka, B.; Podgajny, R.; Przychodzeń, P.; Korzeniak, T. *Coord. Chem. Rev.* **2005**, *249*, 2203. (d) Tanase, S.; Reedijk, J. *Coord. Chem. Rev.* **2006**, *250*, 2501. (e) Dunbar, K. R.; Heintz, R. A. *Prog. Inorg. Chem.* **1997**, *45*, 283.

10. (a) Gruselle, M.; Train, C.; Boubekeur, K.; Gredin, P.; Ovanesyan, N. *Coord. Chem. Rev.* **2006**, *250*, 2491. (b) Pilkington, M.; Decurtins, S. In *Magnetism: Molecules to Materials II*, Miller, J. S., Drillon, M., Eds., Wiley-VCH, New York, **2002**, p. 339.
11. For example: (a) Darriet, J.; Haddad, M. S.; Duesler, E. N.; Hendrickson, D. N. *Inorg. Chem.* **1979**, *18*, 2679. (b) Bordallo, H. N.; Chapon, L.; Manson, J. L.; Hernández-Velasco, J.; Ravot, D.; Reiff, W. M.; Argyriou, D. N. *Phys. Rev. B* **2004**, *69*, 224405. (c) Manson, J. L.; Conner, M. M.; Schlueter, J. A.; Lancaster, T.; Blundell, S. J.; Brooks M. L.; Pratt, F. L.; Papageorgiou, T.; Bianchi, A. D.; Wosnitza, J.; Wangbo, M. H. *Chem. Commun.* **2006**, 4894.
12. (a) Miller, J. S.; Epstein, A. J. *Coord. Chem. Rev.* **2000**, *206–207*, 651. (b) Kaim, W.; Moscherosch, M. *Coord. Chem. Rev.* **1994**, *129*, 157–193. (c) Zhao, H. H.; Bazile, M. J., Jr.; Galán-Mascarós, J. R.; Dunbar, K. R. *Angew. Chem., Int. Ed.* **2003**, *42*, 1015. (d) Clérac, R.; O'Kane, S. Cowe, J.; Ouyang, X.; Heintz, R.; Zhao, H. H.; Bazile, M. J., Jr.; Dunbar, K. R. *Chem. Mater.* **2003**, *15*, 1840. (e) Miyasaka, H.; Izawa, T.; Takahashi, N.; Yamashita, M.; Dunbar, K. R. *J. Am. Chem. Soc.* **2006**, *128*, 11358.
13. (a) Moriya, T. *Phys. Rev.* **1960**, *120*, 91. (b) Moriya, T. *Phys. Rev.* **1960**, *117*, 635. (c) Dzialoshinski, I. *J. Phys. Chem. Solids* **1958**, *4*, 241.
14. (a) Ribas, J.; Escuer, A.; Monfort, M.; Vicente, R.; Cortés, R.; Lezama, L.; Rojo, T. *Coord. Chem. Rev.* **1999**, *193–195*, 1027. (b) Escuer, A.; Aromí, G. *Eur. J. Inorg. Chem.* **2006**, 4721. (c) Bai, S. Q.; Fang, C. J.; Yan, C. H. *Chin. J. Inorg. Chem.* **2006**, *22*, 2123.
15. Harris notation describes the binding mode as $[X{\cdot}Y1Y2Y3 \cdots Yn]$, where X is the overall number of metals bound by the whole ligand and each Y value refers to the number of metal atoms attached to the different donor atoms. The ordering of Y is listed by the Cahn–Ingold–Prelog priority rules. Although there are three nitrogen atoms in azido, the middle one has never been observed to act as a donor atom, so we simply use $Y1$ and $Y2$ for azido. For details, see Coxall, R. A.; Harris, S. G.; Henderson, D. K.; Parsons, S.; Tasker, P. A.; Winpenny, R. E. P. *Dalton Trans.* **2000**, 2349.
16. (a) Mehrothra, R. C.; Bhora, R. *Metal Carboxylates*, Academic Press, New York, **1983**. (b) Oldham, C. *Prog. Inorg. Chem.* **1968**, *10*, 223. (c) Doedens, R. J. *Prog. Inorg. Chem.* **1976**, *21*, 209. (d) Carrell, C. J.; Carrell, H. L.; Erlebacher, J.; Glusker, J. P. *J. Am. Chem. Soc.* **1988**, *110*, 8651.
17. (a) Kageyama, H.; Khomskii, D. I.; Levitin, R. Z.; Vasil'ev, A. N. *Phys. Rev. B* **2003**, *67*, 224422. (b) Tokita, M.; Zenmyo, K.; Kubo, H.; Takeda, K.; Yamagata, K. *Physica B* **2000**, *284–288*, 1497. (c) Rettig, S. J.; Thompson, R. C.; Trotter, J.; Xia, S. H. *Inorg. Chem.* **1999**, *38*, 1360, and references therein.
18. (a) Viertelhaus, M.; Henke, H.; Anson, C. E.; Powell, A. K. *Eur. J. Inorg. Chem.* **2003**, 2283. (b) Sapiña, F.; Burgos, M.; Escrivá, E.; Folgado, J. V.; Marcos, D.; Beltrán, A.; Beltrán, D. *Inorg. Chem.* **1993**, *32*, 4337, and references therein.
19. Cornia, A.; Caneschi, A.; Dapporto, P.; Faberetti, A. C.; Gatteschi, D.; Malevasi, W.; Sangregorio, C.; Sessoli, R. *Angew. Chem., Int. Ed.* **1999**, *38*, 1780.
20. (a) Kabešová, M.; Boča, R.; Melník, M.; Valigura, D.; Dunaj-Jurčo, M. *Coord. Chem. Rev.* **1995**, *140*, 115. (b) Burmeister, J. L. *Coord. Chem. Rev.* **1990**, *105*, 77. (c) Bailey, R. A.; Kozak, S. L.; Michelsen, T. W.; Mills, W. N. *Coord. Chem. Rev.* **1971**, *6*, 407.

21. (a) Escuer, A.; Vicente, R.; El Fallah, M. S.; Solans, X.; Font-Bardía, M. *Dalton Trans.* **1996**, 1013. (b) Liu, J. P.; Meyers, E. A.; Cowan, J. A.; Shore, S. G. *Chem. Commun.* **1998**, 2043. (c) White, C. A.; Yap, G. P. A.; Raju, N. P.; Greedan, J. E.; Crutchley, R. J. *Inorg. Chem.* **1999**, *38*, 2548. (d) Ferlay, S.; Francese, G.; Schmalle, H. W.; Decurtins, S. *Inorg. Chim. Acta* **1999**, *286*, 108. (e) Vicente, R.; Escuer, A.; Ribas, J.; Solans, X.; Font-Bardía, M. *Inorg. Chem.* **1993**, *32*, 6117. (f) Talukder, P.; Datta, A.; Mitra, S.; Rosair, G.; El Fallah, M. S.; Ribas, J. *Dalton Trans.* **2004**, 4161. (g) Escuer, A.; Font-Bardía, M.; Peñalba, E.; Solans, X.; Vicente, R. *Inorg. Chim. Acta* **1999**, *286*, 189. (h) Duggan, D. M.; Hendrickson, D. N. *Inorg. Chem.* **1974**, *13*, 2929, and references therein.

22. (a) Serna, Z. E.; Urtiaga, M. K.; Barandika, M. G.; Cortés, R.; Martin, S.; Lezama, L.; Arriortua, M. I.; Rojo, T. *Inorg. Chem.* **2001**, *40*, 4550. (b) Barandika, M. G.; Serna, Z.; Cortés, R.; Lezama, L.; Urtiaga, M. K.; Arriortua, M. I.; Rojo, T. *Chem. Commun.* **2001**, 45.

23. (a) McElearney, J. N.; Balagot, L. L.; Muir, J. A.; Spence, R. D. *Phys. Rev. B* **1979**, *19*, 306. (b) DeFotis, G. C.; Remy, E. D.; Scherrer, C. W. *Phys. Rev. B* **1990**, *41*, 9074. (c) DeFotis, G. C.; McGhee, E. M.; Echols, K. R.; Wiese, R. S. *J. Appl. Phys.* **1988**, *63*, 3569. (d) Dockum, B. W.; Eisman, G. A.; Witten, E. H.; Reiff, W. M. *Inorg. Chem.* **1983**, *22*, 150.

24. Żurowska, B.; Mroziński, J.; Julve, M.; Lloret, F.; Maslejova, A.; Sawka-Dobrowolska, W. *Inorg. Chem.* **2002**, *41*, 1771.

25. Hitchman, M. A.; Rowbottom, G. *Coord. Chem. Rev.* **1982**, *42*, 55.

26. (a) Gleizes, A.; Meyer, A.; Hitchman, M. A.; Kahn, O. *Inorg. Chem.* **1982**, *21*, 2257. (b) McKee, V.; Zvagulis, M.; Reed, C. A. *Inorg. Chem.* **1985**, *24*, 2914. (c) Ribas, J.; Diaz, C.; Monfort, M.; Vilana, J.; Solans, X.; Font-Altaba, M. *Transition Met. Chem.* **1985**, *10*, 340. (d) Costes, J. P.; Dahan, F.; Ruiz, J.; Laurent, J. P. *Inorg. Chim. Acta* **1995**, *239*, 53. (e) El Fallah, M. S.; Rentshler, E.; Caneschi, A.; Sessoli, R.; Gatteschi, D. *Inorg. Chem.* **1996**, *35*, 3723. (f) Rajendiran, T. M.; Mathonière, C.; Golhen, S.; Ouahab, L.; Kahn, O. *Inorg. Chem.* **1998**, *37*, 2651.

27. (a) Meyer, A.; Gleizers, A.; Girerd, J. J.; Verdaguer, M.; Kahn, O. *Inorg. Chem.* **1982**, *21*, 1729. (b) Renard, J. P.; Verdaguer, M.; Regnault, L. P.; Erkelens, W. A. C.; Rossat-Mignod, J.; Ribas, J.; Stirling, W. G.; Vettier, C. *J. Appl. Phys.* **1988**, *63*, 3538. (c) Chou, L. K.; Abboud, K. A.; Talham, D. R.; Kim, W. W.; Meisel, M. W. *Chem. Mater.* **1994**, *6*, 2051. (d) Esuer, A.; Vicenter, R.; Solans, X. *Dalton Trans.* **1997**, 531. (e) Kahn, O.; Bakalbasis, E.; Mathnière, C.; Hagiwara, M.; Katsumata, K.; Ouahab, L. *Inorg. Chem.* **1997**, *36*, 1530. (f) Rajendiran, T. M.; Kahn, O.; Golhen, S.; Ouahab, L.; Honda, Z.; Katsumata, K. *Inorg. Chem.* **1998**, *37*, 5693. (g) Costes, J. P.; Dahan, F.; Laurent, J. P.; Drillon, M. *Inorg. Chim. Acta* **1999**, *294*, 8. (h) Diza, C.; Ribas, J.; Costa, R.; Tercero, J.; El Fallah, M. S.; Solans, X.; Font-Bardía, M. *Eur. J. Inorg. Chem.* **2000**, 675.

28. (a) Gadet, V.; Verdaguer, M.; Briois, V.; Gleizes, A.; Renard, J. P.; Beauvillain, P.; Chappert, C.; Goto, T.; Le Dang, K.; Veillet, P. *Phys. Rev. B* **1991**, *44*, 705. (b) Chou, L. K.; Abboud, K. A.; Talham, D. R.; Kim, W. W.; Meisel, M. W. *Physica B* **1993**, *194–196*, 311.(c) Renard, J. P.; Regnault, L. P.; Verdaguer, M. In *Magnetism: Molecules to Materials*, Vol. *I*, Miller, J. S., Drillon, M. Eds., Wiley-VCH, Weinheim, Germany, **2002**, p. 49.

29. (a) Srinivasan, R.; Ströbele, M.; Meyer, H. J. *Inorg. Chem.* **2003**, *42*, 3406. (b) Liao, W. P.; Hu, C. H.; Kremer, R. K.; Dronskowski, R. *Inorg. Chem.* **2004**, *43*, 5884. (c) Liao, W. P.; Dronskowski, R. *Inorg. Chem.* **2006**, *45*, 3828. (d) Liu, X. H.; Krott, M.; Müller, P.; Hu, C. H.; Lueken, H.; Dronskowski, R. *Inorg. Chem.* **2005**, *44*, 3001. (e) Liu, X. H.; Müller, P.; Kroll, P.; Dronskowski, R. *Inorg. Chem.* **2002**, *41*, 4259. (f) Cao, R.; Tatsumi, K. *Chem.*

Commun. **2002**, 2144. (g) Tanabe, Y.; Kuwata, S.; Ishii, Y. *J. Am. Chem. Soc.* **2002**, *124*, 6528, and references therein.

30. (a) Batten, S. R.; Murray, K. S.; *Coord. Chem. Rev.* **2003**, *246*, 103. (b) Manson, J. L.; Huang, Q. Z.; Lynn, J. W.; Koo, H. J.; Whangbo, M. H.; Bateman, R.; Otsuka, T.; Wada, N.; Argyriou, D. N.; Miller, J. S. *J. Am. Chem. Soc.* **2001**, *123*, 162. (c) Lappas, A.; Wills, A. S.; Green, M. A.; Prassides, K.; Kurmoo, M. *Phys. Rev. B* **2003**, *67*, 144406, and references therein.
31. Zhang, J. P.; Chen, X. M. *Chem. Commun.* **2006**, 1689.
32. For example: (a) Coughlin, P. K.; Lippard, S. J. *Inorg. Chem.* **1984**, *23*, 1446. (b) Bencini, A.; Benelli, C.; Gatteschi, D.; Zanchini, C. *Inorg. Chem.* **1986**, *25*, 398. (c) Chaudhuri, P.; Karpenstein, I.; Winter, M.; Lengen, M.; Butzlaff, C.; Bill, E.; Trautwein, A. X.; Flörke, U.; Haupt, H. J. *Inorg. Chem.* **1993**, *32*, 888. (d) Rettig, S. J.; Storr, A.; Summers, D. A.; Thompson, R. C.; Trotter, J. *J. Am. Chem. Soc.* **1997**, *119*, 8675. (e) Liu, Y. L.; Kravtsov, V.; Walsh, R. D.; Poddar, P.; Srikanth, H.; Eddaoudi, M. *Chem. Commun.* **2004**, 2806. (f) Patrick, B. O.; Reiff, W. M.; Sánchez, V.; Storr, A.; Thompson, R. C. *Inorg. Chem.* **2004**, *43*, 2330.
33. Hagrman, P. J.; Hagrman, D.; Zubieta, J. *Angew. Chem., Int. Ed.* **1999**, *38*, 2638.
34. For example: (a) Han, S. J.; Manson, J. L.; Kim, J.; Miller, J. S. *Inorg. Chem.* **2000**, *39*, 4182. (b) Martín, S.; Barandika, M. G.; Lezama, L.; Pizarro, J. L.; Serna, Z. E.; de Larramendi, J. I. R.; Arriortua, M. I.; Rojo, T.; Cortés, R. *Inorg. Chem.* **2001**, *40*, 4109. (c) Fu, A. H.; Huang, X. Y.; Li, J.; Yuen, T.; Lin, C. L. *Chem. Eur. J.* **2002**, *8*, 2239. (d) Ghosh, A. K.; Ghoshal, D.; Zangrando, E.; Ribas, J.; Chaudhuri, N. R. *Inorg. Chem.* **2005**, *44*, 1786. (e) Manson, J. L.; Lecher, J. G.; Gu, J. Y.; Geiser, U.; Schlueter, J. A.; Henning, R.; Wang, X. P.; Schultz, A. J.; Koo, H. J.; Whangbo, M. H. *Dalton Trans.* **2003**, 2905.
35. Cundy, C. S.; Cox, P. A. *Chem. Rev.* **2003**, *103*, 663.
36. Cheetham, A. K.; Férey, G.; Loiseau, T. *Angew. Chem., Int. Ed.* **1999**, *38*, 3268.
37. De A. A. Soler-Illia, G. J.; Sanchez, C.; Lebeau, B.; Patarin, J. *Chem. Rev.* **2002**, *102*, 4093.
38. Pilkington, M.; Decurtins, S. In *Crystal Design: Structure and Function*, Desiraju, G. R., Ed., Wiley, Chichester, UK, **2003**, p. 306.
39. Zhang, Y. Z.; Wernsdorfer, W.; Pan, F.; Wang, Z. M.; Gao, S. *Chem. Commun.* **2006**, 3302.
40. Liu, T. F.; Fu, D.; Gao, S.; Zhang, Y. Z.; Sun, H. L.; Su, G.; Liu, Y. J. *J. Am. Chem. Soc.* **2003**, *125*, 13976.
41. Sun Hao Ling. *Doctoral dissertation*, Peking University, China, **2005**.
42. (a) Sun, Z. M.; Prosvirin, A. V.; Zhao, H. H.; Mao, J. G.; Dunbar, K. R. *J. Appl. Phys.* **2005**, *97*, 10B305. (b) Bernot, K.; Luzon, J.; Sessoli, R.; Vindigni, A.; Thion, J.; Richeter, S.; Leclercq, D.; Larionova, J.; Van der Lee, A. *J. Am. Chem. Soc.* **2008**, *130*, 1619.
43. Liu, X. T.; Wang, X. Y.; Zhang, W. X.; Cui, P.; Gao, S. *Adv. Mater.* **2006**, *18*, 2852.
44. Glauber, R. J. *J. Math. Phys.* **1963**, *4*, 294.
45. Li, X. J.; Wang, X. Y.; Gao, S.; Cao, R. *Inorg. Chem.* **2006**, *45*, 1508.
46. (a) Clérac, R.; Miyasaka, H.; Yamashita, M.; Coulon, C. *J. Am. Chem. Soc.* **2002**, *124*, 12837. (b) Ferbinteanu, M.; Miyasaka, H.; Wernsdorfer, W.; Nakata, K.; Sugiura, K.; Yamashita, M.; Coulon, C.; Clérac, R. *J. Am. Chem. Soc.* **2005**, *127*, 3090. (c) Lecren, L.; Roubeau, O.; Coulon, C.; Li, Y. G.; Le Goff, X. F.; Wernsdorfer, W.; Miyasaka, H.; Clérac, R. *J. Am. Chem. Soc.* **2005**, *127*, 17353.

47. Xu, H. B.; Wang, B. W.; Pan, F.; Wang, Z. M.; Gao, S. *Angew. Chem., Int. Ed.* **2007**, *46*, 7388.

48. Georges, R.; Borrás-Almenar, J. J.; Coronado, E.; Curély, J.; Drillon, M. In *Magnetism: Molecules to Materials II*, Miller, J. S., Drillon, M., Eds., Wiley-VCH, New York, **2002**, p. 1.

49. Abu-Youssef, M.; Drillon, M.; Escuer, A.; Goher, M. A. S.; Mautner, F. A.; Vicente, R. *Inorg. Chem.* **2000**, *39*, 5022.

50. Zhang, Y. Z.; Wei, H. Y.; Pan, F.; Wang, Z. M.; Chen, Z. D.; Gao, S. *Angew. Chem., Int. Ed.* **2005**, *44*, 5841.

51. Wang, X. Y.; Li, B. L.; Zhu, X.; Gao, S. *Eur. J. Inorg. Chem.* **2005**, 3277.

52. Wang, X. Y.; Wei, H. Y.; Wang, Z. M.; Gao, S.; Chen, Z. D. *Inorg. Chem.* **2005**, *44*, 572.

53. Liu, T.; Zhang, Y. J.; Wang, Z. M.; Gao, S. *Inorg. Chem.* **2006**, *45*, 2782.

54. Liu, T.; Chen, Y. H.; Zhang, Y. J.; Wang, Z. M.; Gao, S. *Inorg. Chem.* **2006**, *45*, 9148.

55. (a) Yuan, M.; Gao, S.; Sun, H. L.; Su, G. *Inorg. Chem.* **2004**, *43*, 8221. (b) Yuan, M.; Zhao, F.; Zhang, W.; Pan F.; Wang, Z. M.; Gao, S. *Chem. Eur. J.* **2007**, *13*, 2937. (c) Lu, Z. L.; Yuan, M.; Pan, F.; Gao, S.; Zhang, D. Q.; Zhu, D. B. *Inorg. Chem.* **2006**, *45*, 3538.

56. Yuan, M.; Zhao, F.; Zhang, W.; Wang, Z. M.; Gao, S. *Inorg. Chem.* **2007**, *46*, 11235.

57. Escuer, A.; Mautner, F. A.; Goher, M. A. S.; Abu-Youssef, M. A. M.; Vicente, R. *Chem. Commun.* **2005**, 605.

58. Wang, X. Y.; Wang, L.; Wang, Z. M.; Su, G.; Gao, S. *Chem. Mater.* **2005**, *17*, 6369.

59. Wang Xin Yi. *Doctoral dissertation*, Peking University, China, **2006**.

60. Ma, B. Q.; Sun, H. L.; Gao S.; Su, G. *Chem. Mater.* **2001**, *13*, 1946.

61. (a) Gao, E. Q.; Wang, Z. M.; Yan, C. H. *Chem. Commun.* **2003**, *14*, 1748. (b) Escuer, A.; Vicente, J. R.; Goher, M. A. S.; Mautner, F. A. *Inorg. Chem.* **1995**, *34*, 5707. (c) Escuer, A.; Vicente, J. R.; Goher, M. A. S.; Mautner, F. A. *Inorg. Chem.* **1996**, *35*, 6386. (d) Escuer, A.; Vicente, J. R.; Goher, M. A. S.; Mautner, F. A. *Inorg. Chem.* **1997**, *36*, 3440. (e) Escuer, A.; Vicente, J. R.; Goher, M. A. S.; Mautner, F. A. *Dalton Trans.* **1997**, 4431. (f) Goher, M. A. S.; Abu-Youssef, M. A. M.; Mautner, F. A.; Vicente, J. R.; Escuer, A. *Eur. J. Inorg. Chem.* **2000**, 1819.

62. Sun, H. L.; Ma, B. Q.; Gao, S. Su, G. *Chem. Commun.* **2001**, 2586.

63. Wang, X. Y.; Wang, L.; Wang, Z. M.; Gao, S. *J. Am. Chem. Soc.* **2006**, *128*, 674.

64. (a) Greedan, J. E. *J. Mater. Chem.* **2001**, *11*, 37. (b) Ramirez, A. P. *Annu. Rev. Mater. Sci.* **1994**, *24*, 453. (c) Harrison, A. *J. Phys.: Condens. Matter* **2004**, *16*, S553.

65. (a) Verdaguer, M.; Gleizes, A.; Renard, J. P.; Seiden, J. *Phys. Rev. B* **1984**, *29*, 5144. (b) Gleizes, A.; Verdaguer, M. *J. Am. Chem. Soc.* **1984**, *106*, 3727. (c) Kahn, O.; Pei, Y.; Verdaguer, M.; Renard, J. P.; Sletten, J. *J. Am. Chem. Soc.* **1988**, *110*, 782.

66. (a) Feyerherm, R.; Loose, A.; Ishida, T.; Nogami, T.; Kreitlow, J.; Baabe, D.; Litterst, F. J.; Süllow, S.; Klauss, H. H.; Doll, K. *Phys. Rev. B* **2004**, *69*, 134427. (b) Nakayama, K.; Ishida, T.; Takayama, R.; Hashizume, D.; Yasui, M.; Iwasaki, F.; Nogami, T. *Chem. Lett.* **1998**, 497.

67. (a) Wang, X. Y.; Wang, Z. M.; Gao, S. *Inorg. Chem.* **2008**, *47*, 5720. (b) Abu-Youssef, M. A. M.; Mautner, F. A.; Vicente, R. *Inorg. Chem.* **2007**, *46*, 4654.

68. Wang, X. Y.; Wang, Z. M.; Gao, S. *Chem. Commun.* **2007**, 1127.

69. (a) Schmidt, G. M. J. *Pure Appl. Chem.* **1971**, *27*, 647. (b) Toh, N. L.; Nagarathinam, M.; Vittal, J. J. *Angew. Chem., Int. Ed.* **2005**, *44*, 2237. (c) Varshney, D. B.; Gao, X. C.;

Friščić, T.; MacGillivray, L. R. *Angew. Chem., Int. Ed.* **2006**, *45*, 646, and references therein.

70. (a) Sun, H. L.; Ma, B. Q.; Gao, S.; Su, G. *Inorg. Chem.* **2003**, *42*, 5399. (b) Sun, H. L.; Wang, Z. M.; Gao, S. *Inorg. Chem.* **2005**, *44*, 2169. (c) Kutasi, A. M.; Batten, S. R.; Moubaraki, B.; Murray, K. S. *Dalton Trans.* **2002**, 819.
71. Wang, X. T.; Wang, Z. M. Gao, S. *Inorg. Chem.* **2007**, *46*, 10452.
72. Zhang, Y. Z.; Gao, S.; Sun, H. L.; Su, G.; Wang, Z. M.; Zhang, S. W. *Chem. Commun.* **2004**, 1906.
73. Zhang, Y. Z.; Wang, Z. M.; Gao, S. *Inorg. Chem.* **2006**, *45*, 5447.
74. Zhang, Y. Z.; Wang, Z. M.; Gao, S. *Inorg. Chem.* **2006**, *45*, 10404.
75. Yin, P.; Gao, S.; Zheng, L. M.; Wang, Z. M.; Xin, X. Q. *Chem. Commun.* **2003**, 1076.
76. (a) Mautner, F. A.; Hanna, S.; Cortés, R.; Lezama, L.; Barandika, M. G.; Rojo, T. *Inorg. Chem.* **1999**, *38*, 4647. (b) Mautner, F. A.; Cortés, R.; Lezama, L.; Rojo, T. *Angew. Chem., Int. Ed. Engl.* **1996**, *35*, 78.
77. Wang, Z. M.; Zhang, B.; Inoue, K.; Fujiwara, H. Otsuka, T.; Kobayashi, H.; Kurmoo, M. *Inorg. Chem.* **2007**, *46*, 437.
78. (a) Wang, Z. M.; Zhang, B.; Otsuka, T.; Inoue, K.; Kobayashi, H.; Kurmoo, M. *Dalton Trans.* **2004**, 2209. (b) Wang, X. Y.; Gan, L.; Zhang, S. W.; Gao, S. *Inorg. Chem.* **2004**, *43*, 4615.
79. Hu, K. L.; Wang, Z. M.; Gao, S. *Manuscript in preparation.*
80. Wang, Z. M.; Zhang, X. Y.; Batten, S. R.; Kurmoo, M.; Gao, S. *Inorg. Chem.* **2007**, *46*, 8439.
81. (a) Wang, Z. M.; Zhang, B.; Fujiwara, H.; Kobayashi, H.; Kurmoo, M. *Chem. Commun.* **2004**, 416. (b) Zhang, B.; Wang, Z. M.; Kurmoo, M.; Gao, S.; Inoue, K.; Kobayashi, H. *Adv. Funct. Mater.* **2007**, *17*, 577. (c) Wang, Z. M.; Zhang, B.; Kurmoo, M.; Green, M. A.; Fujiwara, H.; Otsuka, T.; Kobayashi, H. *Inorg. Chem.* **2005**, *44*, 1230. (d) Wang, Z. M.; Zhang, B.; Zhang, Y. J.; Kurmoo, M.; Liu, T.; Gao, S.; Kobayashi, H. *Polyhedron* **2007**, *26*, 2207, and references cited therein. (e) Wang, Z. M.; Zhang, Y. J.; Liu, T.; Kurmoo, M.; Gao, S. *Adv. Funct. Mater.* **2007**, *17*, 1523.
82. Viertelhaus, M.; Adler, P.; Clérac, R.; Anson; C. E.; Powell, A. K. *Eur. J. Inorg. Chem.* **2005**, 692.
83. Dybtsev, D. N.; Chun, H.; Yoon, S. H.; Kim, D.; Kim, K. *J. Am. Chem. Soc.* **2004**, *126*, 32.
84. Rood, J. A.; Noll, B. C.; Henderson, K. W. *Inorg. Chem.* **2006**, *45*, 5521.
85. Wang, Z. M.; Zhang, Y. J.; Kurmoo, M.; Liu, T.; Vilminot, S.; Zhao, B.; Gao, S. *Aust. J. Chem.* **2006**, *59*, 617.
86. Tian, Y. Q.; Cai, C. X.; Ren, X. M.; Duan, C. Y.; Xu, Y.; Gao, S.; You, X. Z. *Chem. Eur. J.* **2003**, *9*, 5673.

9

STRUCTURES AND PROPERTIES OF HEAVY MAIN-GROUP IODOMETALATES

LI-MING WU AND LING CHEN

State Key Laboratory of Structural Chemistry, Fujian Institute of Research on the Structure of Matter, Chinese Academy of Sciences, Fuzhou, Fujian, People's Republic of China

9.1 INTRODUCTION

The *p*-block heavy element iodometalates (AM_xI_y; A = a cation, M = Pb or Bi) are a class of interesting complexes that are derivatives of binary PbI_2 or BiI_3. They are remarkable for their diversity of structures [1–4] and interesting optical and electronic properties, such as luminescence [5,6], nonlinear optical activity [7,8], semiconductivity (even metallic conductivity) [9,10], and ferroelectricity [11].

The fascinating chemistry of iodometalates lies in the great structural diversity of the inorganic moieties, that is, the anionic Pb/I and Bi/I clusters and polymers. Meanwhile, the tunable properties originate from the structural modifications of the anions via adjustment of the dimensionality, aggregation, distortion, and hetero components, heterometallic bonding interactions beyond the Bi–I or Pb–I bonds. The effects of ligands coordinated directly with the anionic moiety are discussed in this chapter.

The $MI_6{}^{n-}$ (M = Pb or Bi) octahedra are the most common building units in iodometalates. The Pb–I and Bi–I covalent bonds exhibit low directional correlations and the Pb and Bi atoms are soft. These two facts allow the distortion and aggregation of discrete M/I clusters. The covalency of the Pb–I and Bi–I bonds are relatively greater than those of M–O or M–S bonds, which makes possible the distortion,

Design and Construction of Coordination Polymers, Edited by Mao-Chun Hong and Ling Chen

aggregation, or extension of the building units to a large degree so as to satisfy the charge balance requirements of the cations.

Both Pb and Bi have large atomic numbers and radii, and the energies of their outer s and p orbitals are greatly different; therefore, these two atoms can lose only p electrons and yield as Pb^{2+} and Bi^{3+} ions, respectively. Comparatively, the lighter Sn^{2+} and In^{+} ions are notably less stable with respect to the fully oxidized species Sn^{4+} and In^{3+}. On the other hand, the nuclei of Pb^{2+} or Bi^{3+} weakly bind the outer s electrons, so the Pb^{2+} and Bi^{3+} ions are soft or polarizable, which essentially allows a great degree of distortion and aggregation of MI_x polyhedra.

The structural chemistry of iodometalates has nicely demonstrated evolvement from discrete building units to polymeric 1D, 2D, or 3D compounds. Some of these compounds show novel properties, so that the iodometalates can be ideal systems in which to study structure–property relationships, which may shed useful insights on the rational syntheses of target compounds with desired properties.

9.2 STRUCTURAL FEATURES OF IODOBISMUTHATES AND IODOPLUMBATES

The major structural characteristic of iodometalates are the diverse anionic structure motifs, which range from discrete mono- or polynuclear species to infinite varieties with higher dimensionality (1D, 2D, or 3D).

The conceptional stepwise aggregations of $MI_6{}^{n-}$ (M = Pb or Bi) octahedra are shown impressively by 0D complexes. One-dimensional chains are common in iodometalates that are formed from MI_6 octahedra through sharing I-I-I faces, I-I edges, or I apexes. Similar connections in two or three different directions lead to 2D polymers or 3D networks.

Two types of I atoms, terminal I (μ_t-I) and bridging I (μ_n-I, $n = 2, 3, 4, 5$), are observed. The μ_t-I is defined as a peripheral atom that serves to isolate the M/I cluster (i.e., to terminate the polymerization of the building units). The bridging I atoms (μ_n-I, $n = 2, 3, 4, 5$) are responsible for the aggregation and polymerization.

9.2.1 Binary BiI_3 and PbI_2

Binary PbI_2 and BiI_3 are interesting semiconductors. Hexagonal PbI_2 has a layered structure with $a = b = 4.557$ Å and $c = 6.979$ Å (Fig. 9.1). Each Pb atom has an octahedral PbI_6 coordination sphere that is condensed to layers by sharing edges with six neighboring octahedra. Bulk PbI_2 is yellow powder, melts at 405°C, and has an optical energy gap of about 2.3 eV [12]. BiI_3 crystallizes in a trigonal system with $a = b = 7.525$ Å and $c = 20.703$ Å that is PbI_2-like, in which one-third of the metallic site is now empty (Fig. 9.2). Bulk BiI_3 is black powder, melts at 408°C, sublimates at lower temperature, and has an energy gap of 1.73 eV [13]. Both PbI_2 and BiI_3 show interesting properties, such as electroluminescence [14], photoluminescence [15,16], and nonlinear optical effects [17], and are candidates for room-temperature x-ray or γ-ray detectors [18–20] and thin-film transistors [21].

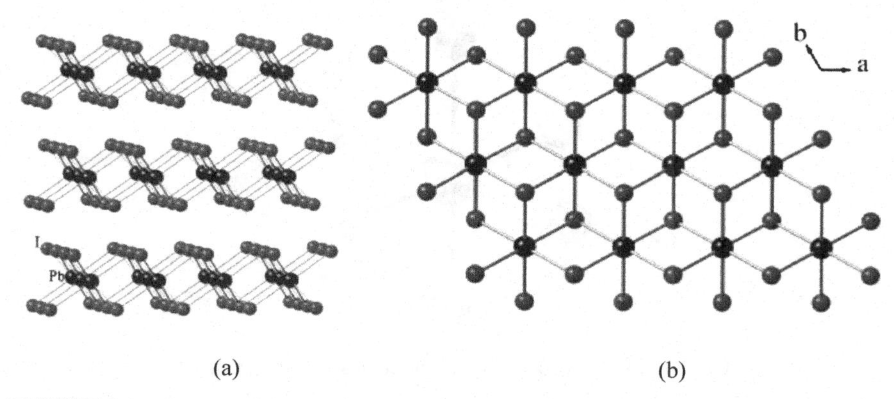

FIGURE 9.1 Structure of PbI_2: (a) packing diagram; (b) layered structure of (001) view.

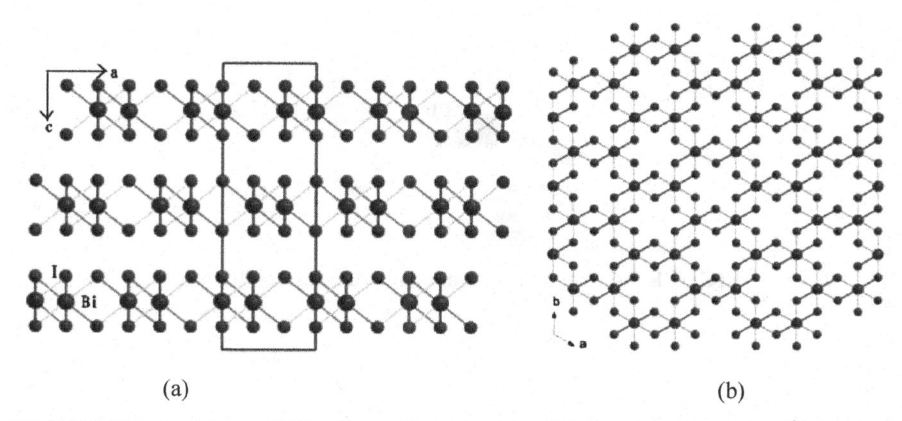

FIGURE 9.2 Structure of BiI_3: (a) packing diagram; (b) layered structure along (001) view.

9.2.2 Zero-Dimensional M/I Clusters

Mononuclear Clusters: $[BiI_6]^{3-}$ and $[PbI_6]^{4-}$ The simplest iodometalates are the mononuclear clusters $[BiI_6]^{3-}$ and $[PbI_6]^{4-}$. The geometry about M atoms is a nearly ideal octahedron, as shown in Figure 9.3. The representative compounds are $[PhCH_2CH_2NH_3]_4[BiI_6][I]\cdot 2H_2O$ [22] and $[CH_6N]_4[PbI_6]\cdot 2H_2O$ [23]. One unusual mononuclear complex is $[Pr_4N]_2[PbI_4]$ [24], in which the tetrahedrally coordinated geometry about Pb atom is rare (Fig. 9.4). The different charges of each cluster unit have dominant roles in determining the configurations of the anionic structures of higher polynuclear clusters or the polymeration of such building units, as discussed below.

Binuclear Clusters: $[Bi_2I_8]^{2-}$, $[Bi_2I_9]^{3-}$, $[Bi_2I_{10}]^{4-}$, and $[Pb_2I_6]^{2-}$ Three different binuclear cluster types are found; the first is the dimer of a unique fivefold-coordinated

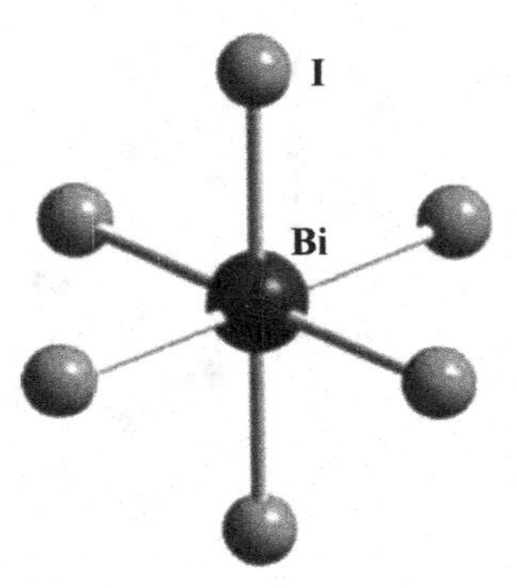

FIGURE 9.3 Structure of mononuclear cluster $[BiI_6]^{3-}$.

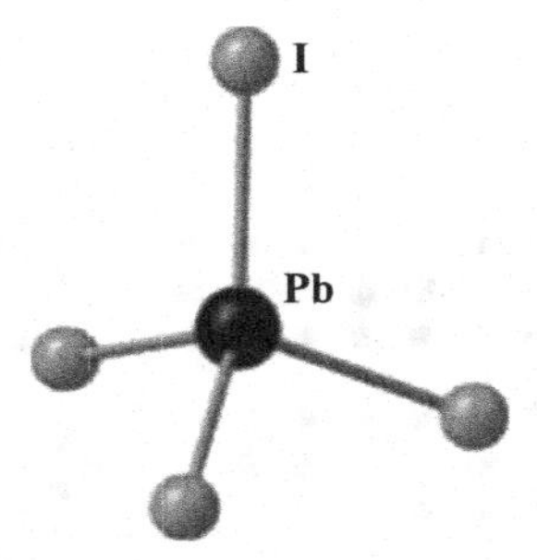

FIGURE 9.4 Structure of mononuclear cluster $[BiI_4]^{2-}$.

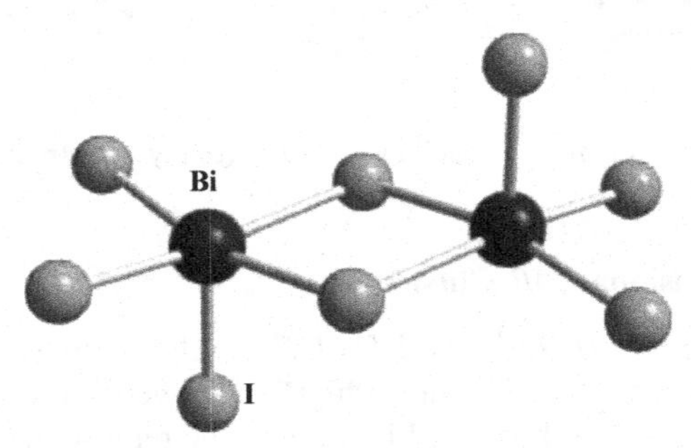

FIGURE 9.5 Structure of binuclear cluster $[Bi_2I_8]^{2-}$.

Bi as in $[(PhCH_2)_4P]_2[Bi_2I_8]$ [25] (Fig. 9.5). The other two, shown in Figs. 9.6 and 9.7, are both dimers of BiI_6 octahedra via either sharing an I–I–I face or an I–I edge. In each case, the geometry of BiI_6 is close to the ideal octahedron, and the Bi–I_{term} bond (3.0 Å) is 0.2 Å shorter than the Bi–μ_2–I bond. The known compounds are $Cs_3Bi_2I_9$ [26,27], $[Me_4N]_3[Bi_2I_9]$, [28] $[Et_2NH_2]_3[Bi_2I_9]$ [29], and $[bpyH]_4[Bi_2I_{10}]$ [30].

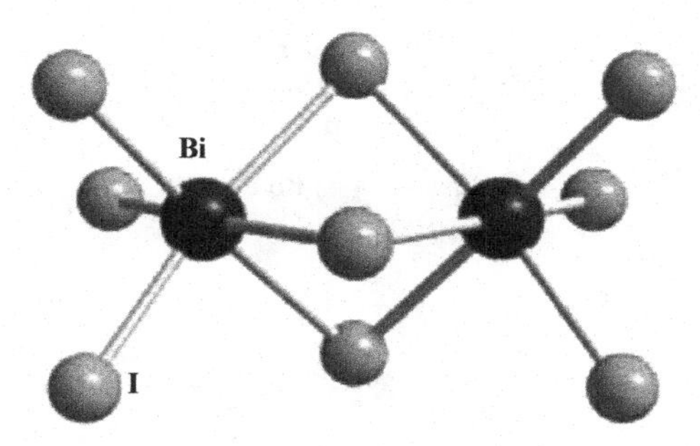

FIGURE 9.6 Structure of binuclear cluster $[Bi_2I_9]^{3-}$.

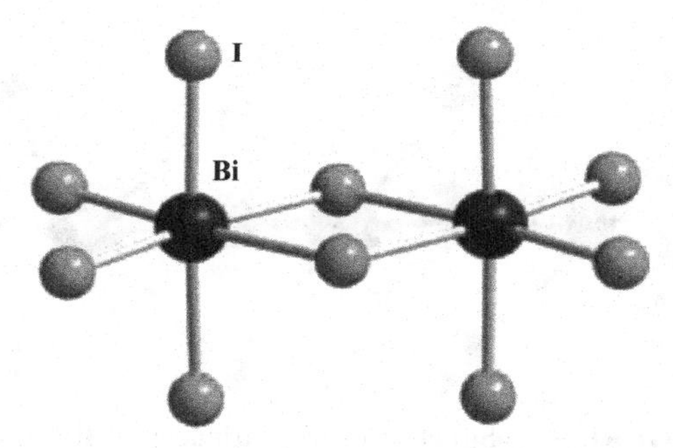

FIGURE 9.7 Structure of binuclear cluster $[Bi_2I_{10}]^{4-}$.

However, only one Pb/I binuclear cluster that is not similar to the aforementioned cluster is found as $[PPh_4]_2[Pb_2I_6]$ [24], in which the unique fourfold-coordinated PbI_4 polyhedron is dimerized (Fig. 9.8).

Trinuclear Clusters: $[Bi_3I_{12}]^{3-}$ and $[Pb_3I_{10}]^{4-}$ Two arrangements of three BiI_6 octahedra give $[Bi_3I_{12}]^{3-}$ anions in *trans*-motif $[n\text{-}Bu_4N]_3[Bi_3I_{12}]$ [31] (Fig. 9.9) or *cis*-motif $[N(CH_3)(n\text{-}C_4H_9)_3]_3[Bi_3I_{12}]$ [32] (Fig. 9.10). The central octahedron in the former shares two opposite I–I–I faces, and in the latter, two *cis* I–I–I faces. The *trans*-$[Bi_3I_{12}]^{3-}$ cluster possesses C_{3v} symmetry with three Bi atoms on a threefold axis. Such an assembly could also be seen as a combination of a binuclear $[Bi_2I_9]^{3-}$ and a BiI_6 octahedron.

The trinuclear $[Bu_3N(CH_2)_4NBu_3]_2[Pb_3I_{10}]$ [33] complex possesses a closed motif of two normally sixfold-coordinated PbI_6 octahedra and one unusual PbI_5 square pyramid (Fig. 9.11). The variation in coordination number (CN) of MI_x ($x = 6, 5, 4$) indicates the considerable tolerance of MI_6 octahedron to endure the distortion. Such tolerance arises from the flexibility of the central Bi or Pb ions and the less direction-correlation nature of the covalent M–I bonds.

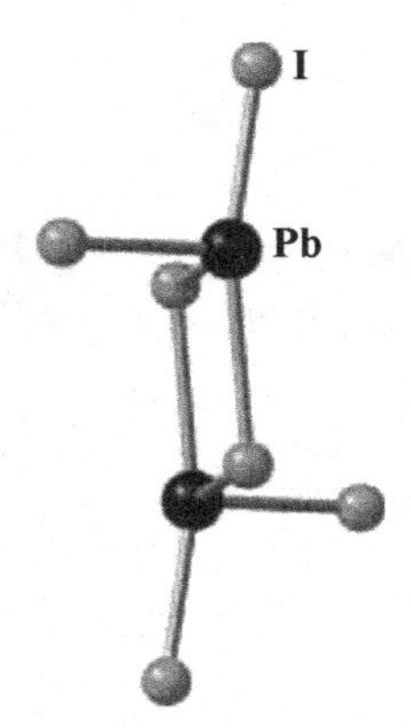

FIGURE 9.8 Structure of binuclear cluster $[Pb_2I_6]^{2-}$.

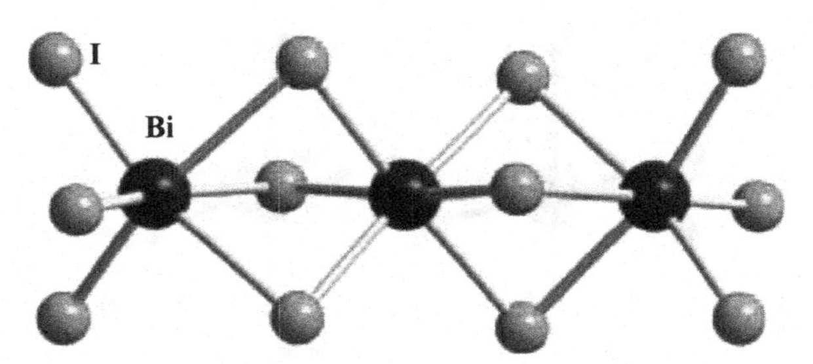

FIGURE 9.9 Structure of *trans*-trinuclear $[Bi_3I_{12}]^{3-}$.

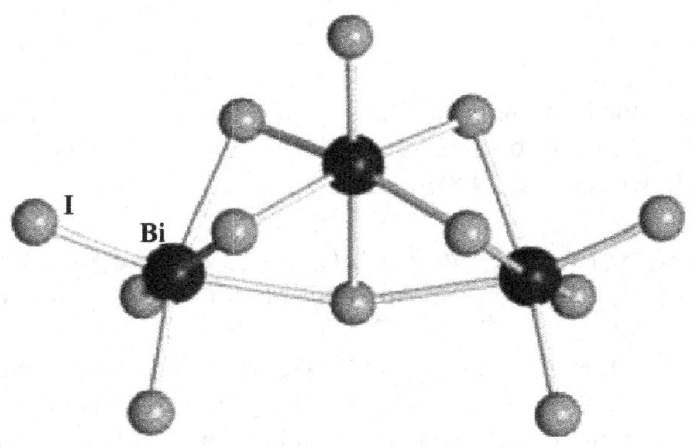

FIGURE 9.10 Structure of *cis*-trinuclear $[Bi_3I_{12}]^{3-}$.

Tetranuclear Cluster: $[Bi_4I_{16}]^{4-}$ A tetranuclear $[Bi_4I_{16}]^{4-}$ cluster (Fig. 9.12) is a dimer of a binuclear $[Bi_2I_{10}]^{4-}$ (Fig. 9.7) via two common μ_2- and two μ_3-I atoms. Compared with the binuclear species, no obvious changes in Bi-I bond distances or I-Bi-I angles are found. Two examples are compounds $[(BiP_c)_4]$ $[Bi_4I_{16}]$ and $[C_3H_9COSCH_4N(CH_3)_3]_4[Bi_4I_{16}]$, [34,35].

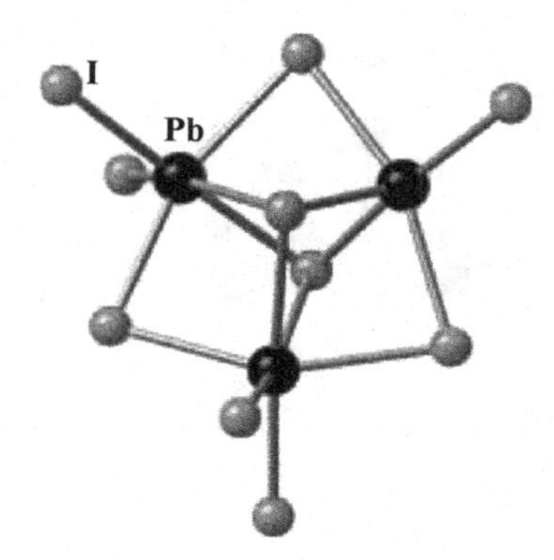

FIGURE 9.11 Structure of closed trinuclear $[Pb_3I_{10}]^{4-}$.

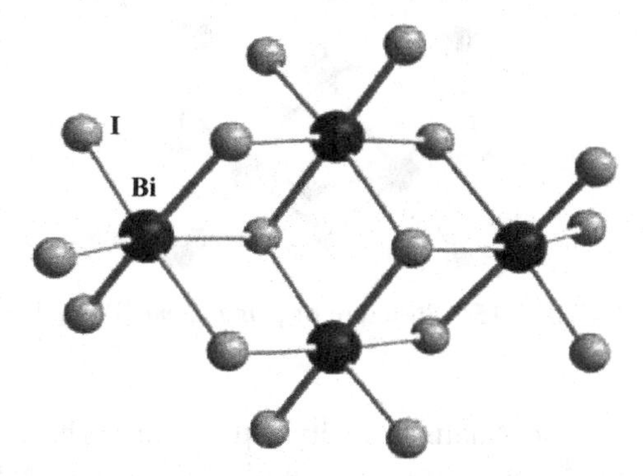

FIGURE 9.12 Structure of tetranuclear $[Bi_4I_{16}]^{4-}$.

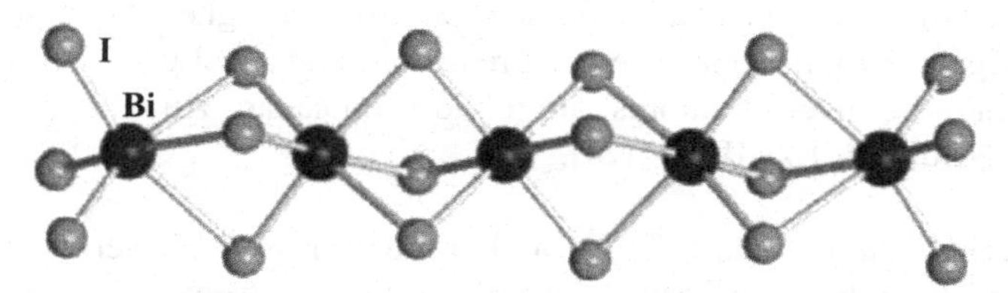

FIGURE 9.13 Structure of pentanuclear $[Bi_5I_{18}]^{3-}$.

Pentanuclear Clusters: $[Bi_5I_{18}]^{3-}$, $[Bi_5I_{19}]^{4-}$, and $[Pb_5I_{16}]^{6-}$ As found in $[Ph_4P]_3[Bi_5I_{18}]$ [36] and $[Ph_4Sb]_3[Bi_5I_{18}]$ [37], the linear pentanuclear cluster shown in Figure 9.13 can be viewed as a combination of a binuclear cluster $[Bi_2I_9]^{3-}$ (Fig. 9.12) and a trinuclear cluster $[Bi_3I_{12}]^{3-}$ (Fig. 9.9) via a shared I-I-I

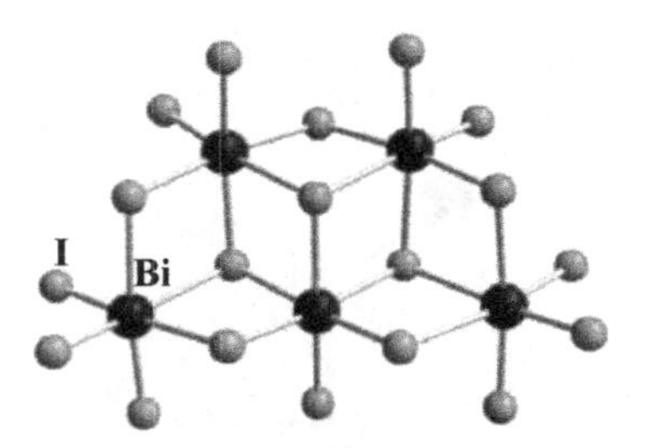

FIGURE 9.14 Structure of pentauclear $[Bi_5I_{19}]^{4-}$.

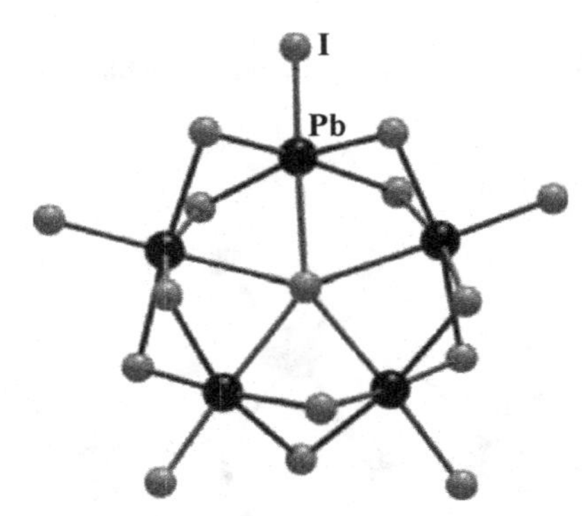

FIGURE 9.15 Structure of pentauclear $[Pb_5I_{16}]^{6-}$.

face. This illustrates the coordinating activity of μ_t-I atoms in bi- or trinuclear species. Interestingly, a different aggregation pattern of five BiI_6 octahedra is found in $[Li(THF)_4]_4[Bi_5I_{19}]$ (Fig. 9.14) [38], which is a combination of a tetranuclear cluster $[Bi_4I_{16}]^{4-}$ (Fig. 9.12) and a $[BiI_6]^{3-}$ octahedron cluster through one μ_3–I and two μ_2–I atoms.

In a third way, five PbI_6 octahedra are arranged in a higher symmetric motif as shown in Figure 9.15, in which the occurrence of an unusual μ_5-I is crucial for the structure and the five Pb atoms are almost coplanar. An example is $[BuN(CH_2CH_2)_3NBu]_3[Pb_5I_{16}]\cdot 4DMF$ [24].

Hexanuclear Clusters: $[Bi_6I_{22}]^{4-}$ No hexanuclear Pb/I cluster is known. On the contrary, three different hexanuclear Bi clusters are formed. The planar type I (Fig. 9.16) exists in $[Ph_4P]_4[Bi_6I_{22}]$, $[Na(THF)_6]_4[Bi_6I_{22}]$ [38], $[Et_4P]_4[Bi_6I_{22}]$ [39], and $[PhEt(Me)_2N]_4[Bi_6I_{22}]$ [40]. Type II, $[PhCH_2NEt_3]_4[Bi_6I_{22}]$ [36], has no planarity (Fig. 9.17), and an unusual μ_4–I atom in type III (Fig. 9.18) occurred in $[Ru(C_{10}H_8N_2)_3]_2[Bi_6I_{22}]$ [41].

Heptanuclear Cluster: $[Pb_7I_{22}]^{8-}$ Only one heptanuclear cluster is known as $[Bu_3N(CH_2)_4NBu_3]_4[Pb_7I_{22}]$ [33], and none are known for Bi/I. The $[Pb_7I_{22}]^{8-}$

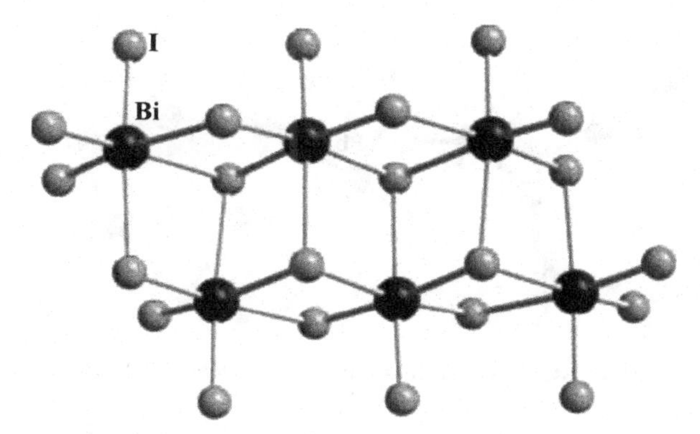

FIGURE 9.16 Structure of type I hexanuclear $[Bi_6I_{22}]^{4-}$.

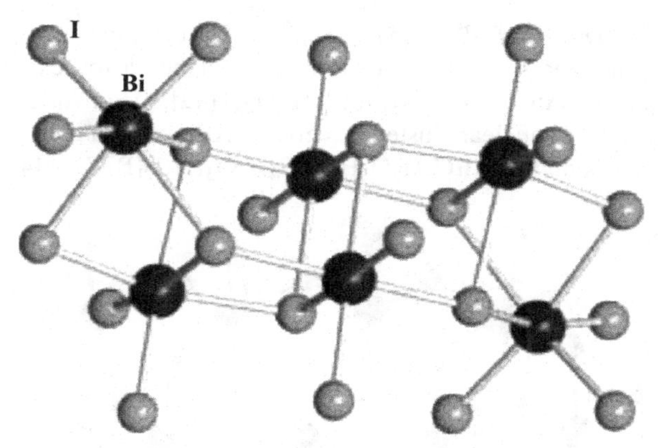

FIGURE 9.17 Structure of type II hexanuclear $[Bi_6I_{22}]^{4-}$.

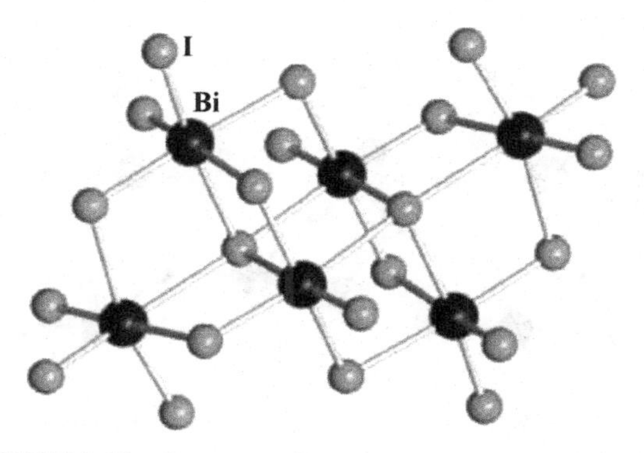

FIGURE 9.18 Structure of type III hexanuclear $[Bi_6I_{22}]^{4-}$.

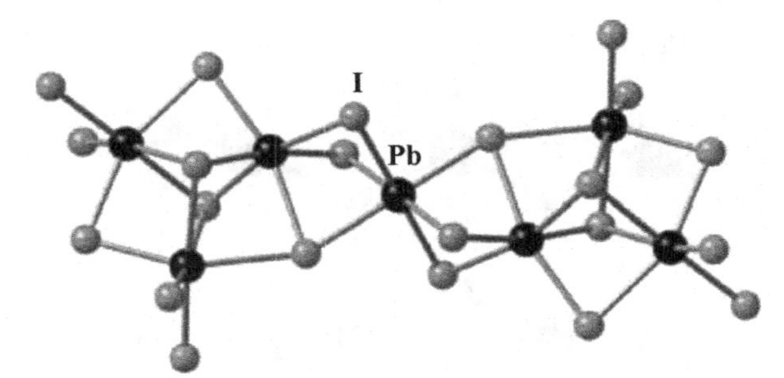

FIGURE 9.19 Structure of heptanuclear $[Pb_7I_{22}]^{8-}$.

(Fig. 9.19) can be constructed from two trinuclear $[Pb_3I_{10}]$ (Fig. 9.11) connected by a PbI_6 octahedron via shared opposite I–I–I faces. Such an example also substantiates the stability of the unusual $[Pb_3I_{10}]^{4-}$.

Octanuclear Clusters: $[Bi_8I_{28}]^{4-}$ and $[Bi_8I_{30}]^{6-}$ The octanuclear cluster $[Bi_8I_{28}]^{4-}$ (Fig. 9.20) is generated by addition of two BiI_6 octahedra to hexanuclear cluster $[Bi_6I_{22}]^{4-}$ (Fig. 9.16). An example is $[Ph_4P]_4[Bi_8I_{28}]$ [42]. Meanwhile, the conceptual dimerization of the tetranuclear cluster also gives an octanuclear cluster (Fig. 9.21) as found in the compound $[Bi_3I(C_4H_8O_3H_2)_2(C_4H_8O_3H)_5]_2[Bi_8I_{30}]$ [43]. These two

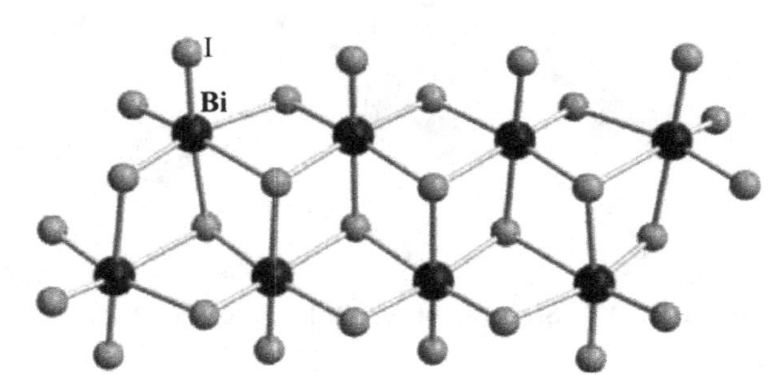

FIGURE 9.20 Structure of octanuclear $[Bi_8I_{28}]^{4-}$.

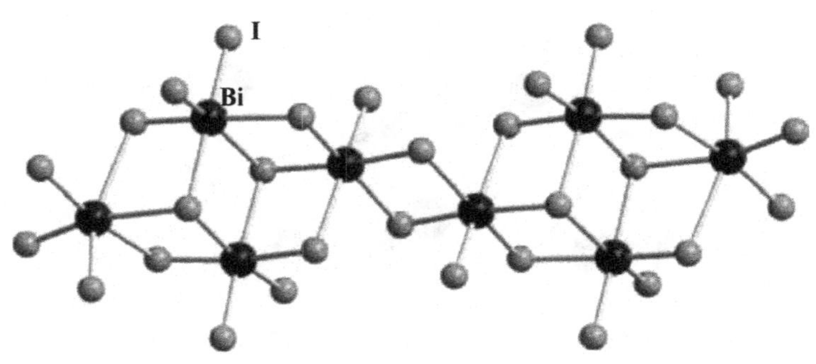

FIGURE 9.21 Structure of octanuclear $[Bi_8I_{30}]^{6-}$.

examples suggest that beyond the BiI_6 octahedron, some smaller clusters can also serve as a building unit to construct polynuclear clusters.

Decanuclear and Octadecanuclear Clusters: $[Pb_{10}I_{28}]^{8-}$ and $[Pb_{18}I_{44}]^{8-}$ $[Pb_{10}I_{28}{}^{8-}]$ [33] (Fig. 9.22) and $[Pb_{18}I_{44}{}^{8-}]$ [1] (Fig. 9.23) are two crystalline examples constructed by more than 10 metal ions, in which the geometry of the PbI_6 octahedron is well kept.

In contrast to 0D Pb/I clusters, Bi/I species tend to aggregate more into discrete polynuclear clusters. The essential reason is that Bi carries a higher positive charge, which allows more terminal I atoms to "protect" the discrete cluster and prevent extension of the building unit.

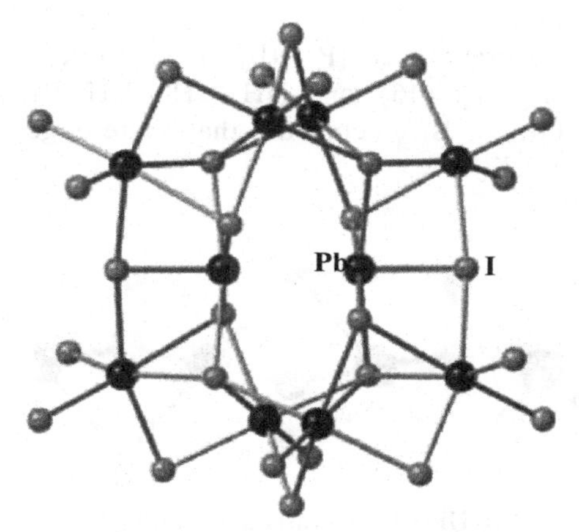

FIGURE 9.22 Structure of $[Pb_{10}I_{28}]^{8-}$.

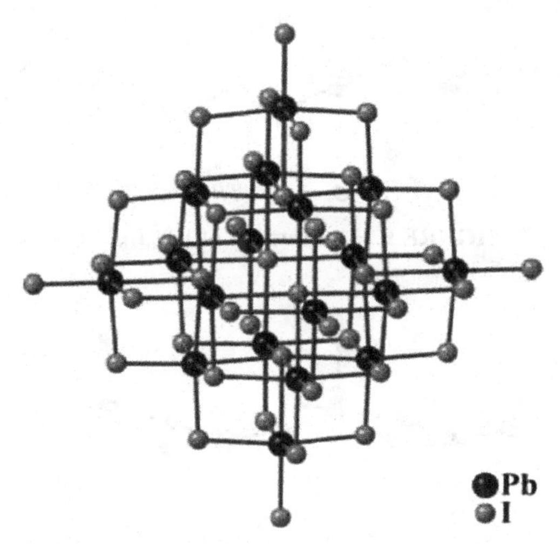

FIGURE 9.23 Structure of $[Pb_{18}I_{44}]^{8-}$.

9.2.3 One-Dimensional M/I Polymeric Chains

1D polymeric chains are common in iodometalates, which are built from MI_6 octahedra via different connection motifs or by different building units, such as Pb_3I_{10} or Bi_4I_{14}.

Linear Chains Constructed by Mononuclear Cluster $[BiI_6]^{3-}$ or $[PbI_6]^{4-}$

Via Shared Faces The most common chain motif for iodoplumbates is found in [7,10] by the infinite connection of PbI_6 octahedron via shared I–I–I faces (Fig. 9.24). The octahedron (O) of PbI_6 can to a large degree be deformed to trigonal prism (TP) in the complex 1,1′-dimethyl-4,4′-bipyridinium (Fig. 9.25) [44].

Via Shared Edges Compounds $[Pr_4N]_n[PbI_3]_n$ [30], $[C_9H_8N][BiI_4]$ [45], $[(C_2H_4N_3S)(C_2H_3N_3S)[BiI_4]$ [46], and $NH_3(CH_2)_2NH_3(BiI_4)_2 \cdot 4H_2O$ [47] are examples in which PbI_6 or BiI_6 octahedra that share edges constitute a linear chain (Figs. 9.26 and 9.27).

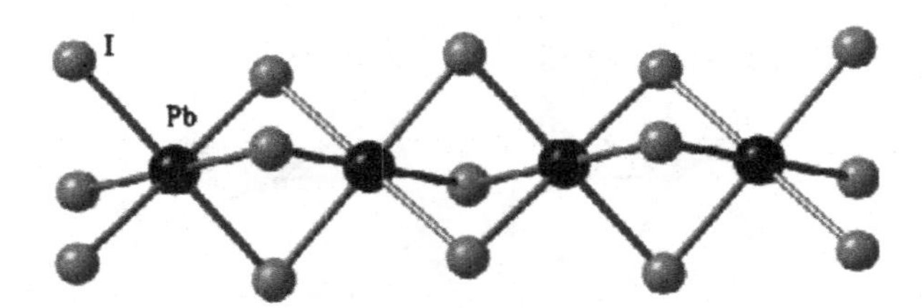

FIGURE 9.24 Structure of $[PbI_3^-]_n$.

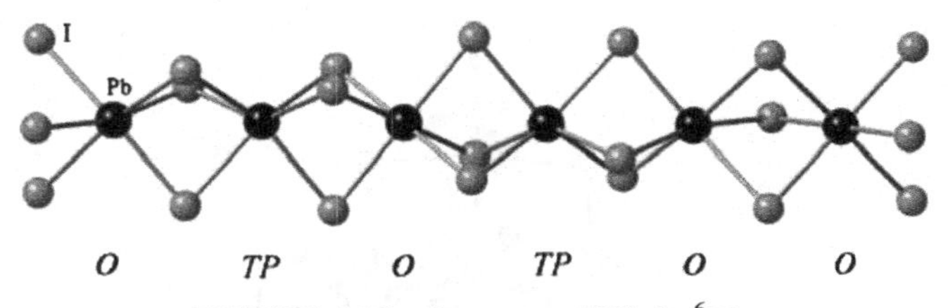

FIGURE 9.25 Structure of $[Pb_6I_{18}^{6-}]_n$.

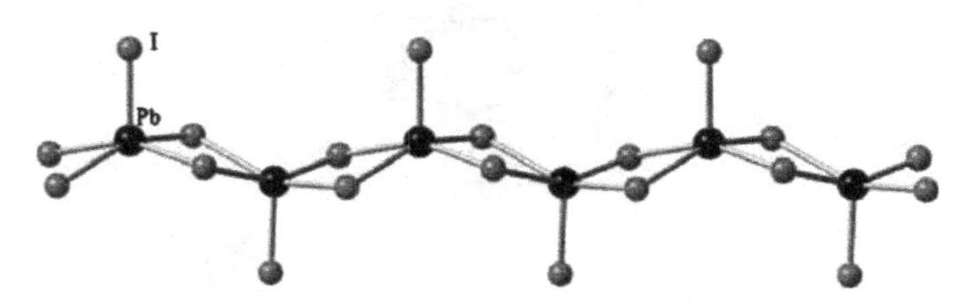

FIGURE 9.26 Structure of $[Pr_4N]_n[PbI_3]_n$.

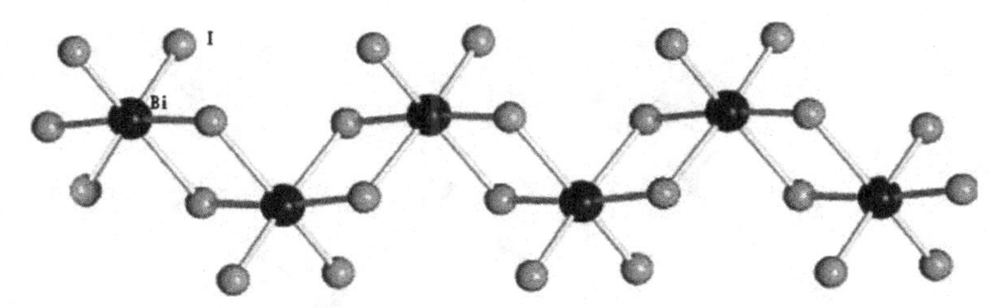

FIGURE 9.27 Structure of $[BiI_4^-]_n$.

Via Shared Apexes The PbI_6 octahedron may share *cis* or *trans* apexes to give $[NH_2C(I){=}NH_2]_{3n}[PbI_5]_n$ [48], as shown in Figure 9.28, or $[C_6H_5CH_2CH_2S{=}C(NH_2)_2]_{3n}[PbI_5]_n$ [49] (Fig. 9.29). This pattern is also adopted by BiI_6 octahedra in $[H_3N(CH_2)_6NH_3]_n[BiI_5]_n$ [50] (Fig. 9.30). In both structures, the metal atoms are coplanar.

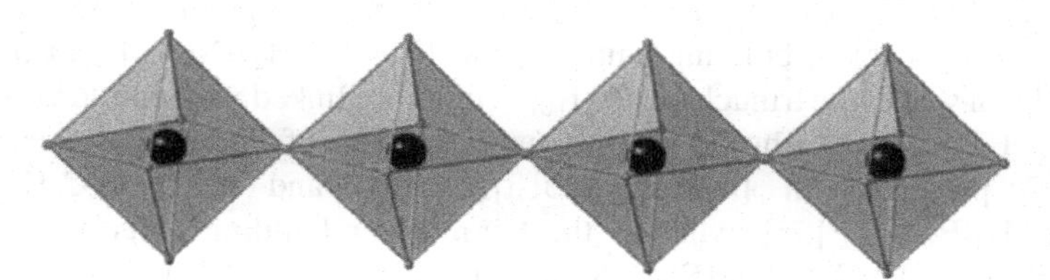

FIGURE 9.28 Structure of $[NH_2C(I){=}NH_2]_{3n}[PbI_5]_n$.

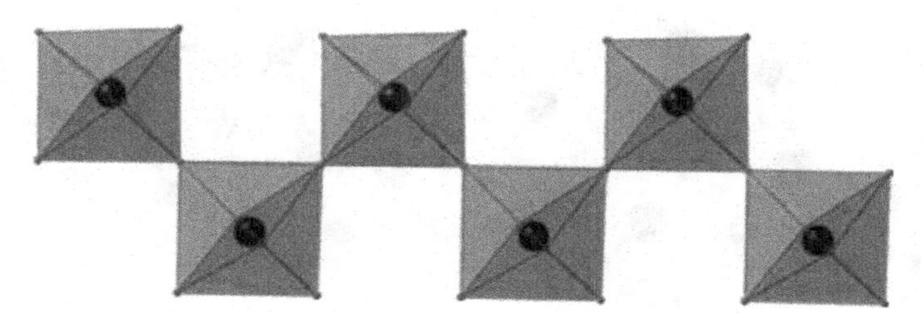

FIGURE 9.29 Structure of $[C_6H_5CH_2CH_2S{=}C(NH_2)_2]_{3n}[PbI_5]_n$.

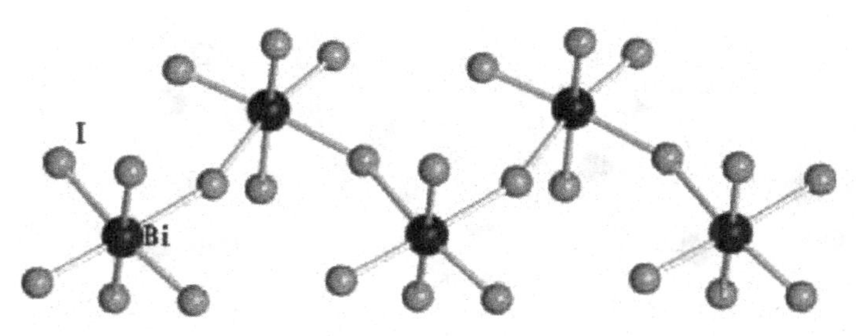

FIGURE 9.30 Structure of $[H_3N(CH_2)_6NH_3]_n[BiI_5]_n$.

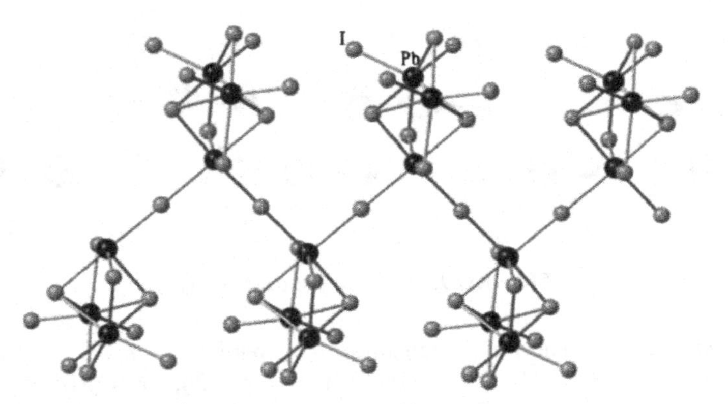

FIGURE 9.31 Structure of $[Me_3N(CH_2)_2NMe_3]_{2n}[Pb_3I_{10}]_n$.

Linear Chains Constructed by Trinuclear Cluster $[Pb_3I_{10}]^{4-}$ or Tetranuclear Cluster $[Bi_4I_{14}]^{2-}$ Except for the MI_6 octahedron, some polynuclear clusters can also serve as conceptual building units, such as $[Me_3N(CH_2)_2NMe_3]_{2n}[Pb_3I_{10}]_n$ [51] (Fig. 9.31), in which the trinuclear $[Pb_3I_{10}]^{4-}$ units are linked via twofold coordinated I atoms. Figure 9.32 shows the polymeric chain of the tetranuclear cluster $[Bi_4I_{14}]^{2-}$ presented in $[n\text{-}Bu_4N]_{2n}[Bi_4I_{14}]_n$ [29] and $[CH_2{=}C(C_6H_4\text{-}4\text{-}NO_2)CH_2NMe_3]_{2n}[Bi_4I_{14}]_n$ [35], whereas the tetranuclear building units are linked by binuclear cluster $[Bi_2I_8]^{2-}$ in $[Et_3PhN]_{4n}[Bi_6I_{22}]_n$ [42] (Fig. 9.33).

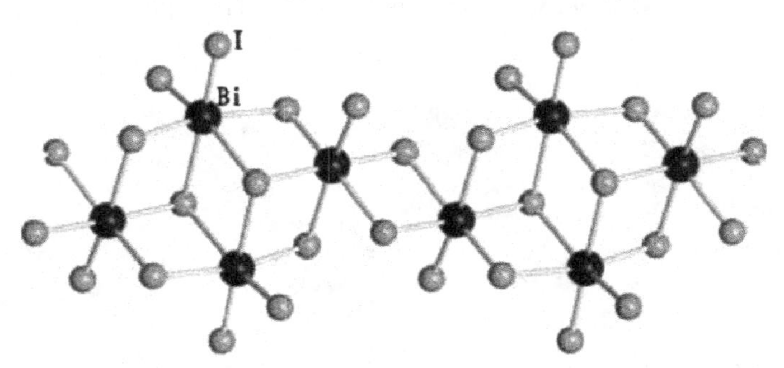

FIGURE 9.32 Structure of $[Bi_4I_{14}{}^{2-}]_n$.

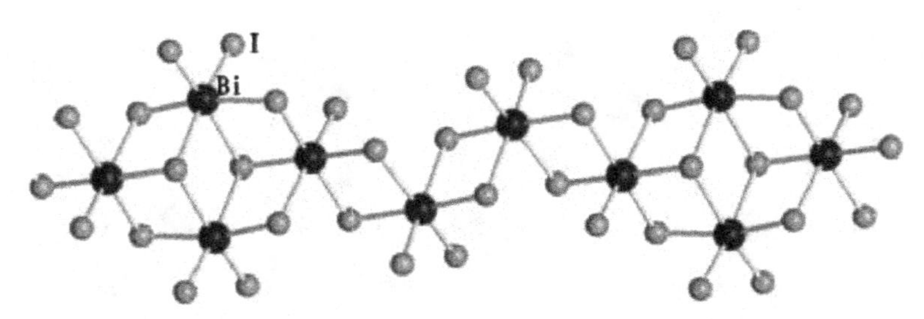

FIGURE 9.33 Structure of $[Et_3PhN]_{4n}[Bi_6I_{22}]_n$.

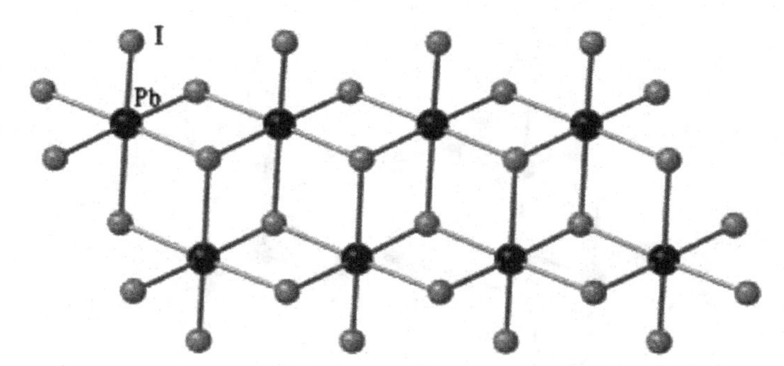

FIGURE 9.34 Structure of $[PbI_3^-]_n$.

Wider Chains Most of the chains described above are of a single string of MI_6 octahedra, and there are some wider chains formed by several strings of MI_6 octahedra, as follows. In $[C_{10}H_7CH_2NH_3]_n[PbI_3]_n$ [49] the anionic chains are two condensed linear edge-sharing PbI_6 chains (Fig. 9.34). In another case, the anionic chain in $[PPh_4]_{2n}[Pb_5I_{12}]_n$ [3] is made of two strings together with some individual PbI_6 octahedra (Fig. 9.35). Additionally, three strings of MI_6 octahedra form the anionic chain in $[PPh_4]_{2n}[Pb_6I_{14}(DMF)_2]_n$ [52], as shown in Figure 9.36, in which half of the peripheral Pb octahedra have a coordination sphere of five I atoms and one DMF solvent molecule. Interestingly, the middle string of PbI_6 octahedra is reminiscent of the PbI_2 binary. Furthermore, an anionic chain of six strings of MI_6 octahedra exists in $[M_4(2,2'\text{-bipy})_{12}Pb_{11}I_{28}S]_n$ (M = Ni, Co) [53], as shown in Figure 9.37. Again, the inner Pb atoms surrounded by six neighboring Pb atoms have the same motif as in binary PbI_2.

If the number of strings of MI_6 octahedra would reach infinity, the anionic structure would be a 2D layer. However, experimental data illustrating such a stepwise development are still lacking.

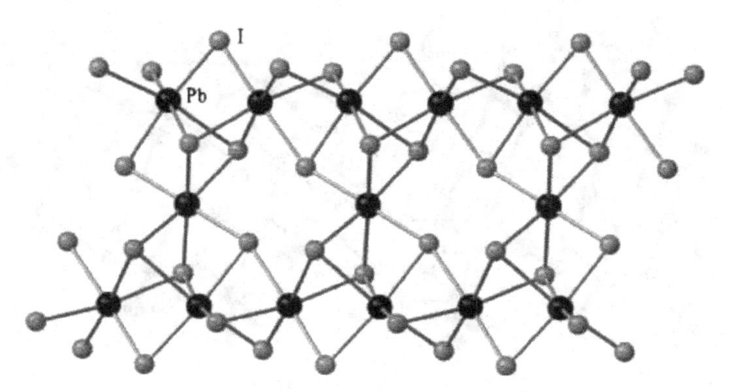

FIGURE 9.35 Structure of $[Pb_5I_{12}^{2-}]_n$.

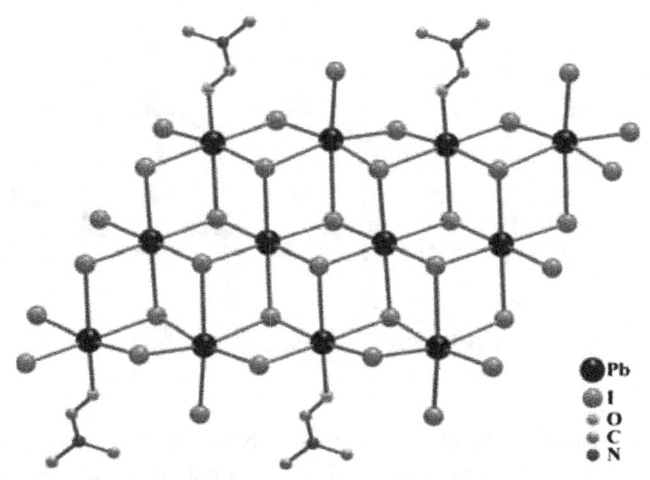

FIGURE 9.36 Structure of $[Pb_6I_{14}(DMF)_2{}^{2-}]_n$.

9.2.4 Two-Dimensional Polymeric M/I Layers

The majority of M/I layers are of perovskite-type construction. Different numbers of layers (m) in the 3D perovskite network could be separated by different organic cations to give 2D layered compounds. The simplest example of this family is $(CH_3C_6H_4CH_2NH_3)_2PbI_4$ [49,54–57], in which single layers of PbI_6 octahedra are separated by organoammonium cations $(CH_3C_6H_4CH_2NH_3)^+$ (Fig. 9.38).

A new layer with double-connected PbI_6 layers is formed with the help of a small cation $(CH_3NH_3)^+$ to fit inside the inorganic PbI_6 layers as shown in Figure 9.39. The thicker layers are separated by the same organic cations $(CH_3C_6H_4CH_2NH_3)^+$.

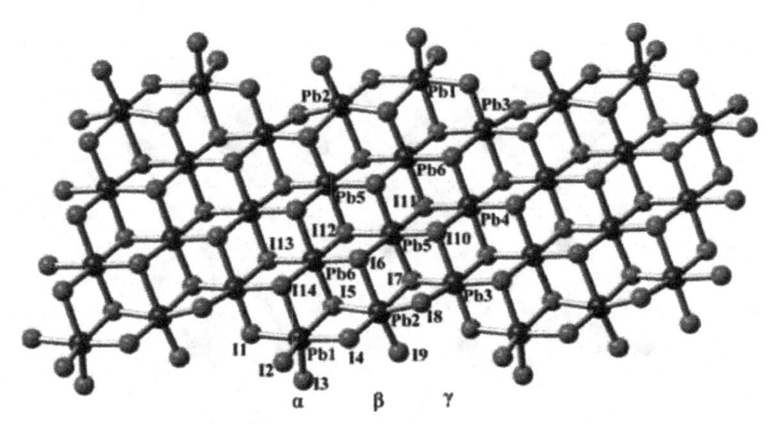

FIGURE 9.37 Structure of $[Pb_{11}I_{28}S^{8-}]_n$.

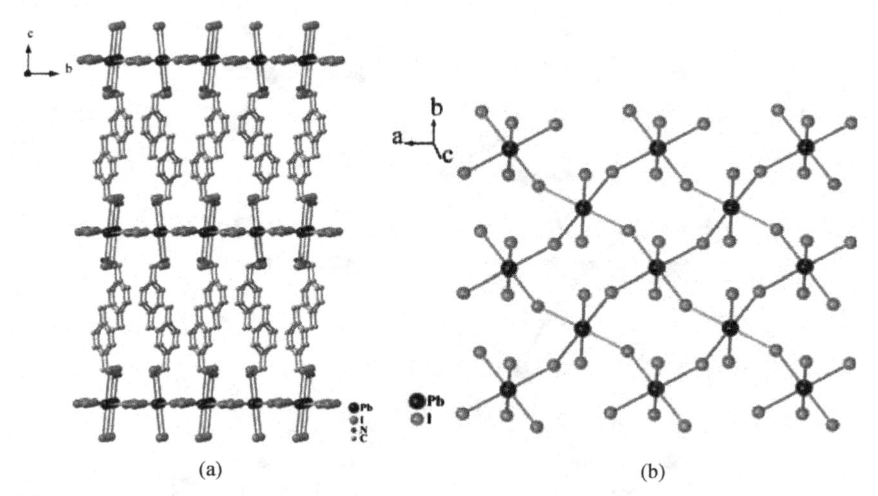

FIGURE 9.38 (a) Structure of $[(CH_3C_6H_4CH_2NH_3)]_{2n}[PbI_4]_n$; (b) single layer of a perovskite-type PbI_6 layer.

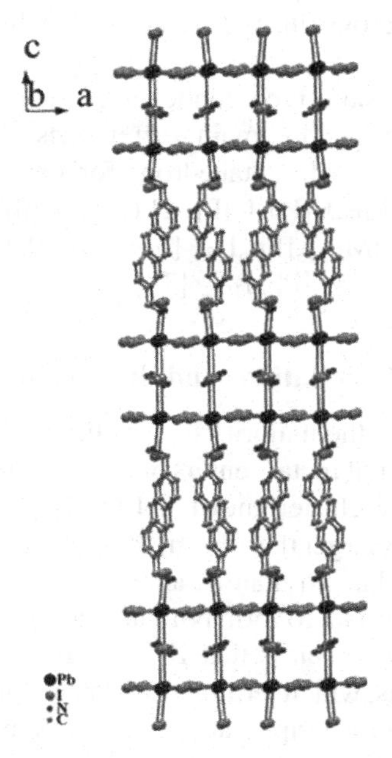

FIGURE 9.39 Structure of $[(CH_3C_6H_4CH_2NH_3)_2(CH_3NH_3)]_n[Pb_2I_7]_n$, and Pb_2I_7 layer with double-connected perovskite-type PbI_6 layers accommodating small $(CH_3NH_3)^+$ cations.

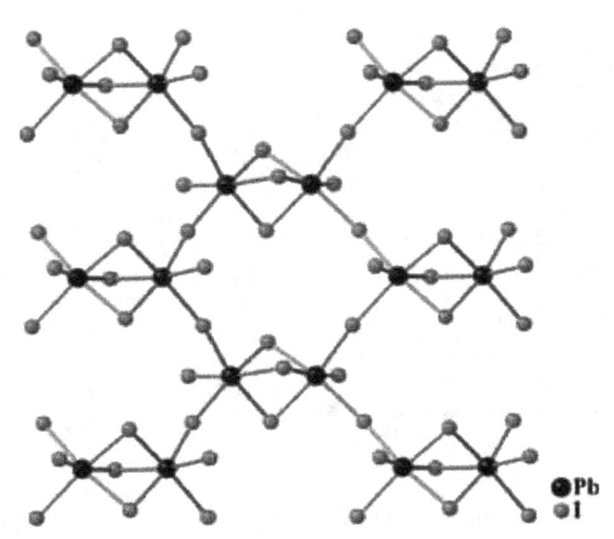

FIGURE 9.40 Structure of $[Pb_2I_7]^{3-}$ anion layer.

Studies have shown that the number (n) of single perovskite-type PbI_6 layers can be 1, 2, 3, 4, 5, and ∞, in the $(RNH_3)_2(CH_3NH_3)_{n-1}M_nX_{3n+1}$ family (M = Pb, Sn, or Cu; X = Cl, Br, or I; and R = C_4H_9, $C_6H_5C_2H_4$) [58–60]. The compound $(H_2AEQT)Bi_{2/3}I_4$ [61] also adopts a perovskite-type structure in which the Bi/I layer is a single perovskite layer.

In addition to the perovskite-type connections, there are patterns in other M/I (M = Pb or Bi) compounds. The main differences come from building units other than the mononuclear MI_6 octahedron: for example, the dinuclear cluster $[Pb_2I_7]^{3-}$ in $[Me_3NC_2H_4NMe_3]_2[Pb_2I_7]I$ [62] (Fig. 9.40) or the hexanuclear cluster $[Pb_6I_{18}]^{6-}$ in $[Me_3NC_3H_6NMe_3]_3[Pb_6I_{18}]$ [51] (Fig. 9.41), or the stepwise ribbon $[Pb_4I_{10}{}^{2-}]_n$ in $[Ni(opd)_2(acn)_2]_n$ $[Pb_4I_{10}]_n$ [78].

9.2.5 Some Structural Correlations and the *r* Value

We have so far summarized the majority of structural types of iodobismuthates and iodoplumbates. The number of metal centers of the discrete cluster can be 1 to 8, 10, or 18. The highest polynuclear cluster known is $[Pb_{18}I_{44}]^{8-}$, but the number of known Bi-containing 0D clusters is larger than that of Pb-containing 0D compounds, whereas more 1D Pb polymers are known than Bi analogs.

Why do Bi/I compounds tend to form polynuclear 0D clusters with respect to Pb/I compounds? The essential reason is that Bi^{3+} carries a higher positive charge, to allow more terminal I atoms, which terminate extension of the building block. In fact, the terminal I atoms serve as peripheral species to terminate extension of the MI_6 building block, so that the more μ_t-I atoms a cluster wears, the more examples of discrete polyclusters there are The μ_2-I atoms correspond to the bridging of two neighboring octahedra; μ_3-I, to the aggregation of three octahedra; and so on.

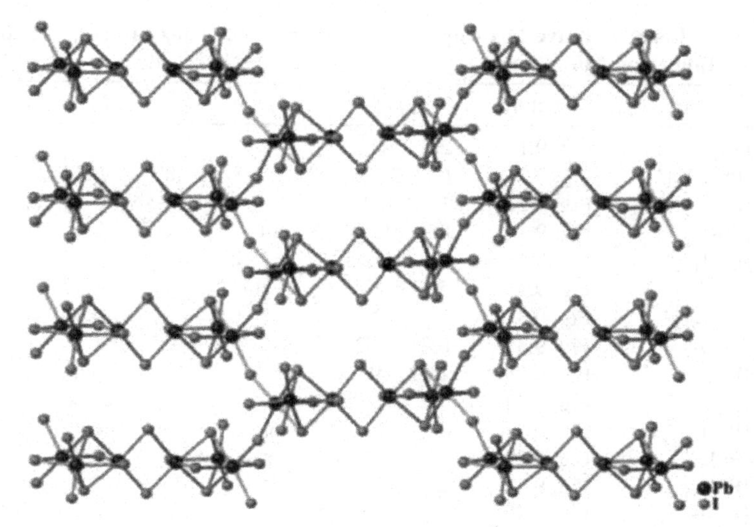

FIGURE 9.41 Structure of $[Pb_6I_{18}]^{6-}$ anion layer.

The types of I atoms reflect the character of the structural motif, and this can be reflected in terms of an r value defined as $r = \Sigma N(\mu_n\text{-I})/n$ (N = the number of μ_n-I atoms and n = 1, 2, 3, 4, 5, 6). In addition, the M/r ratio indicates empirically the density of the inorganic moiety of the iodometalates. Some representative iodometalates, their relevant structural features, and the corresponding r and M/r values are provided in Table 9.1.

The M/r limit in the Pb/I system is 1.5; since binary PbI_2 is the most condensed Pb/I compound and any introduction of cations will decrease such a ratio. Similarly, the M/r ratio of 0.67 for binary BiI_3 is the limit of the Bi/I system. As shown clearly in Figures 9.3–9.23, the higher the number of metal centers, the higher the density of a discrete cluster. As shown in Table 9.1, the density of the inorganic moiety decreases as the M/r ratio decreases. For example, BiI_3 is less condensed than PbI_2 binary because it is a derivative of a PbI_2-type structure with a $\frac{1}{3}$ vacancy; coherently, the M/r ratio of BiI_3 (0.67) is less than that of PbI_2 (1.5). On the other hand, the Bi^{3+} ion has a higher charge versus Pb^{2+}, which requires more negative I^- ions; therefore, the M/r ratios in a Bi/I system are systematically lower than those of the comparable Pb/I system. Take some 0D clusters: for example, the M/r ratios increase monotonously from 0.4 of $[Pb_2I_6]^{2-}$ to 0.82 of $[Pb_{18}I_{44}]^{8-}$. Three complexes of the same building units have a nice dimensional development from the 0D $[Pb_3I_{10}]^{4-}$ cluster (Fig. 9.11) though the 1D $[Pb_3I_{10}{}^{4-}]_n$ chain (Fig. 9.31) to the 2D $[Pb_3I_{10}{}^{4-}]_n$ layer (Fig. 9.42); the density increases as the dimension increases, as does the M/r ratio, from 0.42 to 0.45 to 0.50. Note that there are several related factors in a consideration of the density of 1D or 2D polymers, such as the type of building unit, the extension motif, and so on, that need to be taken into account; therefore, only comparisons within polymers that are made by the same building unit are in direct relevance to the development of the dimension.

TABLE 9.1 Representative M/I Compounds, Their Structural Features, and the Corresponding *r* and M/*r* Values

M	Compound	Structure	$r = \Sigma N(\mu_n\text{-I})/n$	M/*r* Ratio	E_g
Pb	PbI_2	Fig. 9.1	2/3 = 0.67	1.5	
0D	$[Pb_{18}I_{44}]^{8-}$	Fig. 9.23	6 + 24/2 + 8/3 + 6/5 = 21.87	0.82	
	$[Pb_{10}I_{28}]^{8-}$	Fig. 9.22	12 + 6/2 + 6/3 + 4/4 = 18	0.56	
	$[Pb_7I_{22}]^{8-}$	Fig. 9.19	8 + 8/2 + 6/3 = 14	0.5	
	$[Pb_5I_{16}]^{6-}$	Fig. 9.15	5 + 10/2 + 1/5 = 10.2	0.49	
	$[Pb_3I_{10}]^{4-}$	Fig. 9.11	5 + 3/2 + 2/3 = 7.17	0.42	
	$[Pb_2I_6]^{2-}$	Fig. 9.8	4 + 2/2 = 5	0.4	
1D	$[Pb_5I_{12}{}^{2-}]_n$	Fig. 9.35	6/3 + 6/2 = 5	1	
	$[Pb_6I_{18}{}^{6-}]_n$	Fig. 9.25	18/2 = 9	0.66	
	$[PbI_3{}^{-}]_n$	Fig. 9.24	3/2	0.66	
	$[Pb_3I_{10}{}^{4-}]_n$	Fig. 9.31	4 + 4/2 + 2/3 = 6.67	0.45	
	$[PbI_5{}^{3-}]_n$	Fig. 9.29	4 + 1/2 = 4.5	0.22	
2D	$[Pb_3I_{10}{}^{4-}]_n$	Fig. 9.42	2 + 8/2 = 6	0.5	
	$[Pb_2I_7{}^{3-}]_n$	Fig. 9.40	2 + 5/2 = 4.5	0.44	
	$[PbI_4{}^{2-}]_n$	Fig. 9.38	2 + 2/2 = 3	0.33	
Bi	BiI_3	Fig. 9.2	3/2 = 1.5	0.67	1.73
0D	$[Bi_5I_{18}]^{3-}$	Fig. 9.13	6 + 12/2 = 12	0.42	
	$[Bi_8I_{28}]^{4-}$	Fig. 9.20	14 + 8/2 + 6/3 = 20	0.4	
	$[Bi_6I_{22}]^{4-}$	Fig. 9.16	12 + 6/2 + 4/3 = 16.33	0.37	
	$[Bi_8I_{30}]^{6-}$	Fig. 9.21	16 + 10/2 + 4/3 = 22.33	0.36	
	$[Bi_5I_{19}]^{4-}$	Fig. 9.14	11 + 5/2 + 3/3 = 14.5	0.344	
	$[Bi_4I_{16}]^{4-}$	Fig. 9.12	10 + 4/2 + 2/3 = 12.667	0.32	2.16
	$[Bi_3I_{12}]^{3-}$	Fig. 9.9	6 + 6/2 = 9	0.33	
		Fig. 9.10	7 + 4/2 + 1/3 = 9.33	0.32	
	$[Bi_2I_8]^{2-}$	Fig. 9.5	6 + 2/2 = 7	0.29	2.19
	$[Bi_2I_9]^{3-}$	Fig. 9.6	6 + 3/2 = 7.5	0.27	
	$[Bi_2I_{10}]^{4-}$	Fig. 9.7	8 + 2/2 = 9	0.22	
	$[BiI_6]^{2-}$	Fig. 9.3	6	0.17	
1D	$[Bi_4I_{14}{}^{2-}]_n$	Fig. 9.32	6 + 6/2 + 2/3 = 9.67	0.41	2.02
	$[Bi_6I_{22}{}^{4-}]_n$	Fig. 9.33	10 + 10/2 + 2/3 = 15.67	0.38	
	$[BiI_4{}^{-}]_n$	Fig. 9.27	2 + 2/2 = 3	0.33	
	$[BiI_5{}^{-}]_n$	Fig. 9.30	4 + 1/2 = 4.5	0.22	

Among the known iodobismuthates and iodoplumbates, the higher density usually measures a smaller energy gap. We have shown via several examples that the M/*r* ratio decreases with an increase in the energy gap, as noted in Table 9.1. The M/*r* ratios for BiI_3 (2D), $[Bi_4I_{14}{}^{2-}]$ (1D), $[Bi_4I_{16}]^{2-}$ (0D), and $[Bi_2I_9]^{2-}$ (0D) decrease from 0.67 to 0.41, 0.32, and 0.27 as the energy gaps increase from 1.73 to 2.02, 2.16, and 2.19 eV [63], respectively. Such a tendency is also found in the heterometallic Bi/M/I (M = Cu, Ag) complexes and is discussed in Section 9.4. Hence, the structural factor M/*r* might be the best empirical parameter to predict the energy gap of a complex, which decreases with an increase in M/*r*. However, the available experimental data are

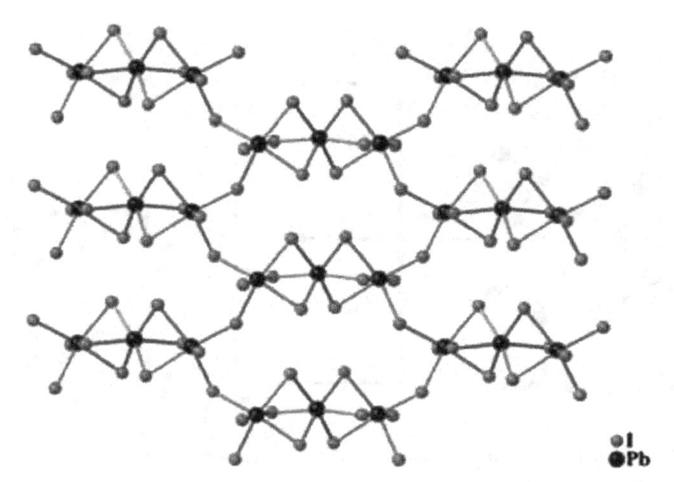

FIGURE 9.42 Structure of the $[Pb_3I_{10}{}^{4-}]_n$ anionic layer.

insufficient to date, and such a rule of thumb needs to be tested. Meanwhile, such an empirical relevance is expected to be suitable for other systems, such as Sn/I, M/X (M = Pb, Bi, Sn and X = F, Cl, Br, I; or M = transition metal and X = S, Se, Te), and so on, in which r is defined as $r = \Sigma N(\mu_m\text{-X})/m$, $m = 1$ to 6.

9.3 STRUCTURAL MODIFICATION

9.3.1 Cation Effect

The cation effects are widely reported in both Pb/I and Bi/I compounds. For example, starting with the same reactant, the cation determines the anionic structure to be $[Pb_{18}I_{44}]^{8-}$ by $[n\text{-}Bu_4N]^+$, or mononuclear $[PbI_4]^{2-}$ by $[Pr_4N]^+$, or binuclear $[Pb_2I_6]^{2-}$ by $[PPh_4]^+$. In another case, anionic $[PbI_5{}^{3-}]_n$ is a linear chain (Fig. 9.28) when it crystallizes with $[NH_2C(I){=}NH_2]^+$, a wavy chain with $[CH_3SC({=}NH_2)NH_2]^+$, or a zigzag chain with $[C_6H_5(CH_2)_2S{=}C(NH_2)_2]^+$ (Fig. 9.29). The anionic $[Pb_3I_{10}{}^{4-}]_n$ adopts a 2D layered structure (Fig. 9.42) when the counter cation is $[C_6H_5CH_2SC(NH_2)_2]^+$ or $[(C_4S)_2SC_2H_4NH_3]^+$, but adopts a zigzag chain motif with $[Me_3N(CH_2)_2NMe_3]^{2+}$, as shown in Figure 9.31. The size and charge of the cation play the most important roles, and the details of the packing of ionic types are the ultimate reasons for stability.

As illustrated in Figure 9.43, a discrete cluster $[n\text{-}Bu_4N]_2[Cu_2(CH_3CN)_2Bi_2I_{10}]$ and polymers of $[Et_4N]_{2n}[Cu_2Bi_2I_{10}]_n$ and $[Cu(CH_3CN)_4]_{2n}[Cu_2Bi_2I_{10}]_n$ are obtained under similar conditions but with different cations [64]. These three compounds exhibit a nice stepwise modification of the anionic structures from a 0D cluster to a linear polymer to a ladderlike chain with a more-or-less gradual decrease in cation size. The larger $n\text{-}Bu_4N^+$ diminish the contacts sufficiently between the isolated anionic cluster units, but the smaller cations allow condensed chains.

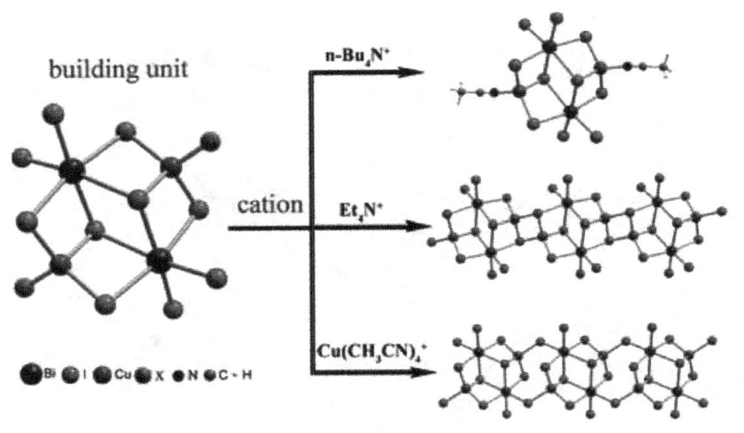

FIGURE 9.43 Different anionic structures have been generated with the guidance of different cations.

9.3.2 Ligand Effects

Organic ligands such as S–, O–, and N–, which coordinate directly to Pb or Bi metal ions with replacement of some of the I atoms, can alter the anionic structures to a great degree. For example, the polymer $[PbI_2(4\text{-MPD})_2]_n$ [65] shows an edge-sharing motif with a building unit of PbI_xZ_{6-z} (Fig. 9.44). More examples may be found in $[PbI_2(TMT)]_n$ [66], $[PbI_2(PYD)_2]_n$ [65], and $[PbI_2(2\text{-MPD})_2]_n$ [67].

Beyond acting merely as a coordination ligand on Pb or Bi centers, the organic ligand can also contribute to the skeleton to generate an inorganic–organic hybrid structure. As shown in Figure 9.45, the organic ligand bipy also affords linkage of the hybrid layer [68]. By changing the length and flexibility of an organic ligand, different hybrid organic/M/I compounds are formed [69].

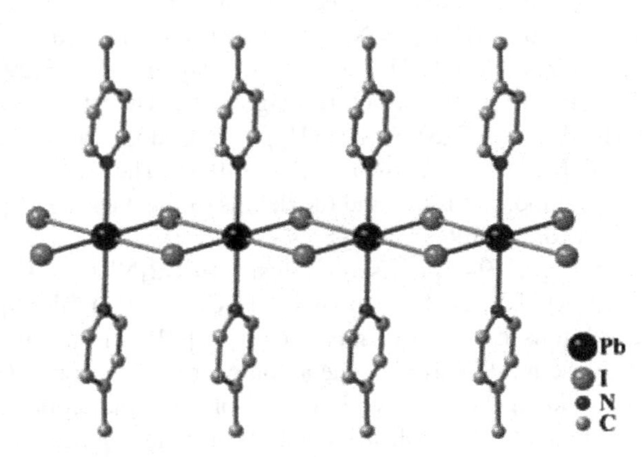

FIGURE 9.44 Structure of $[PbI_2(4\text{-MPD})_2]_n$.

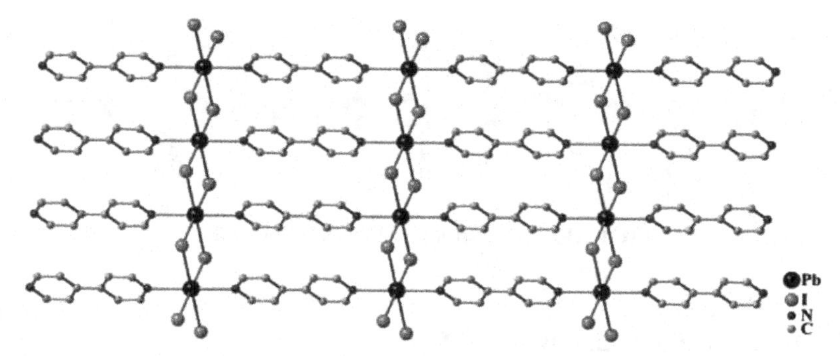

FIGURE 9.45 Structure of layered $[PbI_2(4,4'\text{-BIPY})]_n$.

Recently, we discovered several novel Pb/I/dialkyldithiocarbamates with different dimensionalities: $[PbI(S_2CNR_2)]_n$ [$R_2 = Me_2$, $(CH_2)_4$, and $(CH_2)_5$]. As the length of the alkyl group increases, the Pb/I anionic structure decreases from a 2D layer (Fig. 9.46a) to a 1D chain (Fig. 9.46b). A simple structural relationship between $[PbI(S_2CNMe_2)]_n$ and $[PbI(S_2CNC_4H_8)]_n$ is illustrated in Figure 9.46c; the stepped chain in $[PbI(S_2CNC_4H_8)]_n$ (Fig. 9.46c-i) forms a distorted hypothetical chain by displacement of some Pb–I bonds (Fig. 9.46c-ii), and the aggregation of such distorted chains gives the 2D layer in $[PbI(S_2CNMe_2)]_n$ [70].

By choice of suitable ligands (i.e., those with a bifunctional group), a new 2D layered structure or 3D network is expected. Several aspects of the ligand, such as coordination ability, stereo effects, stacking of aromatic rings, hydrogen bonding, and so on, will influence the M/I structure greatly. Therefore, these features might yield useful access to modification or design a new complex. At the same time, the organic ligand selected may introduce some desired functions into the inorganic Pb/I or Bi/I systems. Explorations are worth trying.

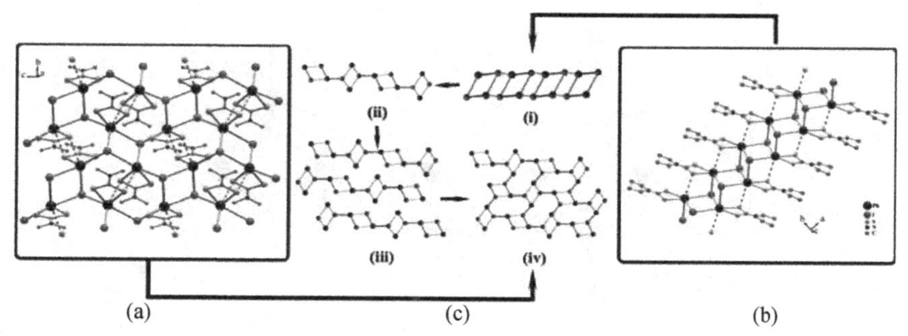

FIGURE 9.46 (a) 2D layered structure in $[PbI(S_2CNMe_2)]_n$; (b) 1D step chain structure in $[PbI(S_2CNC_4H_8)]_n$; (c) structural relationship between step chain in (b) and layer in (a).

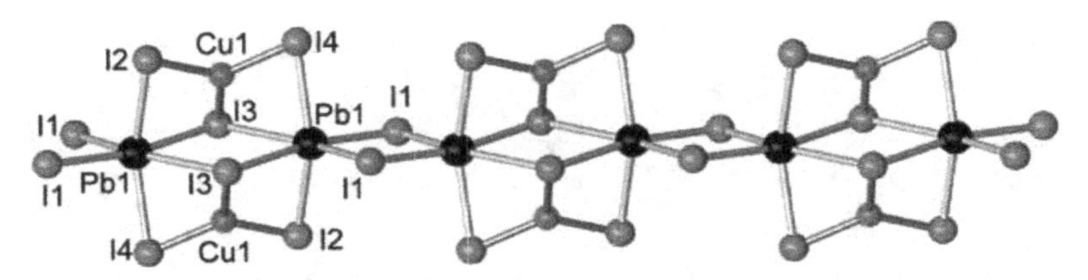

FIGURE 9.47 Structure $[Bu_4N]_n[PbCuI_4]_n$.

9.3.3 Heterometal–Iodine Bonding Effect

Both the cation and ligand effects described above are indirect influences on the structure of the Pb/I or Bi/I inorganic moiety. The direct effects in this sense would be the inorganic bonding interactions within the Pb(Bi)/I moiety. Several recent efforts have focused on introduction of the d^{10} transition metal–iodine interactions into Pb (Bi)/I systems [63,64,71–73]. The heterometallic products obtained have proven that TM-I bonding is an effective means to change the connection of the MI_6 octahedron, and in some cases even the coordination number of Pb/Bi atoms. The unprecedented heterometallic iodometalates also exhibit new and novel properties.

For example, the new polymeric anion structure of compound $[Bu_4N]_n$ $[PbCuI_4]_n$ [71] is shown in Figure 9.47. The PbI_6 octahedra become quite distorted. Two neighboring octahedra are now joined by two trigonally coordinated Cu^+ ions to form a dimer in which adjacent units are connected by a common edge (I1–I1). Because of the Cu-I bonding interaction, the Pb1–I4 and Pb1–I2 bonds are bent toward the Cu^+ ions by 19° from linear. Usually, Cu^+ tends to adopt a fourfold coordination polyhedron; therefore, when a weak ligand (e.g., CH_3CN) coordinates a Cu^+ ion it yields a tetrahedrally coordination sphere and takes Cu^+ ions away from the PbI_6 linear chain. Such a variation is seen in compound $[PbI_4Cu_2(CH_3CN)_2]_n$, shown in Figure 9.48 [74].

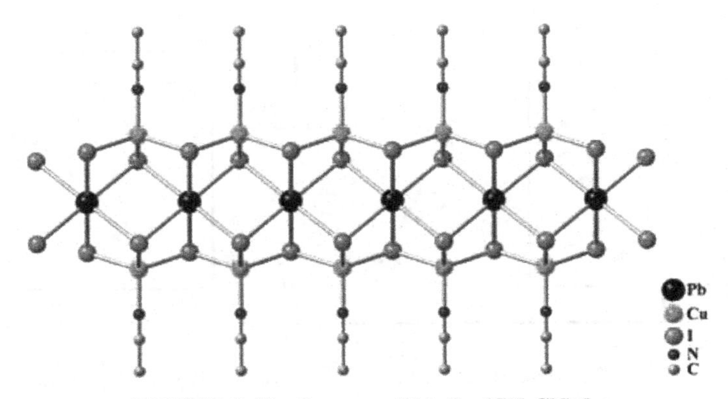

FIGURE 9.48 Structure $[PbI_4Cu_2(CH_3CN)_2]_n$.

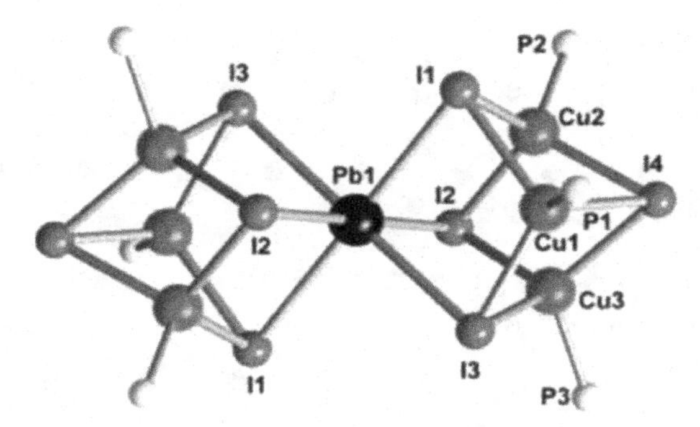

FIGURE 9.49 Molecular structure of compound $PbCu_6I_8(PPh_3)_6$. For clarity, the PPh_3 molecules are omitted.

If a stronger ligand PPh_3^+ is presented, a very different structure is produced, as shown in Figure 9.49. The molecular structure of $[PbCu_6I_8(PPh_3)_6]$ can be described as a bicubane-like cluster formed by two vertex-sharing $[PbCu_3I_4]^+$ units. The large difference in radius between Pb^{2+} and Cu^+ ions leads to distortion of the $[PbCu_3I_4]^+$ cubane. Interestingly, when a larger Ag^+ ion is utilized instead, a somewhat expected cubane unit with less distortion in a 1D chain occurs in $[PbAg_2(PPh_3)_2I_4]_n[PbI_2(DMF)_2]_n$ (Fig. 9.50).

The connection of PbI_6 octahedra as altered by the heavier Ag^+ differs from that by the lighter Cu^+, as illustrated by compound $[Bu_4N]_n[PbAgI_4]_n$ (see Fig. 9.51). The

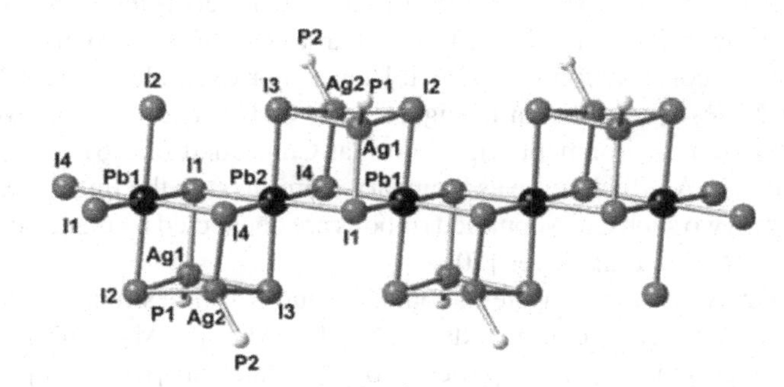

FIGURE 9.50 Cubic chain $[PbAg_2(PPh_3)_2I_4]_n$ along the *a*-axis.

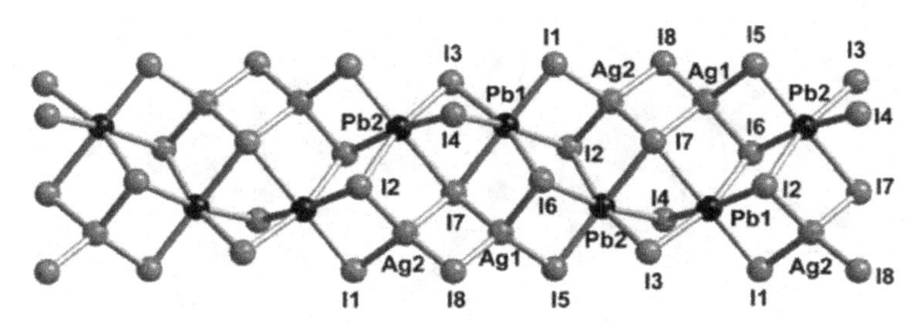

FIGURE 9.51 Section of 1D ribbon in $[Bu_4N]_n[PbAgI_4]_n$ along the *b*-axis.

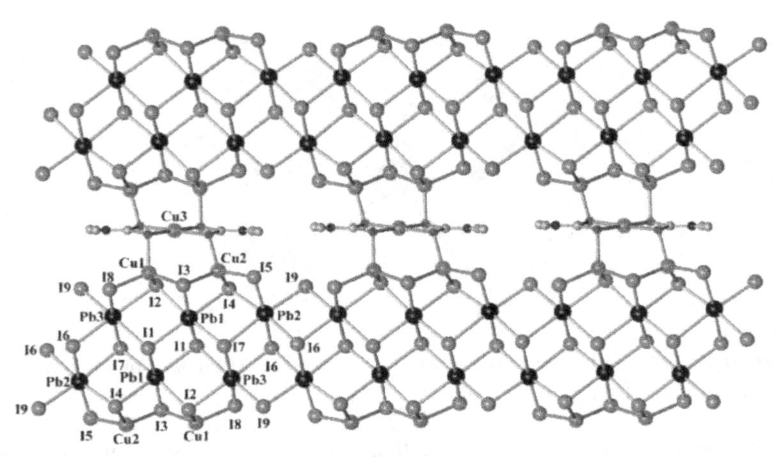

FIGURE 9.52 Section of anionic layer $[Bu_4N]_{2n}[Pb_6I_{18}Cu^I_4Cu^{II}(S_2CNMe_2)_2]_n$. The unmarked small balls represent S, C, and N, respectively.

$[PbAgI_4^-]_n$ double chain becomes a heterometallic polymer constructed for Ag_2I_6 dimers and Pb_4I_{17} tetramers. The face-shared PbI_6 octahedra appear in a long/short-stepped Pb1–Pb2 wavy chain motif.

Although both Ag^+ and Cu^+ ions are fourfold coordinated by I atoms, the structural modifications they produce in Pb/TM/I moiety are different because of the different sizes of Ag^+ and Cu^+ and the different bond energies of Ag–I and Cu–I bonds. In addition, metal ions with different charges (e.g., Cu^+ vs. Cu^{2+}) contribute differently to the construction of a Pb/Cu/I layer. For example, in $[(Bu_4N)_2Pb_6I_{18}$-$Cu^I_4Cu^{II}(S_2CNMe_2)_2]_n$, as shown in Figure 9.52, $Cu1^{1+}$ and $Cu2^{1+}$ have fourfold CuI_3S coordination environment with a normal Cu–I bond of 2.61 to 2.70 Å and a Cu–S bond of 2.34 Å, which serve as boundary atoms to stop the further extension of PbI_6 octahedra. Two such Cu^+-confined ribbons are connected by planar-coordinated $Cu3^{2+}$ cations to give a 2D layer [70].

On the contrary, in the Bi/I system, both Cu^+ and Ag^+ are structurally similar, as shown in Figure 9.53. As in compounds $[Et_4N]_{2n}[Bi_2M_2I_{10}]_n$ ($M = Cu^+$ or Ag^+), the heterometallic polymers are constructed by tetranuclear $[Bi_2M_2I_{10}]^{2-}$ building units [63,64]. In the Cu derivative, the Bi–I bonds lie in the range 2.87 to 3.39 Å,

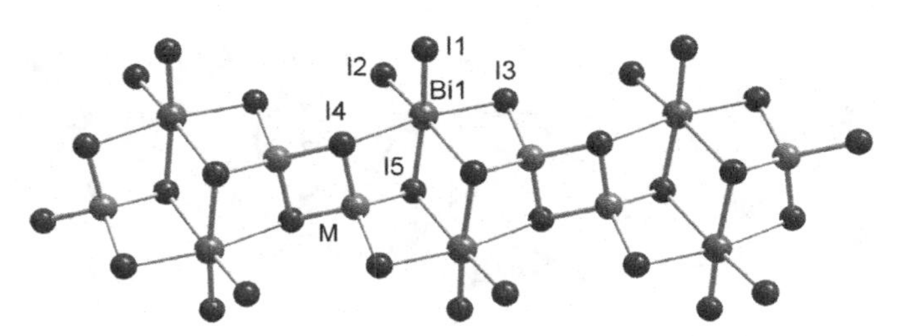

FIGURE 9.53 1D chain of $[Bi_2M_2I_{10}{}^{2+}]_n$. M = Cu or Ag.

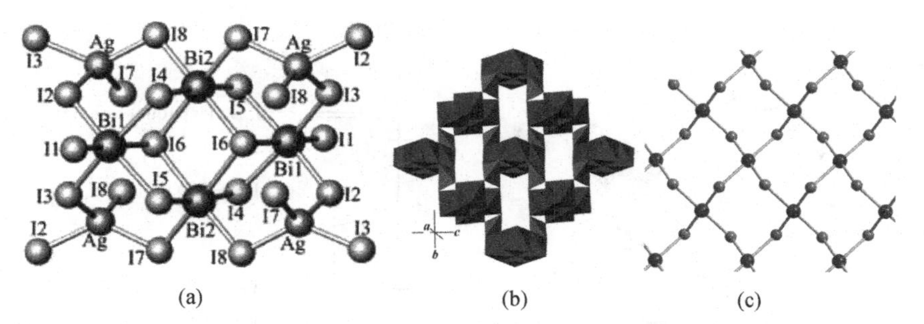

FIGURE 9.54 (a) Structure of the anionic building unit in the compound $[Et_4N]_{2n}$ $[Bi_4Ag_2I_{16}]_n$; (b) polyhedral representation of the 4^4 grid of the anion with BiI_6 octahedra and AgI_4 tetrahedra; (c) topological scheme of the layer in which the Bi_4I_{16} cluster is treated as a node, and the Ag atom as a linkage.

with 3.11 Å as the average, whereas the average Cu–I = 2.65 Å. Although in the corresponding Ag derivative, the anionic polymer is almost identical, except that the Ag–I average distance is 2.84 Å, the BiI_6 octahedra are more distorted, with Bi–I= 2.87 to 3.42 Å and $(Bi–I)_{av} = 3.12$ Å. Again, the high flexibility of BiI_6 octahedron allows accommodation to Cu^+ or Ag^+ ions.

In another case, $[Et_4N]_{2n}[Bi_4Ag_2I_{16}]_n$, the common Bi_4I_{16} cluster can be modified by four AgI_4 tetrahedra (Fig. 9.54). Such heterometallic blocks are extended further to form a 2D layer with Bi_4I_{16} cluster as the node and Ag as the linkage. Such structural motifs have not been observed in a monometallic Bi/I system or Pb/I system, which suggests further that the Ag–I bonding interactions are effective to create unprecedent structures.

9.4 OPTICAL AND THERMAL PROPERTIES

The structural diversity of the Pb(Bi)/I/M system is of great fundamental interest, while a search for interesting properties of the unprecedented heterometallic compounds is another driving force for research in this field.

9.4.1 Optical Change with the Addition of M Ions

The feature of the Pb coordination sphere in $PbI_4Cu_2(PPh_3)_4$ [72] (Fig. 9.55a) is the unusual incomplete octahedron (with two vacant equatorial sites). The axial (ax)I(2)–Pb–I(4) angle is about 168°, approximately a 12° deviation from the ideal octahedral symmetry. Such geometry is comparable to that in mononuclear $[Bu_3N–(CH_2)_2–NBu_3][PbI_4]$, in which $I_{ax}–Pb–I_{ax} = 153.56°$ and $I_{eq}–Pb–I_{eq} = 95.48°$. The DFT calculations have indicated that the 14° axial angular difference of these two comes from increasing the electronegativity of iodine via bonding interaction with copper ions. The axial X–M–X angle is increased with the increasing of electronegativity of halogen, from I to Cl, and becomes closer to the ideal linear geometry.

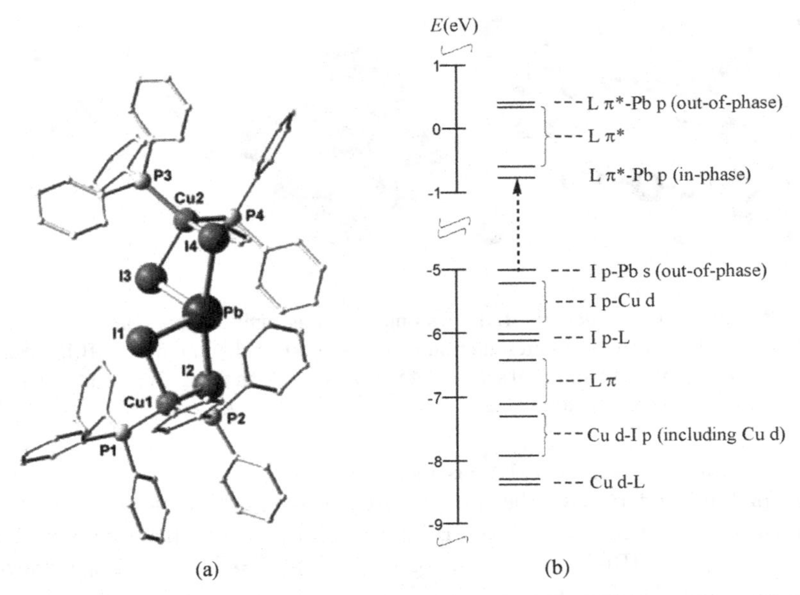

FIGURE 9.55 (a) Molecular structure of the isolated $PbI_4Cu_2(PPh_3)_4$ complex; (b) partial molecular orbital correlation diagram of $PbI_4Cu_2(PPh_3)_4$. The arrow indicates a possible transition from the HOMO.

The diffuse reflectance spectroscopy of $PbI_4Cu_2(PPh_3)_4$ gives an optical bandgap of 2.69 eV, consistent with its yellow color. The solid-state emission spectrum of $PbI_4Cu_2(PPh_3)_4$ at 10 K is shown in Figure 9.56. Excitation of the polycrystalline sample at $\lambda = 358$ nm produces an intense red-infrared emission with a peak maximum at 732 nm ($\tau = 24$ μs). The frontier molecular orbitals (MOs) of $PbI_4Cu_2(PPh_3)_4$ at the B3LYP level of DFT (see Fig. 9.55b) show that the 11 highest-energy MOs among the occupied MOs are mainly iodine 5p in character, with marked mixing with ligand orbitals (PPh_3) and copper 3d orbitals. The 26 lowest unoccupied MOs are aromatic π^* in essence, and five of them have more than 10% lead 6p components. Therefore, the 732-nm emission band can be assigned as an iodine 5p–lead 6s to PPh_3–lead 6p charge transfer: that is, halide and metal-to-ligand and metal charge transfer (XM-LM-CT).

The cluster $PbCu_6I_8(PPh_3)_6$ shown in Figure 9.49 exhibits interesting optical properties. The photoluminescence (PL) spectrum at 10 K (Fig. 9.57) indicates that the emission centered at 690 nm with excitation wavelength (λ_{ex}) < 380 nm and a pair of peaks at 542 and 780 nm with λ_{ex} > 400 nm. The PL peak at 690 nm has the lifetime $\tau = 25$ μs with $\lambda_{ex} = 363$ nm, and the luminescence peak at 542 nm has $\tau = 10$ μs (Fig. 9.57). The interesting near-infrared luminescence emission ($\lambda_{max} = 780$ nm, $\tau = 17$ μm) has been assigned to be an I(5p)–Pb(6s)-to-PPh_3–Pb(6p) charge transition. The higher energy emissions (690 and 542 nm) originated over the cubic cluster core, and the hetero metal ions in such cubes and the relatively higher

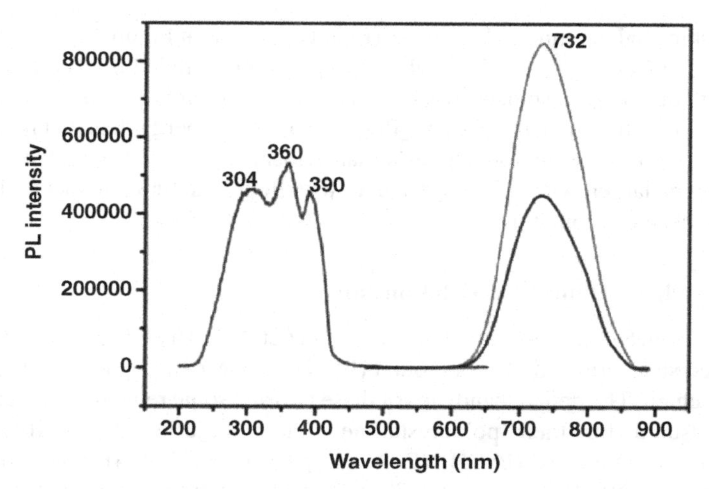

FIGURE 9.56 Solid-state emission and excitation spectra of polycrystalline [$PbI_4Cu_2(PPh_3)_4$] at 10 K. Emission spectrum (top, right) $\lambda_{max} = 732$ nm; the excitation wavelength is at 358 nm, with the spectrum (bottom, right) excited at 390 nm, with maximum = 732 nm and $\tau = 25\ \mu s$, and the excitation spectrum (left) emits at 732 nm.

delocalized Pb–Cu and Cu–Cu interactions over the bicube cluster should have a significant influence on such emissions. The 542-nm emission, which depends on the excitation energy, might be an exciton emission originating from an octahedral coordinated Pb^{2+} ion.

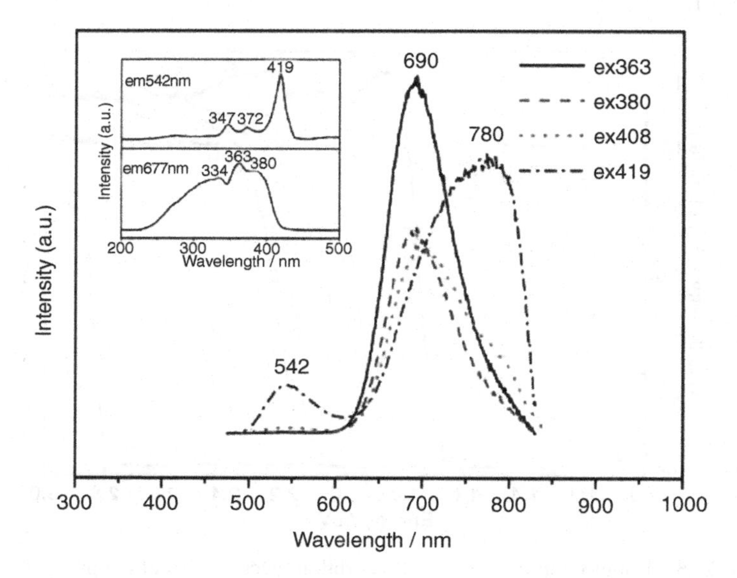

FIGURE 9.57 Solid-state emission (EM) and excitation (EX) spectra of the compound [$PbCu_6I_8(PPh_3)_6$] at 10 K, with representative peaks marked.

The cubic chain compound $[PbAg_2I_4(PPh_3)_2]_n$ shown in Figure 9.50 has a yellow emission at 566 nm with $\tau = 12\,\mu s$, which is about 100 nm red-shifted compared with that of the cubic isolated cluster $Ag_4I_4(PPh_3)_4$ [75]. This emission might be assigned to a transition in the cubic moiety $[Ag_2Pb_2I_4{}^{2+}]$ at lower energy than that in the Ag_4I_4 cube. This simple correlation suggests that delocalization over the silver and lead atoms may be larger, which is in agreement with their structural character (1D cubic chain vs. discrete cubic cluster).

9.4.2 Bandgap–Dimension Relationship

As described in Sections 9.2 and 9.3, compound $(Et_4N)_{2n}[Bi_4Ag_2I_{16}]_n$ (Fig. 9.54) is a 2D layered structure, and the compound $(Et_4N)_{2n}[Bi_2Ag_2I_{10}]_n$ (Fig. 9.53) features a 1D linear chain. The optical bandgaps of these two and some related compounds have been measured for their polycrystalline states: dimeric $(PhCH_2NEt_3)_3(Bi_2I_9)$ (Fig. 9.6) [8], tetrameric $(PhCH_2NEt_3)_4(Bi_4I_{16})$ (Fig. 9.12) [34], tetranuclear-unit 1D chain $(n\text{-}Bu_4N)_{2n}[Bi_4I_{14}]_n$ (Fig. 9.32) [42], and the binary iodides BiI_3 and β-AgI (Fig. 9.58). The optical bandgaps (E_g) for these five compounds are located between those of β-AgI (2.81 eV) [76] and BiI_3 (1.73 eV) [13] and decrease as follows: dimeric $(Bi_2I_9)^{3-}$ (2.19 eV) > tetrameric $(Bi_4I_{16})^{4-}$ (2.16 eV) > tetrameric unit chain $[Bi_2Ag_2I_{10}{}^{2-}]_n$ (2.05 eV) > monometallic tetrameric unit chain $[Bi_4I_{14}{}^{2-}]_n$ (2.02 eV) > 2D layer $[Bi_4Ag_2I_{16}{}^{2-}]_n$ (1.93 eV) (Fig. 9.58). Such a gap reduction apparently follows an increase in the dimensions of the anion structures. Besides, the E_g reduction of the four monometallic compounds also agrees with the increase in the corresponding M/r ratio from 0.27 for a binuclear $[Bi_2I_{10}]^{4-}$ cluster to 0.67 for BiI_3 binary, as shown in Table 9.1.

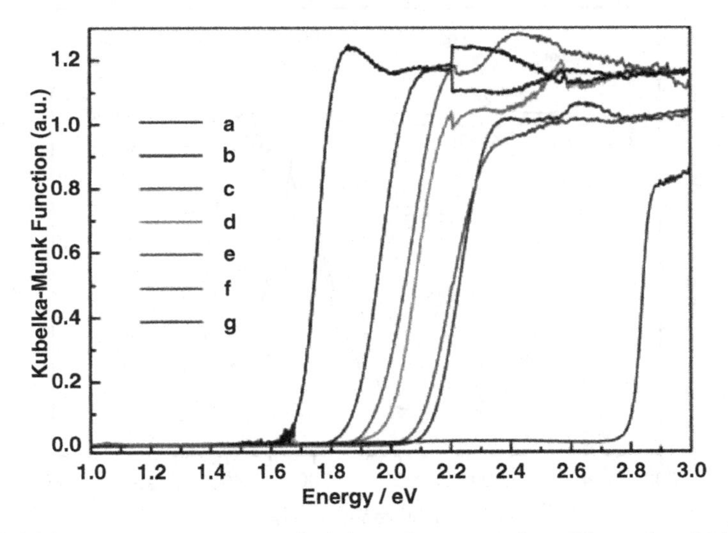

FIGURE 9.58 Room-temperature optical absorption spectra for solid samples of (a) BiI_3, (b) $(Et_4N)_{2n}[Bi_4Ag_2I_{16}]_n$, (c) $(n\text{-}Bu_4N)_{2n}[Bi_4I_{14}]_n$, (d) $(Et_4N)_{2n}[Bi_2Ag_2I_{10}]_n$, (e) $(PhCH_2NEt_3)_4(Bi_4I_{16})$, (f) $(PhCH_2NEt_3)_3(Bi_2I_9)$, and (g) β-AgI.

Since the coordination number of Cu or Ag is less than that of Bi, for a fixed number of total metal atoms the higher the CuI_4 (or AgI_4) percentage corresponds to the higher M/r values. Therefore, comparison of M/r values should be confined to the same system, either monometallic species or heterometallic species.

For a 1D $[Bi_2Ag_2I_{10}{}^{2-}]_n$ chain and a 2D $[Bi_4Ag_2I_{16}{}^{2-}]_n$ layer, $r = 4 + 2/2 + 4/3 = 6.33$, M/r = 0.63, and $r = 2 + 12/2 + 2/3 = 8.67$, M/$r$ = 0.69, respectively. Their energy gaps decrease with increased M/r values.

Similar observations have been found in Bi/I/Cu heterometallic compounds [64]. Values for tetranuclear $[n\text{-}Bu_4N]_2[Bi_2Cu_2(CH_3CN)_2I_{10}]$ are $E_{g(obs)} = 2.06$ eV, $E_{g(cal)} = 2.02$ eV, $r = 4 + 4/2 + 2/3 = 6.67$, and M/r = 0.60; the linear chain $[Et_4N]_{2n}[Bi_2Cu_2I_{10}]_n$ has $E_{g(obs)} = 1.89$ eV, $E_{g(cal)} = 1.86$ eV, $r = 4 + 4/2 + 2/3 = 6.67$, and M/r = 0.63; and the thicker linear chain $[Cu(CH_3CN)_4]_{2n}[Bi_2Cu_2I_{10}]_n$ has $E_{g(obs)} = 1.80$ eV, $E_{g(cal)} = 1.90$ eV, $r = 4 + 4/2 + 2/3 = 6.67$, and M/r = 0.71.

The clear decrease in E_g in these compounds is consistent with their darkening from red to black, and such a trend fits nicely with the increase in M/r values. Note that the E_g values calculated for the two linear compounds deviate considerably from the values observed because of the inaccuracy of the DFT calculation method in the estimation of E_g. In these cases M/r values seem to fit experiments well.

The DFT calculations have indicated that the top of the valence band for each compound is generated by I and Cu, and the bottom of the conduction band consists of the partial DOS of Bi and I. The effective transition edges of these compounds can all be assigned as charge transfer from occupied 3d orbitals of Cu and 5p states of I to empty 6p orbitals of Bi and 5p states of I.

9.4.3 Contribution of Cu^+ or Ag^+ to the Band Structure

The silver iodobismuthate $[Et_4N]_{2n}][Bi_2Ag_2I_{10}]_n$ is structural with $[Et_4N]_{2n}[Bi_2Cu_2I_{10}]_n$ (Fig. 9.53), and both compounds have the same M/r value (0.63). However, the different bandgaps observed (2.05 vs. 1.89 eV) suggest that Cu^+ and Ag^+ play different roles in the electronic structures, and this is supported by the DFT calculation results. In such a case, the M/r value is insufficient to distinguish the difference. As shown in Figure 9.59a, the band structure of the monometallic $[n\text{-}Bu_4N]_{2n}[Bi_4I_{14}]_n$, which possesses a chain motif similar to that of $[Et_4N]_{2n}[Bi_2M_2I_{10}]_n$, exhibits a distribution of Bi–I bonding states, I nonbonding states [which make up the HOCO (highest occupied crystal orbital)], and Bi–I antibonding states [which give the LUCO (lowest unoccupied crystal orbital)] around the Fermi level. With the introduction of Cu^+ (Fig. 9.59b) the corresponding I nonbonding and Bi–I antibonding states show no significant energetic changes, but the Cu–I antibonding states have now been inserted between them. Accordingly, the tops of the valence bands of the Cu analogs now derive nearly entirely from I and Cu orbitals, and consequently, the energy gaps have decreased. In contrast, the results for the Ag/Bi analog (Fig. 9.59c) suggest that the introduction of Ag^+ does not influence the components of the HOCO because the Ag-related states lie at lower energies. Thus, the energy gap is 2.05 eV, 0.03 eV larger than that for $[n\text{-}Bu_4N]_{2n}[Bi_4I_{14}]_n$, an increase that might derive from the distortion of the Bi–I skeleton by the involvement of Ag^+

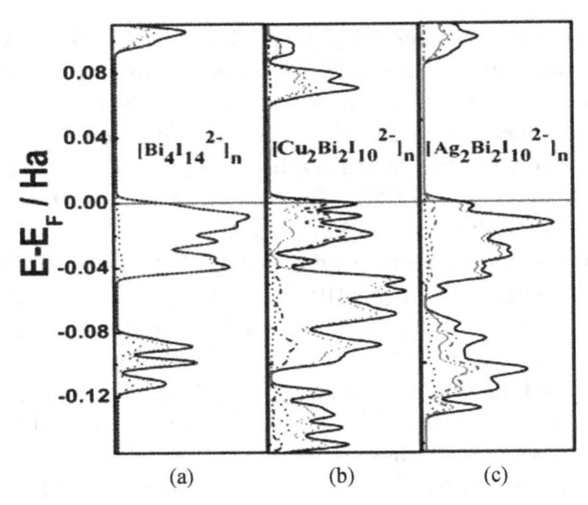

FIGURE 9.59 Total and partial DOS plots of (a) $[n\text{-}Bu_4N]_{2n}[Bi_4I_{14}]_n$, (b) $[Et_4N]_{2n}[Bi_2Cu_2I_{10}]_n$, and (c) $[Et_4N]_{2n}[Bi_2Ag_2I_{10}]_n$ with the Fermi levels set at zero. Solid lines represent total DOS and other lines represent partial DOS, as follows: dashed, Bi; dot, I; dashed-dotted, Cu [in (b)]; dashed-dotted, Ag [in (c)].

ions. These opposed E_g changes with respect to the monometallic $[Bi_4I_{14}{}^{2-}]_n$ come from the fact that Cu^+ contributes to the top of the valance band, pushes it up, and thus leads to a smaller E_g value, but Ag^+ hardly contributes to the HOCO, and its lower energy influences E_g only slightly, by altering the overall electronic structure.

9.4.4 Distinct Thermal Stabilities

Three Bi/Cu/I compounds with structural similarities (Fig. 9.43) have distinct thermal stabilities. The isolated cluster $[Bu_4N]_2[Bi_2Cu_2(CH_3CN)_2I_{10}]$ exhibits a weight loss of about 3.67% (calculated: 3.45%) around 85°C, corresponding to the loss of CH_3CN ligands, and this compound has changed to an unknown product that melts congruently at 145°C and decomposes above 230°C (Fig. 9.60). By contrast, the linear and solvent-free $[Et_4N]_{2n}[Bi_2Cu_2I_{10}]_n$ is stable up to 230°C, and no phase transition or decomposition is found (Fig. 9.60). The endothermic peak for the linear $[Cu(CH_3CN)_4]_{2n}[Bi_2Cu_2I_{10}]_n$ at about 80°C corresponds to a weight loss of about 14% of the CH_3CN molecules, and its decomposition, to trigonal BiI_3 plus cubic CuI. Although these compounds contain the same $[Bi_2Cu_2I_{10}]^{2-}$ building units, the markedly low decomposition temperature of the third compound must be caused by the instability of $Cu(CH_3CN)_4{}^+$ through solvent loss, the resulting "$Bi_2Cu_4I_{10}$" composition being unstable with respect to simple binary iodides. However, the release of CH_3CN in the isolated cluster $[Bu_4N]_2[Bi_2Cu_2(CH_3CN)_2I_{10}]$ results in a "$(Bu_4N)_2(Cu_2Bi_2I_{10})$" composition, which is still a thermally stable phase.

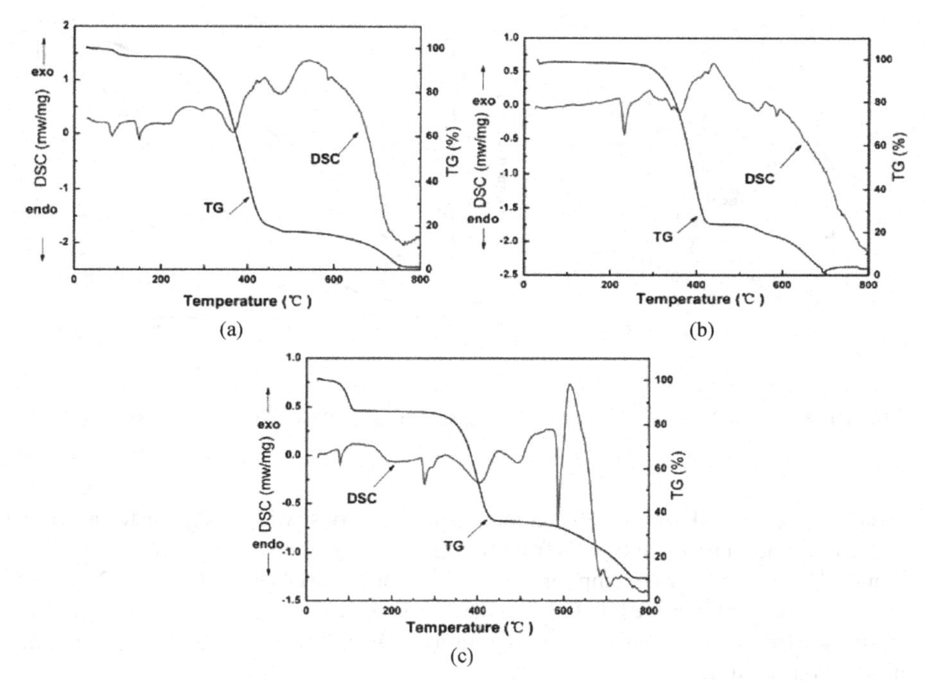

FIGURE 9.60 TG/DSC curves of compounds (a) $[Bu_4N]_2[Bi_2Cu_2(CH_3CN)_2I_{10}]$, (b) $[Et_4N]_{2n}[Bi_2Cu_2I_{10}]_n$, and (c) $[Cu(CH_3CN)_4]_{2n}[Bi_2Cu_2I_{10}]_n$.

9.4.5 Interesting Ferroelectric Property

The structure of $BiI_6Cu_3(PPh_3)_6$ [77] (Fig. 9.61) is that of a Bi atom with an extremely distorted octahedral geometry of six I atoms in which the maximum axial-to-equatorial angular deviation is 29° (the ideal is 90°). Each of the nonadjacent edges

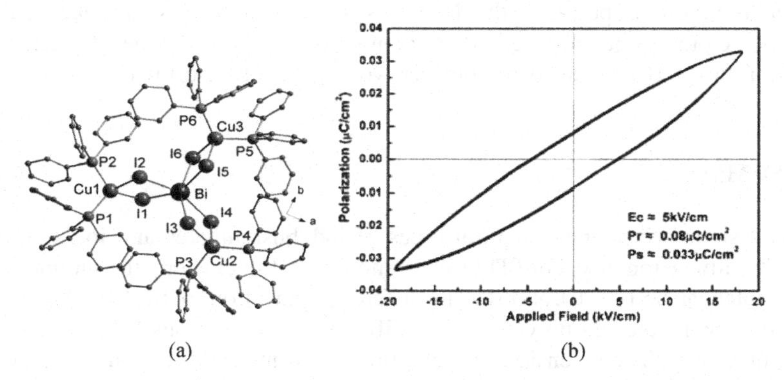

FIGURE 9.61 (a) Molecular structure of $BiI_6Cu_3(PPh_3)_6$; (b) electric hysteresis loop.

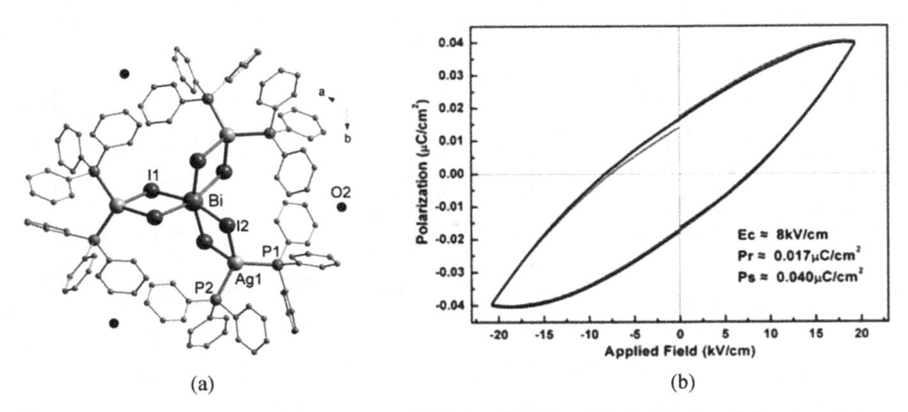

FIGURE 9.62 (a) Molecular structure of $BiI_6Ag_3(PPh_3)_6 \cdot 3H_2O$; (b) electric hysteresis loop.

is capped by crystallographically independent Cu^+ ions, with a regular tetrahedral four-coordination mode of two I's from the $[BiI_6]^{3-}$ center and two P's from the PPh_3 ligand. The isostructural compound $BiI_6Ag_3(PPh_3)_6 \cdot 3H_2O$ [77] (Fig. 9.62) has a similar molecular structure but with less distortion (i.e., a 17° deviation of the axial-to-axial angle from an ideal 180°). The distortion is driven by the unmatched bond length of Ag/Cu–I to Bi–I.

These two isolated clusters crystallize in orthorhombic ($Pna2_1$) and trigonal ($R3c$) systems, respectively, which belong to the polar point groups ($mm2$ and $3m$) required for ferroelectric materials. Their ferroelectric hysteresis loops measured on pellets of the powdered sample at room temperature are shown in Figure 9.62b. The remnant polarization (P_r, μC cm^{-2}) and coercive field (E_c, kV cm^{-1}) are about 0.008, 0.017; and 5, 8, respectively.

Recent efforts have led to more Bi/M/I compounds that exhibit ferroelectric properties, such as $[Cu_3(p\text{-bix})_3]_n[Bi_2I_9]_n$ [77] ($P1$ space group), and the polycrystalline sample shows a remnant polarization (P_r, μC cm^{-2}) of 0.0085, a saturated polarization (P_s, μC cm^{-2}) of 0.017, and a coercive field (E_c, kV cm^{-1}) of about 6.6. The polarization property of the Bi/M/I system may have originated with the flexibility of the large, soft central Bi atoms and the high degree of tolerance of distortion in the BiI_6 octahedron. Further work should be worthwhile.

9.5 SUMMARY

We have summarized the structural types of iodobusmuthates and iodoplumbates sorted by dimensionality. For 0D discrete monometallic clusters, the number of the metal center can be 1–8, 10, and 18. The different connections of the MI_6 (M = Pb, Bi) octahedron generate 1D polymers with different configurations via sharing faces, edges, or apexes. The extension of the building unit in two directions generates 2D polymers or hybrid layers when organic linkage is involved.

Three important factors regarding the structural modification of Pb(Bi)/I compounds are discussed: cation effects, ligand effects, and Ag–I or Cu–I bonding interactions.

An empirical r parameter defined as $\Sigma N(\mu_n\text{–I})/n$ ($n = 1$ to 5) and the M/r value of 0D clusters increase with increases in density (i.e., the number of the metal center). The M/r values for a series of compounds formed by the same building unit increase with increased dimensions in the order 0D to 1D to 2D. Also, the M/r values of different compounds invariably increase with a decrease in the energy gap (E_g), which is in good agreement with their apparent color changes and with electronic structure analyses. The M/r values might be a key to the design and rational synthesis of novel iodometalates, and such empirical rules are expected to be applied to other systems, such as Sn/I, M/X (X = F, Cl, Br), and TM/Ch (TM = transition metal, Ch = chalcogen).

Several property–structure relationships are discussed. Luminescence is related to structural configuration, distortion, and heterometallic bonding interactions. Thermal stability variations with dimensionality and the overall metal-to-iodine ratio are presented. The ferroelectric properties arise from the flexibility and distortion of the BiI_6 octahedron. The various contributions of Ag^+ and Cu^+ to the band structures are discussed. The energy gap changes with dimensionality and density are also presented.

A thorough understanding of the structural chemistry is the basis of structural modification and rational synthesis. Detailed study of the property–structure relationship is a first step in equiping target compounds with the desired properties. Explorations are ongoing, and a rich chemistry is anticipated.

Acknowledgments

This research was supported by the National Natural Science Foundation of China under Projects 20773130, 20521101, and 20733003, the Key Project from CAS (KJCX2-YW-H01), and the Key Project from FJIRSM (SZD08002 and SZD07004).

REFERENCES

1. Krautscheid, H.; Vielsack, F. *Angew. Chem., Int. Ed. Engl.* **1995**, *34*, 2035–2037.
2. Krautscheid, H.; Vielsack, F. *J. Chem. Soc., Dalton Trans.* **1999**, 2731–2735.
3. Krautscheid, H.; Lode, C.; Vielsack, F.; Vollmer, H. *J. Chem. Soc., Dalton Trans.* **2001**, 1099–1104.
4. Fisher, G. A.; Norman, N. C. *Adv. Inorg. Chem.* **1994**, *41*, 233–271.
5. Hattori, T.; Taira, T.; Era, M.; Tsutsui, T.; Saito, S. *Chem. Phys. Lett.* **1996**, *254*, 103–108.
6. Era, M., Morimoto, S., Tsutsui, T., Saito, S. *Appl. Phys. Lett.* **1994**, *65*, 676–678.
7. Guloy, A. M.; Tang, Z.-J.; Miranda, P. B.; Srdanov, V. I. *Adv. Mater.* **2001**, *13*, 833–837.
8. Eickmeier, H.; Jaschinski, B.; Hepp, A.; Nuss, J.; Reuter, H.; Blachnik, R. *Z. Naturforsch. B* **1999**, *54*(3), 305–313.

9. Devic, T.; Evain., M.; Moëlo, Y.; Canadell, E.; Senzier, P. A.; Fourmigué, M.; Batail, P. *J. Am. Chem. Soc.* **2003**, *125*, 3295–3301.
10. Devic, T.; Canadell, E.; Auban-Senzier, P.; Batail, P. *J. Mater. Chem.* **2004**, *14*, 135–137.
11. Medycki, W.; Jakubas, R.; Pislewski, N.; Lefebvre, J. *Z. Naturforsch. A* **1993**, *48*(7), 748–752.
12. Zhu, X. H.; Wei, Z. R.; Jin, Y. R.; Xiang, A. P. *Cryst. Res. Technol.* **2007**, *42*(5), 456–459.
13. Goforth, A. M.; Gardinier, J. R.; Smith, M. D.; Peterson, L.; Loye, H. C. Z. *Inorg. Chem. Commun.* **2005**, *8*(8), 684–688.
14. Artemyev, M. V.; Rakovich, Yu. P.; Yablonski, G. P. *J. Cryst. Growth* **1997**, *171*, 447–452.
15. Baibarac, M.; Preda, N.; Mihut, L.; Baltog, I.; Lefrant, S.; Mevellec, J. Y. *J. Phys., Condens. Mater.* **2004**, *16*, 2345–2356.
16. Lifshitz, E.; Yassen, M.; Bykov, L.; Dag, I. *J. Phys. Chem.* **1994**, *98*, 1459–1463.
17. Sarid, D.; Phee, B. R.; McGinnis, B. P. *Appl. Phys. Lett.* **1986**, *49*(16), 1196–1198.
18. Deich, V.; Roth, M. *Phys. Res. A* **1996**, *380*, 169–172.
19. Cuna, A.; Noguera, A.; Saucedo, E.; Fornaro, L. *Cryst. Res. Technol.* **2004**, *39*(10), 912–919.
20. Jellison, G. E.; Ramey, J. O.; Boatner, A. A. *Phys. Rev. B* **1999**, *59*(15), 9718–9721.
21. Cuna, A.; Aguiar, I.; Gancharov, A.; Perez, M.; Fornaro, L. *Cryst. Res. Technol.* **2004**, *39*(10), 899–905.
22. Papavassiliou, G. C.; Koutselas, I. B.; Terzis, A.; Raptopoulou, C. P. *Z. Naturforsch. B* **1995**, *50*(10), 1566–1569.
23. Vincent, B. R.; Robertson, K. N.; Cameron, T. S.; Knop, O. *Can. J. Chem.* **1987**, *65*, 1042–1046.
24. Krautscheid, H.; Vielsack, F. *Z. Anorg. Allg. Chem.* **1999**, *625*, 562–566.
25. Krautscheid, H. *Z. Anorg. Allg. Chem.* **1999**, *625*(2), 192–194.
26. Lindqvist, O. *Acta Chem. Scand.* **1968**, *22*, 2943–2952.
27. Chabot, B.; Parthe, E. *Acta Crystallogr. B* **1978**, *34*, 645.
28. Feldmann, C. *Z. Kristallogr. New Cryst. Struct.* **2001**, *216*(3), 465–466.
29. Lazarini, F. *Acta Crystallogr. C* **1987**, *43*, 875.
30. Bowmaker, G. A.; Junk, P. C.; Lee, A. M.; Skelton, B. W.; White, A. H. *Aust. J. Chem.* **1998**, *51*(4), 293–309.
31. Geiser, U.; Wade, E.; Wang, H. H.; Williams, J. M. *Acta Crystallogr. C* **1990**, *46*, 1547.
32. Okrut, A.; Feldmann, C. *Z. Anorg. Allg. Chem.* **2006**, *632*(3), 409–412.
33. Krautscheid, H.; Vielsack, F. *J. Chem. Soc., Dalton Trans.* **1999**, 2731–2735.
34. Kubiak, R.; Ejsmont, K. *J. Mol. Struct.* **1999**, *474*, 275–281.
35. Carmalt, C. J.; Farrugia, L. J.; Norman, N. C. *Z. Naturforsch. B* **1995**, *50*(11), 1591–1596.
36. Pohl, S.; Peters, M.; Haase, D.; Saak, W. *Z. Naturforsch. B* **1994**, *49*(6), 741–746.
37. Sharutin, V. V.; Egorova, I. V.; Sharutina, O. K.; Dorofeeva, O. A.; Ivanenko, T. K.; Gerasimenko, A. V.; Pushilin, M. A. *Russ. J. Coord. Chem.* **2004**, *30*(12), 874–883.
38. Krautscheid, H. *Z. Anorg. Allg. Chem.* **1994**, *620*(9), 1559–1564.

39. Clegg, W.; Errington, R. J.; Fisher, G. A.; Green, M. E.; Hockless, D. C. R.; Norman, N. C. *Chem. Ber.* **1991**, *124*(11), 2457–2459.

40. Eickmeier, H.; Jaschinski, B.; Hepp, A.; Nuss, J.; Reuter, H.; Blachnik, R. *Z. Naturforsch. B* **1999**, *54*(3), 305–313.

41. Goforth, A. M.; Tershansy, M. A.; Smith, M. D.; Peterson, L.; zur Loye, H. C. *Acta Crystallogr. C* **2006**, *62*, M381–M385.

42. Krautscheid, H. *Z. Anorg. Allg. Chem.* **1995**, *621*(12), 2049–2054.

43. Feldmann, C. *J. Solid State Chem.* **2003**, *172*(1), 53–58.

44. Tang, Z. J.; Guloy, A. M. *J. Am. Chem. Soc.* **1999**, *121*, 452–453.

45. Nagapetyan, S. S.; Arakelova, A. R.; Ziger, E. A.; Koshkin, V. M.; Struchkov, Y. T.; Shklover, V. E. *Russ. J. Inorg. Chem.* (transl. of *Zh. Neorg. Khim.*) **1989**, *34*, 2244.

46. Cornia, A.; Fabretti, A. C.; Grandi, R.; Malavasi, W. *J. Chem. Crystallogr.* **1994**, *24*(4), 277–280.

47. Chaabouni, S.; Kamoun, S.; Jaud, J. *J. Chem. Crystallogr.* **1997**, *27*(9), 527–531.

48. Wang, S. M.; Mitzi, D. B.; Feild, C. A.; Guloy, A. M. *J. Am. Chem. Soc.* **1995**, *117*, 5297–5302.

49. Papavassiliou, G. C.; Mousdis, G. A.; Raptopoulou, C. P.; Terzis, A. *Z. Naturforsch.* **1999**, *54b*, 1405–1409.

50. Mousdis, G. A.; Papavassiliou, G. C.; Terzis, A.; Raptopoulou, C. P. *Z. Naturforsch. B* **1998**, *53*(8), 927–931.

51. Krautscheid, H.; Vielsack, F. *Z. Anorg. Allg. Chem.* **1997**, *623*, 259–263.

52. H.Krautscheid, J. F. Lekieffre, J. Besinger. *Z. Anorg. Allg. Chem.* **1996**, *622*, 1781–1787.

53. Fan, L. Q.; Wu, L. M.; Chen, L. Unpublished work.

54. Calabrese, J.; Jones, N. L.; Harlow, R. L.; Herron, N.; Thorn, D. L.; Wang, Y. *J. Am. Chem. Soc.* **1991**, *113*, 2328–2330.

55. Mitzi, D. B. *Chem. Mater.* **1996**, *8*, 791–800.

56. Mercier, N. *CrystEngComm* **2005**, *7*, 429–432.

57. Zhu, X. H.; Mercier, N.; Riou, A.; Blanchard, P.; Frère, P. *Chem. Commun.* **2002**, 2160–2161.

58. Mitzi, D. B. *J. Chem. Soc., Dalton Trans.* **2001**, 1–12.

59. Mitzi, D. B.; Feild, C. A.; Harrison, W. T. A.; Guloy, A. M. *Nature* **1994**, *369*, 467–469.

60. Ishihara, T. *J. Lumin.* **1994**, *60–61*, 269–274.

61. Harrowfield, J. M.; Miyamae, H.; Skelton, B. W.; Soudi, A. A.; White, A. H. *Aust. J. Chem.* **1996**, *49*, 1157–1164.

62. Krautscheid, H.; Vielsack, F.; Klaassen, N. *Z. Anorg. Allg. Chem.* **1998**, *624*, 807–812.

63. Chai, W. X.; Wu, L. M.; Li, J. Q.; Chen, L. *Inorg. Chem* **2007**, *46*(4), 1042–1044.

64. Chai, W. X.; Wu, L. M.; Li, J. Q.; Chen, L. *Inorg. Chem* **2007**, *46*(21), 8698–8704.

65. Miyamae, H.; Toriyama, H.; Abe, T.; Hihara, G.; Nagata, M. *Acta Crystallogr., C* **1984**, *40*, 1559.

66. Miyamae, H.; Hihara, G.; Hayashi, K.; Nagata, M. *J. Chem. Soc. Jpn.* **1986**, 1501.

67. Engelhardt, L. M.; Patrick, J. M.; Whitaker, C. R.; White, A. H. *Aust. J. Chem.* **1987**, *40*, 2107–2114.
68. Shi, Y. J.; Xu, Y.; Zhang, Y.; Huang, B.; Zhu, D. R.; Jin, C. M.; Zhu, H. G.; Yu, Z.; Chen, X. T.; You, X. Z. *Chem. Lett.* **2001**, 678–679.
69. Shi, Y. J.; Li, L. H.; Li, Y. Z.; Xu, Y.; Chen, X. T.; Xue, Z. L.; You, X. Z. *Inorg. Chem. Commun.* **2002**, *5*, 1090–1094.
70. Fan, L. Q.; Wu, L. M.; Chen, L. Unpublished work.
71. Fan, L. Q.; Wu, L. M.; Chen, L. *Inorg. Chem.* **2006**, *45*, 3149–3151.
72. Fan, L. Q.; Huang, Y. Z.; Wu, L. M.; Chen, L.; Li, J. Q.; Ma, E. *J. Solid State Chem.* **2006**, *179*, 2361–2366.
73. Burns, M. C.; Tershansy, M. A.; Ellsworth, J. M.; Khaliq, Z.; Peterson, L.; Smith, M. D.; zur Loye, H. C. *Inorg. Chem.* **2006**, *45*(26), 10437–10439.
74. Hartl, H.; Hoyer, M. *Z. Naturforsch. B* **1997**, *52*, 766–768.
75. Henary, M.; Zink, J. I. *Inorg. Chem.* **1991**, *30*, 3111–3112.
76. Victora, R. H. *Phys. Rev. B* **1997**, *56*, 4417.
77. Chai, W. X.; Wu, L. M.; Li, J. Q.; Chen, L.Unpublished work.
78. Li, J. P.; Li, L. H.; Wu, L. M.; Chen, L. *Inorg. Chem.* **2009**, *48*, 1260–1262.

10

CLUSTER-BASED SUPRAMOLECULAR COMPOUNDS FROM Mo(W)/Cu/S CLUSTER PRECURSORS

JIAN-PING LANG, WEN-HUA ZHANG, HONG-XI LI, AND ZHI-GANG REN
College of Chemistry, Chemical Engineering and Materials Science, Suzhou University, Suzhou, Jiangsu, People's Republic of China

10.1 INTRODUCTION

In past decades, a modular and topological approach to the rational design and construction of supramolecular arrays has blossomed because the resulting supramolecular compounds are showing very fascinating structural chemistry and various potential applications in advanced materials [1–3]. Among numerous molecular modules, various transition-metal clusters have received much attention as structural and functional building blocks for supramolecular assemblies [4–10]. For example, Cotton et al. reported adopting discrete metal–metal-bonded dimeric species $[M_2(DArF)_2(MeCN)_4]^{2+}$ (M = Mo, Rh) as building units for a diversity of M_2-based aggregates with attracting electronic properties [4a]. Yaghi et al. synthesized porous frameworks which showed promise as hydrogen storage materials by using tetrazinc carboxylate clusters as connection centers [5c]. Zheng et al. utilized octahedral hexarhenium chalcogenide clusters to create a series of $[Re_6Q_8]$ (Q = S, Se, Te)-based supramolecular arrays, some of which showed interesting electrochemical and photophysical perspectives [7d]. Bain et al. employed hexanuclear clusters containing the octahedral $[Mo_6(\mu_3\text{-}Cl)_8]^{4+}$ cores to generate microporous xerogels for size-selective ion exchange [10a]. However, because of the limited number of suitable

Design and Construction of Coordination Polymers, Edited by Mao-Chun Hong and Ling Chen

cluster precursors and especially, the difficulty in isolating the resulting assemblies, methodologies to use transition-metal clusters as connecting nodes to hold them together via multitopic ligands in predefined patterns within self-assembled oligomeric or polymeric aggregates remain a great challenge.

On the other hand, in the last 40 years, the chemistry of molybdenum (tungsten)–copper–sulfur clusters derived from thiomolybdate or thiotungstate has been investigated extensively due to their diverse structural chemistry [11–14] and their relevance to biological systems [15], industrial catalysis processes [16], and photonic materials [12,13,14h]. We have also been interested in the assembly of Mo(W)/Cu/S clusters in the pursuit of new cluster chemistry and new chemical and/or physical properties [13,14]. So far, over 200 Mo(W)/Cu(Ag)/S clusters have been isolated and structurally characterized in our group. Among them, some clusters, especially those with terminal halides or pseudohalides or solvent molecules coordinated at copper(I) centers, display unique cluster frameworks, which may be used as interesting structure-directed species in the construction of cluster-based supramolecular compounds [14]. After careful consideration of cluster geometry and chemical behavior, we chose and/or prepared the following 12 Mo(W)/Cu/S clusters as potential candidates for the construction of cluster-containing supramolecular compounds: $[Et_4N]_2[MoOS_3CuCN]$ (**1**) [14i], $[A]_2[WS_4(CuCN)_2]$ (**2**: A = Et_4N [17]; **3**: A = PPh_4 [14d]), $[(n\text{-}Bu)_4N]_2[MoOS_3(CuX)_3]$ (**4**: X = I [18a]; **5**: X = NCS [18b]), $[PPh_4][Cp^*MS_3(CuX)_3]$ (Cp^* = pentamethylcyclodienyl (1−); (**6**: M = W, X = Br [13e]; **7**: M = W, X = CN [14b]; **8**: M = Mo, X = NCS [14h]), $[Et_4N]_4[WS_4Cu_4I_6]$ (**9**) [18c], $[(Cp^*WS_3)_3Cu_7(MeCN)_9](PF_6)_4$ (**10**) [14a], and $[(n\text{-}Bu)_4N]_4[MS_4Cu_6I_8]$ (**11**: M = Mo [18d]; **12**: M = W [18e]).

Up to now, several dozens of cluster-based arrays have been generated from these cluster precursors. Among these compounds, some are of consummate structural beauty [14b], some are able to perform specific tasks in regard to host–guest chemistry [14c], and some exhibit enhanced photochemical or photophysical properties relative to those of their precursor compounds [14b,14g,14i]. In this chapter we present the design, assembly, and properties of a series of unique Mo(W)/Cu/S-based supramolecular compounds from these 12 preformed Mo(W)/Cu/S clusters.

10.2 STRATEGIES FOR DESIGN AND ASSEMBLY

10.2.1 Strategy Based on Precursor 1

Compound **1** contains a central $MoOS_3$ moiety coordinated with one CuCN group to form a $[MoOS_3Cu]$ core [14i]. Since the terminal O atom of the $[MoOS_3(CuCN)]^{2-}$ dianion of **1** does not react with Cu^+, the incorporation of one or two Cu^+ into the core framework of **1** via the S atoms may result in the formation of a butterfly-shaped $[MoOS_3Cu_2]$ core or incomplete cubane-like $[MoOS_3Cu_3]$ core. These cluster cores may work as multiconnecting nodes to self-aggregate into larger clusters or polymeric clusters through the bridging S atoms or cyanides. Therefore, reactions of **1** in aniline with equimolar $[Cu(MeCN)_4](PF_6)$ in MeCN or 2 equivalents (equiv.) of CuCN under the presence of KCN in H_2O followed by a standard workup afforded a $[MoOS_3Cu_2]$-based polymeric cluster $[Et_4N]_2[MoOS_3Cu_2(\mu\text{-}CN)]_2\cdot 2$aniline (**13**) and a $[MoOS_3Cu_3]$-based polymeric cluster $[Et_4N]_4[MoOS_3Cu_3CN(\mu'\text{-}CN)]_2(\mu\text{-}CN)_2$ (**14**) in 50% and 54% yields, respectively [14i].

10.2.2 Strategy Based on Precursors 2 and 3

Compound **2** or **3** consists of a central WS_4 unit coordinated symmetrically by two CuCN groups, forming an approximately linear WS_4Cu_2 structure [14d,17]. The two copper centers in the cluster core may have two to four sites available when their coordination geometry is turned into tetrahedral geometry. Therefore, the cluster $[WS_4Cu_2]$ core is expected to serve as a multiconnecting node for the assembly of polydimensional frameworks if fulfillment of the coordination of the Cu atoms and substitution of the terminal CN groups by bridging ligands occurs. For example, it may act as a three-connecting node (a) to link other cluster cores via bridging cyanides, forming a 2D honeycomb-like network (Fig. 10.1). In addition, the four sulfur atoms of the $[WS_4Cu_2]$ framework of **1** or **2** are unsaturated and may bind to one Cu^+ to form a butterfly-type $[WS_4Cu_3]$ cluster core. This core may also serve as another three-connecting node (b) to bridge other cores via bridging cyanides to form a 2D herringbone network (Fig. 10.1).

Reactions of **2** with HOAc in a solution of MeCN which contains 4 vol% HOAc, followed by a workup, produced $[Et_4N]_3[\{WS_4Cu_2(\mu\text{-}CN)\}_2(\mu'\text{-}CN)]\cdot 2MeCN$ (**15**) in 56% yield. On the other hand, reactions of **3** with HOAc in a solution of MeCN which contains 10 vol% HOAc, followed by a workup, produced $[PPh_4]_2[\{WS_4Cu_3(\mu\text{-}CN)\}_2(\mu'\text{-}CN)_2]\cdot 2MeCN$ (**16**) in 12.5% yield. This compound could also be prepared from reactions of **3** with equimolar $[Cu(MeCN)_4](PF_6)$ in MeCN in a high yield [14d].

10.2.3 Strategy Based on Precursors 4 to 8

Compounds **4** to **8** can be viewed as having an incomplete cubane-like $[MS_3Cu_3]$ core structure in which each Cu atom has a trigonal–planar coordination geometry with a terminal halide or pseudohalide [13e,14b,14h,18a,18b]. The three copper centers in

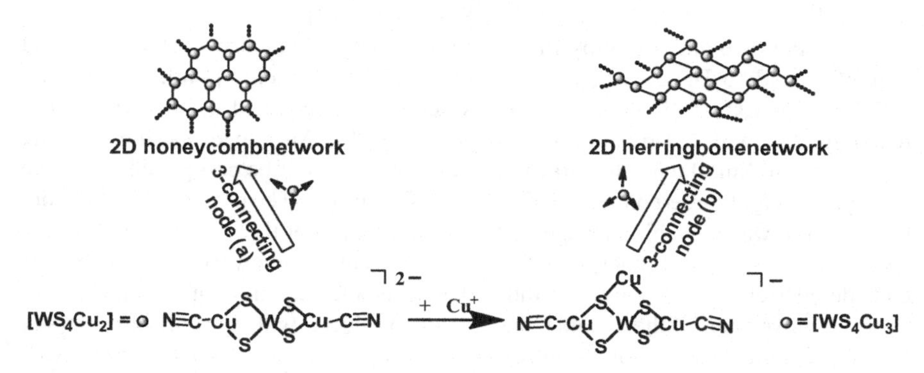

FIGURE 10.1 Possible topological nodes and frameworks derived from **2** or **3**.

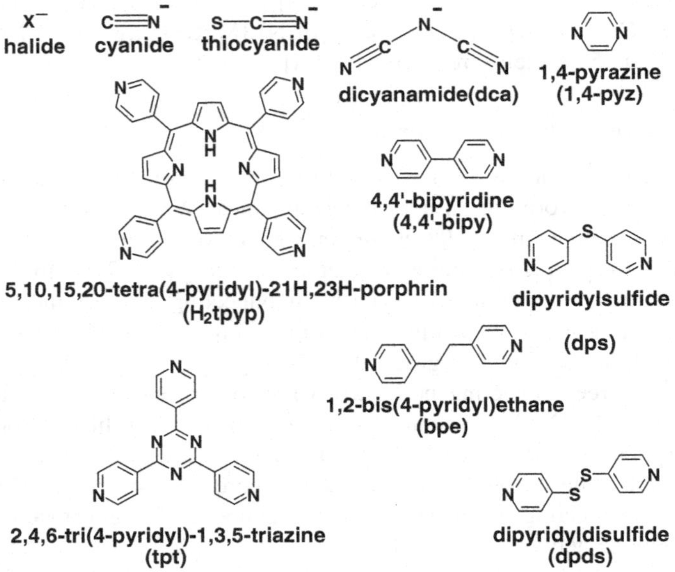

X^- halide; $C{\equiv}N^-$ cyanide; $S{-}C{\equiv}N^-$ thiocyanide; dicyanamide(dca); 1,4-pyrazine (1,4-pyz); 5,10,15,20-tetra(4-pyridyl)-21H,23H-porphrin (H_2tpyp); 4,4'-bipyridine (4,4'-bipy); dipyridylsulfide (dps); 1,2-bis(4-pyridyl)ethane (bpe); 2,4,6-tri(4-pyridyl)-1,3,5-triazine (tpt); dipyridyldisulfide (dpds)

FIGURE 10.2 Some bridging ligands used to connect cluster cores.

the cluster core of **4**–**8** may hold up to six coordination sites if these terminal halides or pseudohalides are replaced by strong donor ligands and the coordination geometries of the copper atoms are changed into a tetrahedral geometry. Topologically, with the assistance of a series of rigid or flexible bridging ligands (Fig. 10.2), each $[MS_3Cu_3]$ cluster core in **4** to **8** is anticipated to serve as an intriguing multiconnecting node to generate colorful cluster-based topological frameworks (Fig. 10.3).

Treatment of a solution of **4** in DMF/MeCN with 2 equiv. of dpds led to the formation of $[MoOS_3Cu_3I(dpds)_2]{\cdot}0.5DMF{\cdot}2(MeCN)_{0.5}$ (**17**) in 64% yield. Reactions of **5** in DMF/MeCN with a methanol solution containing equimolar sodium dicyanamide and 2 equiv. of 4,4′-bipy followed by a workup afforded $[MoOS_3Cu_3(dca)(4,4'\text{-bipy})_{1.5}]{\cdot}DMF{\cdot}MeCN$ (**18**) in 50% yield [14f].

Treatment of **6** and 4,4′-bipy in DMF followed by a standard workup produced $[Cp^*WS_3Cu_3Br(\mu\text{-}Br)(4,4'\text{-bipy})]{\cdot}Et_2O$ (**19**) in 60% yield [14e]. Compound $[(Cp^*WS_3Cu_3)_8Cl_8(CN)_{12}Li_4]{\cdot}2LiCN{\cdot}solvate$ (**20**) was isolated in 20% yield from reactions of **7** with 1,4-pyz and LiCl in MeCN [14b]. We hoped that the bridging ligands would link the clusters to form a supramolecular cube comprised of twelve 1,4-pyz and eight three-connecting Cp^*WS_3CuCl clusters. The product isolated from this reaction was a cyanide-bridged cube structure but not the one expected. In the formation of **20**, the introduction of Li^+ or Cl^- ions into frameworks of the cube was considered to change its charge and thus enhance its solubility in solution. The 1,4-pyz ligand removed part of the cyanide, but did not coordinate with any copper centers of **7**. The combined utilization of nitrogen donor ligands (e.g., 1,4-pyz) and/or LiX (X = halides) may be a good approach to the construction of cluster-based

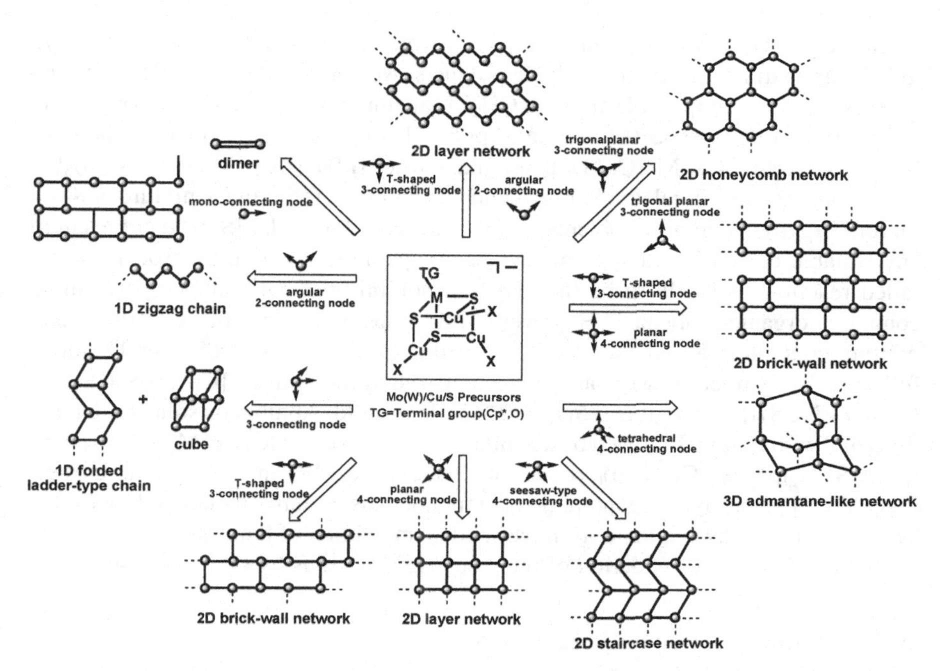

FIGURE 10.3 Possible topological nodes and frameworks derived from **4** to **8**.

supramolecular compounds. Therefore, an analogous reaction of **7** with LiBr and 1,4-pyz in MeCN afforded an unusual one-dimensional (1D) ladder-type chain polymer $[(Cp^*WS_3Cu_3Br)_2(\mu\text{-}CN)_3Li]\cdot$benzene (**21**) in a relatively low yield [14e]. In this case, 1,4-pyz also did not bind at any copper centers, but did remove part of the cyanide in **7**. The Br^- ion from the LiBr used capped the void of the incomplete $[Cp^*WS_3Cu_3]$ cubane-like core, while the Li^+ ion acted as a counterion for the $\{(Cp^*WS_3Cu_3(\mu_3\text{-}Br))_2(\mu\text{-}CN)_3\}^-$ anion. The different outcomes between **7**/LiCl/1,4-pyz and **7**/LiBr/1,4-pyz implied that such an assembly system might proceed in a rather complicated way.

Compound $[Cp^*WS_3Cu_3(\mu\text{-}CN)_2(py)]$ (**22**) was isolated in 70% yield from reactions of **7** with excess pyridine, $[PPh_4][Cp^*WS_3Cu_3(\mu_3\text{-}Cl)(\mu\text{-}CN)(CN)]\cdot$py (**23**) was obtained in 65% yield from reactions of **7** with pyridine in the presence of LiCl, and $[PPh_4]_2[Cp^*WS_3Cu_3(CN)_2]_2(\mu\text{-}CN)_2\cdot0.5(4,4'\text{-bipy})$ (**24**) was generated in 41% yield from reaction of **7** with 4,4′-bipy [14e].

Since many Mo(W)/Cu/S cluster-based assemblies exhibited low solubility in common organic solvents such as MeCN, DMSO, and DMF, it always resulted in the formation of products in very low yields and, particularly, difficulty in their crystallization and subsequent spectral and structural characterization [14b,14c,14d]. To tackle these problems, aniline was finally adopted because it has excellent capability to dissolve Mo(W)/Cu/S cluster-based assemblies. Therefore, compounds $[Cp^*MoS_3Cu_3(1,4\text{-pyz})(\mu\text{-}NCS)_2]$ (**25**) and $[(Cp^*MoS_3Cu_3)_2(NCS)_3(\mu\text{-}NCS)(bpe)_3]\cdot$3aniline (**26**) were isolated from reactions of **8** with excess 1,4-pyz or bpe in aniline in relatively high yields [14h]. However, reactions of **8** in aniline with solid tpt or H_2tpyp did not form any isolable products. Analogous reactions of **8** in aniline with a suspension of tpt (or H_2tpyp) in CH_2Cl_2 at ambient temperature for 3 days also did not produce any expected Mo/Cu/S/tpt (or H_2tpyp) supramolecular compounds. Instead, a tetranuclear Mo/Cu/S cubane-like cluster $[PPh_4]_2[Cp^*MoS_3(CuNCS)_3Cl]$ was always isolated in relatively high yields. In addition, refluxing a mixture of **8** and tpt (or H_2tpyp) in common solvents such as MeCN, DMF, or DMSO did not result in any isolable cluster-based supramolecular compounds. The main reason for these failed reactions is likely to be the very low solubility of tpt and H_2tpyp in these common organic solvents. However, solid-state reactions of a well-ground mixture containing **8** and excess tpt in a sealed Pyrex tube at 100°C for 12 hours, followed by extraction with aniline, led to the formation of $[Cp^*MoS_3Cu_3$(tpt)(aniline)$(NCS)_2]\cdot$0.75aniline$\cdot 0.5H_2O$ (**27**) in 45% yield. Analogous solid-state reactions of **8** with H_2tpyp followed by a similar workup gave rise to $[Cp^*MoS_3Cu_3$(NCS)(μ-NCS)$(H_2\text{tpyp})_{0.4}(\text{Cu-tpyp})_{0.1}]\cdot$2aniline$\cdot$2.5benzene (**28**) in 25% yield. Intriguingly, the free porphyrin ligand in this compound was partially metallated by Cu^{2+}. The Cu^{2+} ions may originate from partial decomposition of **8** during the solid-state reactions followed by oxidation of the resulting Cu^+ ions by O_2 in air [14h].

10.2.4 Strategy Based on Precursor 9

Compound **9** contains a $[WS_4Cu_4]$ core structure in which two Cu atoms have a tetrahedral coordination geometry with two terminal iodides, while the other two have a trigonal planar coordination geometry with one terminal iodide [18c]. If these iodides are all removed by strong donor ligands, the four copper centers may have up to eight coordination sites available when their coordination geometries are all made tetrahedral. Therefore, the $[WS_4Cu_4]$ core in **9** is expected to be a complicated multiconnecting node for the assembly of cluster-based supramolecular frameworks.

The compound $[WS_4Cu_4(4,4'\text{-bipy})_4][WS_4Cu_4I_4(4,4'\text{-bipy})_2]\cdot$8MeCN (**29**) was isolated by reactions of **9** with 2.5 equiv. of 4,4′-bipy in a DMF/MeCN solution [14c]. When crystals of **29** were transferred directly from the mother liquor to a CCl_4 solution of I_2, they darkened considerably within a matter of minutes to a deep reddish brown, consistent with absorption of iodine. A suitable crystal was transferred directly to protective oil and placed on a diffractometer, where it was cooled to 193 K under a stream of nitrogen. Crystal structure analysis indicated that the crystal was essentially identical to the parent crystal except for the fact that I_2 guest molecules were now included in the channels. The crystal formula was found to be $[WS_4Cu_4(4,4'\text{-bipy})_4]$ $[WS_4Cu_4I_4(4,4'\text{-bipy})_2]\cdot 2I_2$ (**30**) [14c].

Reactions of an acetonitrile solution of **9** with 3 equiv. of dps followed by a standard workup produced $[WS_4Cu_4I_2(\text{dps})_3]\cdot$2MeCN (**31**) in 75% yield, while those of **9** in MeCN with a methanol solution containing 2 Eq. of sodium dicyanamide and 2 Eq. of dpds followed by a similar workup used in the isolation of **31** gave rise to $[WS_4Cu_4(\text{dca})_2(\text{dpds})_2]\cdot Et_2O\cdot$2MeCN (**32**) in 63% yield [14f].

Considering that the $[WS_4Cu_4]$ core in **9** has an abundance of potential coordination sites, one approach to overcoming it is to keep some of its coordination sites within the cluster to attach bulky, nonbridging organic ligands, which may enhance the solubility of the resulting metal cluster in organic solvents. The cluster core would still be able to use the remaining coordination sites to bind to bridging ligands within an extended network. Therefore, one bulky ligand, bis(3,5-dimethylpyrazolyl)methane (dmpzm), was introduced into the assembly process of the cluster-based supramolecular arrays derived from **9**. Reactions of a solution of **9** in DMF with 2 equiv. of dmpzm followed by 2 equiv. of sodium dicyanoamide (dca) in MeOH produced a soluble $[WS_4Cu_4]$-based polymer $[WS_4Cu_4(dmpzm)_2(dca)_2]$ (**33**) in 49% yield [14g].

10.2.5 Strategy Based on Precursor 10

Compound **10** contains a $[(Cp^*WS_3)_3Cu_7]$ core structure which may be viewed as being built up via three corner-shared incomplete $Cp^*WS_3Cu_3$ fragments. While the two bridging coppers are surrounded tetrahedrally by four sulfurs, a total of nine MeCN molecules coordinate further with the remaining five coppers to complete their coordination geometries [14a]. A total of 10 sites are available if the coordination geometries of five copper atoms are turned into tetrahedral geometry. Therefore, the cluster $[(Cp^*WS_3)_3Cu_7]$ core is expected to provide very complicated multiconnecting nodes. Also, this cluster core may be cleaved into a $[Cp^*WS_3Cu_3]$ cluster core when it is assembled with bridging ligands such as 4,4′-bipy and its derivatives. Reactions of **10** with equimolar LiCl and equimolar 1,4-pyz in MeCN followed by a standard workup gave rise to $[Cp^*WS_3Cu_3Cl(MeCN)(1,4\text{-pyz})](PF_6)$ (**34**) in 67% yield [14a].

10.2.6 Strategy Based on Precursors 11 and 12

Precursor clusters **11** and **12** were prepared from the solid-state reactions of $[NH_4]_2[MS_4]$ (M = Mo, W) with CuI and $[(n\text{-Bu})_4N]I$ at 100°C followed by extraction of the solid products using CH_2Cl_2 [18d,18e]. Although efforts were made to ensure its stoichiometry, determination of its crystal structure always failed. It was proposed that it contains an octahedral $[MS_4Cu_6]$ core in which the M atom is situated at the center while each of the six Cu(I) centers occupies every corner of the octahedron and one or two terminal iodides coordinate at each copper center. When all iodides are replaced by strong donor bridging ligands, 12 coordination sites are available if the coordination geometries of six copper atoms are turned into tetrahedral geometry. Therefore, the cluster $[MS_4Cu_6]$ core is expected to serve as the most complicated multiconnecting nodes among compounds **1** to **12**. The compound $[MS_4Cu_6I_4(Py)_4]$ (**35**: M = Mo; **36**: M = W) was easily isolated in a relatively high yield from treatment of **11** or **12** with pyridine vapor [18d,18e].

10.3 STRUCTURAL FEATURES

In this chapter the resulting cluster-supported supramolecular assemblies are classified by the types of the Mo(W)/Cu/S cluster cores contained in their structures. A total of six types of cluster cores are found among the structures of 26 Mo(W)/Cu/S cluster-supported supramolecular compounds, which include (1) a linear MCu_2 core, (2) a T-shaped MCu_3 core, (3) a nest-shaped MCu_3 core, (4) a saddle-like MCu_4 core, (5) a hexagonal prismatic M_2Cu_4 core; and (6) an octahedral MCu_6 core. In the following subsections we describe their pertinent structural features in the order of the size of the cluster core.

10.3.1 Assemblies with a Linear MCu_2 Core

Supramolecular assemblies consisting of linear MCu_2 cores are limited in number. The only example is $[Et_4N]_3[\{WS_4Cu_2(\mu\text{-}CN)\}_2(\mu'\text{-}CN)]\cdot 2MeCN$ (**15**) [14d]. X-ray analysis revealed that the asymmetric unit of **15** is represented by one half of the formula unit $[\{WS_4Cu_2(\mu\text{-}CN)\}_2(\mu'\text{-}CN)][Et_4N]_3(MeCN)_2$. As indicated in Figure 10.4a, Cu2 is coordinated by two bridging cyanides and two sulfur atoms (from the cluster), resulting in a tetrahedral Cu(I) center. The other Cu center of the cluster is in an approximately trigonal environment formed from a single bridging cyanide and two sulfur atoms. The oxidation states of the W and Cu atoms in **2**, $+6$ and $+1$, respectively, are retained in **15**. From a topological perspective, the WS_4Cu_2 cluster may be considered as a three-connecting node, which is linked to equivalent nodes to form a 2D (6,3) honeycomb-like anionic network that extends in the *bc*-plane (Fig. 10.4b). The networks are about 10 Å apart, with Et_4N^+ cations and MeCN solvent molecules lying between the sheets.

10.3.2 Assemblies with a T-Shaped MCu_3 Core

Cluster-based supramolecular structures with T-shaped MCu_3 cores are also quite rare. The only example is $[PPh_4]_2[\{WS_4Cu_3(\mu\text{-}CN)\}_2(\mu'\text{-}CN)_2]\cdot 2MeCN$ (**16**) [14d]. Compound **16** consists of an anionic 2D polymer formed by linking WS_4Cu_3 clusters with cyanide bridges. A T-shaped WCu_3 core within **16** may be considered as being

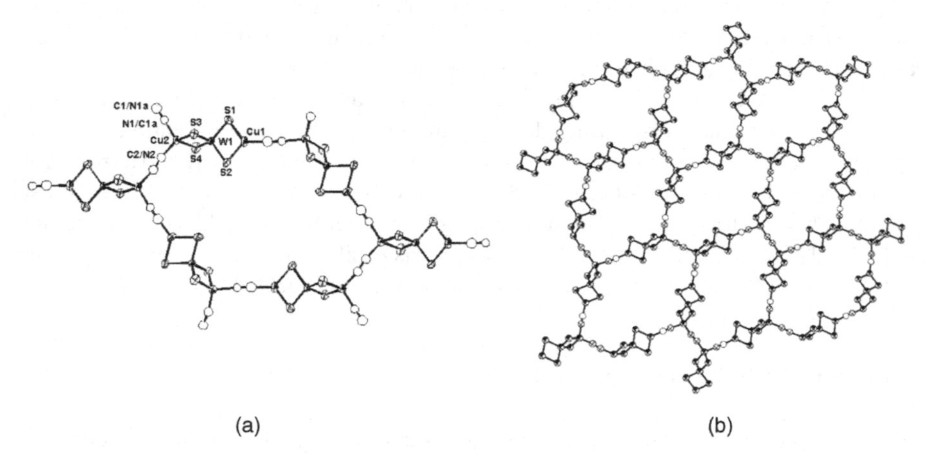

FIGURE 10.4 (a) Repeating unit of the 2D anionic network of **15** with 50% thermal ellipsoids; (b) extended 2D honeycomb network of **15** looking along the *a*-axis.

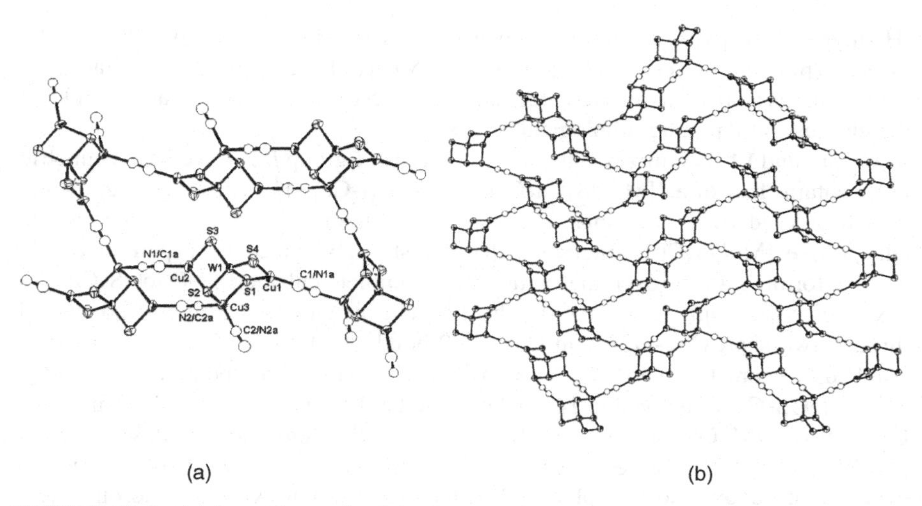

FIGURE 10.5 (a) Repeating unit of the 2D anionic network within **16** with 50% thermal ellipsoids; (b) extended 2D herringbone network of **16** looking along the *c*-axis.

derived from the addition of a Cu atom to the WS_4Cu_2 cluster of **3**. As illustrated in Figure 10.5a, Cu1 and Cu2 are each in approximately trigonal coordination environments formed by two sulfur atoms and a single bridging cyanide. In contrast, Cu3 is tetrahedrally coordinated by two bridging cyanides and a pair of sulfur atoms from the WS_4 unit. Each cluster core is linked to three equivalent clusters through either single or double cyanide bridges. From a topological perspective, each WCu_3 core acts as a three-connecting node to link three equivalent cores through either single or double cyanide bridges, forming a 2D (6,3) herringbone-like anionic network (Fig. 10.5b). The six-membered rings within this network (Fig. 10.5a) resemble the chair conformation of cyclohexane. The average layer-to-layer separation is about 16 Å, with the $[\{WS_4Cu_3(\mu\text{-CN})\}_2(\mu'\text{-CN})_2]_n^{2n-}$ anionic layers separated by $[PPh_4]^+$ cations and MeCN molecules.

10.3.3 Assemblies with a Nest-Shaped MCu_3 Core

Assemblies with nest-shaped MCu_3 cluster cores comprise the biggest family in Mo(W)/Cu/S cluster-supported supramolecular chemistry. A total of 14 Mo(W)/Cu/S-based supramolecules are included: $[Et_4N]_4[MoOS_3Cu_3CN(\mu'\text{-CN})]_2(\mu\text{-CN})_2$ (**14**) [14i], $[MoOS_3Cu_3I(dpds)_2]\cdot0.5DMF\cdot2(MeCN)_{0.5}$ (**17**) [14f], $[MoOS_3Cu_3$ (dca)(4,4′-bipy)$_{1.5}]\cdot$DMF$\cdot$MeCN (**18**) [14f], $[Cp^*WS_3Cu_3Br(\mu\text{-Br})(4,4'\text{-bipy})]\cdot Et_2O$ (**19**) [14e], $[(Cp^*WS_3Cu_3)_8Cl_8(CN)_{12}Li_4]\cdot2LiCN\cdot$solvate (**20**) [14b], $[(Cp^*WS_3Cu_3Br)_2(\mu\text{-CN})_3Li]\cdot$benzene (**21**) [14e], $[Cp^*WS_3Cu_3(\mu\text{-CN})_2(py)]$ (**22**) [14e], $[PPh_4][Cp^*WS_3Cu_3(\mu_3\text{-Cl})(\mu\text{-CN})(CN)]\cdot$py (**23**) [14e], $[PPh_4]_2[Cp^*WS_3Cu_3(CN)_2]_2(\mu\text{-CN})_2\cdot0.5(4,4'\text{-bipy})$ (**24**) [14e], $[Cp^*MoS_3Cu_3(1,4\text{-pyz})(\mu\text{-NCS})_2]$ (**25**) [14h], $[(Cp^*MoS_3Cu_3)_2(NCS)_3(\mu\text{-NCS})(bpe)_3]\cdot3$aniline (**26**) [14h], $[Cp^*MoS_3Cu_3(tpt)(aniline)(NCS)_2]\cdot0.75aniline\cdot0.5H_2O$ (**27**) [14h], $[Cp^*MoS_3Cu_3(NCS)(\mu\text{-NCS})$

$(H_2\text{-tpyp})_{0.4}(\text{Cu-tpyp})_{0.1}]\cdot$2aniline·2.5benzene (**28**) [14h], and $[\text{Cp}^*\text{WS}_3\text{Cu}_3\text{Cl}(\text{MeCN})(\text{pz})](\text{PF}_6)$ (**34**) [14a]. Although the MCu_3 cluster cores are similar, each core works as a different multiconnecting node with the assistance of various linkers, thereby exhibit daunting topological arrays.

Compound **14** crystallizes in the monoclinic space group $P2_1/c$ and its asymmetric unit contains half of a $\{[\text{MoOS}_3\text{Cu}_3\text{CN}(\mu'\text{-CN})]_2(\mu\text{-CN})_2\}^{4-}$ cluster tetraanion and two disordered Et_4N^+ cations. The cluster tetraanion comprises two incomplete cubane-like $[\text{MoOS}_3\text{Cu}_3]$ fragments that are strongly linked by a pair of μ'-CN groups, forming a centrosymmetric double incomplete cubane-like $[\text{MoOS}_3\text{Cu}_3(\mu'\text{-CN})]_2$ core structure (Fig. 10.6a). The two cluster anions are tightly connected through two short Cu1–N2A and Cu1A–N2 bonds. Cu1 and Cu2 adopt a distorted tetrahedral geometry, while Cu3 has a trigonal planar environment. Topologically, each $[\text{MoOS}_3\text{Cu}_3]$ fragment in **14** can be viewed as being built of the addition of two Cu^+ onto the $[\text{MoOS}_3\text{Cu}]$ core of **1**. There are four other bridging cyanides around the $[\text{MoOS}_3\text{Cu}_3(\mu'\text{-CN})]_2$ core. Because two coordination sites (N2, C2A) of two cyanides are above the Cu_4 plane (Cu1, Cu2, Cu1A, Cu2A) while the other two (N2B, C2C) are below the Cu_4 plane, this core works has a unique chairlike four-connecting node to interconnect four equivalent nodes to form another 2D (4,4) network extended along the *bc*-plane (Fig. 10.6b). The average layer-to-layer separation is about. 9.7 Å, with the $[\{\text{MoOS}_3\text{Cu}_3\text{CN}(\mu'\text{-CN})\}_2(\mu\text{-CN})_2]_n^{4n-}$ anionic layers separated by $[Et_4N]^+$ cations.

Compound **17** crystallizes in the monoclinic space group $P2_1/c$ and the asymmetric unit contains one $[\text{MoOS}_3\text{Cu}_3\text{I}(\text{dpds})_2]$ molecule, half of a DMF, and two halves of MeCN solvated molecules. The $[\text{MoOS}_3\text{Cu}_3\text{I}(\text{dpds})_2]$ molecule consists of a nido-shaped $[\text{MoOS}_3\text{Cu}_3]$ core, which closely resembles those in $[\text{MoOS}_3\text{Cu}_3\text{IL}_2]$ (L = 2,2′-bipy [20a], phen [20b]) (Fig. 10.7a). Cu1 and Cu2 adopt a distorted

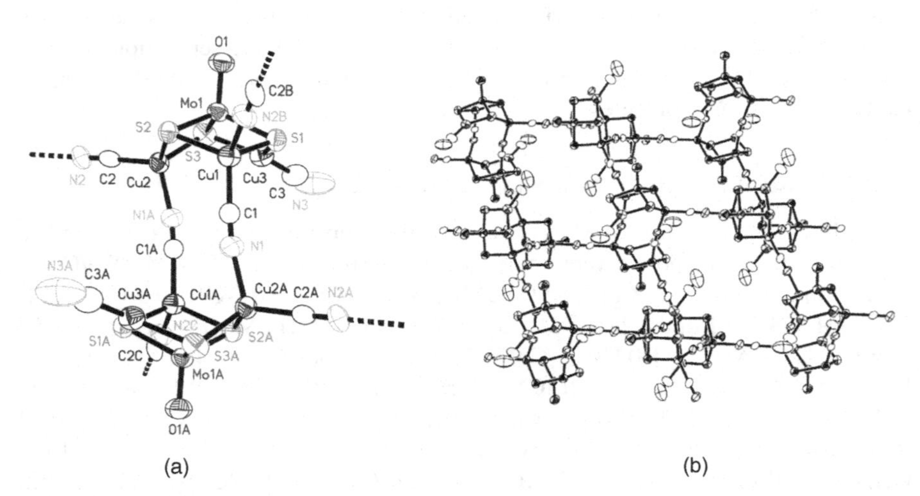

FIGURE 10.6 (a) Perspective view of the repeating unit of **14** with labeling scheme and 50% thermal ellipsoids; (b) extended 2D (4,4) network of **14** viewing along the *a*-axis.

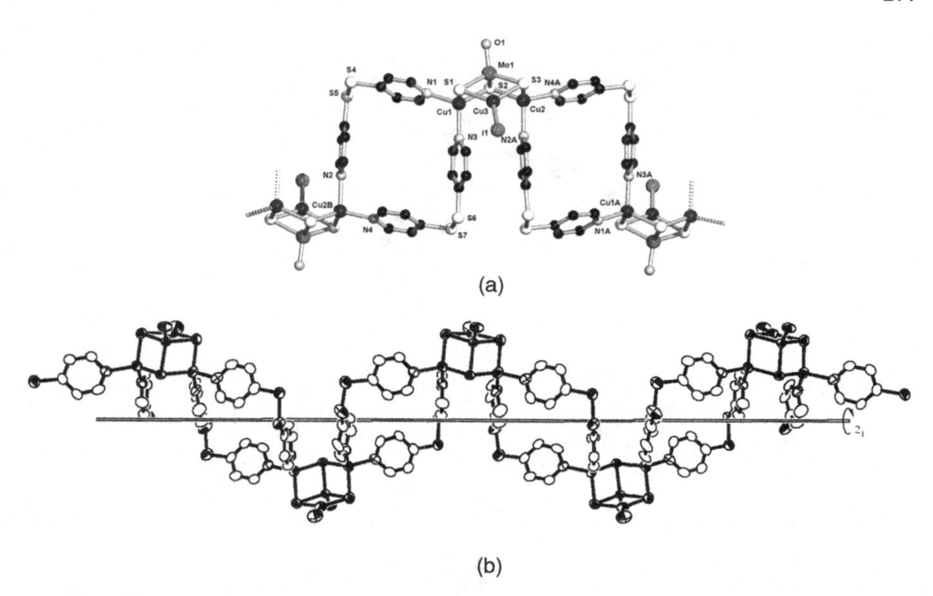

FIGURE 10.7 (a) Perspective view of a $[MoOS_3Cu_3I(dpds)_2]$ molecule of **17** with 50% thermal ellipsoids; (b) portion of the 1D spiral chain of **17** extended along the *b*-axis.

tetrahedral coordination geometry, coordinated by two μ_3-S atoms from the $[MoOS_3]^{2-}$ anion and two N atoms from two dpds ligands. Cu3 has an approximate trigonal planar environment, coordinated by one I and two μ_3-S atoms. The two dtdp ligands in **17** basically keep the gauche conformation of the free dtdp ligand. From a topological perspective, the nido-like $MoCu_3$ core in **17** may be considered as a two-connecting node, which is linked to two equivalent cores via two pairs of dpds ligands to form a 1D spiral chain extended along the *b*-axis (Fig. 10.7b). Each 1D chain is chiral and is represented by either –*M*-$[MoOS_3Cu_3]$-*M*-$[MoOS_3Cu_3]$-*M*-(*M*-chain) or –*P*-$[MoOS_3Cu_3]$-*P*-$[MoOS_3Cu_3]$-*P*-(*P*-chain). This compound consists of a 1 : 1 ratio of *P*- and *M*-chains and is achiral with a centrosymmetric space group $P2_1/c$. The DMF and MeCN solvent molecules lie between the chains, and no evident interaction is observed between the chain and the solvated molecules. Alternatively, the 1D chain structure may be described as being built of cationic $[Cu(dpds)]_2^{2+}$ squares interconnected by $[MoOS_3(CuI)]^{2-}$ cluster anions via three μ_3-S atoms. There is a 2_1 axis running through the center of each $[Cu(dpds)]_2^{2+}$ square within the chain. The internal cavity for each square has an approximate dimension of 9.77 Å (Cu1 ··· Cu2B separation) × 10.15 Å (S4 ··· S6 separation).

Compound **18** crystallizes in the triclinic space group $P\bar{1}$, and the asymmetric unit consists of half of the dimeric $\{[MoOS_3Cu_3]_2(dca)_2(4,4'\text{-bipy})_3\}$ molecule, one DMF molecule, one MeCN solvent molecule. Two nido-like $[MoOS_3Cu_3]$ cores (Fig. 10.8a) are linked by a pair of parallel 4,4′-bipy bridges to form a centrosymmetrically related $\{[MoOS_3Cu_3]_2(dca)_2(4,4'\text{-bipy})_3\}$ dimer (Fig. 10.8b). The resulting narrow mesh within the dimer is estimated to be 3.5 × 11.2 Å in size. Each $[MoOS_3Cu_3]$ core structure in **18** closely resembles those of **14** and **17**. Cu3 is three-coordinated by one N

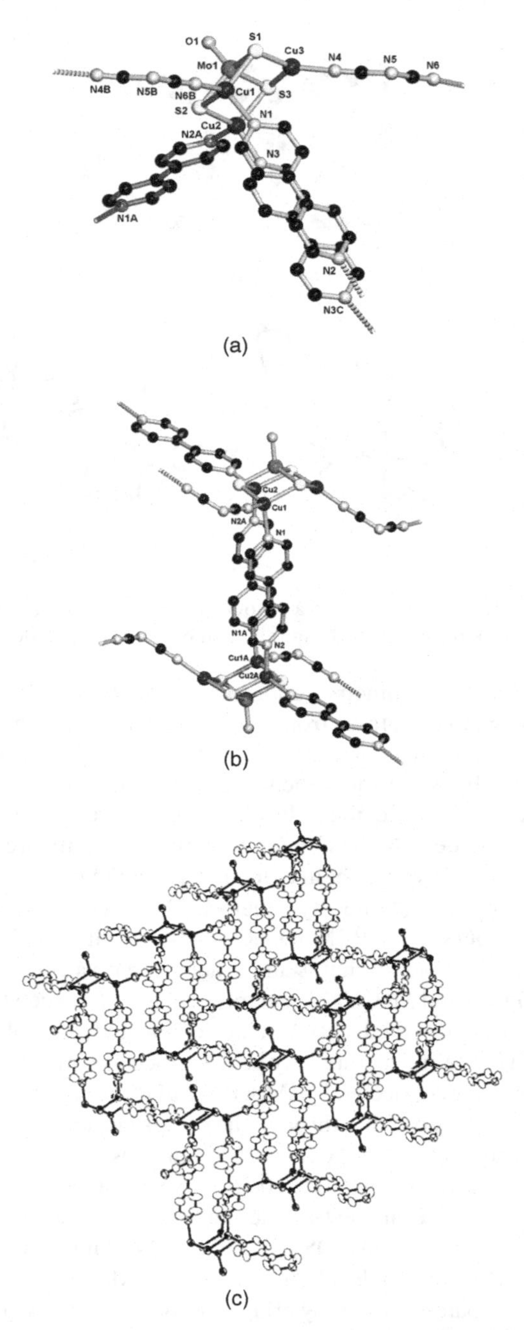

FIGURE 10.8 (a) Perspective view of the $[MoOS_3Cu_3(dca)(4,4'\text{-bipy})_{1.5}]$ molecule of **18** with 50% thermal ellipsoids; (b) dimeric $\{[MoOS_3Cu_3]_2(dca)_2(4,4'\text{-bipy})_3\}$ species of **18**; (c) extended 2D staircase network of **18** looking along the *c*-axis.

from a dca anion and two μ_3-S atoms, while Cu1 is tetrahedrally coordinated by an N atom (4,4′-bipy), an N (dca) atom, and two μ_3-S atoms, while Cu2 is four-coordinated by two N atoms (4,4′-bipy) and two μ_3-S atoms. The dca ligands serve as a $\mu_{1,5}$ coordination mode. Topologically, each [$MoOS_3Cu_3$] core in **18** serves as a rare seesaw-shaped four-connecting node and is linked by a pair of dca anions at Cu1 and Cu3 to form a 1D [$MoOS_3Cu_3$(dca)]$_n$ chain extended along the *a*-axis. Such a chain and its centrosymmetrically related chain are then bridged by a pair of parallel 4,4′-bipy ligands to form a 1D {[$MoOS_3Cu_3$(4,4′-bipy)]$_2$(dca)$_2$}$_n$ double chain in which the orientation of the [$MoOS_3Cu_3$] core is alternating. Each resulting double chain is further interconnected by another pair of symmetry-related 4,4′-bipy ligands to form a unique 2D staircase network extended along the *ab*-plane (Fig. 10.8c).

Compound **19** consists of the repeating dimeric [Cp*WS_3Cu_3Br(4,4′-bipy)]$_2$ units (Fig. 10.9a), which are further interconnected by four Cu–μ-Br–Cu bridges to form a 2D brick-wall layer extended along the *ab*-plane (Fig. 10.9b). Each bricklike cavity is estimated to have dimensions 11.2 × 11.5 Å and is occupied by two [Cp*WS_3Cu_3Br] fragments from the other two dimers. Each dimer is centrosymmetrically related and consists of two [Cp*WS_3Cu_3Br] fragments connected by a pair of parallel 4,4′-bipy bridges. The size of the resulting narrow mesh is estimated to be 3.5 × 11.2 Å. The core structure of each [Cp*WS_3Cu_3Br] fragment closely resembles that of **6**. Cu1 is trigonally coordinated by one terminal Br and two μ_3-S atoms, while Cu2 and Cu3

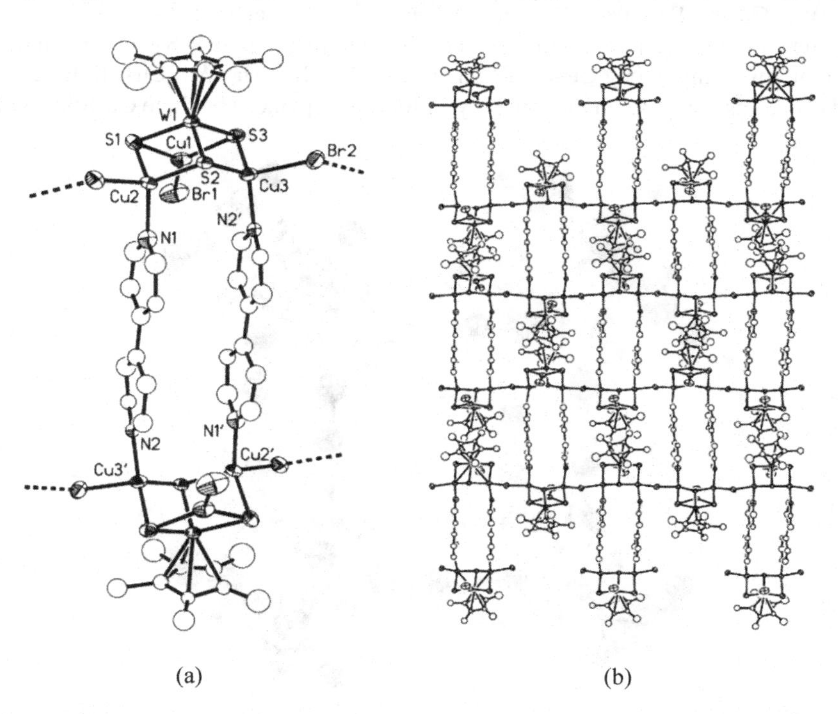

FIGURE 10.9 (a) Perspective view of the repeating unit of **19** with 50% thermal ellipsoids; (b) extended 2D brick-wall network of **19** looking along the *c*-axis.

are tetrahedrally coordinated by an N atom of 4,4′-bipy, a μ-Br atom, and two μ_3-S atoms. Topologically, each [Cp*WS_3Cu_3] core in **19** acts as a T-shaped three-connecting node. The [Cp*WS_3Cu_3Br] fragments are interconnected by μ-Br at Cu3 and Cu2B to form 1D zigzag chains extending along the *b*-axis, where the orientation of the [Cp*WS_3Cu_3Br] fragment is alternating. Alternatively, the 2D brick-wall layer structure of **20** can be viewed as being built up of [Cp*WS_3Cu_3Br(μ-Br)] chains linked by pairs of parallel 4,4′-bipy bridges along the *a*-axis.

Compound **20** consists of a supramolecular cube comprised of 12 cyanides and 8 three-connecting Cp*WS_3Cu_3 cluster fragments (Fig. 10.10). Although the structure possesses the topology of the cubic, the molecule is actually tetragonally distorted. The cubic dimensions as judged by the separation of W centers are 9.64×10.10 10.10 Å. Inside this cube are eight Cl^- and four Li^+ ions. The core structure of each [Cp*WS_3Cu_3] fragment closely resembles that of **7**. As each chloride ion shows significant positional disorder and is weakly interacted with three Cu atoms of each Cp*WS_3Cu_3 cluster core, each Cu center thus adopts a nearly trigonal planar coordination with a fourth weak Cu–Cl bond. At the heart of the cube, four Li^+ ions are arranged in a square whose plane is perpendicular to the tetragonal axis.

Compound **21** crystallizes in the orthorhombic space group *Cmcm*, and the asymmetric unit contains one-fourth of the [{Cp*$WS_3Cu_3(\mu_3$-Br)}$_2$(μ-CN)$_3$]$^-$ anion, one-fourth of a Li^+ ion, and one-fourth of a solvated benzene molecule. The [Cp*$WS_3Cu_3(\mu_3$-Br)] fragment may be viewed as having a distorted cubane-like structure in which one bromide fills into the void of the [Cp*WS_3Cu_3] incomplete cube of **7** with three long Cu–Br distances (Fig. 10.11a). The W1, Cu1, Br1, S2, C1, and C4 atoms are lying on the same crystallographic plane. The three Cu atoms in **21**

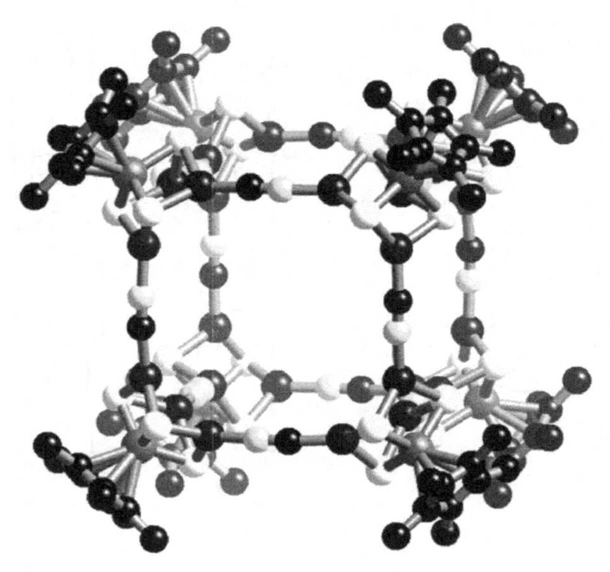

FIGURE 10.10 Perspective view of the cube structure of **20**. All the Cl^- and Li^+ ions are omitted for clarity.

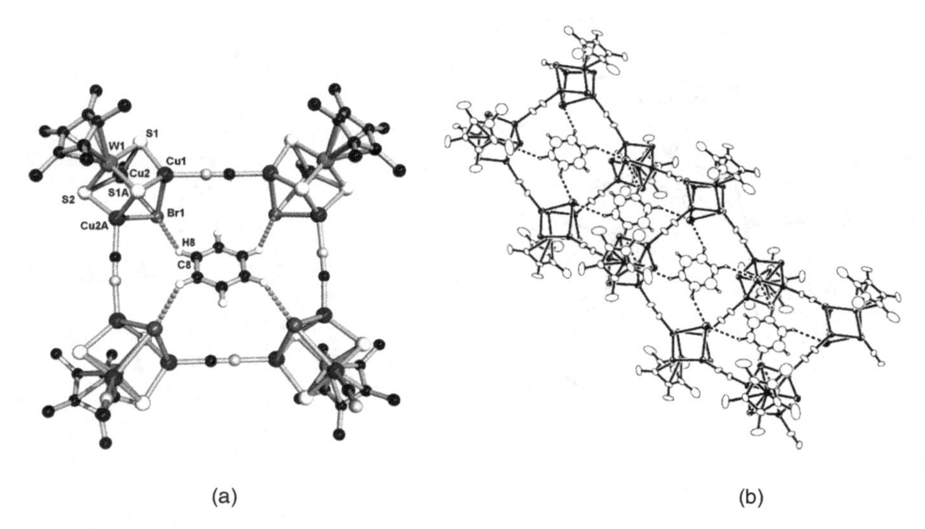

FIGURE 10.11 (a) Perspective view of the repeating square-like unit in **21**; (b) portion of the 1D ladder-shaped chain of **21** extended along the *c*-axis.

may be viewed as having a nearly trigonal planar coordination, with a fourth weak Cu–Br interaction. The four [Cp*$WS_3Cu_3(\mu_3$-Br)] fragments in **21** are linked via four cyanide bridges, forming a repeating square-like [{Cp*$WS_3Cu_3(\mu_3$-Br)(μ-CN)}$_4$](μ-CN)$_2^{2-}$ unit with a crystallographic C_{2v} symmetry (Fig. 10.11a). Each square-like unit is further interconnected by another four cyanides to form a 1D ladder-shaped anionic chain extended along the crystallographic *c*-axis (Fig. 10.11b). Alternatively, the [Cp*$WS_3Cu_3(\mu_3$-Br)] fragment in **21** may be considered topologically as a three-connecting node which is linked to three equivalent nodes via Cu–μ-CN–Cu bridges to form this 1D chain polymer. The average chain-to-chain separation is about 10.4 Å, with the [{Cp*$WS_3Cu_3(\mu_3$-Br)}$_4$(μ-CN)$_6$]$_n^{2n-}$ anionic layers separated by Cp* groups. In addition, a solvated benzene molecule lies at the center of each square-like unit, and its four hydrogen atoms interact with four Br atoms of the square-like unit to form four symmetrical C-H $\cdots$ Br [3.54(4) Å, 140.8°] hydrogen bonds, which may stabilize the entire ladder-shaped chain framework.

Compound **22** crystallizes in the orthorhombic space group $P2_12_12_1$ and the asymmetric unit contains one discrete molecule [Cp*$WS_3Cu_3(\mu$-CN)$_2$(py)]. Although the repeating unit [Cp*$WS_3Cu_3(\mu$-CN)$_2$(py)] (Fig. 10.12a) assumes an incomplete cubane-like [Cp*WS_3Cu_3] core structure similar to that of **7**, its three Cu centers have different coordination environments. Cu1 remains a trigonal planar coordination geometry, while Cu2 (or Cu3) is tetrahedrally bound by two μ_3-S and two C atoms [or one N (py), one C, and two μ_3-S atoms]. The repeating unit [Cp*$WS_3Cu_3(\mu$-CN)$_2$(py)] is coordinated by four bridging cyanide groups, which link four crystallographically equivalent clusters that lie at the corners of a very distorted tetrahedron (Fig. 10.12b). From a topological perspective, each cluster [Cp*WS_3Cu_3] core serves as a tetrahedral node, to form a single adamantine type

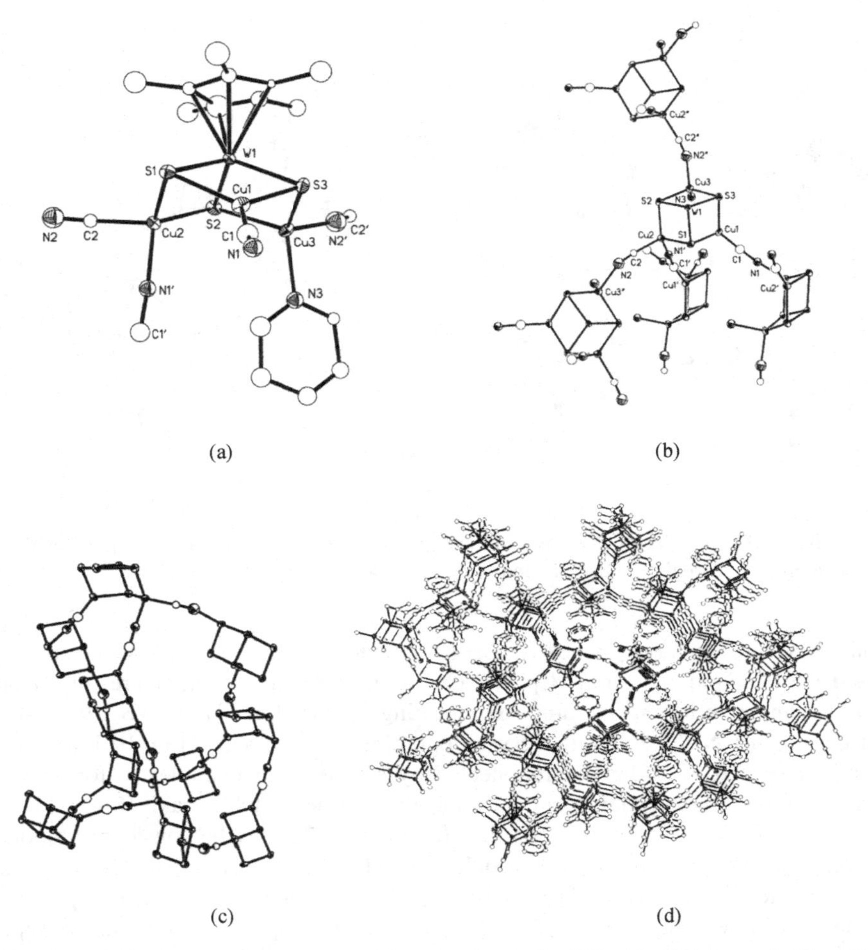

FIGURE 10.12 (a) Perspective view of the repeating unit of **22** with 50% thermal ellipsoids; (b) interactions of a central cluster core with four equivalent cluster centers via four cyanide bridges in **22**; (c) adamantine-type unit within the network in **22**; (d) 3D structure of **22** looking along the *c*-axis.

unit (Fig. 10.12c), which is interconnected via μ-CN anions to form a 3D network (Fig. 10.12d). It should be noted that the pyridyl groups coordinated at Cu2 centers and the bulky Cp* groups on W1 atoms occupy the resulting adamantine cavities, thereby preventing interpenetration of networks.

Compound **23** crystallizes in the monoclinic space group $P2_1/c$ and the asymmetric unit contains one discrete $[Cp^*WS_3Cu_3(\mu_3\text{-}Cl)(CN)(\mu\text{-}CN)]^-$ anion, one $[PPh_4]^+$ cation, and one solvated pyridine molecule. Compound **23** has a one-dimensional zigzag anionic chain, which is composed of the repeating $[Cp^*WS_3Cu_3(\mu_3\text{-}Cl)(CN)]$ units, and is stacked along the crystallographic *c*-axis via Cu–μ-CN–Cu bridges (Fig. 10.13). The anionic chains are separated by $[PPh_4]^+$ cations and pyridine

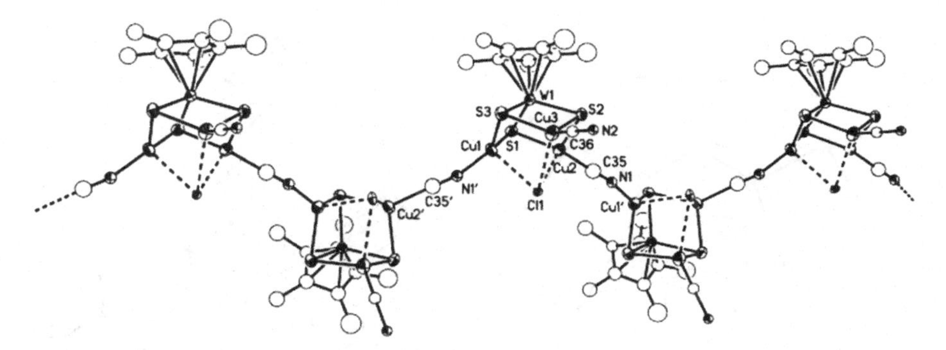

FIGURE 10.13 Section of the 1D zigzag chain of **23** extending along the *c*-axis.

solvent molecules. Topologically, the [Cp*WS_3Cu_3] core in **23** acts as a two-connecting node. The [Cp*WS_3Cu_3(μ_3-Cl)(CN)] unit may be viewed as having a severely distorted cubane-like structure in which one chloride fills into the void of the incomplete cube of **7** with one long and two short Cu-Cl distances. The three Cu atoms in **23** are not equivalent, and their coordination variability ranges from a strongly distorted tetrahedron (Cu1 and Cu3) to a nearly trigonal-planar coordination (Cu2) with a long Cu2–Cl1 interaction.

Compound **24** crystallizes in the triclinic space group $P\bar{1}$, and the asymmetric unit consists of one-half of the {[Cp*$WS_3Cu_3(CN)_2]_2(\mu$-$CN)_2\}^{2-}$ dianion, one $[PPh_4]^+$ cation, and one 4,4′-bipy molecule. Compound **24** consists of two [Cp*$WS_3Cu_3(CN)_3]^-$ anions that are tightly bridged via a pair of Cu-μ-CN-Cu bridges, forming a double incomplete cubane-like structure with a crystallographic center of inversion lying at the center of the W1 and W1A line (Fig. 10.14). Topologically, each [Cp*WS_3Cu_3] core in **24** acts as a mono-connecting node. Cu1 adopts a distorted tetrahedral geometry, while Cu2 and Cu3 have a trigonal-planar coordination geometry. The 4,4′-bipy molecule in the crystal of **24** does not coordinate at copper centers but is sandwiched between the two cluster dianions with evident μ–μ interactions [3.616(14) Å] between each pyridyl group and its nearest Cp* group, thereby affording chains of cluster dianion/4,4′-bipy units running along the *c*-axis.

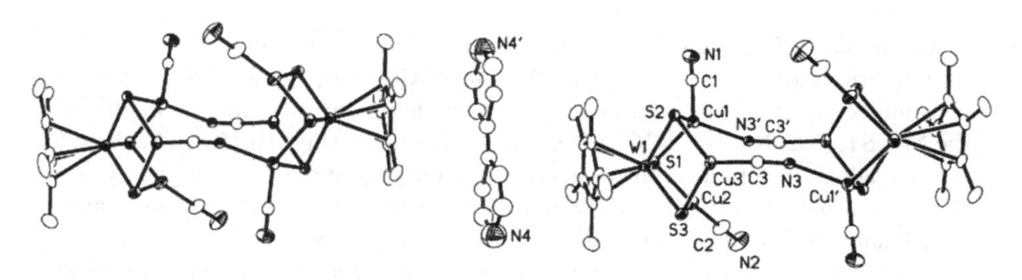

FIGURE 10.14 Dianion structure of **24** along with the sandwiched 4,4′-bipy molecule.

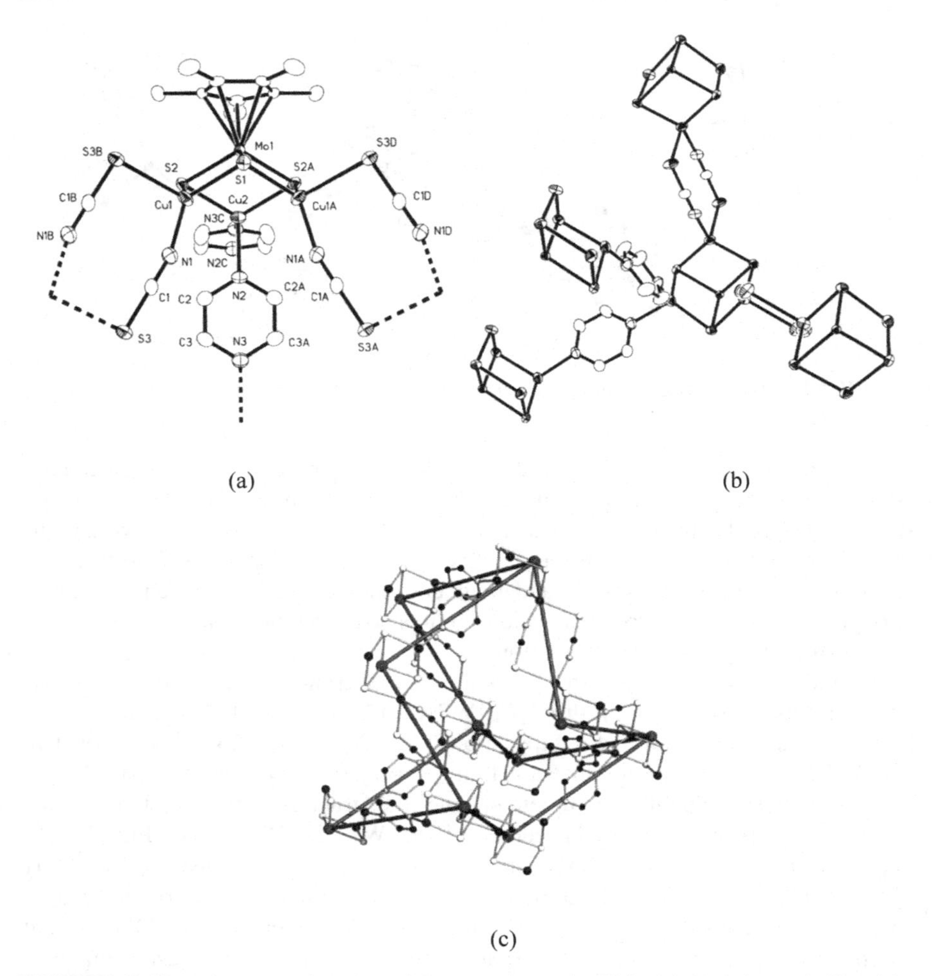

FIGURE 10.15 (a) Perspective view of the repeating unit of **25** with 50% thermal ellipsoids; (b) interactions of a central cluster core with four equivalent cluster centers via a pair of NCS bridges and a pair of 1,4-pyz bridges in **25**; (c) distorted adamantine-type unit within the network in **25**.

Compound **25** crystallizes in the orthorhombic space group *P*nma, and the asymmetric unit contains half of the [$Cp^*MoS_3Cu_3$(1,4-pyz)(μ-$NCS)_2$] molecule. This molecule retains the core structure of **6** and has a mirror plane running through the Mo, S1, Cu2, N2, N3, N2C, N3C, C6, and C9 atoms (Fig. 10.15a). The three Cu centers adopt distorted tetrahedral coordination geometries. Topologically, each [$Cp^*MoS_3Cu_3$] cluster core works as a tetrahedral four-connecting node and is coordinated by two pairs of bridging thiocyanate groups and a pair of bridging 1,4-pyz ligands (Fig. 10.15b), which link four crystallographically equivalent cluster cores that lie at the corners of a very distorted tetrahedron, forming a single adamantine type unit (Fig. 10.15c). This unit is interconnected further via bridging thiocyanate and 1,4-pyz ligands to form a noninterpenetrating 3D network.

Compound **26** crystallizes in the triclinic space group $P\bar{1}$, and the asymmetric unit contains one-half of a $[\{Cp^*MoS_3Cu_3\}_4(NCS)_6(\mu\text{-}NCS)_2(bpe)_6]$ molecule and three aniline solvent molecules. This molecule contains two different dimeric units, $[(Cp^*MoS_3Cu_3)_2(NCS)_2(\mu\text{-}NCS)_2(bpe)_3]$ (unit 1, Fig. 10.16a) and $[(Cp^*MoS_3Cu_3)_2(NCS)_4(bpe)_3]$ (unit 2, Fig. 10.16b), which are interconnected by a single bridging bpe ligand. Each $[Cp^*MoS_3Cu_3]$ core in unit 1 serves as a T-shaped three-connecting node, while that in unit 2 works as an angular two-connecting node. Four angular two-connecting nodes and six T-shaped three-connecting nodes interconnect through two

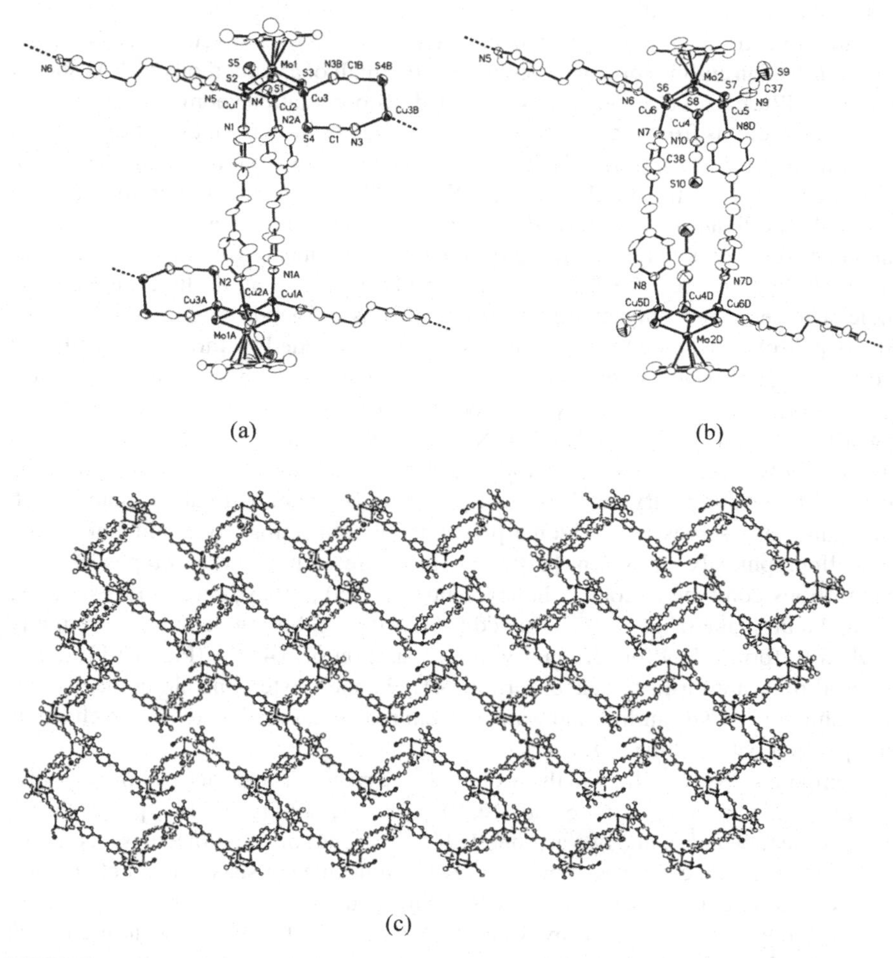

FIGURE 10.16 (a) Perspective view of the dimeric unit (unit 1) of **26** with 50% thermal ellipsoids; (b) perspective view of the dimeric unit (unit 2) of **26** with 50% thermal ellipsoids; (c) extended 2D (6,3) network of **26** parallel to the [104] plane.

pairs of μ-NCS bridges, four single bpe bridges, and four pairs of double bpe bridges, forming a Z-shaped unit $[(Cp^*MoS_3Cu_3)_{10}(bpe)_{12}(NCS)_{14}(\mu\text{-}NCS)_4]$ (Fig. 10.16c). This unit has two rectangle cavities, each with an area of about 415 $Å^2$. The occurrence of an angular two-connecting node and a T-shaped three-connecting node in the same structure is unprecedented in cluster-based supramolecular chemistry. Furthermore, this Z-shaped unit links the equivalent units by a pair of single bpe bridges, a pair of double bpe bridges, and a pair of μ-NCS bridges to form a 2D (6,3) network, parallel to the [104] plane. The 2D networks are about 6.99 Å apart, and when they are stacked along the *a*-axis, 1D channels are found in the unit cell of the crystal. The channels occupy a total volume of 1010.6 $Å^3$ (23.8% of the total cell volume) and are filled with aniline solvent molecules.

Compound **27** crystallizes in the triclinic space group $P\bar{1}$, and the asymmetric unit contains one discrete $[Cp^*MoS_3Cu_3(tpt)(aniline)(NCS)_2]$ molecule, one-half and one-fourth of an aniline solvent molecule, and one-half of a water molecule. As shown in Fig. 10.17a, each Cu center adopts a tetrahedral coordination geometry coordinated by two μ_3-S atoms from the $[Cp^*MoS_3]$ moiety; one N atom from the tpt ligand; and one S atom from the terminal NCS^- (Cu1), one N atom from an aniline molecule (Cu2), or one N atom from the terminal NCS^- (Cu3). It is uncommon that the two terminal NCS ligands show two different coordination preferences to Cu(I) centers in the same compound. In fact, there are only a limited number of Cu(I)/NCS complexes in which the terminal NCS ligand binds to Cu(I) via the S atom. In **27**, the aniline molecules serve not only as a solvent but also as a ligand, coordinating at Cu2. The $[Cp^*MoS_3Cu_3]$ core of **27** is symmetrically surrounded by three tpt ligands and works as a trigonal-planar three-connecting node, which is the first example in cluster-based supramolecular chemistry. The mean deviation from the least-squares plane consisting of Cu1, Cu2, Cu3, N3, N4, N5, N3A, N4A, N5A, N3B, N4B, and N5B is about 0.27 Å (Fig. 10.17a). The formation of such a topological center may be ascribed to the symmetry requirement of the rigid D_{3h}-symmetrical tpt ligand. Each tpt ligand also serves as a trigonal-planar three-connecting node and links three crystallographically equivalent $[Cp^*MoS_3Cu_3]$ cores. Thus, the two types of interconnections combine to form a honeycomb 2D $(6,3)_{core}(6,3)_{tpt}$ network extending along the *bc*-plane (Fig. 10.17b). In addition, two neighboring 2D layers are tightly stacked to form a Piedfort 22 unit with a separation of 3.424 Å (Fig. 10.17c). The Piedfort units are further stacked, forming 1D channels extending along the crystallographic *b*-axis. The aniline and water solvent molecules are located in the channels or squeezed between the 2D layers.

Compound **28** crystallizes in the triclinic space group $P\bar{1}$, and each asymmetric unit contains one-half of a $[Cp^*MoS_3Cu_3(NCS)(\mu\text{-}NCS)(H_2tpyp)_{0.4}(Cu\text{-}tpyp)_{0.1}]_2$ dimer, two aniline molecules, and two and one-half of benzene solvent molecules. As shown in Figure 10.18a, the centrosymmetric dimeric molecule consists of a double incomplete cubane-like structure in which two $[Cp^*MoS_3Cu_3(NCS)(\mu\text{-}NCS)(H_2tpyp)_{0.4}(Cu\text{-}tpyp)_{0.1}]$ fragments are bridged by two μ-NCS ligands. In the dimeric unit of **28**, each $[Cp^*MoS_3Cu_3]$ core works as a T-shaped three-connecting center to link equivalent cores through a pair of NCS bridges and two pyridyl groups of two H_2tpyp (or Cu-tpyp) ligands (Fig. 10.18b). In addition, each H_2tpyp (or Cu-tpyp) ligand, acting as a

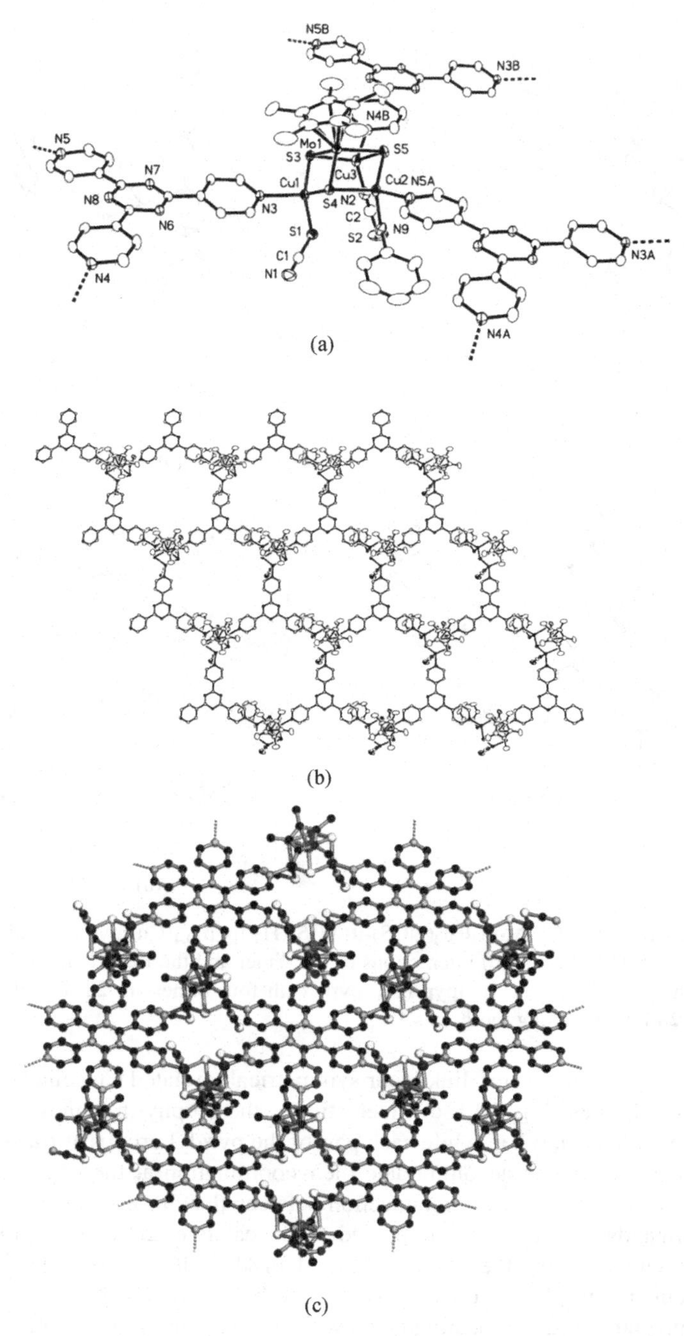

FIGURE 10.17 (a) Perspective view of the repeating unit of **27** with 50% thermal ellipsoids; (b) 2D honeycomb network of **27** extending along the *bc*-plane; (c) 2D sheet pair in a staggered disposition.

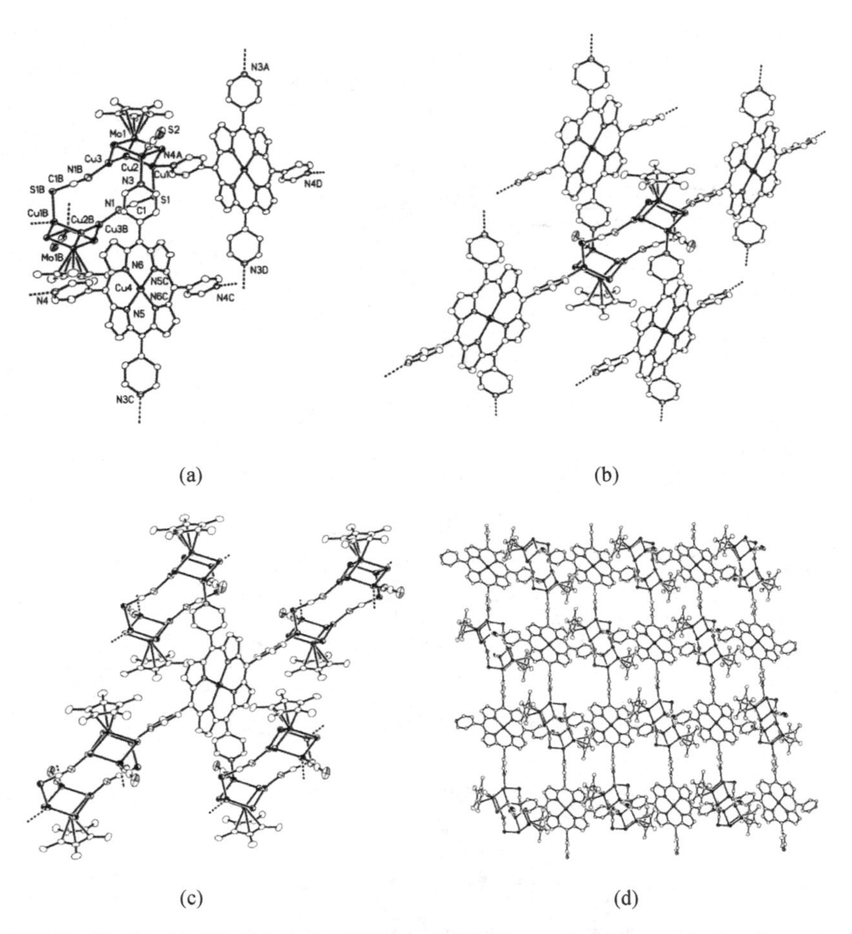

FIGURE 10.18 (a) $[Cp^*MoS_3Cu_3(NCS)(\mu\text{-}NCS)(H_2\text{-tpyp})_{0.4}(Cu\text{-tpyp})_{0.1}]_2$ dimer in **28** with 50% thermal ellipsoids; (b) interactions of the dimer and the four H_2-tpyp (or Cu-tpyp) in **28**; (c) interactions of one H_2-tpyp (Cu-tpyp) with four dimers in **28**; (d) 2D scale-like network of **28** looking along the *b*-axis.

planar four-connecting center, links four symmetrically related Cu centers from four $[Cp^*MoS_3Cu_3]$ cores (Fig. 10.18c). Interestingly, the porphyrin planes are approximately parallel to each other, while each pair of the pyridyl groups at *trans* positions rotates to some extent to accommodate their coordination at the copper cen- ters. Therefore, such a combined interconnection between the cluster core and H_2tpyp (or Cu-tpyp) ligands results in an unprecedented scalelike 2D $(4,6^2)_{core}(4^2,6^2)_{ligand}$ network extending along the *ac*-plane (Fig. 10.18d). Alternatively, the double incomplete cubane-like $[Cp^*MoS_3Cu_3(NCS)(\mu\text{-}NCS)]_2$ core can be visualized as an approximate planar four-connecting node, which links four equivalent nodes through four H_2tpyp (or Cu-tpyp) ligands to form a 2D $(4,4)_{core}(4,4)_{ligand}$ network. The occurrence of such a topological framework is likely to indicate the presence of

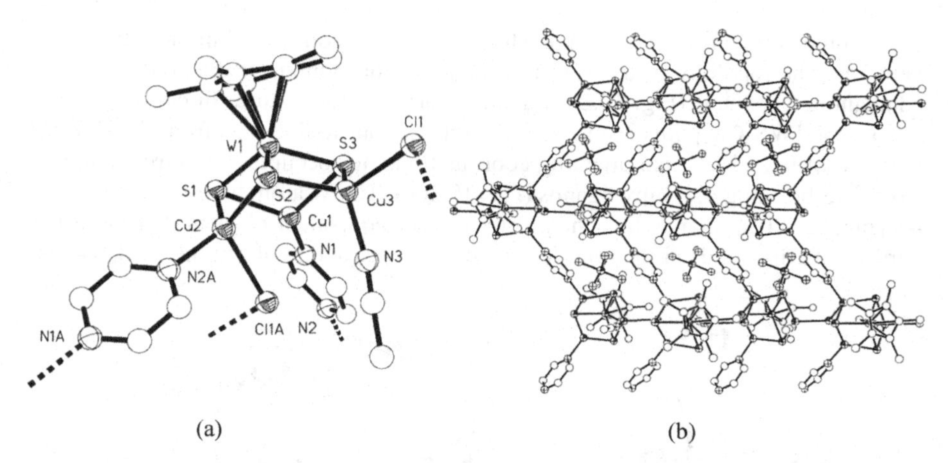

FIGURE 10.19 (a) Perspective view of the repeating unit of **34** with 50% thermal ellipsoids; (b) extended 2D layer network of **34** looking down the *b*-axis.

the four-connecting D_{4h} symmetrical H_2-tpyp (or Cu-tpyp) ligand. When these 2D layers are stacked along the *b*-axis, 1D channels are generated with a volume of 1423.9 Å^3 per unit cell (48.9% of the total cell volume). These channels are occupied by aniline and benzene solvent molecules.

Compound **34** consists of a repeating unit [Cp*WS_3Cu_3(MeCN)] (Fig. 10.19a) that serves as a three-connecting node to link its three equivalent ones by 1,4-pyz to form a zigzag chain extending along the *a*-axis, where the orientation of the [Cp*WS_3Cu_3(MeCN)] unit is alternating. Each chain is held together by μ-Cl bridges along the *c*-axis. The resulting 2D network forms a parallelogrammic mesh, and each cavity is filled by a $[PF_6]^-$ anion (Fig. 10.19b). The surfaces of each layer are covered with Cp* rings, and the thickness of the layers is estimated to be about 9.9 Å. The layers are separated by about 3.5 Å, and no interlayer bonding interactions are observed.

10.3.4 Assemblies of a Saddle-like MCu_4 Core

There are five saddle-like MCu_4-supported supramolecular compounds: [WS_4Cu_4(bpy)$_4$][$WS_4Cu_4I_4$(bpy)$_2$]·8MeCN (**29**) [14c], [WS_4Cu_4(bpy)$_4$][$WS_4Cu_4I_4$(bpy)$_2$]·2I_2 (**30**) [14c], [$WS_4Cu_4I_2$(dps)$_3$]·2MeCN (**31**) [14f], [WS_4Cu_4(dca)$_2$(dpds)$_2$]·Et_2O·2MeCN (**32**) [14f], and [WS_4Cu_4(dmpzm)$_2$(dca)$_2$] (**33**) [14g]. The structures of these five compounds consist of the same MCu_4 cores. However, they show different topological frameworks, due to the complicated multiconnecting nodes presented by the [MCu_4] core.

Compound **29** consists of one-fourth of the [WS_4Cu_4(4,4′-bipy)$_4$]$^{2+}$ dication and one-fourth of the [$WS_4Cu_4I_4$(4,4′-bipy)$_2$]$^{2-}$ dianion in its asymmetric unit. As indicated in Figure 10.20a, the cationic cluster is coordinated by eight bridging 4,4′-bipy ligands. Four pairs of ligands extend to four crystallographically equivalent clusters that lie at the corners of a distorted tetrahedron. The W atom lies on a site of

222 symmetry. Each cluster serves as a four-connecting node in an infinite cationic 3D network. The topology of the net is the same as that found for diamond, with the adamantine-type unit (Fig. 10.20b) as the characteristic feature. In contrast to the cation, the [WS_4Cu_4] cluster in the anion retains four iodide ions from the starting material of **9**, and has its remaining coordination sites occupied by nitrogen atoms from four bridging 4,4′-bipy ligands (Fig. 10.20c). The W atom here lies on a site of $\bar{4}$ symmetry. The ligands link the cluster to four equivalent clusters that lie at the vertices of a distorted tetrahedron. As with the cation, a diamond-type network results,

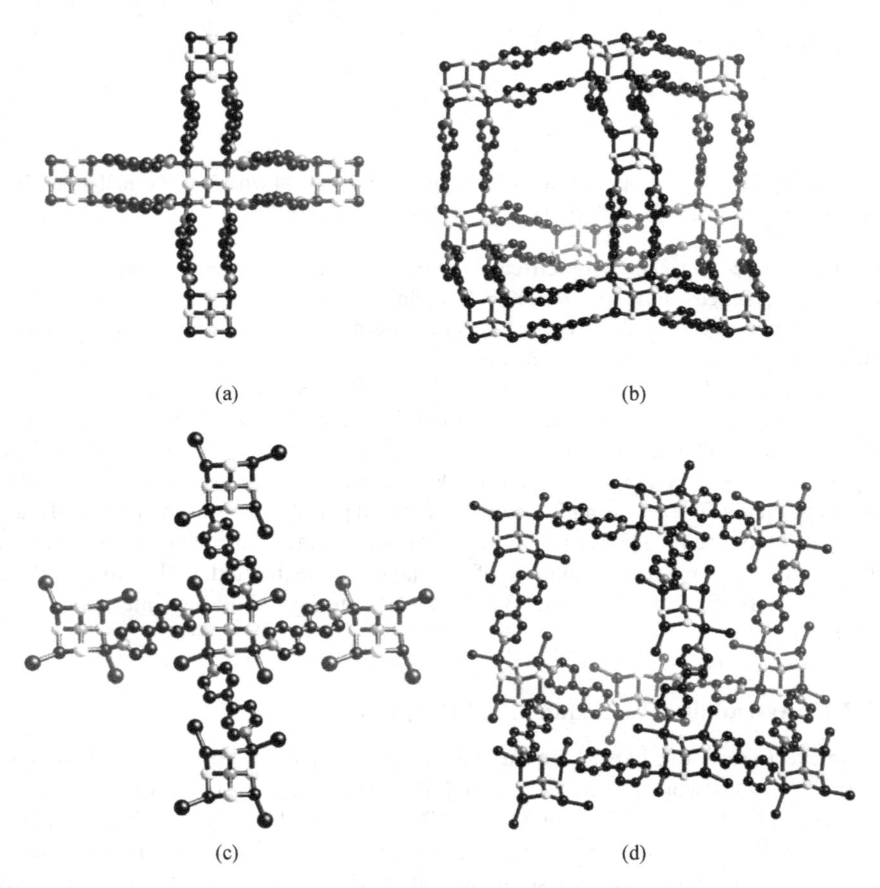

FIGURE 10.20 (a) Part of the cationic net within **29**, showing the double 4,4′-bipy connections between a central WS_4Cu_4 cluster and four equivalent clusters; (b) adamantane-type unit within the cationic diamond network of **29**; (c) part of the anionic net within **29** showing the single 4, 4′-bipy connections between a central $WS_4Cu_4I_4$ cluster and four equivalent clusters; (d) adamantane-type unit in the anionic network of **29**; (e) side-on view of the interpenetration of four adamantane units in **29**; (f) structure of **29** formed from the interpenetration of four diamond networks (two cationic and two anionic) with the channels that result from interpenetration extending along the *c*-axis.

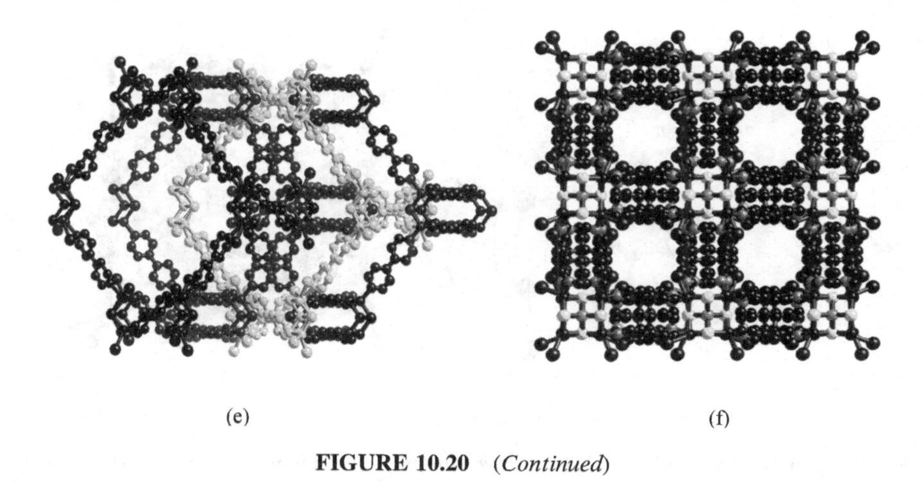

(e) (f)

FIGURE 10.20 (*Continued*)

but the bridges between clusters are now single 4,4′-bipy ligands (Fig. 10.20d). In the anion, the 4,4′-bipy ligand is inclined to the W · · · W vector of the linked clusters, whereas in the cation, the 4,4′-bipy ligands are parallel to the W · · · W vector. In the anionic network the 4,4′-bipy makes close contact with a coordinated iodide ion bound to an adjacent Cu. It is suggested that this steric clash leads to disorder and slight distortion of the 4,4′-bipy ligands belonging to the anionic network. As depicted in Figure 10.20e and 10.20f, interpenetration of the diamond nets is remarkable given that the cationic and anionic nets are so different. For interpenetration to occur, the spacing of network voids in one network needs to match perfectly with the spacing of network voids in the other, and as a consequence, interpenetration is normally observed in crystals containing identical open networks.

In **29**, the independent diamond-like networks stack evenly along a direction parallel to the unique *c*-axis in such a manner that channels of approximate circular cross section extending in the *c*-direction are produced. Allowing for the van der Waals size of the bpy atoms that line the surface of the channel, it is estimated that each channel has a volume of approximately 880 $\mathring{A}^3$ over the *c*-cell length of the tetragonal unit cell. This corresponds to an average cross-sectional area of about 31 $\mathring{A}^2$. The channel volume (van der Waals free space) represents 20% of the crystal volume. Thermogravimetric analysis (TGA) of the compound reveals a weight loss of 2.3% in the region of 30 to 110°C, consistent with the loss of a small amount of residual acetonitrile from the channels. The compound is stable up to about 180°C, after which a series of decomposition steps begin. When the crystals of **29** were dried in vacuo, they failed to diffract as single crystals. However, analysis of the powder diffraction pattern revealed that the framework remained intact upon loss of solvent from the channels.

Compound **30** was essentially identical to the parent crystal of **29** except for the fact that I_2 guest molecules were now included in the channels. Crystal structure analysis revealed that the iodine is associated with the coordinated iodide (Fig. 10.21a). The I_2 molecule slots neatly into a space between coordinated iodide

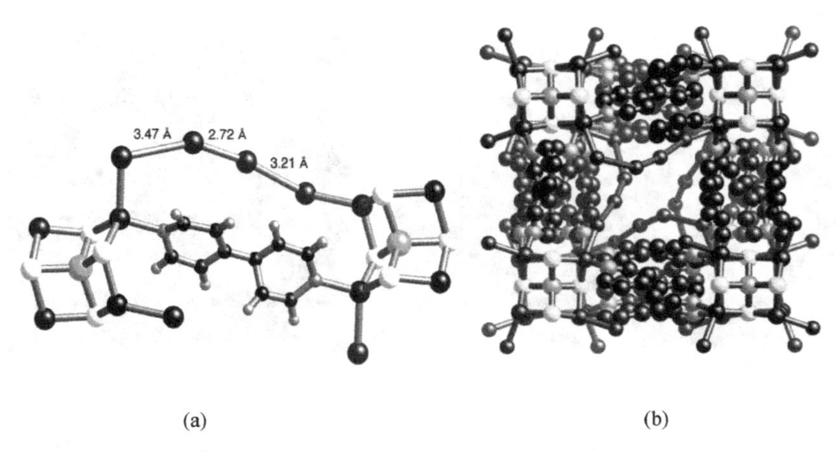

FIGURE 10.21 (a) I_4^{2-} bridge between two clusters of the anionic network of **30**; (b) single channel in **30** extending along the *c*-direction.

ions belonging to two bridged clusters of an anionic net to form a Cu–I–I–I–I–Cu bridge that runs parallel to the 4,4′-bipy bridge. As indicated in Fig. 10.21b, the iodine atoms in this bridge are not evenly spaced and in fact may be considered as an I_2 molecule forming interactions with a pair of coordinated iodide ions.

Compound **31** crystallizes in the orthorhombic space group *Ccc*2 and the asymmetric unit contains half of the $[WS_4Cu_4I_2(dps)_3]$ molecule and two MeCN solvated molecules. The $[WS_4Cu_4I_2(dps)_3]$ molecule consists of a saddle-shaped $[WS_4Cu_4]$ core structure similar to that of **9** (Fig. 10.22a). The W1 atom is located at a twofold axis, while the S4 atom is lying at another twofold axis. Four Cu atoms in **31** adopt a distorted tetrahedral coordination geometry. From a topological perspective, the saddle-shaped $[WS_4Cu_4]$ core in **31** acts as a four-connecting node, which is remarkably different from the tetrahedral four-connecting nodes observed in **29** and **30**. In **31**, the six flexible dsp ligands surround the $[WS_4Cu_4]$ core such that the mean deviation from the least-squares plane consisting of N2, N1B, N3C, N2A, N1D, and N3E atoms is about 1.16 Å. This suggests that the six coordination sites of the six dps ligands lie approximately on the plane highlighted in Figure 10.22b. The $[WS_4Cu_4]$ core in **31** under the presence of the six dps ligands may be viewed as a planar four-connecting node, which is unprecedented in cluster-based supramolecular chemistry. Such a four-connecting node no doubt prefers to generate a 2D network. Therefore, each $[WS_4Cu_4]$ core in **31** is interlinked by a pair of dps ligands to form a 1D $[WS_4Cu_4I_2(dps)_2]_n$ linear chain extended along the *a*-axis. Such a chain is further interconnected via dps ligands to form a 2D layer structure with a parallelogrammic mesh that extends along the *ab*-plane (Fig. 10.22c). The MeCN solvent molecules lie between layers, and there is no evident interaction between the layers and/or the solvated molecules.

Compound **32** crystallizes in the orthorhombic space group $C222_1$, and the asymmetric unit contains one-half of the $[WS_4Cu_4(dca)_2(dpds)_2]$ molecule, one-half of the Et_2O molecule, and two MeCN solvated molecules. The $[WS_4Cu_4]$ unit in **32** again retains the core structure of **9** (Fig. 10.23a). The central W1 atom is lying on a

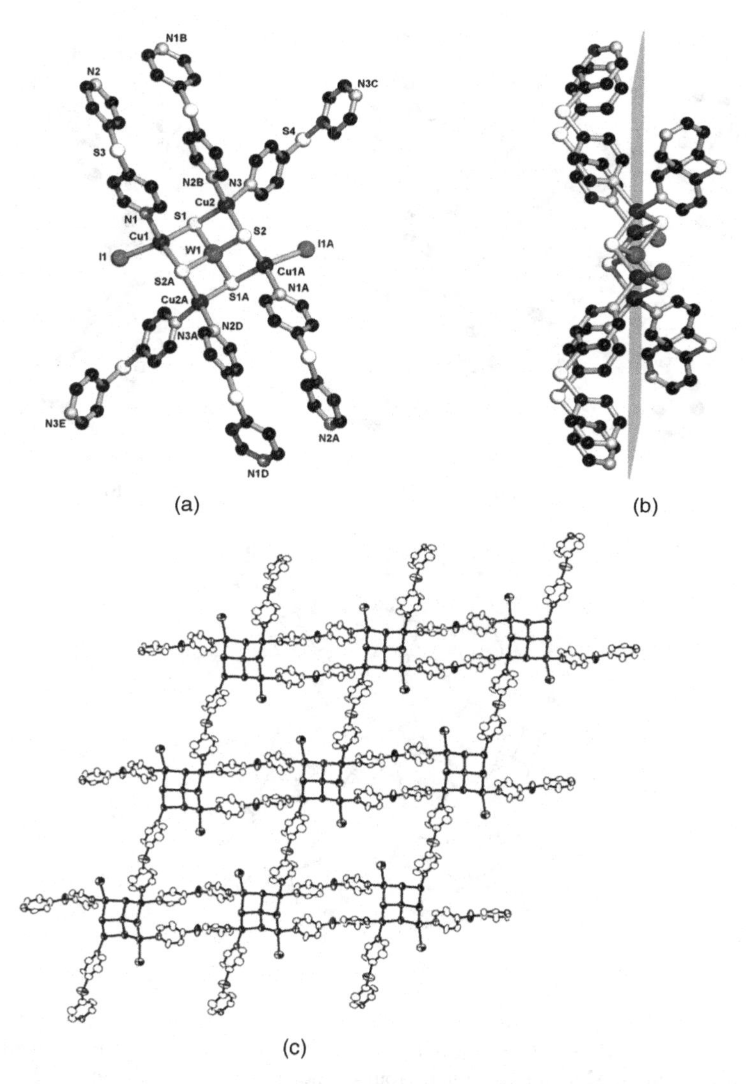

FIGURE 10.22 (a) Perspective view of the repeating unit of **31** with 50% thermal ellipsoids; (b) $[WS_4Cu_4]$ core in **31** acting as a planar four-connecting node; (c) extended network of **31** looking down the *a*-axis.

twofold axis. Each copper center also adopts a distorted tetrahedral coordination geometry. Topologically, each $[WS_4Cu_4]$ core in **32** also serves as a four-connecting node that is similar to that of **31**. The mean deviation from the least-squares plane containing N2, N5, N3B, N1B, N5A, N2A, N3C, and N1C atoms of the four flexible dpds and four dca ligands surrounding the $[WS_4Cu_4]$ core in **32** is about 1.17 Å, suggesting that the eight coordination sites of the dpds and dca ligands are located on approximately the same plane, as highlighted in Figure 10.23b. The $[WS_4Cu_4]$ core in

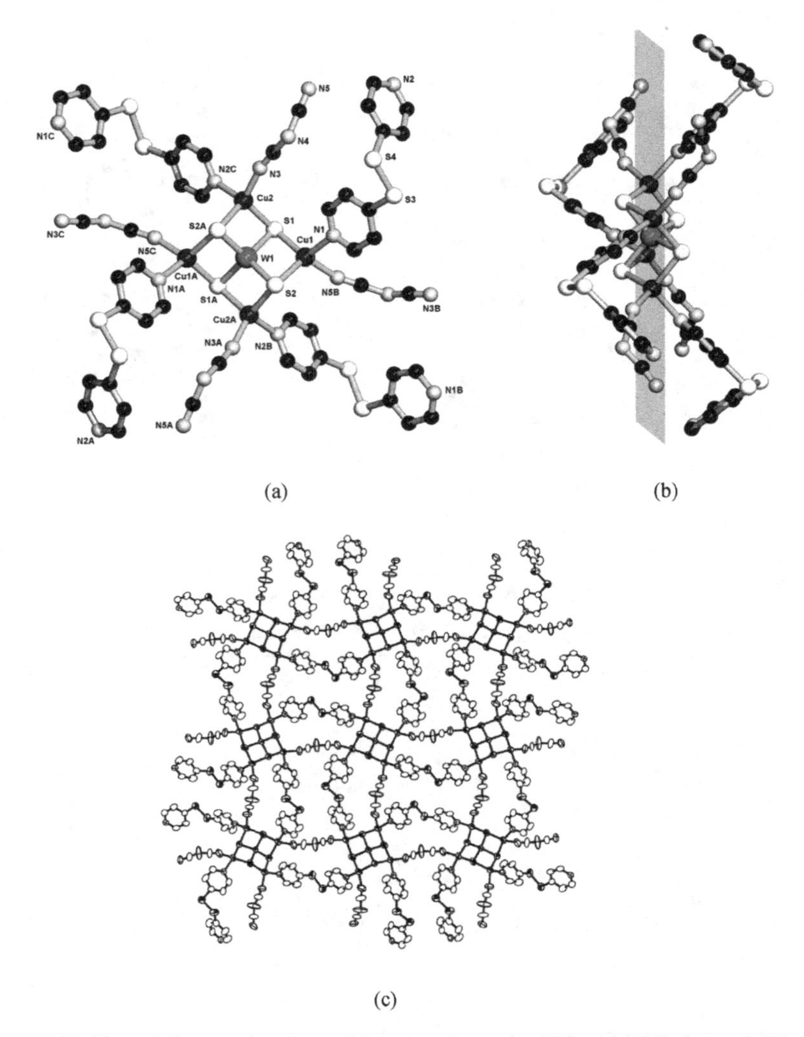

FIGURE 10.23 (a) Perspective view of the repeating unit of **32** with 50% thermal ellipsoids; (b) $[WS_4Cu_4]$ core in **32** acting as a planar four-connecting node; (c) extended network of **32** looking down the *a*-axis.

32 under the presence of four dpds and four dca ligands may be visualized as another type of planar four-connecting node. Each $[WS_4Cu_4]$ core in **32** is interconnected via four pairs of Cu–$\mu_{1,5}$-dca–Cu and Cu–dpds–Cu bridges to form a 2D $[WS_4Cu_4(dca)_2(dpds)_2]_n$ layer network extended along the *bc*-plane (Fig. 10.23c). The assembled 2D sheet is achiral, as the repeating $[WS_4Cu_4]$ core has four dca ligands and four dpds ligands of either *M*- or *P*-form coordinated symmetrically at four Cu centers. Between the layers lie Et_2O and MeCN solvent molecules, and there exists no evident interaction between the layers and the solvated molecules.

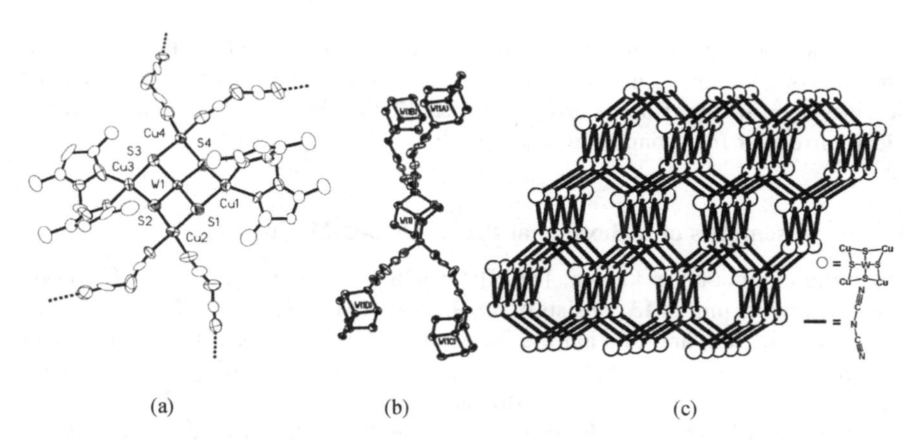

FIGURE 10.24 (a) Perspective view of the repeating units of **33** with thermal ellipsoids drawn at 50% probability level; (b) $[WS_4Cu_4]$ core serving as a tetrahedral four-connecting node, with dca extending to four crystallographically equivalent cluster cores (all dmpzm ligands are omitted for clarity); (c) 3D network topology of **33** viewed along a direction close to the *c*-axis. The hollow spheres represent the $[WS_4Cu_4]$ units, and the black sticks represent the connections provided by the dca anions.

Compound **33** crystallizes in the orthorhombic space group $Pna2_1$, and the asymmetric unit contains one discrete $[WS_4Cu_4(dmpzm)_2(dca)_2]$ unit (Fig. 10.24a). The saddle-like shape of the cluster is also preserved in **9**, but the Cu centers have two distinct coordination environments. All four Cu centers in **33** are coordinated by two μ_3-S atoms from $[WS_4]^{2-}$ anions and two N atoms from dmpzm or dca ligands, which are arranged in a distorted tetrahedral arrangement. Both dmpzm ligands adopt an extended and twisted *exo-anti* conformation and act as bidentate chelating ligands, and together they block four coordination sites on opposite sides of the WS_4Cu_4 core. The other four coordination sites are occupied by terminal nitrogen atoms of four bridging dca anions. Each dca ligand acts as a normal $\mu_{1,5}$ coordination mode. Two pairs of dca ligands extend to four crystallographically equivalent clusters that lie at the vertices of a very distorted tetrahedron (Fig. 10.24b). The separation of tungsten atoms linked through dca1 is 12.39 Å, while those linked by dca2 are 11.82 Å apart. Throughout the crystal the WS_4Cu_4 clusters are in four distinct orientations, but in each case the mean plane involving W and Cu atoms is very closely parallel to the *ac*-plane. From a topological perspective, each $[WS_4Cu_4]$ cluster core serves as a strongly distorted tetrahedral four-connecting node in an infinite 3D network. The topology of this net, although similar to that of the diamond net, which was found in **29** and **30**, is distinctly different. As with the diamond net, parallel (6,3) sheets are linked by connections to equivalent sheets, but the connections between sheets are different from that found in the diamond net. The 3D network in this case has the Schläfli symbol $(6^5 8^1)$. In contrast, the Schläfli symbol for diamond is (6^6). The difference in connectivity between the diamond net and this net may be attributed to the introduction of bulky dmpzm ligands coordinated at two copper sites of the $[WS_4Cu_4]$ cluster core in **9**. The topology of the extended network is represented in Figure 10.24c. The network has the same topology as a network within a copper coordination polymer described by Bourne and co-workers [23a], which is also based on geometrically irregular four-connecting nodes [23b].

10.3.5 Assemblies of a Hexagonal Prism-Shaped M_2Cu_4 Core

Only one compound is known: $[Et_4N]_2[MoOS_3Cu_2(\mu\text{-}CN)]_2 \cdot 2$aniline (**13**) [14i]. The asymmetric unit of **13** consists of one-half of a $[MoOS_3Cu_2(\mu\text{-}CN)]_2^{2-}$ dianion, one Et_4N^+ cation, and one aniline solvated molecule. The cluster dianion has a hexagonal prismatic $[MoOS_3Cu_2]_2$ cage structure with a crystallographic inversion center lying on the midpoint of the Mo1 and Mo1A contact (Fig. 10.25a). The cage has two butterfly-shaped $[MoOS_3Cu_2]$ fragments interconnected via four Cu–S bonds, and resembles those in $[M_2E_2S_6Cu_4(PR_3)_4]$ (M = Mo, W; E = O, S) [24]. Each $[MoOS_3Cu_2]$ fragment in **36** can be viewed as being built of the addition of one Cu^+ atom onto the $[MoOS_3Cu]$ core of **1**. In each $[MoOS_3Cu_2]$ fragment, Cu1 or Cu2 is coordinated by one bridging cyanide and three sulfur atoms, resulting in a tetrahedral coordination geometry. As the four coordination sites (C1, C1A, N1B, N1C) of the four cyanides surrounding the prismatic $[MoOS_3Cu_2]_2$ cage core lie nearly on the plane composed of the four Cu atoms (their mean deviation from this plane being 0.16 Å), this core serves topologically as an uncommon planar four-connecting node to link four equivalent nodes, forming a 2D (4,4)-layer network (extending along the *bc*-plane) with parallelogrammatic meshes (Fig. 10.25b). The networks are about 12 Å apart, with Et_4N^+ cations and aniline solvent molecules lying between the sheets. The terminal O atoms interact with amine groups of aniline to afford N–H···O intra- or intermolecular hydrogen bonds, thereby forming a 3D network.

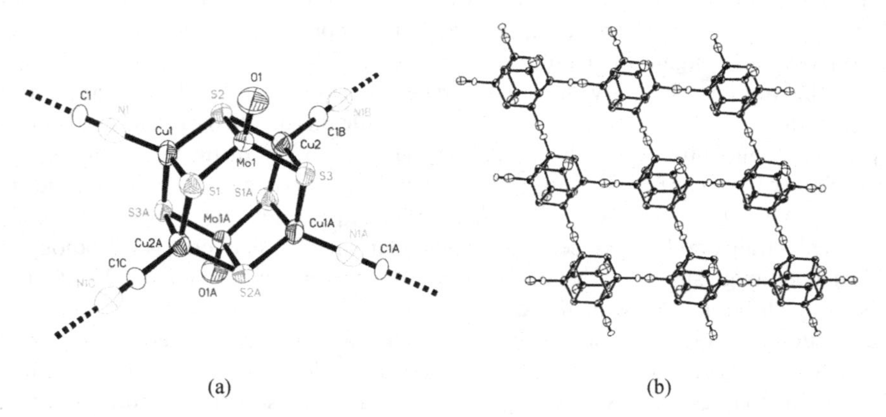

FIGURE 10.25 (a) Perspective view of the repeating unit of **13** with a labeling scheme and 50% thermal ellipsoids; (b) 2D (4,4) network of **13** extending along the *bc*-plane.

10.3.6 Assemblies of an Octahedral MCu_6 Core

Compounds **35** [18d] and **36** [18e] are isostructural and crystallize in the tetragonal space group $I\bar{4}$ 2d, and their asymmetric units contain one-fourth of $[MS_4Cu_6I_4(py)_6]$ molecule. Each $[MS_4Cu_6]$ cluster core (Fig. 10.26a) is coordinated by four pyridine ligands and six iodine atoms, of which four work as bridges (Fig. 10.26b). With the assistance of the four bridging iodides, this cluster core acts as a tetrahedral four-connecting node to link four crystallographic equivalent cores (Fig. 10.26c), forming a 3D diamond-like network (Fig. 10.26d).

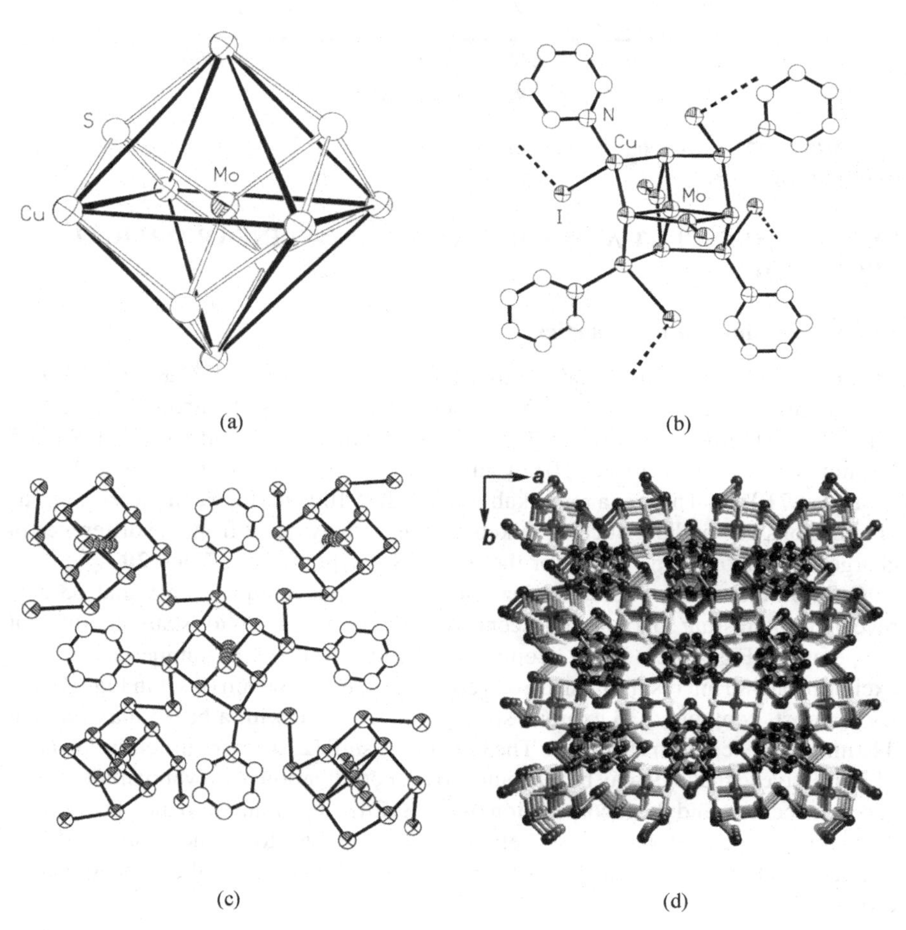

FIGURE 10.26 (a) Octahedral $[MS_4Cu_6]$ core in **35** or **36**; (b) perspective view of the repeating units in **35** or **36** with thermal ellipsoids drawn at the 50% probability level; (c) MS_4Cu_6 core serving as a tetrahedral four-connecting node with iodides extending to four crystallographically equivalent cluster cores; (d) 3D structure of **35** or **36** viewed along a direction close to the *c*-axis.

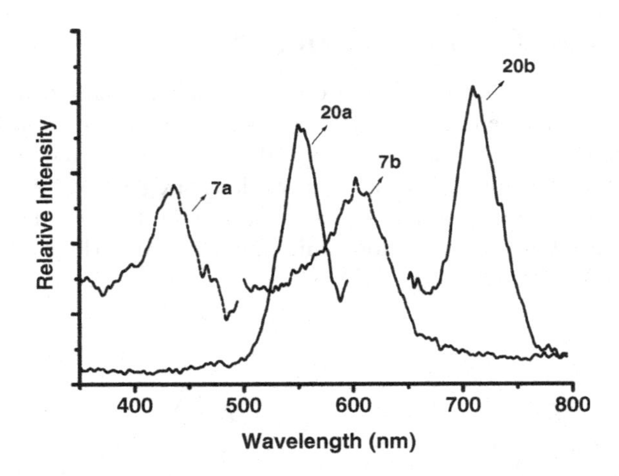

FIGURE 10.27 (a) Excitation and (b) emission spectra of **7** and **20** in the solid state at ambient temperatures.

10.4 LUMINESCENT AND THIRD-ORDER NONLINEAR OPTICAL PROPERTIES

10.4.1 Luminescent Properties

Preliminary photochemical and photophysical investigations of **7** and **20** revealed very interesting photoluminescent properties in the solid state at ambient temperature (Fig. 10.27) [14b]. Excitation of **7** at 435 nm resulted in a broad emission band at 600 nm, while that of **20** at 550 nm gave a relatively narrow emission with λ_{max} located at 710 nm. There is a remarkable red shift (>100 nm) between the excitations or emissions of **7** and **20**. Such a marked difference may result from the difference in charge and donor atoms found for the $[Cp^*WS_3Cu_3]$ cluster in **7** and **20**.

On the other hand, photochemical and photophysical studies on **9** and **33** also revealed interesting photoluminescent properties in the solid state at ambient temperature (Fig. 10.28) [14g]. Compound **9** shows very weak luminescence, with excitation at 480 nm resulting in a broad emission band at 660 nm (inset in Fig. 10.28). By contrast, compound **33** exhibits strong luminescence with a broad maximum at 440 nm upon excitation at 360 nm. The larger blue shift (220 nm) of the emission band of **33** relative to that of **9** may be attributed to the coordination of dicyanoamide anions at copper centers and the incorporation of a WS_4Cu_4 cluster unit into the 3D network. The emission peak at 440 nm is likely to be due to metal-to-ligand charge transfer, with an electron being transferred from the copper(I) center to the unoccupied π^* orbital of the CN group.

10.4.2 Third-Order Nonlinear Optical Properties

Because of the low solubility of the Mo(W)/Cu/S cluster-based supramolecular compounds in common organic solvents, only a limited number of the compounds

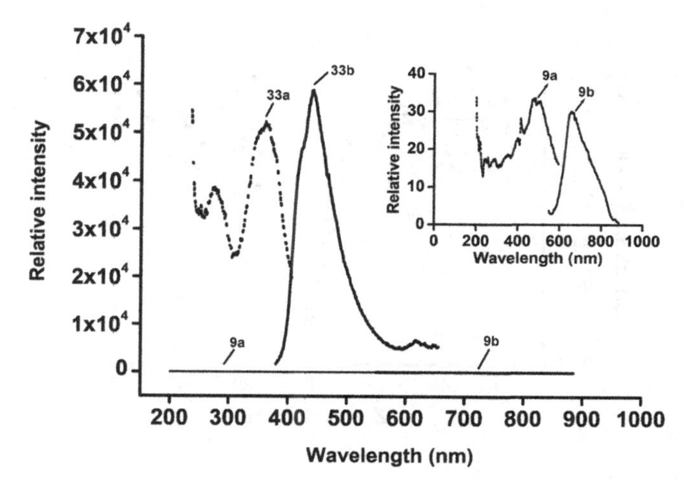

FIGURE 10.28 Solid-state excitation (dashed line) and emission (solid line) spectra of **9** and **33** at room temperature. *Inset*: Amplified excitation (dashed line) and emission (solid line) spectra of **9**.

noted above were investigated for their third-order nonlinear optical (NLO) properties. For example, the NLO absorption performances of the aniline solutions of **8** and **25–28** along with pure aniline solvent were evaluated by the Z-scan technique under an open-aperture configuration with 4.5-ns laser pulses at 532 nm (Fig. 10.29) [14h]. For the pure aniline solvent, the linear transmittance for aniline was found to be very large, amounting to 97.2%, which suggest that it has weak NLO refraction and absorption and is not enough to affect the NLO properties of complexes in aniline. Although the detailed mechanism is still unknown, the NLO absorption data obtained under the conditions used in this work can be evaluated by literature methods [25]. The nonlinear absorptive indexes α_2 for **8** and **25–28** were calculated to be $7.58 \times 10^{-11}\,\mathrm{mW}^{-1}$ (**8**), $6.86 \times 10^{-11}\,\mathrm{mW}^{-1}$ (**25**), $6.78 \times 10^{-11}\,\mathrm{mW}^{-1}$ (**26**), $6.11 \times 10^{-11}\,\mathrm{mW}^{-1}$ (**27**), and $7.79 \times 10^{-11}\,\mathrm{mW}^{-1}$ (**28**), respectively, implying that they have good NLO absorption properties. Intriguingly, **8** and **25–28** were not detected to have nonlinear refractive effects, implying that their nonlinear refractive effects are weak in our experimental conditions. In accordance with the α_2 values observed , the effective third-order susceptibility $\chi^{(3)}$ [26a] and the corresponding hyperpolarizability γ [26b] values for **8** and **25–28** were calculated to be 2.86×10^{-12} esu/3.09×10^{-29} esu (**8**), 2.78×10^{-12} esu/2.99×10^{-29} esu (**25**), 2.74×10^{-12} esu/2.96×10^{-29} esu (**26**), 2.47×10^{-12} esu/2.67×10^{-29} esu (**27**), and 3.15×10^{-12} esu/6.80×10^{-29} esu (**28**), respectively. These results showed that **8** and **25–28** possess relatively strong third-order optical nonlinearities. In addition, the third-order NLO properties for **28** were slightly better than those of **8** and **25–27**, which may be ascribed to the fact that **28** has relatively lower linear absorption at 532 nm than **8** and **25–27**.

Another example is involved in the third-order NLO properties of **1** and **13–14** [14i]. The electronic spectra of **1** and **13–14** showed relatively low linear absorption in

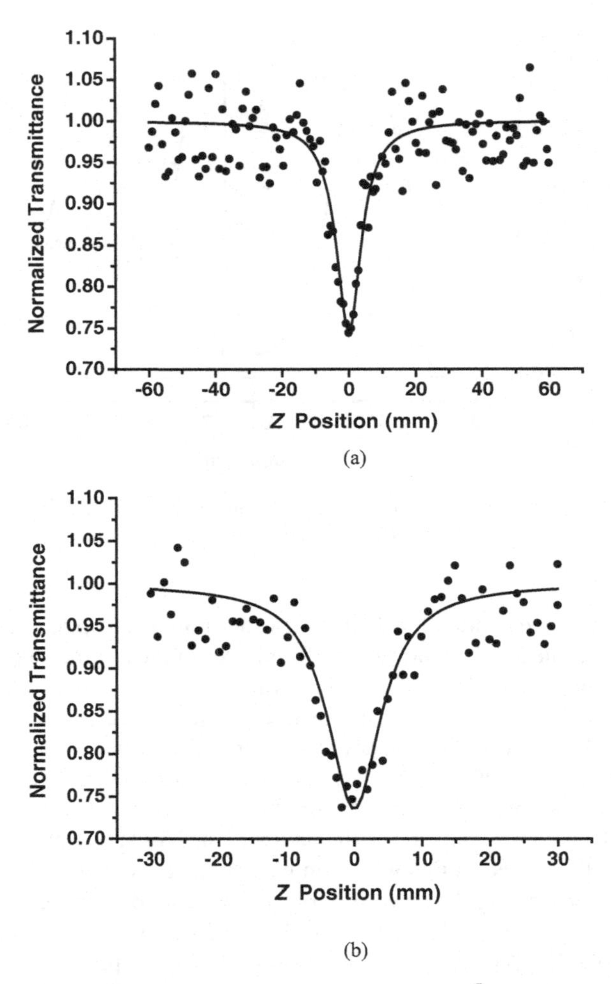

FIGURE 10.29 Z-scan data for aniline solutions of a 3.0×10^{-5} M for **8** and (b) 3.0×10^{-5} M for **28**. The Filled circles represent experimental data, and the solid curves are the theoretical fit.

532 nm, which promises low intensity loss and little temperature change by photon absorption during the NLO measurements. The nonlinear absorption performance of the solutions of these three compounds in DMF was evaluated by the Z-scan technique under an open-aperture configuration (Fig. 10.30). The open-aperture Z-scan experiments revealed that the NLO transmittances for **13** and **14** (84% and 46%, respectively) are much better than that of **1** (94%). Although the detailed mechanism is still unknown, the NLO absorption data obtained under the conditions used in this study can be evaluated from the methods reported. The nonlinear absorptive indexes α_2 for **1** and **13–14** were calculated to be 1.28×10^{-11} mW^{-1} (**1**), 5.10×10^{-11} mW^{-1} (**13**), and 1.58×10^{-10} mW^{-1} (**14**), indicating that **13** and **14** exhibit better nonlinear

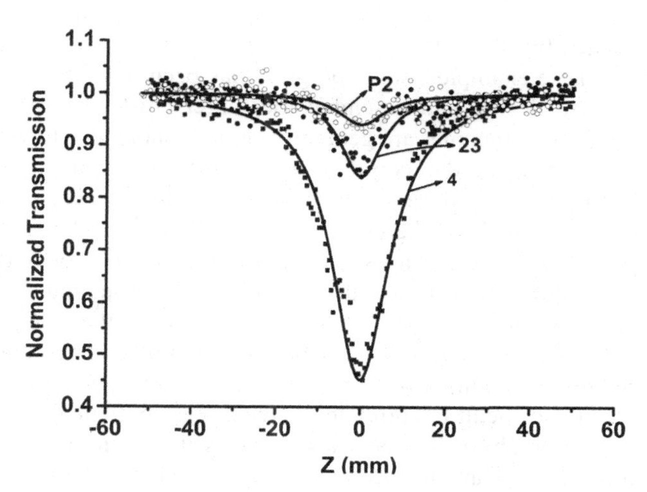

FIGURE 10.30 Z-scan data of the DMF solutions of **1** (5.49×10^{-4} M), **13** (5.00×10^{-4} M), and **14** (4.25×10^{-4} M), at 532 nm. The data were evaluated under an open-aperture configuration. The open circles (**1**), dots (**13**), and squares (**14**) are the experimental data, and the solid curves are the theoretical fit.

absorption properties than those of **1**. Intriguingly, compounds **1** and **13–14** were not detected to have nonlinear refractive effects, implying that their nonlinear refractive effects are weak under our experimental conditions. In accordance with the α_2 values observed, the effective third-order susceptibilities $\chi^{(3)}$ and the corresponding hyperpolarizability γ for **1** and **13–14** were calculated to be 1.66×10^{-11} esu/1.55×10^{-29} esu (**1**), 6.63×10^{-11} esu/6.78×10^{-29} esu (**13**), and 2.06×10^{-10} esu/2.47×10^{-28} esu (**14**). These results showed that **13** and **14** possess better third-order optical nonlinearities than those of **1**. It is noted that when **1** was converted into **13** and **14**, an obvious increase in the γ value was found (i.e., the γ values of **13** and **14** are 4.4 and 15.9 times larger than that of **1**, respectively). Structurally, when one or two Cu^+ atoms were incorporated into the [$MoOS_3Cu$] core of **1**, a [$MoOS_3Cu_2$]$_2$ core for **13** or a [$MoOS_3Cu_3$] core for **14** is formed. According to the previous results [11h,12c], clusters containing [$MoOS_3Cu_3$] cores always showed high NLO activity, while those containing [$MoOS_3Cu$] cores have low NLO activity. Thus, such an increased order of γ values among **1** and **13–14** may be reasonable, and their NLO performances are likely to be cluster core–dependent. Besides this, the enhanced NLO performances of **13** and **14** may be also ascribed to the skeletal expansion of these fragments and the formation of cluster-based supramolecular arrays [12c,26d, 26f].

10.5 CONCLUSIONS

In the work reported here, we demonstrated our efforts to explore the rational design and construction of Mo(W)/Cu/S cluster-based supramolecular arrays. A total of 24 molybdenum (tuingsten)–copper–sulfur cluster-based supramolecular arrays were isolated from reactions of 12 preformed clusters with various bridging ligands. Among these examples, some progress has been made toward an understanding of how to design and assemble the target cluster-based compounds. However, the structure outcomes are generally unpredictable, due to the numerous permutations arising from the coordination geometries of the cluster cores, bonding preferences of the bridging ligands, counterion effects, and synthetic routes. Other factors, such as the solvents, the symmetries of the ligands, the reaction temperatures, and the shapes and sizes of the bridging ligands, may also be responsible for the composition and structure of the compounds generated. In addition, several examples showed that upon assembly, the chemical and/or physical properties of the resulting cluster-based supramolecular compounds were greatly changed relative to those of their cluster precursors. The problem is that at the present time it is difficult to correlate the resulting topological structures with their chemical and/or physical properties, such as their luminescent and NLO properties. Although we have to face these tough problems, the beautiful supramolecular chemistry presented in this chapter so activates and inspires us that a much broader and richer chemistry, with the potential to be rationally designed and assembled, may be anticipated.

Acknowledgments

This work was supported by the National Natural Science Foundation of China (20525101 and 20871088), the State Key Laboratory of Structural Chemistry of the Fujian Institute of Research on the Structure of Matter, the State Key Laboratory of Organometallic Chemistry of Shanghai Institute of Organic Chemistry (08–25), the Qin-Lan Project of Jiangsu Province, the "Soochow Scholar" Program, and the Program for Innovative Research Team of Soochow University.

REFERENCES

1. (a) Lehn, J.-M. *Supramolecular Chemistry Concepts and Perspectives*, VCH, Weinheim, Jermany. 1995, p. 139. (b) Stang, P. J.; Olenyuk, B. *Acc. Chem. Res.* **1997**, *30*, 502. (c) Piguet, C.; Bemardinelli, G.; Hopfgartner, G. *Chem. Rev.* **1997**, *97*, 2005. (d) Robson, R.In *Comprehensive Supramolecular Chemistry*, Vol. 6, Atwood, J. L., Davies, J. E. D., MacNicol, D. D., Vögtle, F., Lehn, J.-M., Eds., Pergamon Press, Oxford, UK, 1997, p. 733. (e) Yaghi, O. M.; Li, H. L.; Davis, C.; Richardson, D.; Groy, T. L. *Acc. Chem. Res.* **1998**, *31*, 474. (f) Caulder, D. L.; Raymond, K. N. *Acc. Chem. Res.* **1999**, *32*, 975. (g) Swiegers, G. F.; Malefeste, T. J. *Chem. Rev.* **2000**, *100*, 3483. (h) Moulton, B.; Zaworotko, M. J. *Chem. Rev.* **2001**, *101*, 1629. (i) Chisholm, M. H.; MacIntosh, A. M. *Chem. Rev.* **2005**, *105*, 2949.
2. (a) Batten, S. R.; Robson, R. *Angew. Chem., Int. Ed.* **1998**, *37*, 1460. (b) Fujita, M. *Chem. Soc. Rev.* **1998**, *27*, 417. (c) Blake, A. J.; Champness, N. R.; Hubberstey, P.; Li, W. S.; Withersby, M. A.; Schröder, M. *Coord. Chem. Rev.* **1999**, *183*, 117. (d) Hagrman, P. J.; Hagrman, D.; Zubieta, J. *Angew. Chem., Int. Ed.* **1999**, *38*, 2638. (e) Leininger, S.; Olenyuk, B.; Stang, P. J. *Chem. Rev.* **2000**, *100*, 853. (f) Holiday, B. J.; Mirkin, C. A.

Angew. Chem., Int. Ed. **2001**, *40*, 2022. (g) Seidel, S. R.; Stang, P. J. *Acc. Chem. Res.* **2002**, *35*, 972–983.

3. (a) Robson, R.; Hoskins, B. F. *J. Am. Chem. Soc.* **1990**, *112*, 1546. (b) Blake, A. J.; Champness, N. R.; Khlobystov, A. N.; Parsons, S.; Schröder, M. *Angew. Chem., Int. Ed.* **2000**, *39*, 2317. (c) Campos-Femández, C. S.; Clérac, R.; Kooman, J. M.; Russell, D. H.; Dunbar, K. R. *J. Am. Chem. Soc.* **2001**, *123*, 773. (d) Xiong, R. G.; You, X. Z.; Abrahams, B. F.; Xue, Z. L.; Che, C. M. *Angew. Chem., Int. Ed.* **2001**, *40*, 4422. (e) Galán-Mascarós, J. R.; Dunbar, K. R. *Angew. Chem., Int. Ed.* **2003**, *42*, 2289. (f) Berlinguette, C. P.; Dragulescu-Andrast, A.; Sieber, A.; Galán-Mascarós, J. R.; Güdel, H. U.; Achim, C.; Dunbar, K. R. *J. Am. Chem. Soc.* **2004**, *126*, 6222. (g) Jiang, H.; Lin, W. B. *J. Am. Chem. Soc.* **2004**, *126*, 7426. (h) Huang, X. C.; Zhang, J. P.; Chen, X. M. *J. Am. Chem. Soc.* **2004**, *126*, 13218.
4. (a) Cotton, F. A.; Lin, C.; Murillo, C. A. *Acc. Chem. Res.* **2001**, *34*, 750. (b) Cotton, F. A.; Lin, C.; Murillo, C. A. *Inorg. Chem.* **2001**, *40*, 5886. (c) Cotton, F. A.; Lin, C.; Murillo, C. A. *Chem. Commun.* **2001**, 11. (d) Cotton, F. A.; Dikarev, E. V.; Petrukhina, M. A.; Schmitz, M.; Stang, P. J. *Inorg. Chem.* **2002**, *41*, 2903. (e) Cotton, F. A.; Donahue, J. P.; Lichtenberger, D. L.; Murillo, C. A.; Villagrán, D. *J. Am. Chem. Soc.* **2005**, *127*, 10808–10809. (f) Cotton, F. A.; Murillo, C. A.; Villagrán, D.; Yu, R. M. *J. Am. Chem. Soc.* **2006**, *128*, 3281–3290.
5. (a) Vodak, D. T.; Braun, M. E.; Kim, J.; Eddaoudi, M.; Yaghi, O. M. *Chem. Commun.* **2001**, 2534. (b) Eddaoudi, M.; Moler, D. B.; Li, H. L.; Chen, B.; Reineke, T. M.; O'Keeffe, M.; Yaghi, O. M. *Acc. Chem. Res.* **2001**, *34*, 319. (c) Rosi, N. L.; Eckert, J.; Eddaoudi, M.; Vadak, D. T.; Kim, J.; O'Keeffe, M.; Yaghi, O. M. *Science* **2003**, *300*, 1127.
6. (a) Ouyang, X.; Campana, C.; Dunbar, K. R. *Inorg. Chem.* **1996**, *35*, 7188. (b) Miyasaka, H.; Campos-Fernández, C. S.; Galán-Mascarós, J. R.; Dunbar, K. R. *Inorg. Chem.* **2000**, *39*, 5870. (c) Miyasaka, H.; Clérac, R.; Campos-Fernández, C. S.; Dunbar, K. R. *Inorg. Chem.* **2001**, *40*, 1663. (d) Conan, F.; Gall, B. L.; Berbaol, J.-M.; Stang, S. L.; Sala-Pala, J.; Mest, Y. L.; Bacsa, J.; Ouyang, X.; Dunbar, K. R.; Campana, C. F. *Inorg. Chem.* **2004**, *43*, 3673. (e) Chifotides, H.; Dunbar, K. R. *Acc. Chem. Res.* **2005**, *38*, 146–156.
7. (a) Zheng, Z. P. *Chem. Commun.* **2001**, 2521. (b) Roland, B. K.; Carter, C.; Zheng, Z. P. *J. Am. Chem. Soc.* **2002**, *124*, 6234. (c) Roland, B. K.; Selby, H. D.; Carducci, M. D.; Zheng, Z. P. *J. Am. Chem. Soc.* **2002**, *124*, 3222. (d) Selby, H. D.; Roland, B. K.; Zheng, Z. P. *Acc. Chem. Res.* **2003**, *36*, 935. (e) Roland, B. K.; Flora, W. H.; Selby, H. D.; Armstrong, N. R.; Zheng, Z. P. *J. Am. Chem. Soc.* **2006**, *128*, 6620–6625.
8. (a) Long, J. R.; McCarty, L. S.; Holm, R. H. *J. Am. Chem. Soc.* **1996**, *118*, 4603. (b) Beauvais, L. G.; Shores, M. P.; Long, J. R. *J. Am. Chem. Soc.* **2000**, *122*, 2763. (c) Bennett, M. V.; Beauvais, L. G.; Shores, M. P.; Long, J. R. *J. Am. Chem. Soc.* **2001**, *123*, 8022. (d) Tulsky, E. G.; Crawford, N. R. M.; Baudron, S. A.; Batail, P.; Long, J. R. *J. Am. Chem. Soc.* **2003**, *125*, 15543.
9. (a) Abrahams, B. F.; Egan, S. J.; Robson, R. *J. Am. Chem. Soc.* **1999**, *121*, 3535. (b) Abrahams, B. F.; Haywood, M. G.; Robson, R. *Chem. Commun.* **2004**, 938. (c) Yan, B. B.; Zhou, H. J.; Lachgar, A. *Inorg. Chem.* **2003**, *42*, 8818. (d) Yan, Z. H.; Day, C. S.; Lachgar, A. *Inorg. Chem.* **2005**, *44*, 4499.
10. (a) Bain, R. L.; Shriver, D. F.; Ellis, D. E. *Inorg. Chim. Acta* **2001**, *325*, 171–174. (b) Jin, S.; DiSalvo, F. J. *Chem. Mater.* **2002**, *14*, 3448–3457. (c) Naumov, N. G.; Cordier, S.; Perrin, C. *Angew. Chem., Int. Ed.* **2002**, *41*, 3002–3004. (d) Yan, B. B.; Zhou, H. J.; Lachgar, A. *Inorg. Chem.* **2003**, *42*, 8818–8822. (e) Mironov, Y. V.; Naumov, N. G.; Brylev, K. A.;

Efremova, Q. A.; Fedorov, V. E.; Hegetschweiler, K. *Angew. Chem. Int. Ed.* **2004**, *43*, 1297–1300. (f) Yan, Z. H.; Day, C. S.; Lachgar, A. *Inorg. Chem.* **2005**, *44*, 4499–4505. (g) Wang, J. Q.; Ren, C. X.; Jin, G. X. *Chem. Commun.* **2005**, 4738–4740. (h) Wang, J. Q.; Ren, C. X.; Weng, L. H.; Jin, G. X. *Chem. Commun.* **2006**, 162–164.

11. (a) Müller, A.; Diemann, E.; Jostes, R.; Bögge, H. *Angew. Chem., Int. Ed.* **1981**, *20*, 934–955. (b) Müller, A.; Bögge, H.; Schimanski, U.; Penk, M.; Nieradzik, K.; Dartmann, M.; Krickemeyer, E.; Schimanski, J.; Römer, C.; Römer, M.; Dornfeld, H.; Wienböker, U.; Hellmann, W. *Monatsh. Chem.* **1989**, *120*, 367–391. (c) Howard, K. E.; Rauchfuss, T. B.; Rheingold, A. L. *J. Am. Chem. Soc.* **1986**, *108*, 297–299. (d) Ansari, M. A.; Ibers, J. A. *Coord. Chem. Rev.* **1990**, *100*, 223–266. (e) Jeannin, Y.; Sécheresse, F.; Bernés, S.; Robert, F. *Inorg. Chim. Acta* **1992**, *198–200*, 493–505. (f) Wu, X. T.; Chen, P. C.; Du, S. W.; Zhu, N. Y.; Lu, J. X. *J. Cluster Sci.* **1994**, *5*, 265–285. (g) Holm, R. H. *Pure Appl. Chem.* **1995**, *67*, 217–224. (h) Hou, H. W.; Xin, X. Q.; Shi, S. *Coord. Chem. Rev.* **1996**, *153*, 25–56. (i) Wu, D. X.; Hong, M. C.; Cao, R.; Liu, H. Q. *Inorg. Chem.* **1996**, *35*, 1080–1082. (j) Coucouvanis, D. *Adv. Inorg. Chem.* **1998**, *45*, 1–73.

12. (a) Chan, C. K.; Guo, C. X.; Wang, R. J.; Mak, T. C. W.; Che, C. M. *J. Chem. Soc., Dalton Trans.* **1995**, 753–757. (b) Zheng, H. G.; Ji, W.; Low, M. L. K.; Sakane, G.; Shibahara, T.; Xin, X. Q. *J. Chem. Soc., Dalton Trans.* **1997**, 2375–2362. (c) Shi, S. In *Optoelectronic Properties of Inorganic Compounds*, Roundhill, D. M., Fackler, J. P., Jr., Eds., Plenum Press, New York, 1998, pp. 55–105. (d) Zhang, C.; Song, Y. L.; Xu, Y.; Fun, H. K.; Fang, G. Y.; Wang, Y. X.; Xin, X. Q. *J. Chem. Soc., Dalton Trans.* **2000**, 2823–2829. (e) Che, C. M.; Xia, B. H.; Huang, J. S.; Chan, C. K.; Zhou, Z. Y.; Cheung, K. K. *Chem. Eur. J.* **2001**, *7*, 3998–4006. (f) Coe, B. J. In *Comprehensive Coordination Chemistry II*, Vol. 9, McCleverty, J. A., Meyer, T. J., Eds., Elsevier Pergamon, Oxford, UK, 2004, pp. 621–687.(g) Wu, X. T. *Inorganic Assembly Chemistry*, Science Press, Beijing, 2004, pp. 1–179.

13. (a) Shi, S.; Ji, W.; Tang, S. H.; Lang, J. P.; Xin, X. Q. *J. Am. Chem. Soc.* **1994**, *116*, 3615–3616. (b) Lang, J. P.; Xin. X. Q. *J. Solid State Chem.* **1994**, *108*, 118–127. (c) Shi, S.; Ji, W.; Lang, J. P.; Xin, X. Q. *J. Phys. Chem.* **1994**, *98*, 3570–3572. (d) Lang, J. P.; Tatsumi, K.; Kawaguchi, H.; Lu, J. M.; Ge, P.; Ji, W.; Shi, S. *Inorg. Chem.* **1996**, *35*, 7924–7927. (e) Lang, J. P.; Kawaguchi, H.; Ohnishi, S.; Tatsumi, K. *Chem. Commun.* **1997**, 405–406. (f) Lang, J. P.; Tatsumi, K. *Inorg. Chem.* **1998**, *37*, 160–162. (g) Lang, J. P.; Tatsumi, K. *Inorg. Chem.* **1998**, *37*, 6308–6316. (h) Yu, H.; Xu, Q. F.; Sun, Z. R.; Ji, S. J.; Chen, J. X.; Liu, Q.; Lang, J. P.; Tatsumi, K. *Chem. Commun.* **2001**, 2614–2615. (i) Lang, J. P.; Ji, S., J.; Xu, Q. F.; Shen, Q.; Tatsumi, K. *Coord. Chem. Rev.* **2003**, *241*, 47–60. (j) Chen, J. X.; Xu, Q. F.; Zhang, Y.; Chen, Z. N.; Lang, J. P. *Eur. J. Inorg. Chem.* **2004**, 4247–4252. (k) Ren, Z. G.; Li, H. X.; Liu, G. F.; Zhang, W. H.; Lang, J. P.; Zhang, Y.; Song, Y. L. *Organometallics* **2006**, *25*, 4351–4357. (l) Yang, J. Y.; Gu, J. H.; Song, Y. L.; Shi, G.; Wang, Y. X.; Zhang, W. H.; Lang, J. P. *J. Phys. Chem. B* **2007**, *111*, 7987–7993. (m) Wang, J.; Sun, Z. R.; Deng, L.; Wei, Z. H.; Zhang, W. H.; Zhang, Y.; Lang, J. P. *Inorg. Chem.* **2007**, *46*, 11381–11389.

14. (a) Lang, J. P.; Kawaguchi, H.; Tatsumi, K. *Chem. Commun.* **1999**, 2315–2316. (b) Lang, J. P.; Xu, Q. F.; Chen, Z. N.; Abrahams, B. F. *J. Am. Chem. Soc.* **2003**, *125*, 12682–12683. (c) Lang, J. P.; Xu, Q. F.; Yuan, R. X.; Abrahams, B. F. *Angew. Chem., Int. Ed.* **2004**, *43*, 4741–4745. (d) Lang, J. P.; Jiao, C. M.; Qiao, S. B.; Zhang, W. H.; Abrahams, B. F. *Inorg. Chem.* **2005**, *44*, 3664–3668. (e) Xu, Q. F.; Chen, J. X.; Zhang, W. H.; Ren, Z. G.; Li, H. X.; Zhang, Y.; Lang, J. P. *Inorg. Chem.* **2006**, *45*, 4055–4064. (f) Lang, J. P.; Xu, Q. F.; Zhang, W. H.; Li, H. X.; Ren, Z. G.; Chen, J. X.; Zhang, Y. *Inorg.*

Chem. **2006**, *45*, 10487–10496. (g) Ding, N. N.; Zhang, W. H.; Li, H. X.; Ren, Z. G.; Lang, J. P.; Zhang, Y.; Abrahams, B. F. *Inorg. Chem. Commun.* **2007**, *10*, 623–626. (h) Zhang, W. H.; Song, Y. L.; Ren, Z. G.; Li, H. X.; Li, L. L.; Zhang, Y.; Lang, J. P. *Inorg. Chem.* **2007**, *46*, 6647–6660. (i) Zhang, W. H., Song, Y. L.; Zhang, Y.; Lang, J. P. *Cryst. Growth Des.* **2008**, *8*, 253–258. (j) Zhang, W. H.; Lang, J. P.; Zhang, Y.; Abrahams, B. F. *Cryst. Growth Des.* **2008**, *8*, 399–401.

15. (a) Holm, R. H. *Adv. Inorg. Chem.* **1992**, *38*, 1–71. (b) Stiefel, E. I.; Coucouvanis, D.; Newton, W. E., Eds., *Molybdenum Enzymes, Cofactors and Model Systems*, ACS Symposium Series 535, American Chemical Society, Washington, DC, 1993. (c) Stiefel, E. I.; Matsumoto, K., Eds., *Transition Metal Sulfur Chemistry, Biological and Industrial Significance*, ACS Symposium Series 653, American Chemical Society, Washington, DC, 1996. (d) George, G. N.; Pickering, I. J.; Yu, Y. E.; Prince, R. C.; Bursakov, S. A.; Gavel, O. Y.; Moura, I.; Moura, J. J. G. *J. Am. Chem. Soc.* **2000**, *122*, 8321–8322. (e) Dobbek, H.; Gremer, L.; Kiefersauer, R.; Huber, R.; Meyer, O. *Proc. Natl. Acad. Sci. USA* **2002**, *99*, 15971–15976.
16. (a) Chianelli, R. R.; Picoraro, T. A.; Halbert, T. R.; Pan, W. H.; Stiefel, E. I. *J. Catal.* **1984**, *86*, 226–230. (b) Curtis, M. D. *J. Cluster Sci.* **1996**, *7*, 247–262.(c) Adams, R. D.; Cotton, F. A., Eds., *Catalysis by Di– and Polynuclear Metal Cluster Complexes*, Wiley-VCH, New York, 1998. (d) Hernandez–Molina, R.; Sykes, A. G. *J. Chem. Soc., Dalton Trans.* **1999**, 3137–3148. (e) Hidai, M.; Kuwata, S.; Mizobe, Y. *Acc. Chem. Res.* **2000**, *33*, 46–52.
17. Gheller, S. F.; Hambley, T. W.; Rodger, J. R.; Brownlee, R. T. C.; O'Connor, M. J.; Snow, M. R.; Wedd, A. G. *Inorg. Chem.* **1984**, *23*, 2519.
18. (a) Hou, H. W.; Long, D. L.; Xin, X. Q.; Huang, X. Y.; Kang, B. S.; Ge, P.; Ji, W.; Shi, S. *Inorg. Chem.* **1996**, *35*, 5363. (b) Lang, J. P.; Bao, S. A.; Zhu, H. Z.; Xin, X. Q.; Cai, J. H.; Weng, L. H.; Hu, Y. H.; Kang, B. S. *Chem. J. Chin. Univ.* **1992**, *13*, 889. (c) Lang, J. P.; Bian, G. Q.; Cai, J. H.; Kang, B. S.; Xin, X. Q. *Transition Met. Chem.* **1995**, *20*, 376. (d) Lang, J. P.; Xin, X. Q. *Acta Chim. Sin.* **1996**, *54*, 461–467. (e) Lang, J. P.; Zhou, W. Y.; Xin, X. Q.; Cai, J. H.; Kang, B. S.; Yu, K. B. *Polyhedron* **1993**, *12*, 1647–1653.
19. Julia, S.; Sala, P.; del Mazo, J.; Sancho, M.; Ochoa, C.; Elguero, J.; Fayet, J. P.; Vertut, M. C. *J. Heterocycl. Chem.* **1982**, *19*, 1141.
20. (a) Zheng, H. G.; Zhang, C.; Chen, Y.; Xin, X. Q.; Leung, N. H. *Synth. React. Inorg. Met., Org. Chem.* **2000**, *30*, 349. (b) Hou, H. W.; Ang, H. G.; Ang, S. G.; Fan, Y. T.; Low, M. K. M.; Ji, W.; Lee, Y. W. *Inorg. Chim. Acta* **2000**, *299*, 147.
21. (a) Raper, E. S.; Clegg, W. *Inorg. Chim. Acta* **1991**, *180*, 239–244. (b) Goher, M. A. S.; Yang, Q. C.; Mak, T. C. W. *Polyhedron* **2000**, *19*, 615–621.
22. (a) Jessiman, A. S.; MacNicol, D. D.; Mallinson, P. R.; Vallance, I. *J. Chem. Soc., Chem. Commun.* **1990**, 1619–1620. (b) Ke, Y. X.; Collins, D. J.; Sun, D. F.; Zhou, H. C. *Inorg. Chem.* **2006**, *45*, 1897–1899. (c) Czugler, M.; Weber, E.; Párkányi, L.; Korkas, P. P.; Bombicz, P. *Chem. Eur. J.* **2003**, *9*, 3741–3747. (d) Saha, B. K.; Nangia, A. *Cryst. Growth Des.* **2007**, *7*, 393–401.
23. (a) Bourne, S. A.; Lu, J.; Moultan, B.; Zaworotko, M. J. *Chem Commun.* **2001**, 861. (b) Batten, S. R. *CrystEngComm* **2001**, *3*, 67.
24. Müller, A.; Bögge, H.; Hwang, T. K. *Inorg. Chim. Acta* **1980**, *39*, 71–74.
25. (a) Sherk-Bahae, M.; Said, A. A.; Wei, T. H.; Hagan, D. J.; Van Stryland, E. W. *IEEE J. Quantum Electron.* **1990**, *26*, 760–769. (b) Sherk-Bahae, M.; Said, A. A.; Van Stryland, E. W. *Opt. Lett.* **1989**, *14*, 955–957.

26. (a) Yang, L.; Dorsinville, R.; Wang, Q. Z.; Ye, P. X.; Alfano, R. R.; Zamboni, R.; Taliani, C. *Opt. Lett.* **1992**, *17*, 323–325. (b) Chen, Z. R.; Hou, H. W.; Xin, X. Q.; Yu, K. B.; Shi, S. *J. Phys. Chem.* **1995**, *99*, 8717–8721. (c) Zhang, C.; Song, Y. L.; Jin, G. C.; Feng, G. Y.; Wang, Y. X.; Rag, S. S. S.; Fun, H. K.; Xin, X. Q. *J. Chem. Soc., Dalton. Trans.* **2000**, 1317. (d) Shi, S.; Lin, Z.; Mo, Y.; Xin, X. Q. *J. Phys. Chem.* **1996**, *100*, 10695. (f) Zhang, Q. F.; Niu, Y. Y.; Lueng, W. H.; Song, Y. L.; Williams, I. D.; Xin, X. Q. *Chem. Commun.* **2001**, 1126.

11

MICROPOROUS METAL–ORGANIC FRAMEWORKS AS FUNCTIONAL MATERIALS FOR GAS STORAGE AND SEPARATION

LONG PAN, KUN-HAO LI, JEONGYONG LEE, DAVID H. OLSON, AND JING LI
Department of Chemistry and Chemical Biology, Rutgers University, Piscataway, New Jersey

11.1 INTRODUCTION

Microporous metal–organic frameworks (MMOFs) comprise a subset of the polymeric metal–organic coordination compounds that have one-, two-, and three-dimensional (1D to 3D) extended network structures [also known as metal–organic frameworks (MOFs)] [1–3]. MMOFs generally consist of metal or metal cluster vertices interconnected by rigid or semirigid di-, tri-, or multifunctional organic linkers, encompassing void space (isolated cavities, or 1D–3D interconnected channels of various size, shape, and curvature) where solvent templates and guests typically reside (with the exception of a few guest-free MMOFs). The accessible microporosity (pore diameter less than 2 nm [4]) associated with MMOFs has attracted intensive attention because of the unique properties and potential applications of MMOFs in gas storage and separation, heterogeneous catalysis, and chemical sensing [2,3,5–17].

The burgeoning research in the past one and half decades has demonstrated that MMOFs, as a new type of adsorbent material, possess several advantageous features compared to the widely used zeolites as well as other adsorbent materials.

Design and Construction of Coordination Polymers, Edited by Mao-Chun Hong and Ling Chen

1. MMOFs are highly crystalline materials whose synthesis and crystal growth are more facile; and the structures of MMOFs can be determined more readily by methods such as single-crystal x-ray diffraction. Thousands of new MOF structures [3], together with information regarding the properties of some of them, have been reported. Accumulation of this information helps deepen our understanding of the chemical principles of their formation and the critical correlation between their structures and properties. A few generalizations have been recognized, among which the concept of secondary building units (SBUs) and "isoretucular" chemistry have been broadly accepted and applied in the rational design and synthesis of MMOFs with predesigned structures and properties [3].
2. MMOFs are modular in nature, which means that both the organic ligands and the metals can be varied to access a potentially unlimited number of structures [3]. This flexibility in their synthesis offers an opportunity to systematically tune the structures for a specific application. For example, different metals and/or functional groups on the ligands can be incorporated into the final structure to enhance the interactions between the adsorbates and the MMOFs or to offer selectivity among the various adsorbates (e.g., gases and hydrocarbons).
3. MMOFs have shown higher porosity than that shown by zeolites as well as by other adsorbent materials, such as the carbon-based adsorbents silica and alumina. The observed specific surface area highest is an order of magnitude higher than the typical surface area of zeolites. High surface areas are believed to be beneficial to gas storage and separation applications [7,18].
4. The pores of MMOFs are perfectly ordered and more readily accessible by adsorbates. The pores of MMOFs are more homogeneous than those of silica and alumina in terms of the pore size and size distribution, pore shape and curvature, and the nature of the pore walls, which offers more predictable sorption behaviors.

Hydrogen and methane storage, and CO_2 (and NO_x) sequestration and storage in MMOFs are areas of intense current interest and are being pursued by researchers around the globe, who are motivated by imminent energy needs and environment concerns [3,8,19–35]. Similar to other adsorbent materials, such as activated carbons, silica, alumina, and zeolites, MMOFs adsorb such adsorbates by fully reversible physisorption as characterized by low isosteric heats of adsorption (Q_{st}). Compared to other competing technologies based on chemisorption (e.g., metal and chemical hydrides), physisorption offers better desorption and readsorption kinetics.

MMOFs have also shown capabilities and potential for separation of gases and hydrocarbons. Our systematic investigation of the hydrocarbon adsorption properties of the MMOFs reported herein has revealed many unique characteristics in their hydrocarbon adsorption and their great potential for hydrocarbon separations by equilibrium or kinetic processes.

11.2 DESIGN, RATIONAL SYNTHESIS, AND STRUCTURE DESCRIPTION

The design and synthesis of microporous metal–organic framework materials (MMOFs) call for insights into the chemistry on both the inorganic and organic sides. In combination with the well-developed coordination chemistry, especially the recently recognized chemistry of SBUs, MMOFs of predictable topology can be designed and synthesized rationally based on the metal coordination habit, and by variation of the size, geometry (directionality of the ligating groups), and functionality of the ligands.

To form MMOFs with desirable pore characteristics, a variety of ligands were employed in the synthesis. These include rigid and flexible, straight and bent, aromatic and aliphatic, short and long, and oxygen- and nitrogen-containing molecules, to name a few. Representative examples include 5-*tert*-butyl isophthalic acid (H_2tbip), 4,4′-(hexafluoroisopropylidene)dibenzoic acid (H_2hfipbb), 4,4′-biphenyldicarboxylic acid (H_2bpdc), 4,4′-bipyridine (bpy or 4,4′-bpy), formic acid, 1,4-benzenedicaroxylic acid (H_2bdc), triethylenediamine (ted), and *N*,*N*-dimethylformamide (DMF). Sources of metal are typically metal salts (e.g., nitrates, halides, acetates). All chemicals used in the studies were purchased from commercial providers and used without further purification. Reactions were typically carried out in Teflon-lined Parr bombs or capped glass vials or bottles.

By judicious selection of the ligands (or ligand combination), metals, and reaction conditions that favor the formation of particular SBUs and topology, we have discovered a series of MMOFs with specific structural features and porosity. We have investigated their adsorption properties systematically over a wide temperature and pressure range. In this chapter we describe our synthetic approach and rational, structure characterization and important findings on the adsorption properties of these materials.

The crystal structures of these MMOFs have been determined by single-crystal x-ray diffraction methods. Full data sets have been collected on a Bruker-AXS smart APEX I CCD diffractometer or a CAD4 Diffractis-586 single-crystal x-ray diffractometer (Enraf-Nonius) with graphite-monochromated Mo K_α radiation ($\lambda = 0.71073$ Å). The structures were solved by direct methods and refined by full-matrix least-squares on F^2. Selected crystal data and structural information are summarized in Table 11.1. The phase purity of all bulk samples was confirmed by matching the PXRD patterns observed for the samples to those simulated from the corresponding single-crystal structures. A brief description of the synthesis and structure for each compound are summarized in the following sections.

11.2.1 [Cu(hfipbb)(H_2hfipbb)] (1)

Reactions of $Cu(NO_3)_2{\cdot}3H_2O$ (0.029 g, 0.12 mmol), H_2hfipbb (0.195 g, 0.5 mmol), and deionized water (5 mL) in the molar ratio of 1.2 : 5 : 2278 at 125°C for 3 days resulted in blue column-like crystals of **1**. The product was washed with DMF (10 mL × 3) to remove unreacted H_2hfipbb and dried in air.

TABLE 11.1 Selected Crystal Data and Structural Information

	1	2	3
Formula	$C_{51}H_{26}Cu_2F_{18}O_{12}$	$C_{12}H_{12}O_4Zn$	$C_{51}H_{39}CoN_3O_5$
f.w. (g mol^{-1})	1299.80	285.59	983.33
Crystal system	monoclinic	trigonal	monoclinic
Space group	*P2/n*	*R-3m*	*C2/c*
a (Å)	18.723(4)	28.863(4)	9.523(2)
b (Å)	7.271(1)	28.863(4)	20.618(4)
c (Å)	20.481(4)	7.977(2)	25.874(5)
α (deg)	90	90	90
β (deg)	107.02(3)	90	96.20
γ (deg)	90	120	90
V (Å^3)	2666.1(9)	5755(2)	5050.5(2)
Z	4	18	4
P_{calcd} (g cm^{-3})	1.619	1.483	1.294
T (K)	293	293	293
λ (Å)	0.71073 (Mo K_{α})	0.71073 (Mo K_{α})	0.71073 (Mo K_{α})
GoF	1.293	1.047	1.220
R_1/wR_2 [$I>2\sigma(I)$]	0.044/0.069	0.033/0.087	0.123/0.263
Free volume[a], [Å^3(%)]	309.3 (11.6%)	1018.6 (17.7%)	1446 (28.6%)
	6	**7**	**9**
Formula	$C_9H_{13}Co_3NO_{13}$	$C_9H_{13}Mn_3NO_{13}$	$C_{64}H_{62}N_6O_{17}Co_3$
f.w. (g mol^{-1})	520.00	508.02	1363.99
Crystal system	monoclinic	monoclinic	orthorhombic
Space group	*P2$_1$/n*	*P2$_1$/n*	*Pbcn*
a (Å)	11.3834(6)	11.683(2)	14.195(3)
b (Å)	9.9292(6)	10.250(2)	25.645(5)
c (Å)	14.4324(8)	15.154(3)	18.210(4)
α (deg)	90	90	90
β (deg)	91.25(10)	91.86(3)	90
γ (deg)	90	90	90
V (Å^3)	1630.88(16)	1813.8(6)	6629(2)
Z	4	4	4
P_{calcd} (g cm^{-3})	2.358	1.860	1.367
T (K)	100	293	293
λ (Å)	0.71073 (Mo K_{α})	0.71073 (Mo K_{α})	0.71073 (Mo K_{α})
GoF	1.066	1.013	1.130
R_1/wR_2 [$I>2\sigma(I)$]	0.018/0.044	0.029/0.080	0.056/0.120
Free volume[a], [Å^3(%)]	503.3 (30.9%)	628.2(34.6%)	2691.4 (40.6%)
	10	**11**	
Formula	$C_{64}H_{61.38}N_6O_{16.69}Zn_3$	$C_{17}H_{24.10}N_3O_{6.05}Zn$	
f.w. (g mol^{-1})	1377.88	432.69	
Crystal system	monoclinic	tetragonal	
Space group	*P2$_1$/n*	*P4/ncc*	
a (Å)	14.608(3)	14.8999(4)	
b (Å)	18.041(4)	14.8999(4)	

TABLE 11.1 ***(Continued)***

	10	**11**
c (Å)	24.803(5)	19.1369(2)
α (deg)	90	90
β (deg)	90.005(1)	90
γ (deg)	90	90
V (Å^3)	6537	4248.4(2)
Z	4	8
P_{calcd} (g cm^{-3})	1.400	1.353
T (K)	100	100
λ (Å)	0.71073 (Mo K_α)	0.71073 (Mo K_α)
GoF	1.007	1.008
R_1/wR_2 [$I>2\sigma(I)$]	0.044/0.104	0.044/0.119
Free volume[a], [Å^3(%)]	2095.7 (32.1%)	2604.3 (61.3%)

[a] Solvent accessible volume, calculated using Platon software [38].

The overall 3D guest-free framework structure of Cu(hfipbb)(H_2hfipbb)$_{0.5}$] (**1**) (Fig. 11.1a) is built on di-copper paddlewheel-shaped SBUs that are formed by two copper atoms sharing four carboxylate groups from four hfipbb ligands (shown as a ball-and-stick model in Fig. 11.1b, Cu–Cu interatomic distance in the paddlewheel: 2.645 Å) [36]. Each paddlewheel SBU binds to four adjacent paddlewheel SBUs through the remaining carboxylate groups of the four hfipbb to form an undulating 2D network with 4^4 topology (Fig. 11.1c) [37]. Two such identical networks interpenetrate to form a 2D layer. The adjacent layers are further interconnected by monodentate carboxylate groups of H_2hfipbb ligands (shown as line models in Fig. 11.1a and b) along the axial directions of the paddlewheel SBUs to generate a 3D network structure as shown in Figure 11.1a. The most remarkable features of the structure are the microchannels (Fig. 11.1a), featuring curved internal surfaces with small pore openings (ca. 3.5 × 3.5 Å, excluding the van der Waals radius of carbon) and larger cages or chambers (ca. 5.1 × 5.1 Å) at about 7.3-Å intervals, which are more obvious in a helium adsorption simulation study (see Section 11.3). The calculated (Platon) [38] solvent accessible volume in **1** is 309.3 Å^3 out of a unit cell volume of 2666.1 Å^3 (or 11.6%).

11.2.2 [Zn(tbip)] (2)

Hydrothermal reaction of $Zn(NO_3)_2 \cdot 6H_2O$ (0.197 g, 0.66 mmol) and H_2tbip (0.148 g, 0.66 mmol) with a mixture solution of 13 mL H_2O and 3 mL ethylene glycol at 180°C for 3 days generated uniform brownish column-shaped crystals. The crystals were vacuum filtered, washed with water (3 × 30 mL) and ethanol (3 × 30 mL), and dried in air (0.082 g, 42% yield based on H_2tbip).

Like **1**, [Zn(tbip)] (**2**) is example, of a guest-free microporous metal–organic framework [39]. Tetrahedral zinc centers are linked by tbip ligands to generate the 3D structure shown in Figure 11.2a. Each tetrahedral Zn(II) cation occupies a two-fold rotation symmetry position. All adjacent zinc nodes are bridged along the c-axis by two carboxylate groups to form a 3_1 helical chain with a pitch of 7.977 Å along the

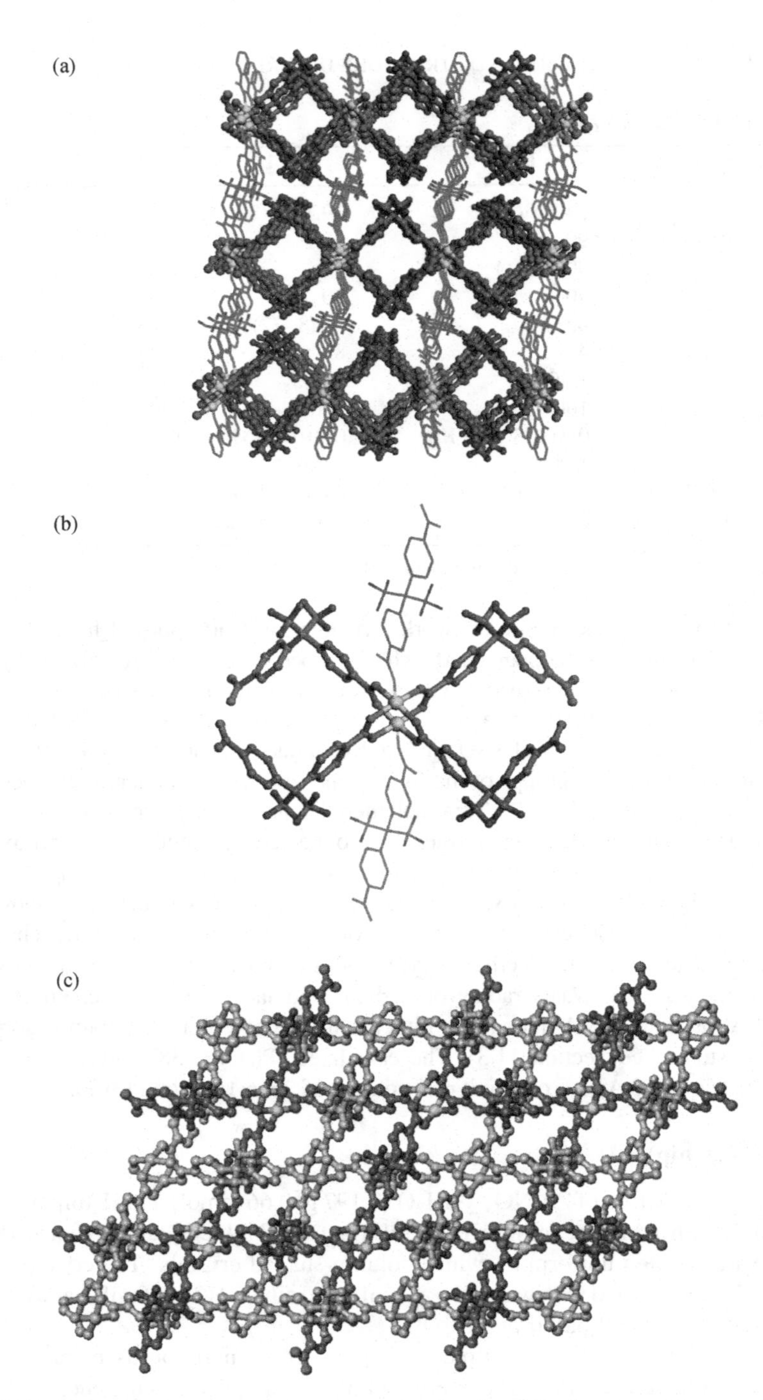

FIGURE 11.1 (a) Perspective view along the *b*-axis of **1**, showing 1D open channels; (b) di-copper paddlewheel SBU in **1**, with the hfipbb shown in a ball-and-stick model and H_2hfipbb in a line model; (c) interpenetration of two identical nets in **1**, with the axial H_2hfipbb ligands removed.

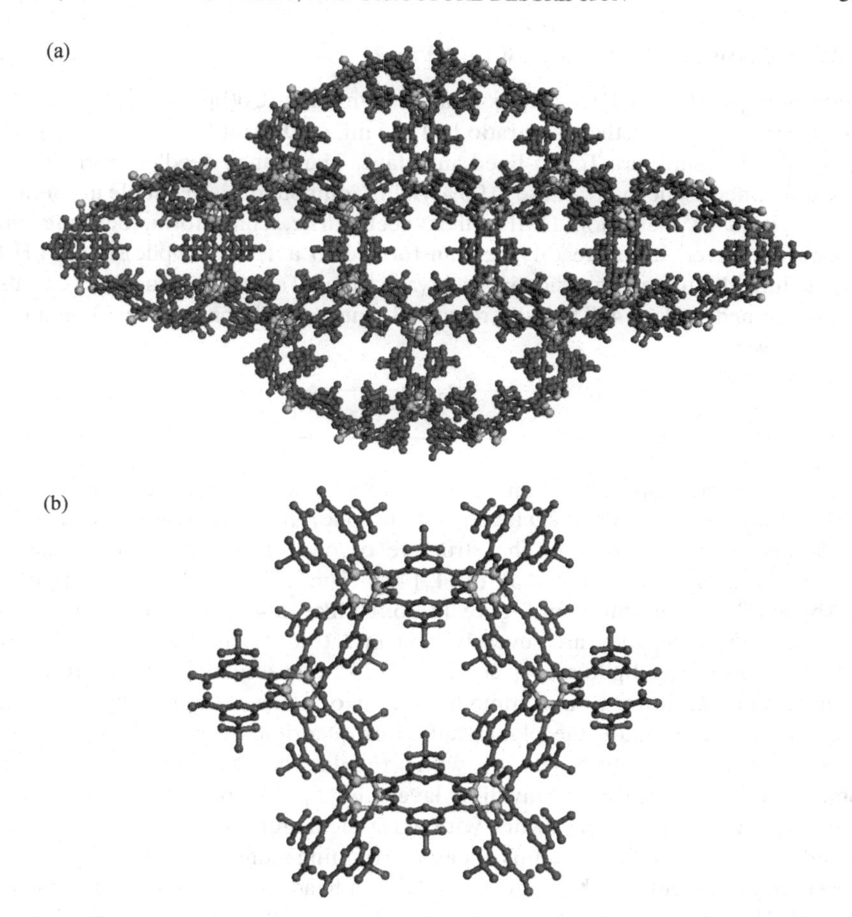

FIGURE 11.2 (a) Perspective view of the overall structure of **2** along the crystallographic *c*-axis; (b) 3_1 chains in **2** that are of opposite handedness.

crystallographic *c*-axis (the screw axis). Each chain connects to three neighboring chains of opposite handedness through tbip ligands (Fig. 11.2b), resulting in a 3D structure containing hexagonal close-packed 1D open (guest-free) channels along the *c*-direction (Fig. 11.2a).

With 3.55-Å (tile) and 3.25-Å (perpendicular) distances between the phenyl rings that form the channel walls, which are comparable with the value found in graphite (3.35 Å), strong π–π interactions via a slipped π stacking mode are anticipated [40]. All three methyl groups of the ligand are disordered and protrude into the channels, significantly reducing the diameter of the tubular channels (along the *c*-axis) to around 4.5 Å (excluding the van der Waals radius of hydrogen). While the aperture of the channels in **2** is on the same order as that of **1**, the freely accessible volume of 17.7% calculated from Platon for **2** is significantly higher as a result of a larger void space and more effective packing of the channels.

11.2.3 [Co(bpdc)(bpy)]·0.5DMF (3, RPM-2)

[Co(bpdc)(bpy)]·0.5DMF (**3**) was assembled by mixing [Co(bpdc)$(py)_2$]·H_2O (**4**), a 2D structure, and bpy in the molar ratio 1 : 4 in 5 mL of DMF at 120°C for 1 day under autogenous pressure in a Teflon-lined autoclave. The orange needlelike crystals of **3** were obtained as the major phase in 90% yield. Compound **3** is insoluble in common solvents such as methanol, DMF, ether, acetonitrile, chloroform, benzene, and toluene. However, it can readily be transformed to a 1D [Co(bpdc)$(H_2O)_2$]·H_2O (**5**) structure [42] upon immersion in a hot water/ethanol solution. Structure **5** can also be transformed back to **4**. The fully recyclable interconversion between **3**, **4**, and **5**, is as follows:

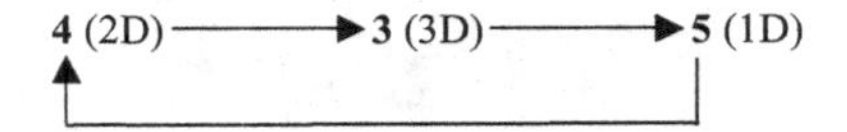

Single-crystal x-ray diffraction analysis of **3** revealed a 3D porous structure that can be viewed as interwoven 2D layers of [Co(bpdc)] pillared by bpy ligands along the *b*-axis (Fig. 11.3a) [41]. The structure of one of the two interwoven 2D networks closely resembles that of **4** [42]. The basic building block of **3**, $Co_2(bpdc)_4(bpy)_4$, is shown in Figure 11.3b. Unlike in **4**, where bpdc ligands are essentially in plane, they are severely bent in **3** (the dihedral angles of the two benzene rings in two bpdc are 41.20° and 17.25°, respectively). This distortion is required to make room to accommodate the two interlocking [Co(bpdc)] layers (Fig. 11.3c). As a result, the planar lattice of quadrilateral shaped grids in **4** is replaced by an undulating network with irregular rhombus (Fig. 11.3c). Interconnection of the adjacent wavelike layers by bpy as pillars gives rise to a noninterpenetrating 3D framework with a 1D rectangular channel (ca. 5.6 × 3 Å, excluding the van der Waals radius of carbon) running along the *a*-axis (Fig. 11.3a). The solvent molecules (DMF) are arranged in a head-to-tail fashion within every channel. Calculation of the pore volume in **3** reveals that 28.6% (1446 $Å^3$ out of 5050 $Å^3$) of the unit cell volume is occupied by guest molecules, which will become accessible after removal of the guests.

11.2.4 Metal (Co, Ni, Mn) Formate (6, 7, 8)

Both single crystals and pure polycrystalline samples of **6** and **7**, as well as the powder sample of **8**, were prepared by solvothermal reactions, which are exemplified by the following synthetic procedure for **6** [43]. Heating a mixture of cobalt(II) nitrate hexahydrate (422 mg, 1.4 mmol) and formic acid (0.1 mL, 2.8 mmol) in *N,N*-dimethylformamide (DMF, 20 mL) in a Teflon-lined acid digestion bomb at 100°C in an oven for 48 hours yielded pink blocklike single crystals of $[Co_3(HCOO)_6]$·DMF (**6**) suitable for x-ray diffraction study. The facile syntheses of **6** were easily scaled up to produce multigrams of products. For example, a pink solution of cobalt(II) nitrate hexahydrate (5.236 g, 17.6 mmol) and formic acid (4.5 mL, 115.7 mmol) in DMF (amine free, 15 mL) was heated at 100°C in an oven for 24 hours before it was cooled to room temperature naturally. The pink crystalline

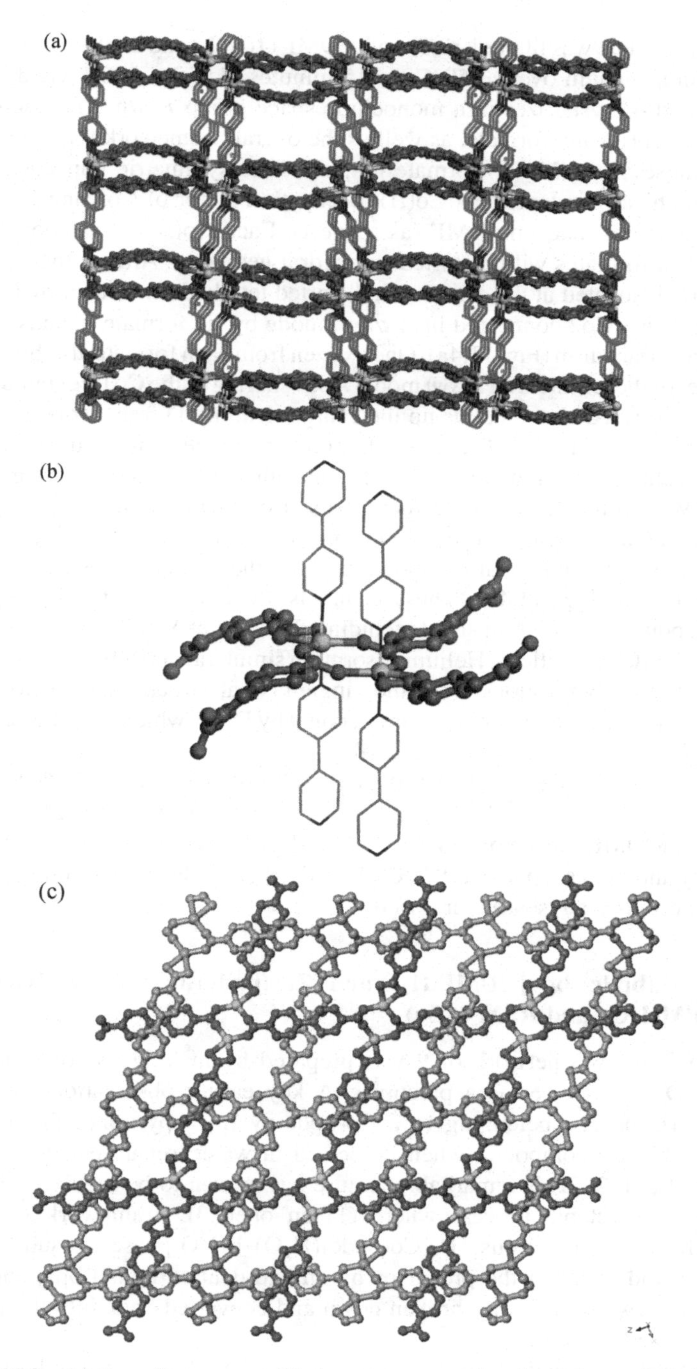

FIGURE 11.3 (a) Overall structure of **3** showing rectangular channels running along the *a*-axis; (b) basic building block found in **3**; (c) 2D layer formed by two interwoven identical nets formed by bpdc and the Co(II) centers.

powdery precipitate was filtered, rinsed with DMF (10 mL) and diethyl ether (10 mL), and dried in a vacuum oven at 50°C for 15 minutes (3.055 g, 100% yield).

Compound **6** crystallizes in a monoclinic space group $P2_1/n$. The coordination geometry of Co(II) and formate, as well as the overall framework topology, closely resemble those of the Mn(II) formate [44,45]. The asymmetric unit contains four crystallographically independent Co(II) (occupation factor of Co3 and Co4 is 0.5), six formate anions, and one DMF (as a guest). Each Co(II) is six-coordinated to oxygens from formates with distorted octahedral geometry. Among the four Co(II) centers, Co1 is located at the center of a distorted tetrahedron formed by Co2, Co3, Co4, and another Co2 connected in *syn/anti* mode by six formate ligands along the edges of the tetrahedron (Fig. 11.4a). One oxygen from each formate also binds to Co1 at the center of the tetrahedron in *syn* mode. Apex sharing of the Co1-centered Co1Co_4 tetrahedra described above builds up the charge-neutral 3D framework structure of Co(II) formate (Fig. 11.4b). The network topology can be simplified by connecting all the Co1 centers (as tetrahedral nodes) through other Co(II) centers (as the midpoint between two tetrahedral nodes) to form a distorted diamond network as shown in Figure 11.4c. One-dimensional zigzag channels roughly 5 to 6 Å in size (excluding the van der Waals radius) can be observed along the *b*-axis where the DMF guest molecules reside (Fig. 11.4d). These channels are interconnected by very small apertures about 1.4 × 5.3 Å in size (excluding the van der Waals radii of the closest atoms) along [101] directions. Helium adsorption simulation results (see Section 11.3) confirmed these observations from the single-crystal structure. As calculated by Platon, 30.9% of the unit cell volume is occupied by DMF, which becomes accessible by other guests upon removal of the DMF.

Due to the bigger size of Mn(II), the unit cell volume of **7** is 11.2% larger than that of **6** (1813.8 $Å^3$ vs. 1630.88 $Å^3$). The accessible pore volume calculated after removal of the DMF guest molecules is 34.6%; slightly more than that (33%) has been reported by another group. $[Ni_3(HCOO)_6]$·DMF (**8**), as judged by its almost identical PXRD pattern, is also isostructural to **6**.

11.2.5 $[Co_3(bpdc)_3bpy]$·4DMF·H_2O and $[Zn_3(bpdc)_3bpy]$·4DMF·H_2O (9, 10, RPM-1-Co and RPM-1-Zn)

Similar to **3**, the 3D network of **9** was prepared by making use of a previously reported 1D structure, **5**, as a precursor. A key earlier observation was that all materials, 1D or 2D, belonging to a Co-bpdc-py (py = pyridine) family readily convert to this 1D compound when immersed in water, regardless of their initial dimensionality [42]. This remarkable feature has proven to be highly desirable for designing a recyclable process where (1) an open 3D framework structure is achieved from the nonporous 1D $[Co(bpdc)(H_2O)_2]$·H_2O phase via substitution of ancillary ligand water in the precursor by an exo-dentate ligand bpy, and (2) the open 3D framework is readily broken down and converted back into the precursor when immersed in water.

Alternatively, recyclable porous materials (RPMs) $[Co_3(bpdc)_3bpy]$·4DMF·H_2O (**9**, RPM-1-Co) and $[Zn_3(bpdc)_3bpy]$·4DMF·H_2O (**10**, RPM-1-Zn) can also be grown

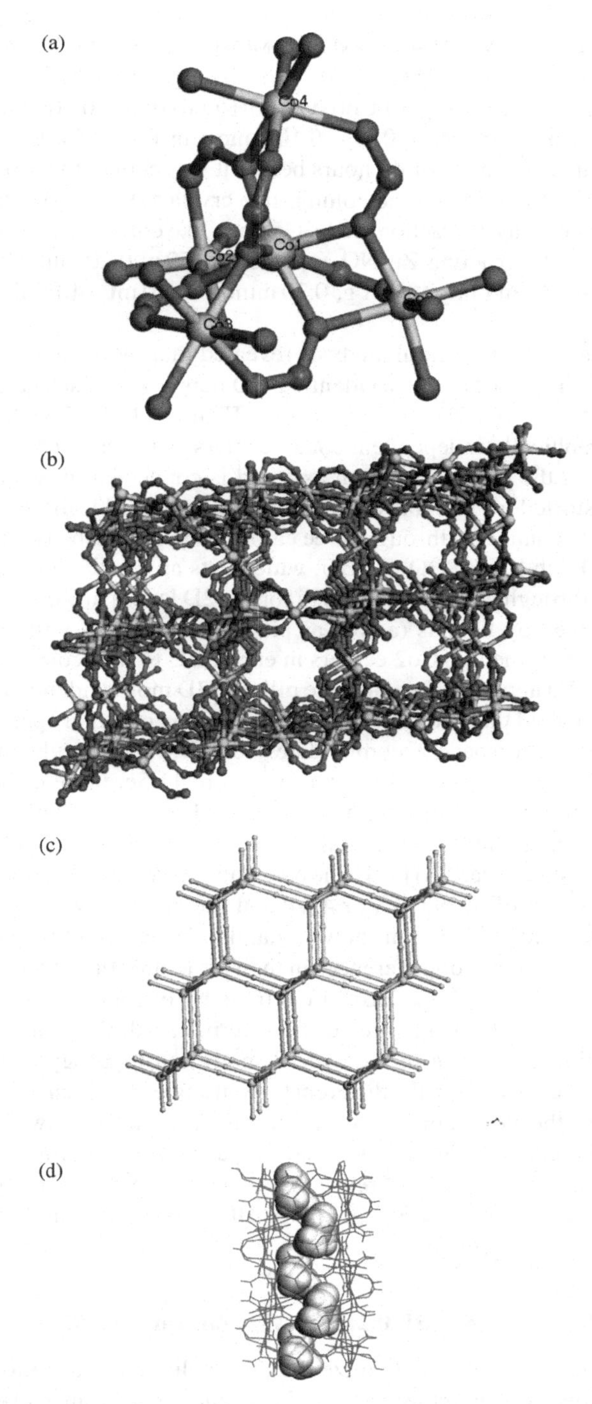

FIGURE 11.4 Crystal structure of **6**: (a) distorted Co_4 tetrahedron with Co1 residing in the center of the tetrahedron; (b) perspective view of the overall structure along the *b*-axis; (c) diamondoid net simplified by connecting the tetrahedral nodes Co(1) through other Co(II) centers; (d) packing DMF inside a segment of the zigzag channel.

as the following [46]. A mixture of $Co(NO_3)_2{\cdot}6H_2O$ (0.020 g, 0.069 mmol), H_2bpdc (0.025 g, 0.13 mmol) and bpy (0.016 g, 0.10 mmol) in DMF (5 mL) was heated at 150°C in an isothermal oven for 48 hours before it was cooled to room temperature naturally. One of the purple-colored column-like crystals (**9**) thus obtained was used for a single-crystal x-ray diffraction study. Column-like colorless crystals of **10** were obtained similarly by reacting $Zn(NO_3)_2{\cdot}6H_2O$ (0.059 g, 0.20 mmol) with H_2bpdc (0.048 g, 0.20 mmol) and bpy (0.031 g, 0.20 mmol) in 10 mL of DMF at 150°C for 48 hours.

Single-crystal x-ray structural analysis revealed that the structure of **9** features interpenetration (Fig. 11.5a) of two identical 3D networks constructed on a unique trinuclear SBU, $[Co_3(COO)_6]$, as shown in Figure 11.5b. The SBU contains two crystallographically independent cobalt centers. The octahedrally coordinated Co1 is located on the twofold rotation axis and is connected to two adjacent Co2 centers with distorted trigonal bipyramidal geometry (Fig. 11.5b). Each of the two Co1–Co2 pairs is connected through three carboxylate groups by two μ_2 and one μ_3 oxygen atoms. Each $[Co_3(COO)_6]$ SBU, which acts as a node, is connected to six adjacent nodes through six bpdc ligands to form a 2D layer parallel to the *ab*-plane. Nitrogen from the bpy ligands (acting as pillars) coordinates with the remaining coordination sites of the two Co2 centers in each SBU to interconnect the adjacent 2D layers into a 3D network. Two of these pillared 3D motifs, identical in structure, interpenetrate to yield the catenated network consisting of large, open, 1D channels (Fig. 11.5a). The free space accommodates four DMF molecules and one H_2O molecule per formula unit. As illustrated in Figure 11.5c, these channels contain large-diameter supercages (approximately $11 \times 11 \times 5$ Å, excluding the van der Waals radius of carbon atoms) and smaller windows (triangular, with an effective maximum dimension of ca. 8 Å) [47]. The calculated solvent accessible pore volume of **9** is 2811.1 Å^3 out of 6629.0 Å^3, or 42.4% of the unit cell volume.

Single-crystal x-ray diffraction showed that the crystal structure of **10** is closely related to **9**, with some minor differences in their structures [48]. One is that the two Zn2 centers in the $[Zn_3(COO)_6]$ SBU are tetrahedrally coordinated instead of the five-coordination in **9**. However, this difference in coordination number does not cause any striking difference in the overall motif and topology of the structure. Another difference arises from the different torsion angles between the benzene rings of the ligands and the metal atoms, as well as the dihedral angles between the benzene rings, so that the channels in **10** have more surface atoms sticking into the voids. A direct consequence of this difference is a reduction in accessible pore volume. As calculated by Platon, the free volume in **10** is 2658.9 Å^3 out of 6537.0 Å^3, or 40.7% of the unit cell volume.

11.2.6 $[M(bdc)(ted)_{0.5}]{\cdot}2DMF{\cdot}0.2H_2O$ (M = Zn, Cu, Co, Ni; 11, 12, 13, 14)

One of the strategies used to maximize the pore volume in our rational design of porous structures was to construct noninterpenetrated frameworks. This was attempted by employing a bulky molecular species, triethylenediamine (ted), as a pillar ligand. Thus, blocklike single crystals of $[Zn(bdc)(ted)_{0.5}]{\cdot}2DMF{\cdot}0.2H_2O$

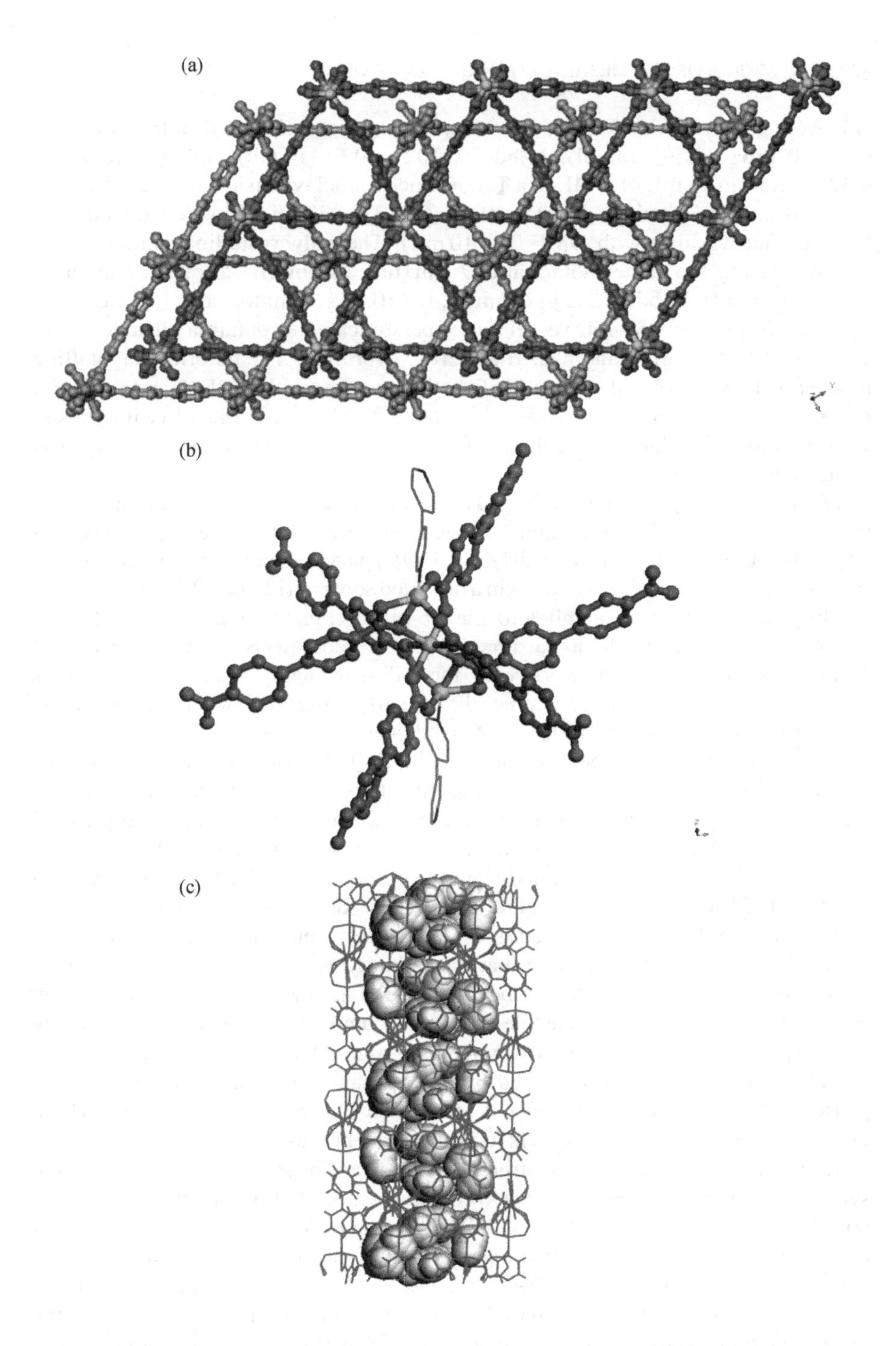

FIGURE 11.5 (a) Perspective view of the channels in **9** that are formed by the interpenetration of two identical nets; (b) SBU found in 9; (c) packing of DMF and water guest molecules, showing the alternate supercages and narrow windows along a segment of the channels of irregular shape in **9**.

(**11**) were grown successfully by solvothermal reactions of zinc(II) nitrate hexahydrate (0.156 g, 0.597 mmol), H_2bdc (0.102 g, 0.614 mmol), and ted (0.036 g, 0.321 mmol) in 15 mL of DMF in a Teflon-lined autoclave inside an oven at 120°C for 2 days. Colorless single crystals of **11** (0.212 g, 83% yield) were isolated after filtering and washing with DMF (3 × 10 mL). The polycrystalline samples were prepared using the same molar ratio. A mixture of zinc(II) nitrate tetrahydrate (1.56 g, 6 mmol), H_2bdc (1.02 g, 6 mmol), ted (0.36 g, 3 mmol) and 150 mL DMF was transferred to a 250-mL vessel and supersonicated to obtain a clear solution. The vessel was covered and heated at 120°C overnight. The colorless crystalline powder of **11** was isolated by vacuum filtration, washed with DMF (10 mL × 3) and dried in air (1.68 g, 65.11% yield). The isostructural **12**, **13**, and **14** (with guests) were prepared similarly in medium to good yields, and their phase purity was confirmed by PXRD.

$[Zn(bdc)(ted)_{0.5}]\cdot 2DMF\cdot 0.2H2O$ (**11**) is a 3D porous structure without interpenetration. It crystallizes in the tetragonal space group *P4/ncc* [49]. The structure is built on paddlewheel-shaped SBUs. Each $[Zn_2(COO)_4]$ paddlewheel SBU is linked by bdc ligands to four other such SBUs to form a distorted square grid like a 2D net (due to the bending of bdc ligands) parallel to the *ab*-plane (Fig. 11.6a), which is further connected by ted along the axial direction of the paddlewheels (along the *c*-axis) to give rise to a 3D framework with three-dimensionally interconnected pores (Fig. 11.6b and c). To a large degree, the structure resembles what was originally proposed by Seki and Mori [50] for a Cu–bdc–ted compound.

The unique portion of the links in **11** is the $Zn_2(\frac{1}{2}\ bdc)_4(\frac{1}{2}\ ted)_2$ vertex along the fourfold symmetry axis at $(\frac{1}{4}\ \frac{1}{4}\ z)$. Thus, the ted ligands, which have threefold molecular symmetry, are necessarily disordered about the fourfold axis to yield 12 partially occupied ethylene sites. At the center of the pore lies the fourfold axis $(-\frac{1}{4}\ \frac{1}{4}\ z)$ and the partially occupied water molecule site, which is surrounded by a quartet of DMF molecules that are additionally disordered about their C1–N1 bond. The disordering of adsorbed guest molecules is quite common in the highly porous structures. It is interesting that the more hydrophobic portion of the DMF molecules point away from the 1,4-bdc linkers and in toward the center of the pore, and further, that a water molecule can be accommodated in the center of the hydrophobic pore, surrounded only by methyl groups. The structural features of the overall network were first described by Dybtsev and co-workers in their *I4/mcm* phase [51], which are similar but not identical to our *P4/ncc* phase of **11**. These two structures apparently represent two different phases and contain some noteworthy differences in molecular structure, as a result of significantly different site symmetries at the vertices. The free volume, as calculated by Platon, is 2482.3 Å^3 out of 4248.4 Å^3, or 58.4% of the unit cell volume of **11**. However, upon removal of the guest molecules, the bdc ligands would become linear from the bent conformation as proven by single-crystal and PXRD studies. There are two types of channels apparent from the guest-free crystal structure. The first type is along the *c*-axis, with a cross section of about 7.5 × 7.5 Å (excluding the van der Waals radius of carbons). The second type is along the *a*- and *b*-axes, with a cross section of about 4.8 × 3.2 Å (excluding the van der Waals radii).

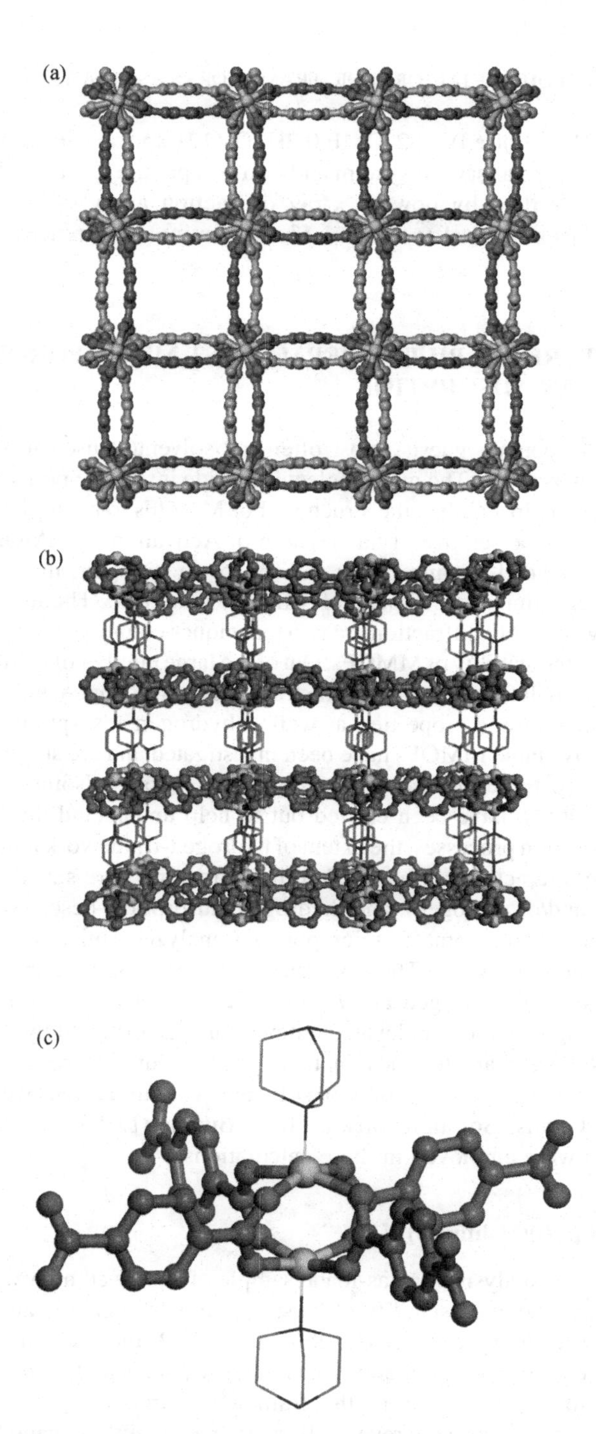

FIGURE 11.6 Structure of **11**: (a) two distorted square-grid-like layers formed by bdc connecting di-zinc paddlewheel SBUs; note the severely bent conformation of the bdc ligands; (b) perspective view of the channels in **11** along the [110] direction; (c) paddlewheel SBU in **11**.

Compound [Cu(bdc)(ted)$_{0.5}$]·2 DMF·0.2H2O (**12**) appears to be isostructural to **11**, with the tetragonal crystal system and unit cell parameters $a = 15.118(6)$ and $c = 19.274(5)$ Å verified by powder x-ray diffraction analysis. Compounds **13** and **14** are also isostructural to **11** and **12**, with minor differences in their unit cell constants.

11.3 STRUCTURE STABILITY, PERMANENT MICROPOROSITY, AND HYDROGEN ADSORPTION

MMOFs typically contain guests, most often the solvent(s) used in the synthesis and crystallization process. The guest molecules reside inside the pores or cavities of the framework and help stabilize the structure. For MMOFs to be used as adsordents, the guests must first be removed (via activation). Activation of MMOFs is usually carried out by extended heating under vacuum or in nitrogen; therefore, thermal stability is a very important aspect of MMOF materials. Thermal gravimetric analysis and powder x-ray diffraction are two techniques commonly used in combination to investigate the stability of MMOFs. Out of the large number of MMOFs reported each year, only a small fraction retain their structural integrity upon moderate heating.

The pore structures and properties, as well as hydrogen adsorption, of a selected group of thermally stable MMOFs have been investigated and are summarized in the following sections. Experimental gas adsorption–desorption isotherms, combined with simulation study, have been carried out to help understand the nature of the adsorption–desorption processes, the extent of hydrogen-framework interactions, and the role in these interactions of material composition and pore size and shape.

The argon (and/or nitrogen) and hydrogen adsorption–desorption isotherms were collected using an automated micropore gas analyzer Autisorb-1 MP (Quantachrome Instruments, Florida). The cryogenic temperatures were controlled using liquid argon and liquid nitrogen at 87 and 77 K, respectively. The samples were activated under high vacuum at elevated temperatures determined by thermal gravimetric analyses. Pore characteristics, including pore volume, pore size, and surface area, were analyzed using Autosorb v1.50 software. The simulations were performed using Accelrys Cerius^2 Sorption software [52]. Burchart1.01-Universal1.02 forcefield parameters were employed in these calculations.

11.3.1 [Cu(hfipbb)(H_2hfipbb)] (1)

Thermogravimetric analysis of an as-made sample of **1** showed no weight loss up to 330°C, suggesting the good stability of its structure, which can be attributed to the twofold interpenetration. As described in Section 11.2, the guest-free ultramicrochannels in **1** are composed of alternating smaller windows and larger cages, which are very difficult to observe from the framework structure. Helium-adsorption simulation [52] results instead reveal alternating oval-shaped cages and narrow "neck" regions (of roughly 3.2 and 7.3 Å in the largest dimensions) along the microchannel (Fig. 11.7) [53].

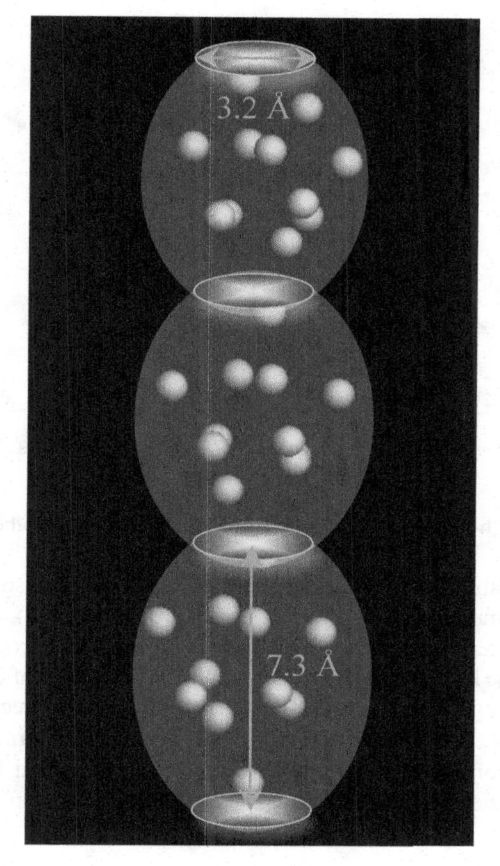

FIGURE 11.7 Helium adsorption simulation reveals the alternating oval-shaped cages and narrow necks along the microchannel in **1**.

The hydrogen adsorption–desorption isotherms were measured on **1** at cryogenic temperatures. Figure 11.8 shows the isotherms obtained at 77 K. Strong hysteresis was observed at all pressure levels. This behavior is probably a result of the small pore dimensions in **1**. The maximum value of hydrogen uptake is 0.23 wt% (1 atm). The density of adsorbed hydrogen at these temperatures is 0.03 to 0.033 $g\,cm^{-3}$ (calculated based on the estimated pore volume, 0.07 $cm^3\,g^{-1}$) [36], comparable to the value at its critical point (33 K and 13 atm).

11.3.2 [Zn(tbip)] (2)

Thermogravimetric analysis of **2** showed no weight loss during prolonged heating at 350°C over 24 hours. PXRD after TGA of the sample verified that it fully retained its structural integrity. This exceptional stability demonstrates that MMOFs of this type

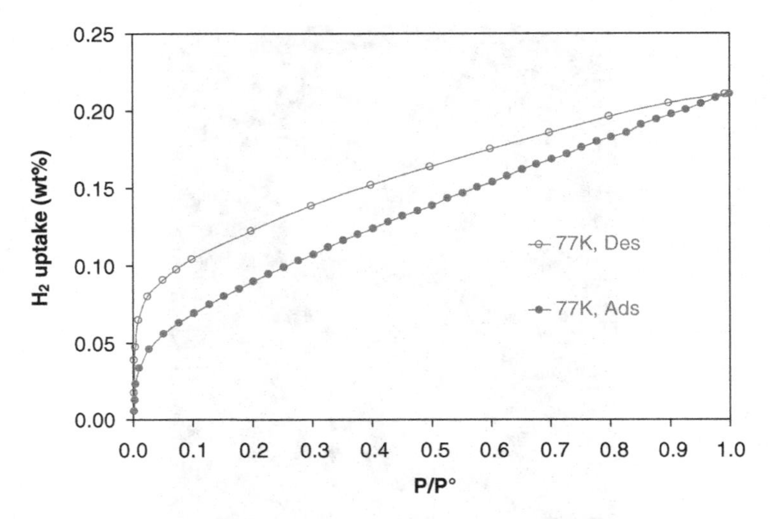

FIGURE 11.8 High-resolution hydrogen adsorption–desorption isotherms at 77 K for **1**.

may indeed be suitable for applications that require frequent adsorption–desorption cycles over long time periods (e.g., on-board hydrogen storage).

The pore characteristics of **2** were characterized by both nitrogen and argon gas sorption studies. The DA (Dubinin–Astakhov) pore volume [54] of **2** was calculated to be 0.14 and 0.15 $cm^3 g^{-1}$ based on the N_2 and Ar sorption data collected at 77 and 87 K, respectively (Fig. 11.9). Both follow a typical type I isotherm. A gravimetric density of 0.75 and 0.52 wt% of hydrogen was achieved at 77 and 87 K and 1 atm, respectively (Fig. 11.10). The density of adsorbed H_2 at 77 K and 1 atm, calculated to be

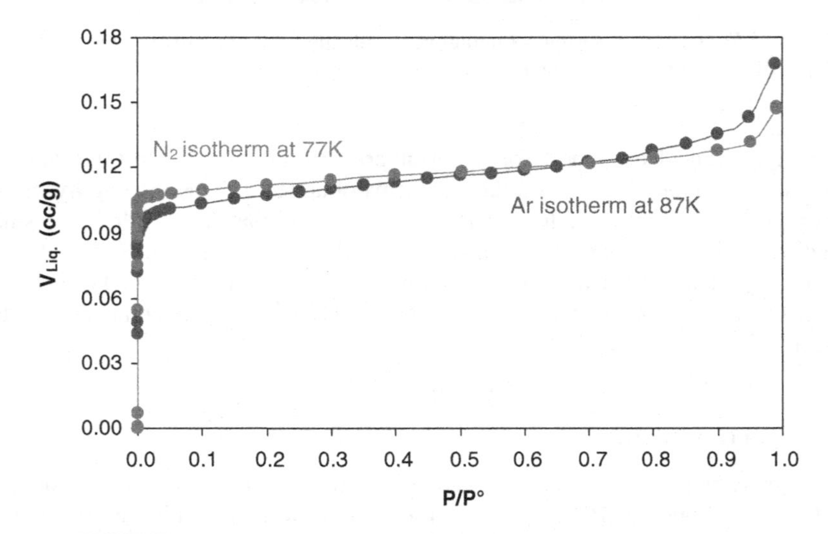

FIGURE 11.9 Adsorption isotherms for **2**: Ar (87 K) and N_2 (77 K).

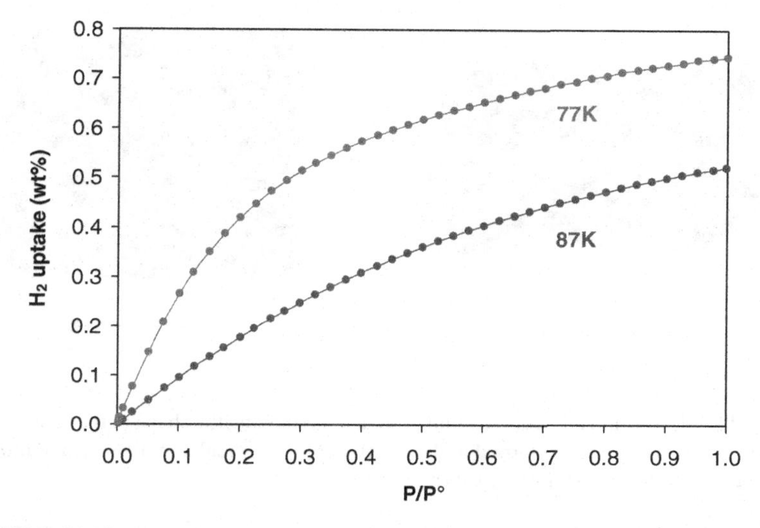

FIGURE 11.10 High-resolution hydrogen adsorption isotherms at 77 and 87 K for **2**.

0.054 g cm^{-3}, is as high as that of our previously reported value, 0.053 g cm^{-3}, which is one of the highest values among the porous metal–organic frameworks reported to date. Multipoint Brunauer–Emmett–Teller (BET) surface area was calculated to be 256 m^2 g^{-1} from N_2 adsorption data at 77 K. In addition, the isosteric heats of hydrogen adsorption Q_{st} of **2** were calculated to be 6.7 to 6.4 kJ mol^{-1} (coverage 0.02 to 0.5 wt%), suggesting relatively strong adsorbent–adsorbate interaction.

11.3.3 [Co(bpdc)(bpy)]·0.5DMF (3)

Thermogravimetric analysis performed on the as-synthesized **3** showed one-step weight loss of DMF in the range 160 to 200°C. The framework structure is stable up to 350°C. PXRD analysis indicated Co_3O_4 as the only residue after the sample was heated to 700°C. For adsorption experiments, the sample was heated to 200°C to remove the guest molecules. The PXRD pattern showed no change compared to the simulated pattern from the single-crystal structure. The hydrocarbon adsorption–desorption behavior of this compound is described in Section 11.4.

11.3.4 Cobalt Formate (6)

Thermal gravimetric analysis of a freshly made sample of **6** indicates a clear weight-loss step between 120 and 175°C (14.00% experimentally vs. 14.06% calculated for DMF loss), followed by another weight-loss step around 300°C corresponding to the decomposition of the compound (decomposition product Co_3O_4 confirmed by PXRD). PXRD of a sample activated at 200°C on TGA showed no obvious shift of the peaks, suggesting that the structure remained essentially intact. We attribute the good stability of the framework to the extensive dative bonds between Co(II) centers and the formate ligands, as well as the small pore size (Fig. 11.4a and b).

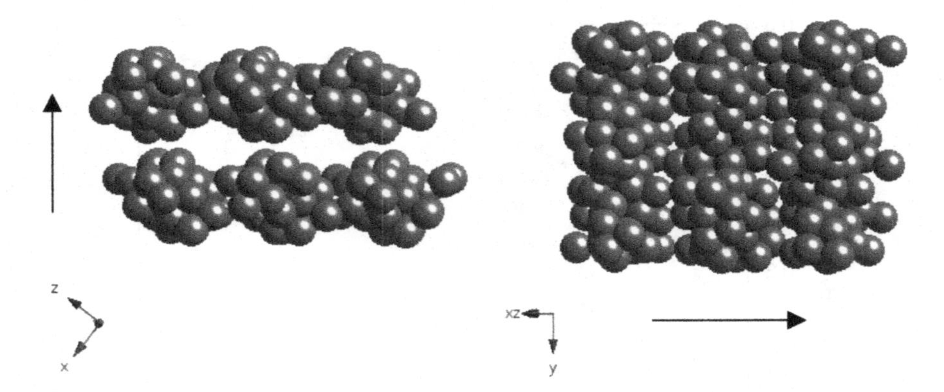

FIGURE 11.11 Helium adsorption simulation results showing the shape of the channels in **6**. No connection in the direction normal to the [−101] plane (vertical direction); small molecule connection within the [−101] plane (normal to the *y*-axis).

The N_2 adsorption isotherm for **6** follows type I behavior, with a hysteresis in its desorption isotherm. The surface area was calculated to be 354 $m^2\,g^{-1}$ (Langmuir) and 304 $m^2\,g^{-1}$ (BET), respectively. The D-A (Dubinin-Astakhov) micropore volume of the sample was 0.14 $cm^3\,g^{-1}$, in good agreement with the value reported. The main pore size was determined to be about 5.0 Å from the Horvath–Kawazoe (H-K) pore size distribution [55], consistent with the size estimated from the single-crystal structure.

Simulated adsorption of He also shed light on the irregular channel architecture, which complements the information obtained from single-crystal structure analysis and the gas adsorption study described above. Three views of the packing of the He atoms, with all structural atoms (Co and formate) removed, are depicted in Figure 11.4. It shows clearly that the main channel running parallel to the [010] direction has a diameter of about 5.5 Å. Within the (−101) plane there appears to be an interconnection of these main channels (along the *b*-axis) for species of the size of He or H_2. There is no channel connection normal to the (−101) planes (Fig. 11.11).

Gravimetric uptake of 0.75 and 0.65 wt% of hydrogen was achieved at 77 and 87 K and 1 atm, respectively (Fig. 11.12). As plotted in Figure 11.12a, the hydrogen adsorption isotherm almost reached saturation at 1 atm and 77 K. The isosteric heats of hydrogen adsorption (Q_{st}) were calculated to be 8.3 to 6.5 kJ mol^{-1} in a low coverage region (0.06 to 0.26 wt%), which are among the highest numbers reported so far for porous metal–organic frameworks, suggesting relatively strong adsorbent–adsorbate interactions in the relatively small pored compound **6** (Fig. 11.12).

11.3.5 $[Co_3(bpdc)_3bpy]\cdot 4DMF\cdot H_2O$ and $[Zn_3(bpdc)_3bpy]\cdot 4DMF\cdot H_2O$ (9, 10)

Both compounds clearly showed two-stage weight losses in thermogravimetric analyses. The weight losses around 200°C were 22% for **9**, and 21% for **10**, in

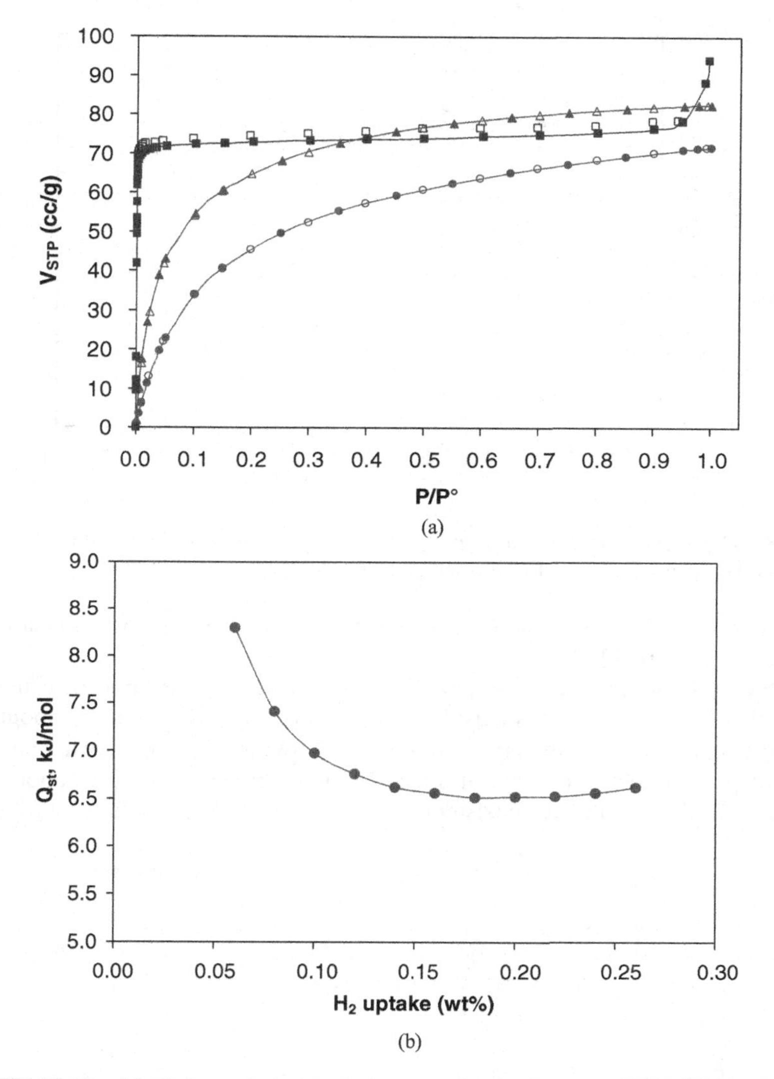

FIGURE 11.12 (a) High-resolution hydrogen sorption isotherms at 77 K (filled triangles, ADS; open triangles, DES) and 87 K (filled circles, ADS; open circles, DES), and a nitrogen isotherm at 77 K (filled squares, ADS; open squares, DES) for **6**; (b) isosteric heats of adsorption calculated based on 77 and 87 K hydrogen isotherms.

agreement with the amount of guest molecules (DMF and H_2O) calculated, while the weight losses at higher temperatures (400 and 384°C, respectively) correspond to decomposition of **9** and **10**. The Ar adsorption–desorption isotherms (Fig. 11.13) of both compounds show a small hysteresis at $P > 0.4$ atm, which is typically a result of intercrystalline voids in the samples. At rather low pressure ($P < 0.05$ atm), hystereses

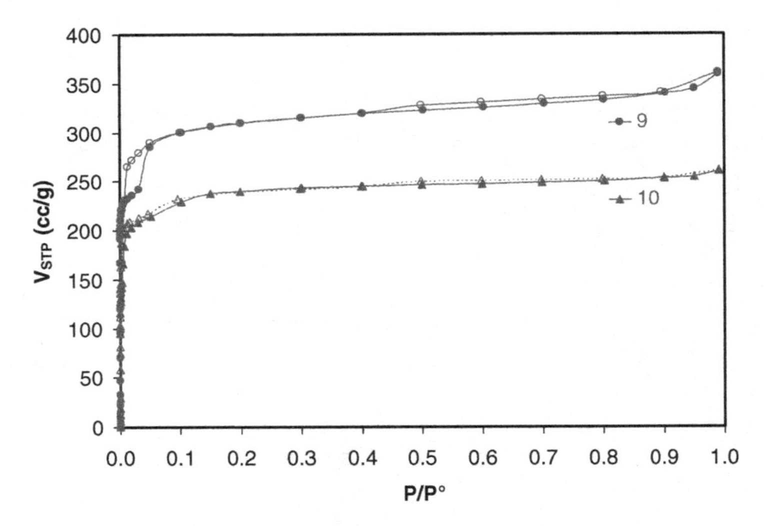

FIGURE 11.13 Ar adsorption (filled circles and triangles) and desorption (open circles and triangles) isotherms for **9** and **10** obtained at 87 K.

are also observed for both **9** and **10**, which could originate from changes in volume of the adsorbents [56,57].

Based on Ar adsorption results, the Horvath–Kawazoe (H–K) pore size analysis shows that two types of pores exist in **9**: ultramicropores (diameters of about 7 Å) and supermicropores (diameters of about 15 Å). Similarly for **10**, there are also two types of pores: ultramicropores of about 7 Å and supermicropores of about 10 Å (Fig. 11.14). These data are consistent with the dimensions estimated based on the

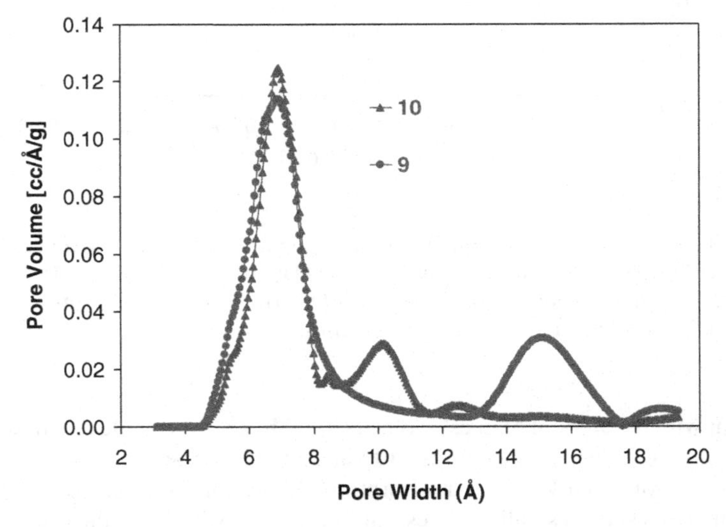

FIGURE 11.14 Horvath–Kawazoe pore size distribution of **9** and **10** based on Ar adsorption data.

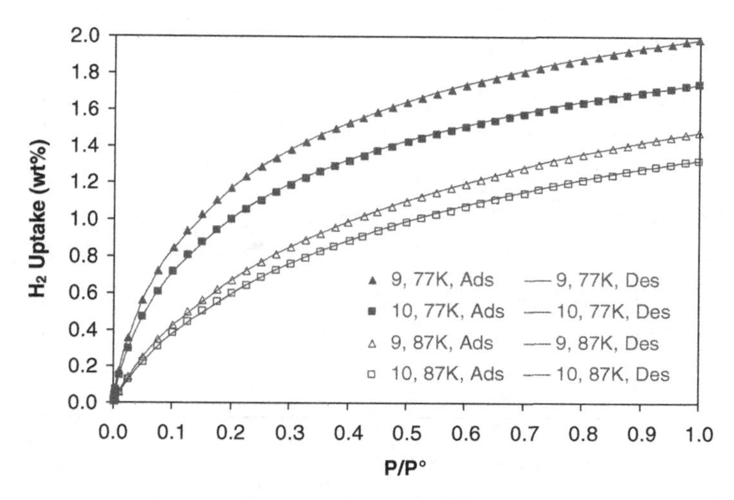

FIGURE 11.15 High-resolution H_2 adsorption and desorption isotherms at 77 and 87 K for **9** and **10**, respectively.

single-crystal structures. The total pore volume estimated based on their Ar isotherm data is consistent with these observations, giving a value of 0.38 cm g^{-1} for **9** and 0.33 cm g^{-1} for **10**.

No hysteresis was observed in the hydrogen adsorption–desorption isotherms for **9** and **10** at 77 and 87 K (Fig. 11.15). It is interesting to see the difference in the hydrogen uptake in the two structures. The Co structure (**9**) adsorbs more hydrogen than does the Zn structure (**10**) at both temperatures and all pressure levels. At 87 K and 1 atm, the values are 1.48 and 1.32 wt% (1.98 and 1.74 wt% at 77 K and 1 atm) for **9** and **10**, respectively. This difference is in agreement with the difference in their estimated total micropore volumes (0.38 and 0.33 $cm^3 g^{-1}$ for **9** and **10**, respectively). The amounts of hydrogen adsorbed in both **9** and **10** are significantly higher than those absorbed in ZSM-5 (0.7 wt%) [58] and H-SSZ-13 (1.28 wt%) [59], the latter of which was recently reported to have the highest uptake value of hydrogen among zeolite materials. It is also worth noting that the densities of the adsorbed H_2 at 77 K and 1 atm are 0.052 to 0.053 g cm^{-3} in **9** and **10**, calculated based on the estimated pore volumes using Ar sorption data at 87 K. These values are comparable to those of liquid H_2 (e.g., 0.053 g cm^{-3} at 30 K and 8.1 atm, or 0.03 g cm^{-3} at critical point 33 K and 13 atm) [60]. The isosteric heat of adsorption (Q_{st}) calculated using hydrogen-adsorption isotherms measured at 77 and 87 K, are higher for **9** than for **10** over the weight range 0.1 to 1.3 wt% (Fig. 11.16), indicative of somewhat stronger adsorbent–adsorbate interactions in **9**. The Q_{st} values for both **9** and **10** decrease as the hydrogen uptake increases, as expected.

11.3.6 [M(bdc)(ted)$_{0.5}$]·2DMF·0.2H$_2$O (M = Zn, Cu, Co, Ni; 11, 12, 13, 14)

The thermogravimetric analysis results indicate that all guest molecules of **11** (32.7%) and **12** (32.6%) can be removed at about 160 to 170°C and the resulting structure was stable up to about 300°C. The permanent porosity of the two guest-free

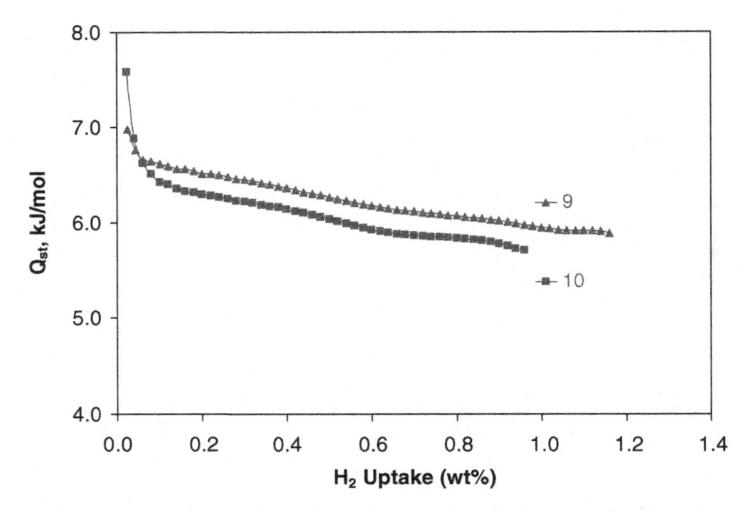

FIGURE 11.16 Isosteric heats of H_2 adsorption for **9** (filled triangles) and **10** (filled squares), respectively, as a function of the amount of hydrogen adsorbed.

frameworks was confirmed by argon adsorption at 87 K. The pore diameters calculated, determined by the Horvath–Kawazoe method, show that both materials have very similar pore size distribution between 6.6 and 8.8 Å, with a peak centered at about 7.8 Å, and the results are in excellent agreement with the values calculated from the single-crystal structure (Fig. 11.17). The BET surface area was estimated, from the type I argon adsorption isotherm, to be 1794 and 1461 $m^2\ g^{-1}$ for **11** and **12**,

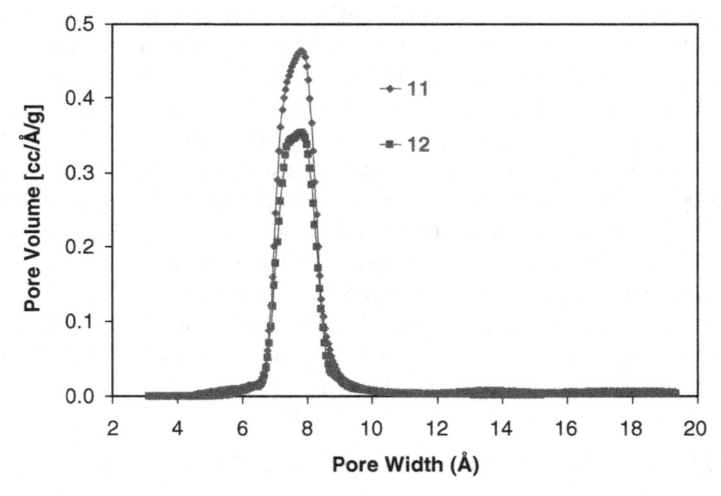

FIGURE 11.17 Horvath–Kawazoe pore size distribution for **11** and **12** (calculated from Ar sorption isotherms at 87 K).

respectively. The pore volume was calculated by the H–K method in the micropore range (i.e., pore size < 20 Å). The values are 0.65 and 0.52 $cm^3\ g^{-1}$ for **11** and **12**, respectively.

The hydrogen adsorption–desorption isotherms were measured for both **11** and **12** at 77 and 87 K as a function of relative pressure ($P/P°$) in the range 10^{-4} to 1 atm (1 atm = 101,325 Pa). No hysteresis was observed (Fig. 11.18). Higher H_2 uptake was observed for the Zn compound (**11**) at 77 and 87 K and throughout the entire

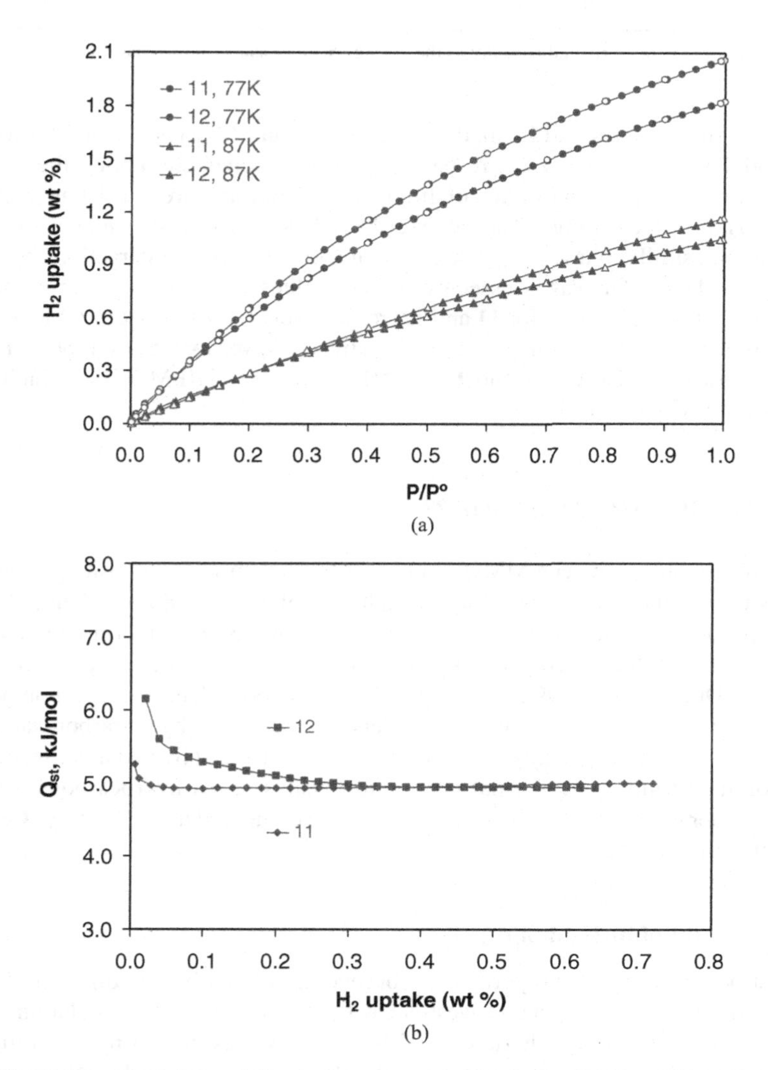

FIGURE 11.18 (a) High-resolution hydrogen sorption isotherms for **11** and **12** at 77 and 87 K (filled triangles and circles for adsorption and open ones for desorption); (b) isosteric heats of adsorption for **11** and **12**.

TABLE 11.2 Heats of Hydrogen Adsorption for Selected MMOFs

	Q_{st} (kJ mol^{-1})	Pore Size (Å)	H_2 Coverage (wt%)
6	8.3–6.5[a]	5.0	0.06–0.26
2	6.7–6.4	4.5	0.02–0.5
9	6.7–6.3	7.15	0.06–0.5
10	6.7–5.8	7.10	0.06–0.5
11	5.3–5.0	7.8	0.01–0.7
12	6.1–4.9	7.8	0.02–0.6

[a] The higher Q_{st} may be attributed partially to the zigzag irregular shape of the channels in **6**.

pressure region. At 77 K and 1 atm, the values are 2.1 and 1.8 wt% (1.2 and 1.1 wt% at 87 K and 1 atm) for **11** and **12**, respectively. These values are among the highest reported thus far under the same conditions [51,61,62] and are consistent with the estimated large pore volumes. The value for **11** at 77 K also matches well with the data reported for a similar phase [51]. The extent of sorbent–sorbate interactions between hydrogen and the MOF was assessed by Q_{st}. The values range roughly from 5.0 to 5.3 and from 4.9 to 6.1 kJ mol^{-1} for **11** and **12**, respectively (H_2 coverage: 0.01 to 0.7 and 0.02 to 0.6 wt%). Note that these values are slightly lower than those reported for **6** (cobalt formate), **2** ([Zn(tbip)], and **9** and **10** (RPM-1-Co and RPM-1-Zn), which have smaller pores (Table 11.2).

11.4 HYDROCARBON ADSORPTION

The close similarity between MMOFs and zeolites in their microporosity prompts an investigation of the hydrocarbon sorption behavior of the MMOFs and their potential to be used for gas and hydrocarbon separation applications either thermodynamically or kinetically. All the hydrocarbon sorption studies were conducted on a computer-controlled DuPont Model 990 thermogravimetric analyzer. The hydrocarbon partial pressure was varied by changing the blending ratios of hydrocarbon-saturated nitrogen and pure nitrogen gas streams. The samples were activated at temperatures based on the results from thermal gravimetric analyses. All hydrocarbons were of 99% or higher purity. Adsorption simulations were computed using Accelrys Cerius2 Sorption software [52].

11.4.1 [Cu(hfipbb)(H_2hfipbb)] (1)

The hydrocarbon sorption properties of **1** for several hydrocarbons of different shapes, sizes, and polarities have been investigated by experiments and simulations [53]. While a very small pore volume of 0.070 cm^3 g^{-1} was estimated by adsorption, **1** exhibits unique adsorption properties. It adsorbs propane and butane rapidly (Fig. 11.19) but does not adsorb pentane or higher normal or branched hydrocarbons (Table 11.3). This type of carbon-number cutoff behavior is rarely observed in other

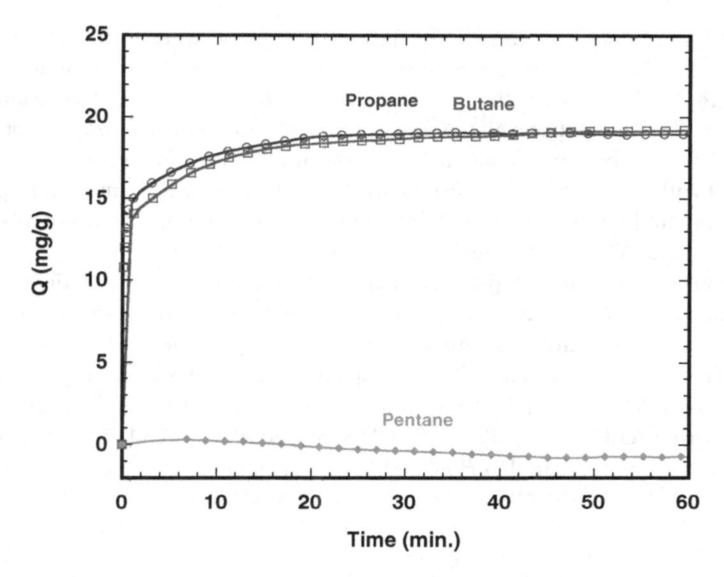

FIGURE 11.19 Adsorption rates of propane (open circles), butane (open squares), and pentane (filled diamonds) at 90°C and 650 torr in **1**.

types of adsorbents, including zeolites. Compound **1** is the first material to be found to have the potential of separating normal C2 to C4 olefins and alkanes from all branched alkanes and all normal hydrocarbons above C4. This unusual property is attributed to a narrowing of the channels at intervals of about 7.3 Å (the length of the large chamber or cage and unit cell repeat distance), which is just greater than the length of *n*-C4 (ca. 6.9 Å), and just less than the length of *n*-C5 (ca. 8.1 Å). The diameter of the neck is approximately 3.2 Å (excluding the van der Waals radii of hydrogen); therefore, the neck is too small for this region to be an equilibrium position for alkanes with diameters of around 3.9 Å but large enough to allow the passage of normal alkanes while excluding branched alkanes.

TABLE 11.3 Hydrocarbon Adsorption Properties of 1 at 25°C

	P/P^{oa}	Sorption (wt%)
Methanol	0.60	2.0
Propane	0.062	2.6
Propene	0.019	2.0
n-Butane	0.33	4.0
2-Methylpropane	~0.1	0
n-Pentane	0.100	0
3-Methylbutane	~0.1	0
n-Hexane	0.071	0
3-Methylpentane	0.078	0

[a] P is the partial pressure of the adsorbate, and P^o is the vapor pressure of the adsorbate at the sorption temperature.

Cerius2 sorption simulations indicate that if pentane were to be located in the large chamber area of the microchannel, the closest H–H interatomic distance between the molecule and the wall would be 1.86 Å, which is too small to be reasonable. The calculations also predict that a maximum of one propane or butane molecule is adsorbed per cage and that essentially no pentane is adsorbed, in agreement with the experimental results. In addition, the sorption simulations indicate that 2,2-dimethylbutane does not adsorb, that is, it does not fit in the chamber area. Thus, the shape of the molecules is critical.

Adsorption isotherms for propane and butane in **1** are shown in Figure 11.20. Values for zero-loading heat of adsorption, calculated from these data using the Clausius–Clapeyron equation, are 47 kJ mol^{-1} for propane and 50 kJ mol^{-1} for butane. The value of 47 kJ mol^{-1} for propane adsorption in **1** is higher than the values of 41 and 34 kJ mol^{-1} reported for propene adsorption in pure-silica eight-membered-ring zeolites, ITQ-12 and Si-CHA, respectively [63,64]. Compound **1** also adsorbs 2.0 wt% of methanol at $P/P^o = 0.6$ (80 torr) and 25°C, which shows that it is capable of gas-phase separation of methanol from water.

11.4.2 [Zn(tbip)] (2)

Compound **2** has the property of being very hydrophobic, exhibiting essentially zero water adsorption (<~1 mg g^{-1} at $P/P^o = 0.65$) but adsorbing high volumes of methanol, 110 mg g^{-1} at 24.5°C and 90 torr MeOH ($P/P^o = 0.73$). The shape of the MeOH adsorption isotherms (Fig. 11.21) is typical for an adsorbent that exhibits relatively weak interactions with the adsorbate (less than ca. 3 mg g^{-1} up to 40 torr, followed by a very sharp increase in adsorption that is believed to be pore filling by capillary condensation). The condensation point shifts about 4.7 torr °C^{-1}, predicting that capillary condensation should occur at very low pressure (e.g., 1 torr MeOH) at about 15°C. The adsorption properties of **2** may be compared with those of pure silica ZSM-5 (4 ppm Al), which is hydrophobic but adsorbs as much as 7 mg g^{-1} of water at $P/P^o = 0.65$ and shows moderate affinity for MeOH. Its MeOH adsorption isotherms are fitted reasonably by the Langmuir equation, with adsorption beginning at low pressures but not exhibiting the sharp increase associated with capillary condensation.

11.4.3 [Co(bpdc)(bpy)]·0.5DMF (3, RPM-2)

RPM-2 displays high sorption capacity for hydrocarbons despite its small 1D channels. Its aromatic hydrocarbon-based pore surface and distinctive pore structure offer some unique features in hydrocarbon adsoprtion. The uptake of large molecules such as cyclohexane, xylenes, and mesitylene is unexpected for **3** because of the small channel opening of about 5.6 × 3 Å (estimated based on its crystal structure). In particular, the uptake of large mesitylene molecules, which are too large for zeolite ZSM-5 with about 5.5 × 5.5 Å channels, is quite surprising, suggesting the higher flexibility of the RPM-2 structure. Triisopropylbenzene

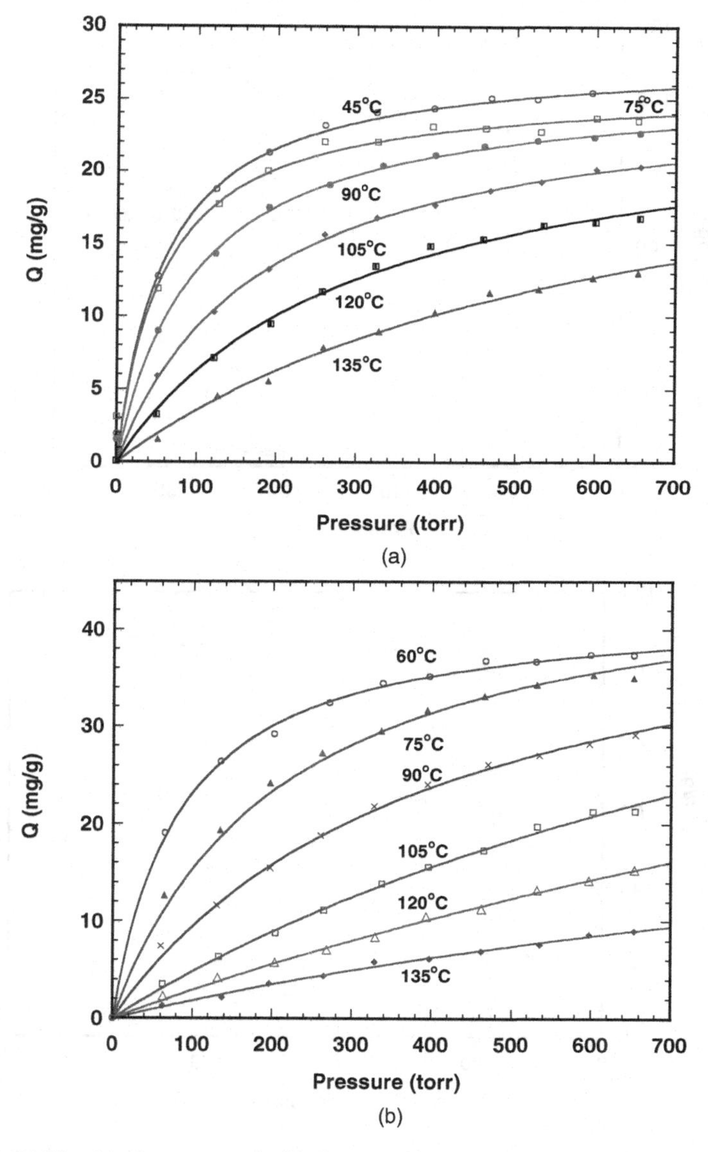

FIGURE 11.20 (a) Propane and (b) butane adsorption isotherms for **1** at various temperatures.

(ca. 8.5 Å) is excluded, consistent with RPM-2′s pore diameter. RPM-2′s adsorption capacity for *ortho*-xylene at 30°C (15 wt%), is unusually higher than for other hydrocarbons (ca. 10 wt%). In addition, an unexpected increase in *ortho*-xylene uptake capacity is observed for 60 → 40°C (ca. 4 wt% vs. ca. 1 wt% for *para*-xylene). This is attributed to more efficient packing of *ortho*-xylene. The

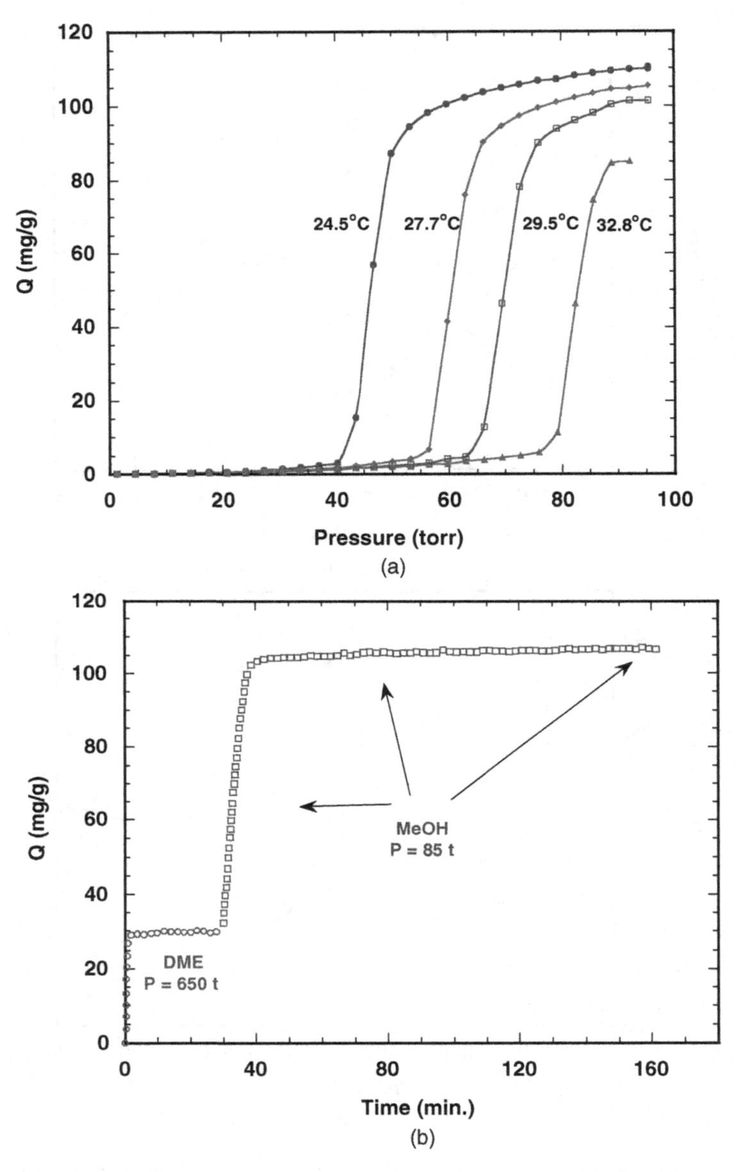

FIGURE 11.21 (a) MeOH adsorption isotherms at different temperatures; (b) adsorption of DME at 30°C followed by MeOH on **2** as a function of time.

isosteric heat of adsorption for *n*-hexane at zero coverage is calculated to be about 68 kJ mol^{-1}, considerably larger than that for zeolite H-Y (45.5 kJ mol^{-1}), and comparable to that for zeolite H-ZSM-5 (68.8 kJ mol^{-1}) [65] and RPM-1s (66 kJ mol^{-1}).

11.4.4 [Co(HCOO)] (6)

The adsorption of a series of light alcohols, light aromatics, and three linear hydrocarbons (i.e., propane, propene, and *n*-hexane) provides insights into the strong control of the channel structure of **6** on its adsorption capacity and selectivity. The lowest temperature adsorption isotherm for benzene (Fig. 11.22) is nearly flat in the high-pressure region, reaching $Q = 174\ \mathrm{mg\,g^{-1}}$, which corresponds to 3.94 molecules of benzene per unit cell. Simulated benzene adsorption gives 3.73 molecules per unit cell (with the amount adsorbed still increasing slowly at the end of the simulation). The benzene positioning and orientation from adsorption simulation correspond well with the experimental data that have been reported for the Mn–FA isotype (Fig. 11.23) [45]. This is a clear example of commensurate channel structure–guest molecule accommodation. In zeolites, this phenomenon, typically referred to as *commensurate freezing* [66–69], has been observed for the adsorption of *p*-xylene in ZSM-5 [70], and subsequently for benzene, toluene, and *n*-hexane and *n*-heptane [66–69,71].

Isotherms for toluene, ethylbenzene, and *p*-xylene demonstrate that the amount adsorbed, in terms of weight and the number of molecules adsorbed per unit cell, are all lower than that of benzene and decrease monotonically for this series. This can be attributed to increasing steric constraints as the molecular size and rigidity increases from toluene to *p*-xylene. This effect is also reflected in the isosteric heats of adsorption, which reach a maximum for toluene and then decline for the two C8 aromatics (Table 11.4).

Similarly, the adsorption isotherms of the homologous series of alcohols, from methanol and to pentanol, also reveal the controlling role of the channel structure.

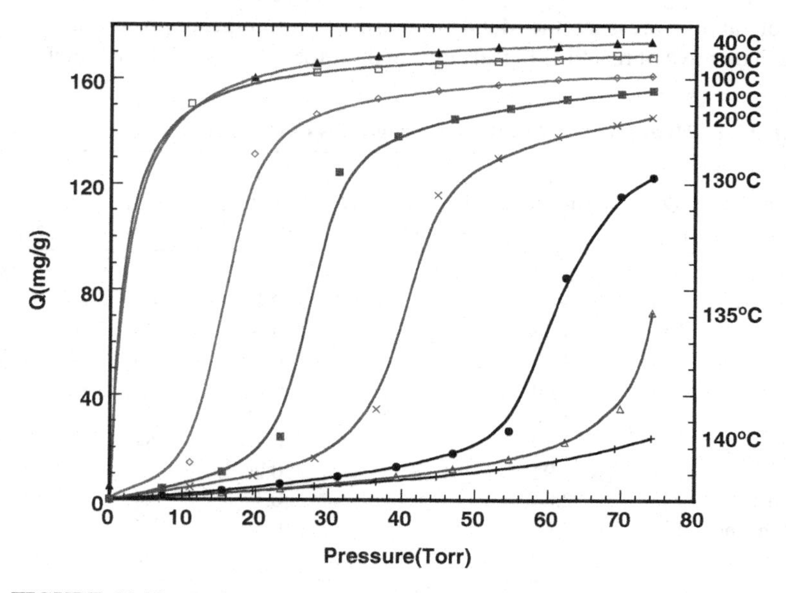

FIGURE 11.22 Benzene adsorption isotherms for **6** at various temperatures.

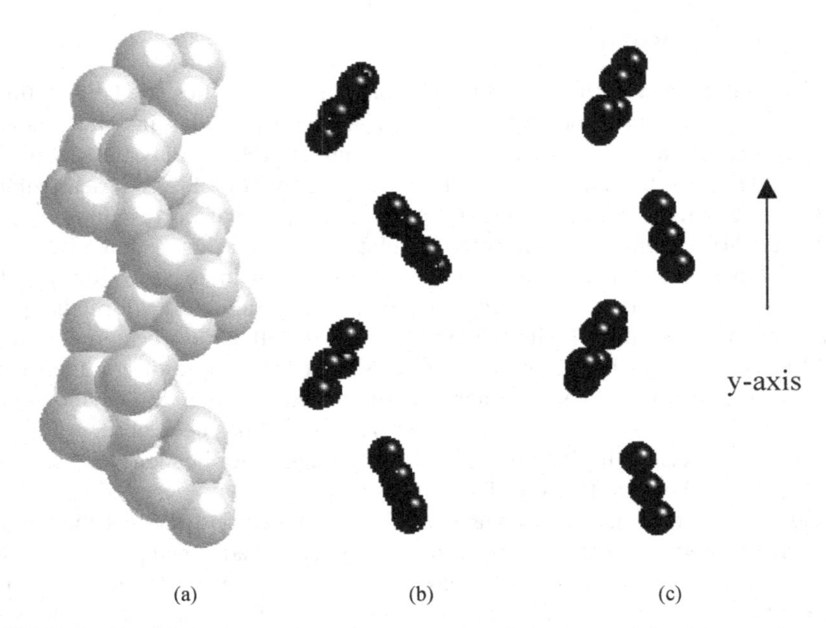

FIGURE 11.23 (a) Packing of He in a segment of the channel in **6** from adsorption simulation; (b) packing of benzene in **6** from benzene adsorption simulation; (c) packing of benzene in a channel in Mn-FA [45]. The framework atoms are removed for clarity.

At a P/P^o of about 0.5 (30°C), both MeOH and EtOH reach similar adsorption levels (99 and 106 mg g^{-1}, respectively) (Fig. 11.24). While the 30°C isotherm for EtOH levels out at high pressure, the MeOH isotherm has a significant positive slope in this region, indicative of reaching the adsorption limit for EtOH but not for MeOH. EtOH,

TABLE 11.4 Hydrocarbon Adsorption Properties of 6

	Q_{obs} (mg g^{-1})	Q_{obs} (molecules/ unit cell)	Q_{sim} (molecules/ unit cell)	Pore Volume (cm^3 g^{-1})	Q_{st} (kJ mol^{-1})	Length (Å)
Methanol	99	5.47	2.96	0.12	58	4.3
Ethanol	106	4.06	3.94	0.13	62	5.7
Propanol	143	4.20	—	0.18	76	6.8
Butanol	168	4.00	—	0.21	56	8.0
Pentanol	108	2.20	—	0.13	60	9.1
Propane	91	3.82	3.74	0.18	38	5.8
n-Hexane	102	2.10	—	0.16	54	10.2
Benzene	174	3.94	3.73*a*	0.20	54	6.7
Toluene	148	2.84	—	0.17	64	7.6
Ethylbenzene	130	2.16	—	0.15	56	8.8
p-Xylene	48	0.71	~0	0.06	62	8.5

[a] Still increasing at the end of simulation.

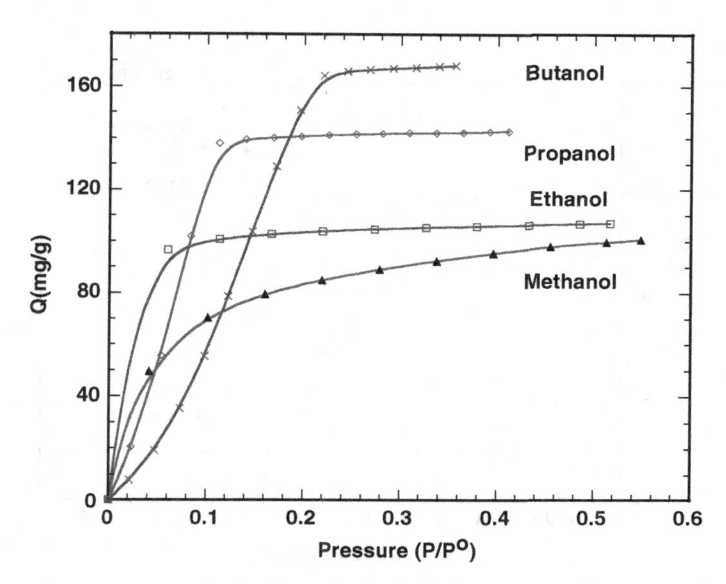

FIGURE 11.24 Light alcohol adsorption isotherms for **6** vs. P/P^o.

as well as the C3 and C4 alcohols, has an apparent adsorption limit of four molecules per unit cell (one per channel segment), indicative of commensurate accommodation of these molecules in the channels of **6**. Due to its small size, MeOH adsorption does not, however, have this loading limit of four molecules per unit cell.

In comparing the MeOH and EtOH adsorption isotherms in a low-pressure region, it appears that **6** has higher affinity for EtOH than for MeOH, which is consistent with the expected increase in heat of adsorption due to one more CH_2 group interacting with **6**. Going on to the isotherms for C3 and C4 alcohols, the affinities appear to decrease as the sharp increase in adsorption shifts to higher pressure. This seeming contradiction can be understood in terms of changes in entropy. The pore volume occupied by these molecules (Table 11.4) increases from $0.134\,cm^3\,g^{-1}$ to $0.207\,cm^3\,g^{-1}$, suggesting that the alcohols, from EtOH to BuOH, are increasingly confined by **6**. This results in a progressively positive $T\Delta S$ contribution to the free energy, decreasing the driving force for adsorption. Thus, P/P^o shifts to a higher value before dQ/dP increases sharply.

The adsorption of propane, a molecule of length comparable to EtOH (5.84 Å vs. 5.64 Å), shows a similar effect of commensurate pore shape–molecule accommodation. At high pressure ($P > 300$ torr), the isotherm at 40°C is nearly flat and the amount of adsorbed propane, $Q = 91\,mg\,g^{-1}$, equals 3.82 molecules per unit cell, again indicating a limiting value of 4, while hexane, whose size is about twice that of propane (10.2 Å), shows an adsorbed amount of $102\,mg\,g^{-1}$, which corresponds to 2.10 molecules per unit cell, roughly one-half that for propane.

The isosteric heats of adsorption for most of the adsorbents examined herein are essentially constant with loading level (Table 11.4). This is consistent with adsorption that is controlled mainly by adsorbate–adsorbent interactions.

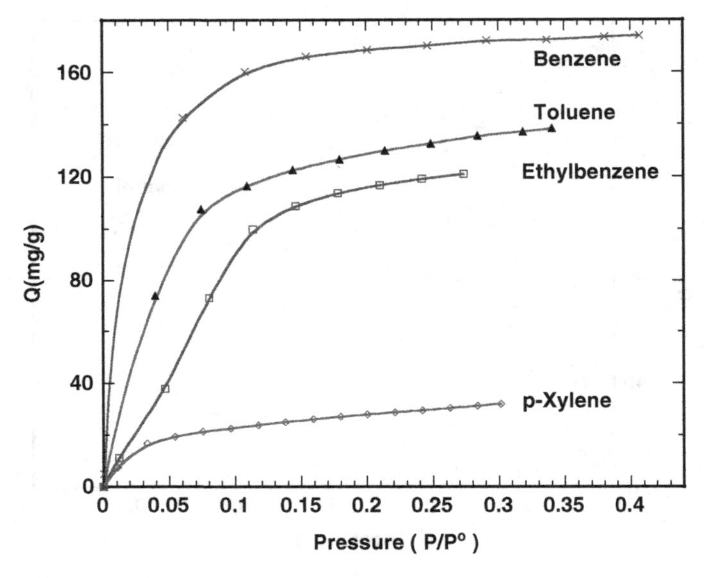

FIGURE 11.25 Aromatic (C6–C8) adsorption isotherms for **6** plotted versus P/P^{o}.

The adsorption isotherms for **6** also indicate a potential to affect separations. A comparison of the isotherms for the aromatics, plotted as the amount adsorbed versus P/P^{o} (Fig. 11.25), as well as real pressure (Fig. 11.26), indicates that the two close-boiling chemicals ethylbenzene and *p*-xylene could be separated via an

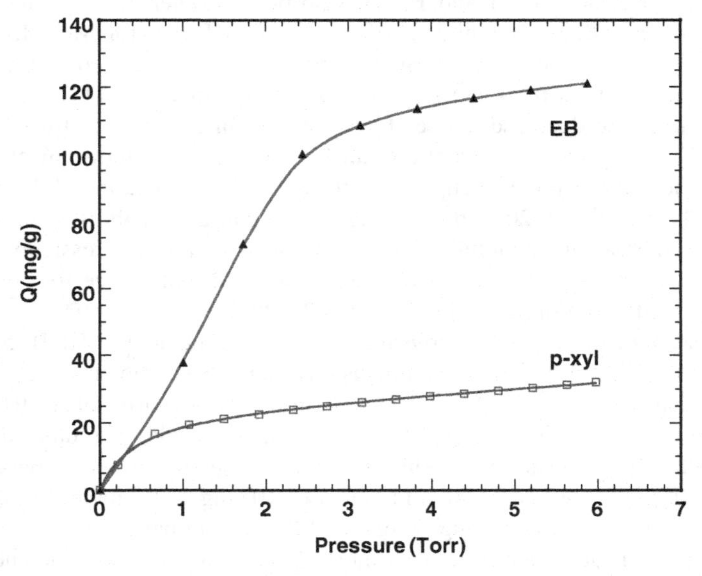

FIGURE 11.26 Comparison of 40°C ethylbenzene (EB) and *p*-xylene (*p*-xyl) adsorption isotherms for **6** under real pressure.

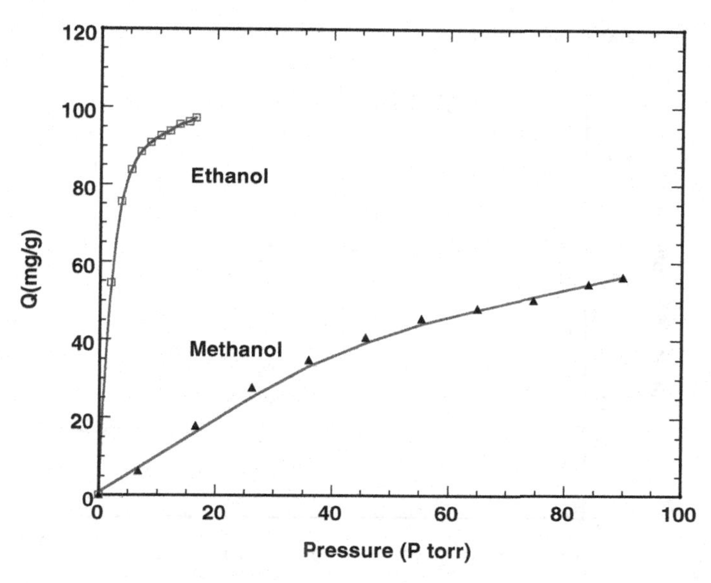

FIGURE 11.27 Comparison of 70°C methanol and ethanol adsorption isotherms for **6** under real pressure.

equilibrium-based process such as pressure swing adsorption (PSA) using **6** as the adsorbent. Commercially, C8 aromatics (i.e., *m*-, *o*-, and *p*-xylenes and ethylbenzene) are currently separated via the PARX process [72] or fractional crystallization.

A similar comparison of the series of homologous alcohols indicates that methanol and ethanol could also be separated, again via a PSA-type process. This potential is much clearer when the isotherms are compared as a function of pressure (Fig. 11.27). Although the data above indicate that separation can be achieved, no measurements or modeling has been conducted that would allow us to make a more realistic assessment of separation efficiency or cost-effectiveness.

11.4.5 [$Co_3(bpdc)_3bpy$]·4DMF·H_2O and (9, RPM-1-Co)

Compound **9** displays a high adsorption capacity for hydrocarbons, in accordance with the large channels observed. A pore volume of 0.25 $cm^3\,g^{-1}$ is estimated based on the *n*-hexane ($P = 490$ torr, $P/P^o = 0.48$) adsorption capacity (17 wt%) at 30°C. This volume is between those for the large-pore zeolite H-Y (0.32 $cm^3\,g^{-1}$, one of the most widely used large-pore zeolites) [11a] and medium-pore zeolite H-ZSM-5 (0.19 $cm^3\,g^{-1}$). At 80°C, the sorption capacity (Fig. 11.28) for cyclohexane ($P = 55$ torr, $P/P^o = 0.074$) of **9** is 19 wt%, which exceeds that of H-Y (17 wt%). The higher uptake capacity, considering its lower pore volume than H-Y, may be attributed to more efficient packing of the cyclohexane molecules in **9** and/or their stronger interaction with the pore walls. Other larger hydrocarbon molecules can also be

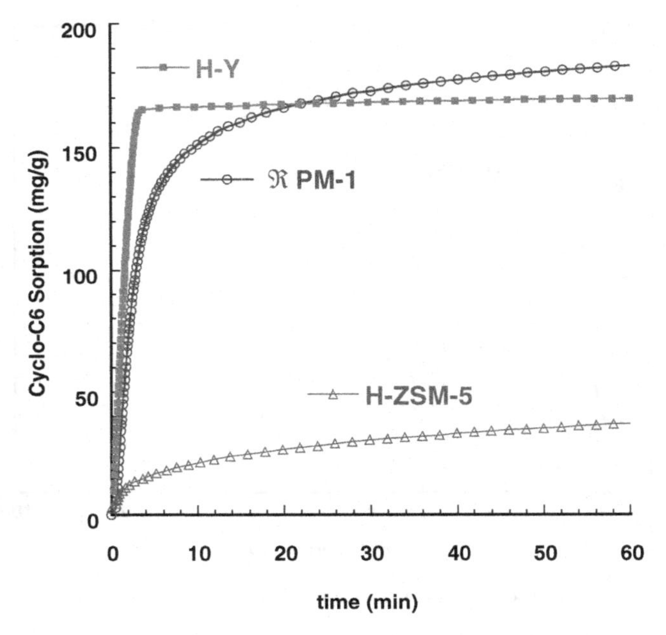

FIGURE 11.28 Cyclohexane uptake at 80°C on H-Y, RPM-1, and H-ZSM-5 as a function of sorption time.

taken up by **9**. The sorption capacities at 80°C for mesitylene (7 Å) and triisopropylbenzene (8.5 Å) are 17 and 12 wt% ($P = 1.4$ torr, $P/P^{o} = 0.27$), respectively.

The heat of *n*-hexane sorption on **9** (66 kJ mol^{-1}) is considerably larger than that of H-Y (45.5 kJ mol^{-1}) and is close to that of H-ZSM-5 (68.8 kJ mol^{-1}) [12] and **3** (68 kJ mol^{-1}). Comparison of Henry constants [13] of *n*-hexane sorption isotherms measured at 250°C: 0.0070, 0.0060, and 0.0038 torr^{-1} for H-ZSM-5, **9**, and H-Y, respectively, also confirms the strong hydrocarbon molecule–pore wall interaction in **9**.

The sorption rate can be used to estimate the size of the effective pore windows in porous materials [2]. The sorption rate of cyclohexane in **9** is slower than that in H-Y, which contains 12-membered ring windows with openings of 7.4 × 7.4 Å, and faster than that in H-ZSM-5, which has 10-membered ring windows with apertures of 5.3 × 5.6 Å. The retardation in rate for the nearly spherical cyclohexane molecule (kinetic diameter 6.0 Å) suggests a window dimension of approximately 6 Å, between those of H-Y and H-ZSM-5.

11.4.6 [Zn(bdc)(ted)$_{0.5}$]·2DMF·0.2H$_2$O (11)

Activated **11** exhibits very high adsorption capacity for MeOH, 520 mg g^{-1}, at 25°C and 60 torr (Fig. 11.29). At low pressure, adsorption is linear with pressure, then rises sharply as capillary condensation commences. A classical adsorption hysteresis loop,

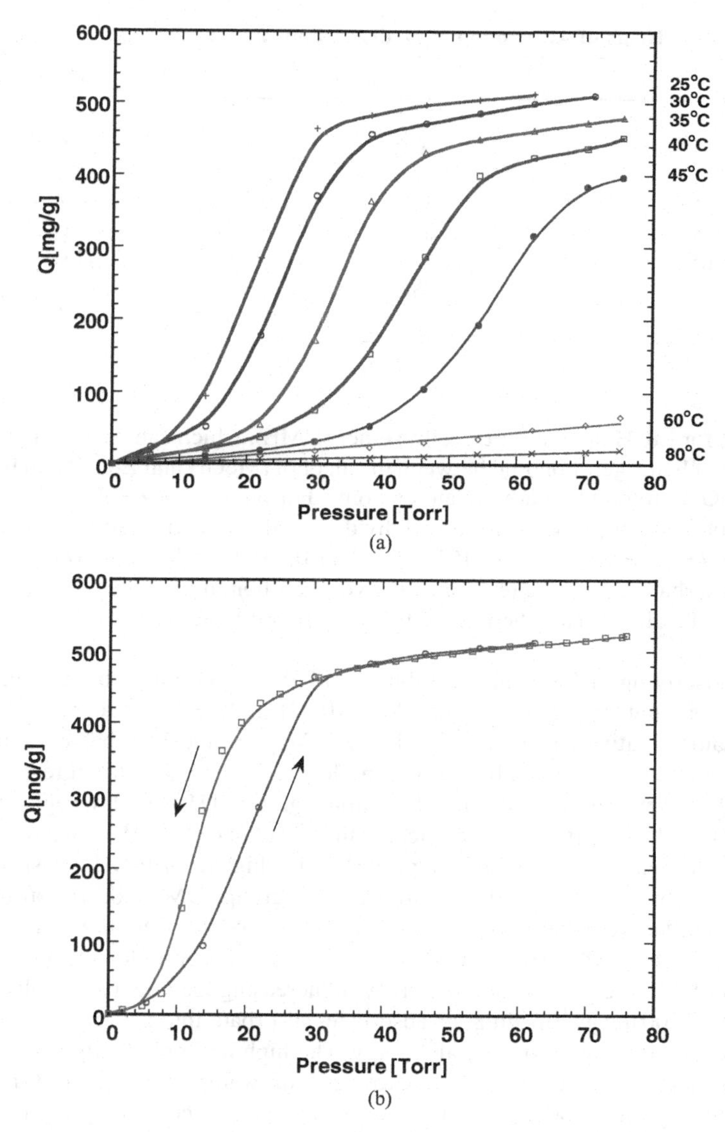

FIGURE 11.29 (a) MeOH adsorption isotherms for **11** at various temperatures (25 to 80°C); (b) adsorption and desorption of MeOH at 25°C.

frequently observed for very large pore adsorbents or where adsorbate–adsorbent interaction energies are not high, is illustrated in Figure 11.29. The isosteric heat of adsorption for MeOH in the capillary condensation region, at $Q = 260\ \mathrm{mg\ g^{-1}}$, is $40.8\ \mathrm{kJ\ mol^{-1}}$ (Table 11.5). In the Henry's law region, Q_{st} is $59.5\ \mathrm{kJ\ mol^{-1}}$, indicating stronger interaction of MeOH with the adsorbent at low loadings. A similar trend is

TABLE 11.5 Heats of Adsorption of Hydrocarbons for 11

	Q_{st} (kJ mol^{-1})	Q_{st} (Low Loading) kJ/mol^{-1})
Propane	32.2	—
Methanol	40.8	59.5
Ethanol	40.8	65.8
Dimethyl ether	45	30.3
n-Hexane	66.7	49.0
Cyclohexane	57.8	41.9
Benzene	58.0	41.8
Water	—	45.4

observed for EtOH but not for dimethyl ether (DME), which can be explained by the presence of hydrogen-bonding between the alcohol (MeOH and EtOH) and the MOF (e.g., an OH proton to a carboxylate O atom). For pure silica ZSM-5 zeolite at low MeOH loadings (the amount absorbed divided by the amount adsorbed at very high pressure, $Q/Q_{\infty} = 0.14$), Q_{st} is 45 kJ mol^{-1} [73]. At high loadings ($Q/Q_{\infty} = 0.88$), where adsorbate–adsorbate (e.g., mostly hydrogen-bonding) interactions increase, it is 60 kJ mol^{-1}. In this case there is no strong hydrogen-bonding at low loading level in ZSM-5.

The adsorption of EtOH also exhibits capillary condensation, but the maximum adsorbed amount of 418 mg g^{-1} (at $P/P^0 = 0.42$) is 20% less than that of MeOH at the same relative pressure (Fig. 11.30). We attribute this behavior to more efficient packing of the smaller MeOH molecule in the pore structure of **11**. For EtOH, Q_{st} values in the capillary condensation region and Henry's law region are 40.8 and 65.8 kJ mol^{-1}, respectively, mimicking the behavior of MeOH, with the exception of Q_{st} in the Henry's law region being 6.3 kJ mol^{-1} higher for EtOH, consistent with the energy contributions of the additional CH_2 group. DME adsorption does not exhibit çapillary condensation over the pressure and temperature range studied (Fig. 11.30). At $P/P^o = 0.13$ (30°C, $P = 520$ torr) it adsorbs 405 mg g^{-1}. The Q_{st} values for DME increase monotonically with increasing loading, rising substantially at less than 200 mg g^{-1} loading, as adsorbate–adsorbate interactions (e.g., stacking of dipoles) contribute to the overall energy. The high hydrophobicity of the surface of **11** is demonstrated, with only <6 mg g^{-1} of water adsorbed at $P/P^o = 0.42$ compared to over 400 mg g^{-1} for the three oxygen-containing hydrocarbons (Fig. 11.30c).

Clearly, **11** is expected to separate these hydrocarbons easily from water. Simplistic estimates of the separation potential of **11** for the first cycle of a pressure swing adsorption (PSA) type of process for the separation of MeOH and EtOH from water are given in Table 11.6. For example, in a first pressure swing cycle, less than 0.3 g of 98.5% EtOH can be produced for each gram of **11**. Other adsorbents, for example, Linde type 4A molecular sieves, should also be capable of removing water from ethanol and methanol. However, as water interacts strongly with this zeolite, a process using this zeolite would require activation at 200 to 300°C to desorb the water.

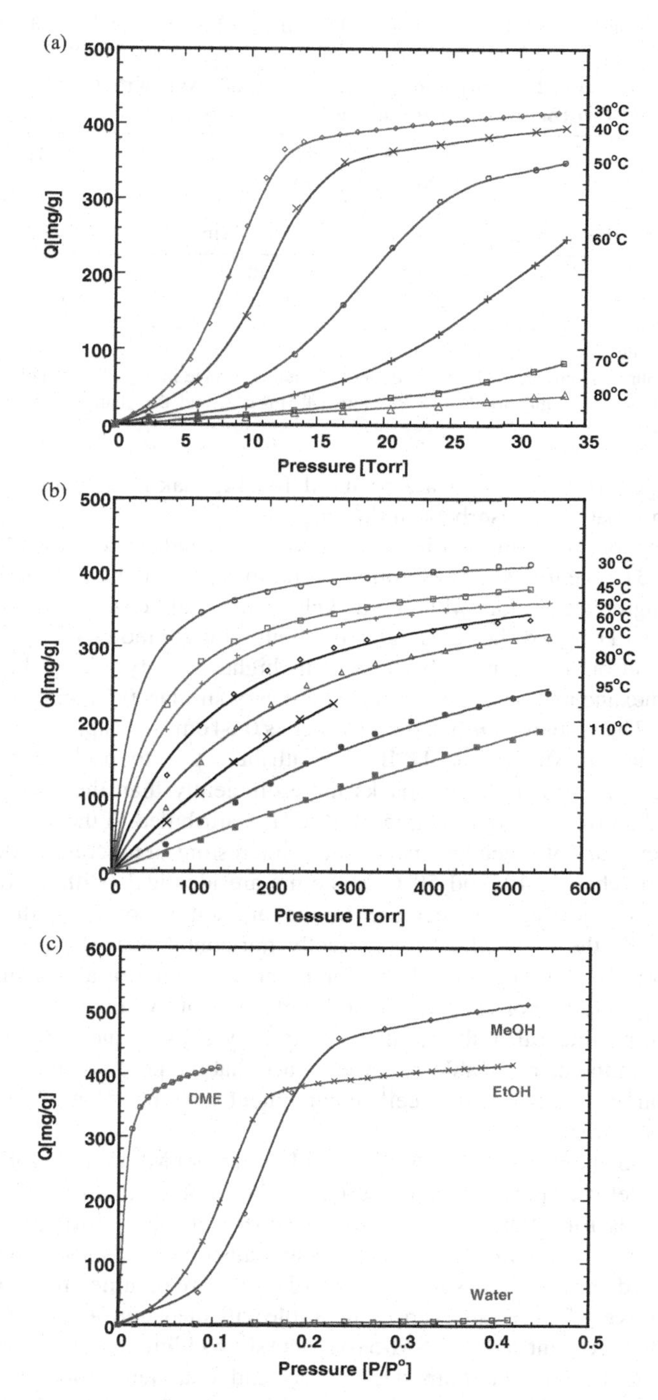

FIGURE 11.30 (a) Ethanol and (b) dimethyl ether adsorption isotherms for **11** at various temperatures; (c) overlay of MeOH, EtOH, DME, and water adsorption isotherms at 30°C.

TABLE 11.6 Effect of Pressure Change on Loading of Oxygenated Hydrocarbons and Water in 11 and Hydrocarbon Purity of the Product Stream

	Hydrocarbon/H_2O		Hydrocarbon Purity[b]	
	Pressure Change (torr)	Loading[a] ($mg\,g^{-1}$)	wt%	mol%
Methanol	45–15	395/22	94.7	87.7
Ethanol	15–5	318/46[c]	98.5	96.2

[a] Maximum loading as sole adsorbate at 40°C.
[b] Hydrocarbon purity of product effluent assuming the loading change in the adjacent column.
[c] The change in water loading for $P = 15$ to 5 torr = 4.2 mg g^{-1}, and the change in EtOH loading is 273 mg g^{-1}.

Lower energies for desorption are required for **11**, making it a more favorable hydrocarbon separation adsorbent in this case.

The adsorption isotherms of *n*-hexane, cyclohexane, and benzene for **11** are shown in Figure 11.31. The *n*-hexane adsorption isotherms appear to be modeled well with a simple Langmuir equation, while both cyclohexane and benzene exhibit capillary condensation. The two hexanes adsorb about equal amounts (weight basis), but benzene adsorbs 16% more because of its higher density. The sorbed volumes of the cyclohexane and benzene molecules are very similar: 0.54 and 0.56 $cm^3\,g^{-1}$, respectively. The volume of *n*-hexane adsorbed is 0.64 $cm^3\,g^{-1}$, approximately equal to the volumes of MeOH and DME. We attribute this to packing efficiencies; for example, the linear *n*-hexane packs more efficiently than the two cyclic C6s. As discussed above, the high volume of MeOH is attributed to the same factor.

The adsorption of cyclohexane has an interesting temperature dependence (Fig. 11.31b). The 30, 40, and 50°C isotherms are identical (within experimental error), that is, showing no decrease in the amount adsorbed with increasing temperature. Furthermore, they are perfectly horizontal over the pressure range $P = 10$ to 60 torr. Although this behavior points to a structural explanation, the $Q = 418$ mg g^{-1} value corresponds to 5.68 molecules of cyclohexane per unit cell, which is not an integral multiple of the symmetry sites of the tetragonal crystal structure. From the data in Table 11.4 we deduce that the pore volume of **11** is very high, 0.64 $cm^3\,g^{-1}$, which is in excellent agreement with the value of 0.65 $cm^3\,g^{-1}$, obtained from Ar sorption data.

Comparison of the heats of adsorption of **11** with other MOFs and zeolites reveals a correlation between pore size and adsorbate–adsorbent interactions. The relatively high Q_{st} values for MOF materials are in general attributed to their relatively small pore diameters, which leads to extensive adsorbate–adsorbent interactions. High heat of adsorption values of 66 and 68 kJ mol^{-1} are obtained for *n*-hexane in **9** and **3**, respectively. These values are comparable with the value 66.7 kJ mol^{-1} for **11**. The high-silica-content zeolite ZSM-5 [65] also shows high Q_{st} values for *n*-hexane, again attributed to its medium-sized pores and channels, allowing extensive adsorbate–adsorbent interactions. Accordingly, the larger-pore zeolite H-Y has a significantly lower Q_{st} value, 46 kJ mol^{-1} [65]. Finally, Q_{st} for propane is similar for

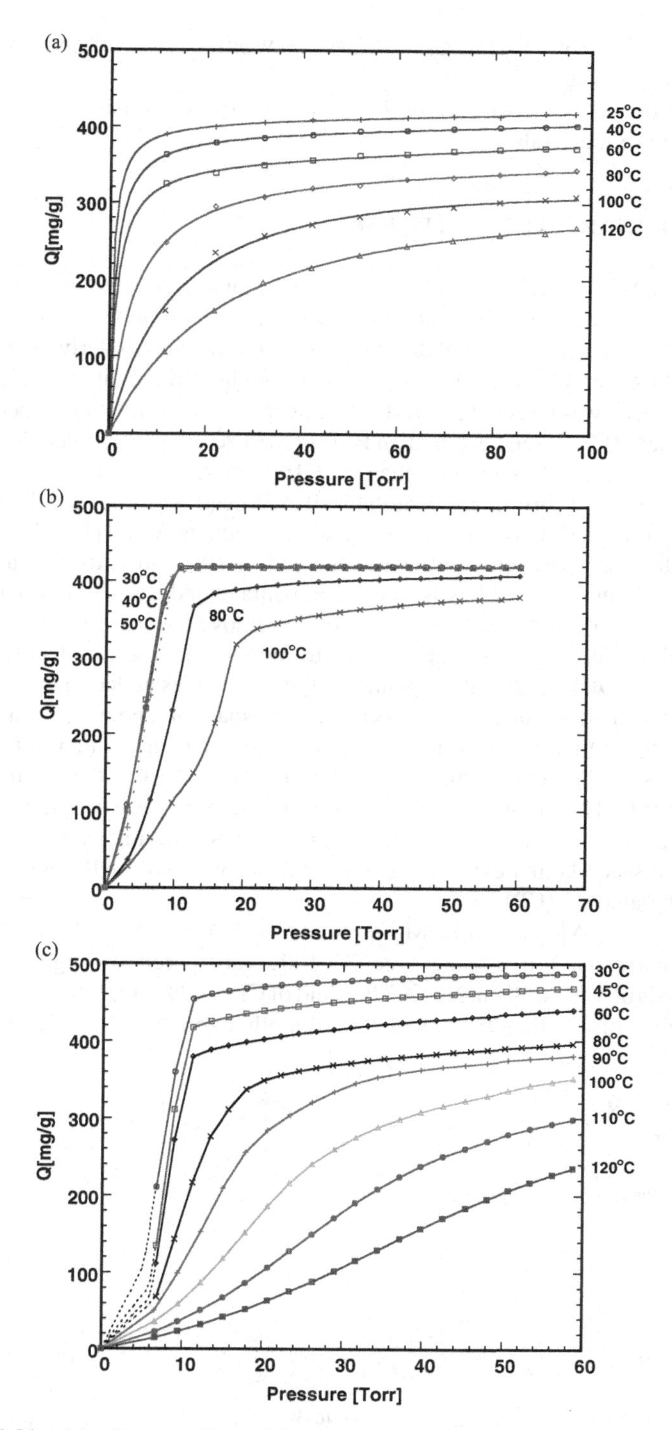

FIGURE 11.31 (a) *n*-Hexane, (b) cyclohexane, and (c) benzene sorption isotherms for **11** at various temperatures.

11 and the pure-silica small-pore zeolite Si-CHA [63,64], consistent with the large cavities in both materials.

11.5 SHIP-IN-BOTTLE SYNTHESIS

The unique 3D (**9**) $\leftrightarrow$ 1D (**5**) recyclable interconverions discovered in the synthesis of **9** prompted investigation of **9**'s ability to serve as a host for photochemical reactions [74]. The photochemistry of dibenzylketone (DBK) and its derivatives adsorbed on classical FAU and MFI zeolites has been thoroughly investigated [75,76]. For our study, *o*-MeDBK was selected as a test substrate in consideration of the specific shape of 1D channels of **9**. When adsorbed on FAU or MFI zeolites this molecule undergoes two photoreactions, as shown in Scheme 11.1: (1) α-cleavage followed by loss of carbon monoxide to form a geminate pair of hydrocarbon radicals, which undergo geminate (product AB) or random combination (products AA, AB, and BB), and (2) an intramolecular hydrogen abstraction followed by cyclization to form a cyclopentanol (CP). About 50 mg of **9** was slurried in pentane and transferred to a branched quartz cell. Argon was blown though the cell to remove the solvent. The sample was then heated to 150°C for an hour at 1 torr to remove the guest molecules. A 2-mg sample of *o*-MeDBK in 0.3 mL of pentane/ether (1 : 1) was added to the cell at room temperature under Ar. The mixture was allowed to soak for 2 hours, then flushed with Ar, and pumped to 2×10^{-5} torr overnight. It was then irradiated with a 500-W medium-pressure mercury lamp for 1 hour. The sample was then extracted with ether (extract 1). The irradiated sample was then soaked in water until the color turned to whitish-gray and was then extracted with an excess amount of ether (extract 2). The first and second ether extracts gave about 30% and ~30% AB, respectively, and >40% of the alcohol (CP).

Photolysis of *o*-MeDBK at RPM-1 produced only AB with 60% yield in reaction (1), which corresponds to a "cage effect" of 100% (compared to a cage effect of 70% for photolysis of *o*-MeDBK in NaX). The yield of CP is 40% in reaction (2). This yield of CP is much higher than the values found in other zeolites (e.g., NaX), where the

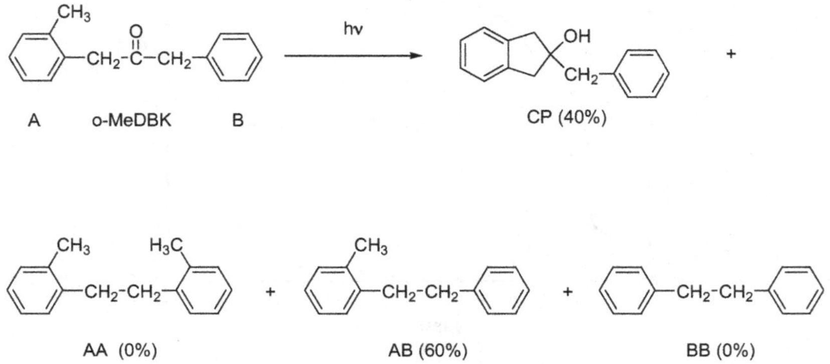

A o-MeDBK B —hν→ CP (40%) +

AA (0%) + AB (60%) + BB (0%)

SCHEME 11.1

maximum yield is about 10%. More significantly, only about 50% of the overall products could be extracted before breaking the RPM-1 framework. The remaining 50% of the products were recovered only after **9** was immersed in water and converted completely to nonporous 1D precursor, giving a 100% mass balance (compared to ca. 60 to 70% mass balance of NaX). The results from the photolysis demonstrate the unique potential of **9** to serve as a "smart" porous host for "ship-in-bottle" photochemistry and other reactions.

11.6 SUMMARY AND CONCLUSIONS

In this chapter we present the rational syntheses, structure characterization, and gas and hydrocarbon adsorption–desorption properties of a series of MMOFs that feature increasing pore sizes and pore volumes. Depending on the size, shape, and functionality of the ligands, as well as the coordination habit of the primary and secondary metals or metal cluster building units, these MMOFs feature very different framework topology and pore structures, which are fully characterized by single-crystal and powder x-ray diffraction, thermogravimetric analysis, and gas adsorption.

These materials all show good thermal stability and retain structural integrity after removal of the solvent guest molecules. Their permanent microporosity has been confirmed by gas adsorption studies. As the pore size and pore volume as well as the specific surface area increases, the H_2 uptake capacity (at low pressure and low temperatures) of these materials also shows monotonical increase, while the isosteric heats of hydrogen adsorption show more complicated correlations. Further considerations as to their framework and pore structures, such as the identity of the metals and ligands, pore shape, curvature, and nature of the pore walls, must also be taken into account to give a reasonable explanation.

Our study also shows that the MMOFs reported herein can adsorb significant amounts of various hydrocarbons. For some of them, significant differences in the adsorption isotherms of various hydrocarbons were observed, pointing to the potential utility of these MMOFs for separation and purification of the corresponding hydrocarbons. Overall, MMOFs have been proven, by our and others' research, to be a new type of microporous adsorbent material that is more versatile than other microporous adsorbents in synthesis, characterization, and applications. MMOFs with new structures and properties will continue to emerge, and their use as practical functional materials is expected in the foreseeable future.

Acknowledgments

We are grateful to the National Energy Technology Laboratory (grant DE-FC26-05NT42446), the Department of Energy (grant DE-FG02-08ER46491), and the New Energy and Industrial Technology Development Organization (NEDO), Japan for partial support of the work discussed in this chapter. We thank Dr. Xiaoying Huang and Dr. Thomas J. Emge for their contributions to structure analysis.

REFERENCES

1. Yaghi, O. M.; Li, H. L.; Davis, C.; Richardson, D.; Groy, T. L. *Acc. Chem. Res.* **1998**, *31*, 474.
2. Kitagawa, S.; Kitaura, R.; Noro, S.-I. *Angew. Chem., Int. Ed.* **2004**, *43*, 2334.
3. Ferey, G. *Chem. Soc. Rev.* **2008**, *37*, 191.
4. Everett, D. H. *Pure Appl. Chem.* **1972**, *31*, 578.
5. Eddaoudi, M.; Moler, D. B.; Li, H.; Chen, B.; Reineke, T. M.; O'Keeffe, M.; Yaghi, O. M. *Acc. Chem. Res.* **2001**, *34*, 319.
6. Seayad, A. M.; Antonelli, D. M. *Adv. Mater.* **2004**, *16*, 765.
7. Rowsell, J. L. C. O.; Yaghi, M. *Micropor. Mesopor. Mater.* **2004**, *73*, 3.
8. Dincă, M.; Yu, A. F.; Long, J. R. *J. Am. Chem. Soc.* **2006**, *128*, 8904.
9. Bradshaw, D.; Claridge, J. B.; Cussen, E. J.; Prior, T. J.; Rosseinsky, M. J. *Acc. Chem. Res.* **2005**, *38*, 273.
10. Kitagawa, S.; Kondo, M. *Bull. Chem. Soc. Jpn.* **1998**, *71*, 1739.
11. Barton, T. J.; Bull, L. M.; Klemperer, W. G.; Loy, D. A.; McEnaney, B.; Misono, M.; Monson, P. A.; Pez, G.; Scherer, G. W.; Vartuli, J. C.; Yaghi, O. M. *Chem. Mater.* **1999**, *11*, 2633.
12. Dinolfo, P. H.; Hupp, J. T. *Chem. Mater.* **2001**, *13*, 3113.
13. Bahr, D. F.; Reid, J. A.; Mook, W. M.; Bauer, C. A.; Stumpf, R.; Skulan, A. J.; Moody, N. R.; Simmons, B. A.; Shindel, M. M.; Allendorf, M. D. *Phys. Rev. B* **2007**, *76*, 7.
14. Wang, X. L.; Qin, C.; Wang, E. B.; Su, Z. M. *Chem. Eur. J.* **2006**, *12*, 2680.
15. Maspoch, D.; Ruiz-Molina, D.; Veciana, J. *J. Mater. Chem.* **2004**, *14*, 2713.
16. Maspoch, D.; Ruiz-Molina, D.; Veciana, J. *Chem. Soc. Rev.* **2007**, *36*, 770.
17. Maspoch, D.; Domingo, N.; Roques, N.; Wurst, K.; Tejada, J.; Rovira, C.; Ruiz-Molina, D.; Veciana, J. *Chem. Eur. J.* **2007**, *13*, 8153.
18. Frost, H.; Snurr, R. Q. *J. Phys. Chem. C* **2007**, *111*, 18794.
19. Rosi, N. L.; Eckert, J.; Eddaoudi, M.; Vodak, D. T.; Kim, J.; O'Keeffe, M.; Yaghi, O. M. *Science* **2003**, *300*, 1127.
20. Li, Y. W.; Yang, R. T. *J. Am. Chem. Soc.* **2006**, *128*, 726.
21. Blomqvist, A.; Araujo, C. M.; Srepusharawoot, P.; Ahuja, R. *Proc. Natl. Acad. Sci. USA.* **2007**, *104*, 20173.
22. Dincă, M.; Dailly, A.; Liu, Y.; Brown, C. M.; Neumann, D. A.; Long, J. R. *J. Am. Chem. Soc.* **2006**, *128*, 16876.
23. Collins, D. J. H.; Zhou, C. *J. Mater. Chem.* **2007**, *17*, 3154.
24. Thomas, K. M. *Catal. Today* **2007**, *120*, 389.
25. Ma, S. Q.; Sun, D. F.; Simmons, J. M.; Collier, C. D.; Yuan, D. Q.; Zhou, H. C. *J. Am. Chem. Soc.* **2008**, *130*, 1012.
26. Thallapally, P. K.; Kirby, K. A.; Atwood, J. L. *New J. Chem.* **2007**, *31*, 628.
27. Wang, S. Y. *Energy Fuels* **2007**, *21*, 953.
28. Frost, H.; Duren, T.; Snurr, R. Q. *J. Phys. Chem. B* **2006**, *110*, 9565.

29. Eddaoudi, M.; Kim, J.; Rosi, N.; Vodak, D.; Wachter, J.; O'Keeffe, M.; Yaghi, O. M. *Science* **2002**, *295*, 469.
30. Duren, T.; Sarkisov, L.; Yaghi, O. M.; Snurr, R. Q. *Langmuir* **2004**, *20*, 2683.
31. Eddaoudi, M.; Kim, J.; Rosi, N.; Vodak, D.; Wachter, J.; O'Keefe, M.; Yaghi, O. M. *Science* **2002**, *295*, 469.
32. Keskin, S.; Sholl, D. S. *J. Phys. Chem. C* **2007**, *111*, 14055.
33. Hayashi, H.; Côté, A. P.; Furukawa, H.; O'Keeffe, M.; Yaghi, O. M. *Nat. Mater.* **2007**, *6*, 501.
34. Loiseau, T.; Lecroq, L.; Volkringer, C.; Marrot, J.; Férey, G.; Haouas, M.; Taulelle, F.; Bourrelly, S.; Llewellyn, P. L.; Latroche, M. *J. Am. Chem. Soc.* **2006**, *128*, 10223.
35. Wang, Q. M.; Shen, D.; Bülow, M.; Lau, M. L.; Deng, S.; Fitch, F. R.; Lemcoff, N. O.; Semanscin, J. *Micropor. Mesopor. Mater.* **2002**, *55*, 217.
36. Pan, L.; Sander, M. B.; Huang, X. Y.; Li, J.; Smith, M.; Bittner, E.; Bockrath, B.; Johnson, J. K. *J. Am. Chem. Soc.* **2004**, *126*, 1308.
37. Batten, S. R.; Robson, R. *Angew. Chem., Int. Ed.* **1998**, *37*, 1460.
38. Spek, A. L.Utrecht University, Utrecht, The Netherlands, available at http://www.cryst.chem.uu.nl/platon/platon.
39. Pan, L.; Parker, B.; Huang, X. Y.; Olson, D. H.; Lee, J.; Li, J. *J. Am. Chem. Soc.* **2006**, *128*, 4180.
40. Janiak, C. *J. Chem. Soc., Dalton Trans.* **2000**, 3885.
41. Pan, L.; Liu, H.; Kelly, S. P.; Huang, X.; Olson, D. H.; Li, J. *Chem. Commun.* **2003**, 854.
42. Pan, L.; Ching, N.; Huang, X. Y.; Li, J. *Inorg. Chem.* **2000**, *39*, 5333.
43. Li, K.; Olson, D. H.; Lee, J.; Bi, W.; Wu, K.; Yuen, T.; Xu, Q.; Li, J. *Adv. Funct. Mater.* **2008**, *18*, 2205.
44. Dybtsev, D. N.; Chun, H.; Yoon, S. H.; Kim, D.; Kim, K. *J. Am. Chem. Soc.* **2004**, *126*, 32.
45. Wang, Z. M.; Zhang, B.; Fujiwara, H.; Kobayashi, H.; Kurmoo, M. *Chem. Commun.* **2004**, 416.
46. Lee, J.; Li, J.; Jagiello, J. *J. Solid State Chem.* **2005**, *178*, 2527.
47. Pan, L.; Liu, H.; Lei, X.; Huang, X.; Olson, D. H.; Turro, N. J.; Li, J. *Angew. Chem., Int. Ed.* **2003**, *42*, 542.
48. Lee, J. Y.; Pan, L.; Kelly, S. P.; Jagiello, J.; Emge, T. J.; Li, J. *Adv. Mater.* **2005**, *17*, 2703.
49. Lee, J. Y.; Olson, D. H.; Pan, L.; Emge, T. J.; Li, J. *Adv. Funct. Mater.* **2007**, *17*, 1255.
50. Seki, K.; Mori, W. *J. Phys. Chem. B* **2002**, *106*, 1380.
51. Dybtsev, D. N.; Chun, H.; Kim, K. *Angew. Chem., Int. Ed.* **2004**, *43*, 5033.
52. *Accelrys Cerius^2 Sorption, Molecular Simulations*, Accelrys, San Diego, CA.
53. Pan, L.; Olson, D. H.; Ciemnolonski, L. R.; Heddy, R.; Li, J. *Angew. Chem., Int. Ed.* **2006**, *45*, 616.
54. Dubinin, M. M.; Astakhov, V. A. *Adv. Chem. Ser.* **1971**, *102*, 69.
55. Horvath, G.; Kawazoe, K. *J. Chem. Eng. Jpn.* **1983**, *16*, 474.
56. Lowell, S.; Shields, J. E.; Thomas, M. *Characterization of Porous Solids and Powders: Surface Area, Pore Size, and Density*; Kluwer Academic, Dordrecht, The Netherlands, **2004**.

57. Sing, K. S. W.; Everett, D. H.; Haul, R. A. W.; Moscou, M.; Pierotti, R. A.; Rouquerol, J.; Siemieniewska, T. *Pure Appl. Chem* **1985**, *57*, 603.
58. Weitkamp, J.; Fritz, M.; Ernst, S. *Int. J. Hydrogen Energy* **1995**, *20*, 967.
59. Zecchina, A.; Bordiga, S.; Vitillo, J. G.; Ricchiardi, G.; Lamberti, C.; Spoto, G.; Bjorgen, M.; Lillerud, K. P. *J. Am. Chem. Soc.* **2005**, *127*, 6361.
60. National Institute of standards and Technology, http://www.nist.gov/.
61. Chun, H.; Dybtsev, D. N.; Kim, H.; Kim, K. *Chem. A Eur. J.* **2005**, *11*, 3521.
62. Wong-Foy, A. G.; Matzger, A. J.; Yaghi, O. M. *J. Am. Chem. Soc.* **2006**, *128*, 3494.
63. Olson, D. H.; Camblor, M. A.; Villaescusa, L. A.; Kuehl, G. H. *Micropor. Mesopor. Mater.* **2004**, *67*, 27.
64. Olson, D. H.; Yang, X. B.; Camblor, M. A. *J. Phys. Chem. B* **2004**, *108*, 11044.
65. Denayer, J. F.; Souverijns, W.; Jacobs, P. A.; Martens, J. A.; Baron, G. V. *J. Phys. Chem. B* **1998**, *102*, 4588.
66. Makowski, W.; Majda, D. *Appl. Surf. Sci.* **2005**, *252*, 707.
67. Vlugt, T. J. H.; Krishna, R.; Smit, B. *J. Phys. Chem. B* **1999**, *103*, 1102.
68. Vanwell, W. J. M.; Wolthuizen, J. P.; Smit, B.; Vanhooff, J. H. C.; Vansanten, R. A. *Angew. Chem., Int. Ed.* **1995**, *34*, 2543.
69. Smit, B.; Maesen, T. L. M. *Nature* **1995**, *374*, 42.
70. Olson, D. H.; Kokotailo, G. T.; Lawton, S. L.; Meier, W. M. *J. Phys. Chem.* **1981**, *85*, 2238.
71. Reischman, P. T.; Schmitt, K. D.; Olson, D. H. *J. Phys. Chem.* **1988**, *92*, 5165.
72. Mowry, J. R. Handbook of Petroleum Refining Processes, McGraw-Hill, New York, **1986**.
73. Akhmedov, K. S.; Rakhmatkariev, G. U.; Dubinin, M. M.; Isirikyan, G. V. *SSR* **1987**, *8*, 1717.
74. Lei, X.; Doubleday, J. C. E.; Zimmt, M. B.; Turro, N. J. *J. Am. Chem. Soc.* **1986**, *108*, 2444.
75. Turro, N. J. *Acc. Chem. Res.* **2000**, *33*, 637.
76. Ramamurthy, V.; Garcia-Garibay, M. In *Comprehensive Supramolecular Chemistry*, Bein, T., Ed., Pergamon Press, Oxford, UK, **1996**.

12

DESIGN AND CONSTRUCTION OF METAL–ORGANIC FRAMEWORKS FOR HYDROGEN STORAGE AND SELECTIVE GAS ADSORPTION

SHENG-QIAN MA, CHRISTOPHER D. COLLIER, AND HONG-CAI ZHOU
Department of Chemistry and Biochemistry, Miami University, Oxford, Ohio

12.1 INTRODUCTION

In the last two decades we have witnessed an explosive growth in the field of metal–organic frameworks (MOFs), a new type of functional material. MOFs (also known as *coordination polymers*) are crystalline inorganic–organic hybrids constructed by assembling metal ions or small metal-containing clusters [known as *secondary building units* (SBUs)] with multitopic organic ligands (such as carboxylates, tetrazolates, and sulfoxolates) via coordination bonds [1–4]. They can be one-, two-, or three-dimensional infinite networks [5–7]. Of those, three-dimensional MOFs with permanent porosity are of the greatest interest because the voids inside the frameworks can accommodate guest molecules for a number of applications. Through judicious selection of the metal ions or SBUs and organic linkers, not only can a variety of topologies and structures be produced [8–10], but the pore sizes can be tuned systematically and the pore walls can be functionalized for specific applications in catalysis [11,12], gas storage [13–15], and gas separation [16,17]. In this chapter we review hydrogen storage and selective gas adsorption studies of MOFs.

Design and Construction of Coordination Polymers, Edited by Mao-Chun Hong and Ling Chen

12.2 HYDROGEN STORAGE IN POROUS METAL–ORGANIC FRAMEWORKS

12.2.1 Hydrogen Storage Targets and Current Storage Methods

The decreasing amount of fossil fuels and increasing threat of global warming have prompted the search for alternative energy carriers to supplement those currently in use. Among various alternatives, hydrogen stands at the forefront, due to its clean combustion and high gravimetric energy density.

In 2003, the U.S. government launched the Hydrogen Fuel Initiative for developing clean, hydrogen-powered automobiles to replace those currently powered by fossil fuels. The success of commercialization of hydrogen fuel cell–powered vehicles, however, relies largely on the development of a safe, efficient, and economic on-board hydrogen storage system. Based on the concept that today's vehicles will be powered by future higher-efficiency hydrogen fuel cell power sources, the U.S. Department of Energy (DOE) has set a number of targets for the hydrogen storage system (including the container and necessary components): 6.0 wt% or 45 kg m^{-3} by the year 2010, and 9.0 wt% or 81 kg m^{-3} by the year 2015 at near-ambient temperatures (-50 to 80°C) and applicable pressures (less than 100 bar). Additionally, the kinetics of hydrogen release and recharging must meet the requirements for practical applications. In other words, hydrogen adsorption and desorption should be totally reversible and the recharging of hydrogen should be completed within minutes [18–20].

Several methods are being explored for on-board hydrogen storage. Although compressed hydrogen gas and cryogenically stored liquid hydrogen are currently utilized in fuel cell–powered vehicles for demonstration, the high pressure (>700 bar) of the compressed hydrogen gas and the large amount of energy input for the cryogenic storage of liquid hydrogen preclude their commercialization for daily use [21].

Metal hydrides and chemical hydrides have also been studied actively as hydrogen carriers in the past decades; however, the irreversibility of hydrogen sorption and poor kinetics of hydrogen recharging necessitate continued investigation to improve the uptake and release kinetics and retention of cycling capacity [22,23].

Compared to chemical means of hydrogen storage, physisorption of hydrogen using porous materials has the advantages of fast charge–recharge kinetics as well as favorable thermodynamics. In the past decade, activated carbons [24], nanotubes [25], and inorganic zeolites [26] have been investigated widely as potential candidates for hydrogen storage. However, the weak interactions (through van der Waals forces) between hydrogen molecules and the aforementioned sorbents lead to very limited hydrogen uptake, even at low temperatures under high pressures, despite their substantial surface areas.

12.2.2 Porous MOFs as Promising Candidates for Hydrogen Storage

Porous MOFs have recently been considered as one of the most promising candidates to approach the DOE targets for on-board hydrogen storage, due to their high specific surface areas, tunable pore sizes, functionalizable pore walls, and well-defined framework–hydrogen interacting sites [27].

In 2003, Rosi and co-workers reported the first measurements of hydrogen adsorption on a MOF, albeit the exceedingly high uptake of 4.5 wt% at 77 K and 1 bar was later revised to 1.32 wt% [28]. Since then, over 70 porous MOFs have been investigated for hydrogen adsorption [27].

Low–Pressure Cryotemperature Hydrogen Adsorption Studies Although the DOE targets for hydrogen storage are set at the condition of near-ambient temperatures and applicable pressures (<100 bar), the uptake values of hydrogen at 77 K and 1 bar have been widely investigated and deemed to be one of the standards to compare the hydrogen adsorption capacities of MOF materials [27]. These values are very useful and instructive at the early stage of exploration for hydrogen storage materials. Several factors influencing the hydrogen uptake of porous MOFs at 77 K and 1 bar, such as specific surface area/pore volume, pore sizes, and catenation have been studied extensively.

Specific Surface Area/Pore Volume Generally speaking, pore volume is proportional to specific surface area. In general, porous MOFs exhibit higher specific surface areas than those of carbon materials and inorganic zeolites. The record for surface area has been broken repeatedly in the past several years. For example, a porous MOF with a specific surface area of 4500 $m^2\,g^{-1}$ and pore volume of 1.61 $cm^3\,g^{-1}$ was reported for MOF-177 in 2004 [29]; the record was eclipsed by MIL-101 with a surface area and pore volume of 5500 $m^2\,g^{-1}$ and 1.9 $cm^3\,g^{-1}$, respectively in 2005 [30].

The parameters of surface area and pore volume influencing hydrogen uptake at 77 K and 1 bar have been studied intensively in MOFs. However, it has been found that MOFs with high specific surface areas (above 1000 $m^2\,g^{-1}$) and large pore volumes (over 1.0 $cm^3\,g^{-1}$) show no direct correlation between specific surface area/pore volume and hydrogen uptake [27]. For example, MOF-177 can adsorb only 1.25 wt% hydrogen at 77 K, 1 atm, despite its high surface area of 4500 $m^2\,g^{-1}$ and pore volume of 1.61 $cm^3\,g^{-1}$; however, IRMOF-8, whose surface area (1466 $m^2\,g^{-1}$) and pore volume (0.52 $cm^3\,g^{-1}$) are less than one-third of those of MOF-177, can adsorb 1.5 wt % hydrogen under similar conditions [31]. The lack of a linear correlation between hydrogen adsorption capacity and surface area/pore volume strongly indicates that low-pressure hydrogen adsorption is controlled by other factors, which are discussed below.

Pore Size The low hydrogen adsorption capacities in porous MOFs with high surface areas and large pore volumes are presumably due to the weak interactions between hydrogen molecules and the frameworks resulting from large pore sizes and spatial free void spaces [27]. Reduction in pore size is known to enhance the interaction energy as the attractive potential fields of opposite walls overlap [32]. This has been explored extensively as a strategy to increase hydrogen–framework interactions, thereby improving hydrogen uptake [33,34]. Systematic investigation of pore sizes on hydrogen uptake was recently exemplified in a series of NbO-type MOFs based on tetracarboxylate organic ligands and di-copper paddlewheel SBUs. Extension of biphenyl-3,3′,5,5′-tetracarboxylate to terphenyl-3,3′,5,5′-tetracarboxylate and quaterphenyl-3,3′,5,5′-tetracarboxylate leads to proportional increase in pore size but decrease of hydrogen uptake at 77 K, 1 atm [35].

The ideal pore size for effective adsorption of hydrogen molecules in MOFs is comparable to the kinetic diameter of dihydrogen, which is 2.89 Å. This leads to optimal interaction between the dihydrogen molecule and the framework, thus maximizing the total van der Waals forces acting on dihydrogen [36].

Catenation Catenation, frequently encountered in porous MOFs, is the intergrowth of two or more identical frameworks [8]. It is favored by the use of longer linkers, and deemed as an alternative strategy for reducing pore sizes in porous MOFs [27,33]. The effect of catenation on hydrogen uptake was illustrated by hydrogen adsorption studies on IRMOFs, which revealed that catenated IRMOF-9, IRMOF-11, and IRMOF-13 showed higher hydrogen adsorption capacities than those of noncatenated IRMOF-1 [37]. This effect is probably related directly to the reduction of pore diameter due to catenation.

Conceptually, a catenated MOF and its noncatenated counterpart can be viewed as a supramolecular pair of stereoisomers. Recently, we developed a templating strategy to synthesize catenation isomer pairs predictably by using copper paddlewheel SBUs and two trigonal-planar ligands (TATB and HTB) (Fig. 12.1). This allowed us to evaluate catenation as an independent criterion in the hydrogen uptake of a MOF. The catenation isomerism is controlled by the presence or absence of oxalic acid. Although the noncatenated form (PCN-6′) has a higher overall porosity, based on the solvent-accessible volume calculated from the single-crystal x-ray structure, its catenated counterpart (PCN-6) exhibits a 41% increase in surface area, 133% increase

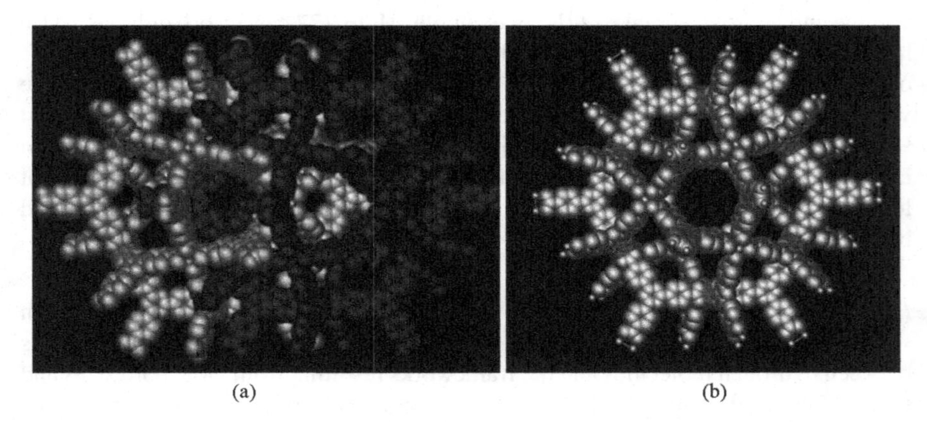

FIGURE 12.1 (a) Catenated and (b) noncatenated frameworks. (From ref. 38, with permission from the American Chemical Society.)

in volumetric hydrogen uptake, and 29% increase in gravimetric hydrogen uptake [38]. This finding is consistent with a recent theoretical simulation suggesting that new adsorption sites and small pores formed as a result of catenation may strengthen the overall interaction between gas molecules and the pore walls, thereby increasing the apparent surface area and hydrogen uptake [39].

Ligand Functionalization The functionalization of organic linkers not only plays a critical role in the construction of MOFs but also plays an important role in further enhancement of hydrogen adsorption [27,37]. Organic linkers with aromatic fragments, such as phenylene, naphthalene, and biphenylene, are used widely in the synthesis of MOFs to form a rigid three-dimensional porous framework [31,37]. Increasing the aromaticity of these organic ligands has been both predicted theoretically [40,41] and proved experimentally to be an effective way to improve hydrogen adsorption capacity [31]. A typical example is the synthesis of a series of IRMOFs which have similar topology based on octahedral Zn_4O SBU. It was found that the hydrogen adsorption per formula unit at 77 K and 1 bar increased with an increasing number of aromatic rings in the organic linkers. The maximum adsorption increased from 4.2 molecules of H_2 per formula unit in IRMOF-18 (2,3,5,6-tetramethylphenylene-1,4-dicarboxylate) to 9.8 in IRMOF-13 (pyrene-2,7-dicarboxylate). Meanwhile, the gravimetric hydrogen capacity of IRMOF-13 (1.73 wt%) is almost double that of IRMOF-18 (0.89 wt%) formed by 2,3,5,6-tetramethylphenylene-1,4-dicarboxylate [31,37]. These results indicate that the more aromatic rings in the organic ligands, the stronger the interactions between hydrogen and the MOFs. Recently, we prepared a nanoscopic-cage-based porous MOF, PCN-14, built from an anthracene derivative, 5,5′-(9,10-anthracenediyl)di-isophthalate. PCN-14 exhibits significant enhancement of both hydrogen adsorption capacity and hydrogen affinity compared to those of the NbO-type MOFs, which are also based on tetracarboxylate organic ligands and di-copper paddlewheel SBUs [42]. The improved hydrogen uptake in PCN-14 can be ascribed to the central anthracene aromatic rings.

In addition to the increase of aromaticity in the organic ligands, chemical modification of the organic linkers by introducing an electron-donating group (or groups) has been suggested, based on *ab initio* calculations, as another way to further enhance framework affinity for the dihydrogen molecule [43]. This was illustrated in the hydrogen adsorption studies of the IRMOF series. Adding one –Br or one $-NH_2$ or four methyl groups to the central benzene ring of the linker in IRMOF-1 affords IRMOF-2, IRMOF-3 and IRMOF-18, respectively [31], while replacing the phenyl ring of bdc with a thieno-[3,2*b*]thiophene moiety affords IRMOF-20 [37]. The increased polarizability of the heteropolycyclic ligand essentially improves the hydrogen sorption on a molar basis in IRMOF-20 due to a stronger interaction of hydrogen with the organic linker, despite a reduction in the gravimetric capacity due to the heavy sulfur atom. Little enhancement, however, was found in IRMOF-2, IRMOF-3, or IRMOF-18 [37], although MP2 computational studies suggest that functionalizing phenylene ring with electron-donor groups such as NH_2 or Me can improve hydrogen affinity by about

15% [43]. A similar lack of hydrogen adsorption enhancement was found in some pillared MOFs constructed by ligands with all phenyl H atoms replaced with either –F or $-CH_3$ [44]. This may be attributed to the smaller pore sizes or blocking of some high-affinity binding sites by the larger ligand, thus canceling out the benefit derived from electronic enhancement of the ligand. It has also been proposed that N-heterocyclic ligands may have a higher hydrogen affinity than purely graphitic ligands, based on some studies of hydrogen adsorption in carbon, carbon nitride, and boron nitride nanotube structures [45]. This was illustrated in some porous MOFs constructed from triazine ligands developed in our laboratory [38,46].

The versatility of organic ligands has provided almost infinite possibilities for the construction of MOFs with various topologies. Instead of simply modifying the rigid organic ligands to build MOFs with similar topology, utilizing flexible organic ligands having different stereoisomers under external stimuli can result in porous MOFs with quite different topologies. This phenomenon is referred to as *supramolecular isomerism*. As structure determines property, supramolecular isomers are expected to exhibit different hydrogen adsorption capacities. Recently, we designed a tetracarboxylate ligand, the tetra-anion of *N,N,N′,N′*-tetrakis(4-carboxyphenyl)-1,4-phenylenediamine (tcppda), which has three stereoisomers with a pair of enantiomers and a diastereomer [47]. The diastereomer has a C_{2h} symmetry with three phenyl rings oriented as left- and right-handed propellers around the two N atoms and a plane of symmetry through the central phenyl ring, reflecting one N-centered propeller to the other. The pair of enantiomers possesses D_2 point group symmetry, with the two N-centered propellers being either right-handed (${}^{\delta}D_2$) or inverted (${}^{\lambda}D_2$). Under solvothermal conditions, the reaction between Cu $(NO_3)_2{\cdot}2.5H_2O$ and $H_4TCPPDA$ in DMSO at 115°C gave rise to a porous MOFs with NbO topology, wherein only the C_{2h} isomer of the tcppda ligand exists. When raising the temperature to 120°C, another porous MOF with PtS topology was obtained, wherein only the D_2 tcppda isomer was found in a racemic combination with the ratio of ${}^{\delta}D_2$ and ${}^{\lambda}D_2$ 1 : 1 (Fig. 12.2). The temperature-dependent supramolecular isomerism of the two MOFs can be attributed to the interconversion of the D_2 and C_{2h} isomers of tcppda from low temperature to high temperature. Nitrogen and hydrogen adsorption studies at 77 K revealed that both surface area and hydrogen uptake of the porous MOFs with PtS topology are about 20% higher than those of the NbO-type porous MOF. These studies suggest that designing flexible organic linkers is a promising way to construct porous MOFs with high hydrogen uptake.

Unsaturated Metal Centers The impregnation of unsaturated metal centers (UMCs) into porous MOFs is very attractive for hydrogen adsorption. One of the advantages in porous MOFs compared to carbon materials is that metal ions incorporated in porous MOFs have higher hydrogen binding energy than carbon [33,34]. Recent neutron studies have revealed that hydrogen adsorption is very dependent on the nature of the metal cation or oxide of the SBUs in porous MOFs [28,48]. Coordinatively unsaturated metal centers are usually very reactive and are known to play an important role in catalysis; consequently, their use in MOFs is a promising strategy to reach high hydrogen uptake due to their significantly high

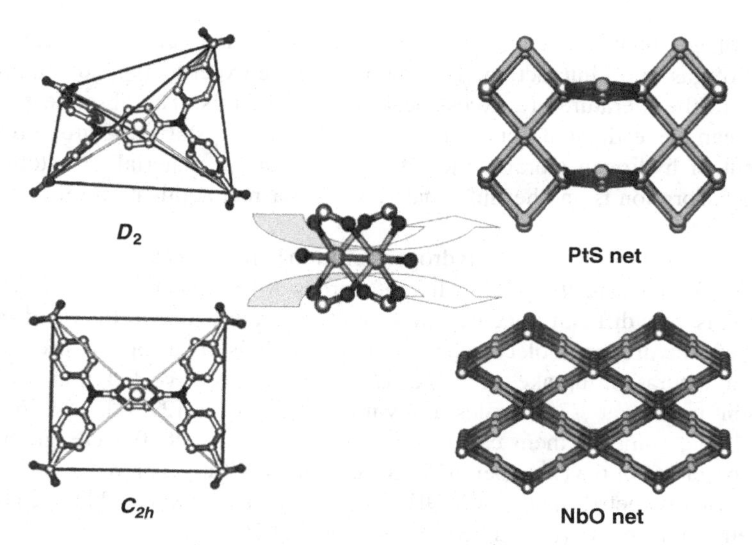

FIGURE 12.2 Different conformations of tcppda ligand lead to two porous MOFs with different topologies. (From ref. 47, with permission from the Royal Society of Chemistry.)

hydrogen affinity [34]. An effective way to achieve coordinative unsaturation of the metal ions is to liberate the terminal bound labile solvent (aqua) ligands by evacuation at elevated temperature (usually, 100 to 200°C), provided that the porous framework integrity is retained after the process.

The contribution from such UMCs to hydrogen adsorption capacity is quite remarkable [49]. This is well illustrated by the fact that MOF-505 has a hydrogen uptake as high as 2.47 wt% at 77 K and 1 atm after removal of axial aqua ligands from di-copper paddlewheel SBUs via thermal activation, generating the coordinatively unsaturated copper centers in MOF-505 [49]. Our recent studies on catenation isomers revealed that UMCs can lead to about a 20% increase in gravimetric hydrogen uptake in PCN-6 at 1 atm and 77 K [38].

Hydrogen Saturation at Cryotemperatures and High-Pressure Hydrogen Adsorption Studies at Room Temperature Although extensive studies have been focused on hydrogen uptake at low temperature and low pressure (usually 77 K, 1 atm), increasing attention is being drawn to high-pressure hydrogen-adsorption studies because of their application potential in practical on-board hydrogen storage. In addition to evaluating high-pressure gravimetric adsorption capacity in porous MOFs, volumetric adsorption capacity has also been assessed widely as it is another important criterion for on-board hydrogen storage.

Excess Adsorption and Absolute Adsorption In high-pressure studies, two concepts, excess and absolute adsorption, are frequently used to describe hydrogen adsorption in porous MOFs. In brief, *excess adsorption* is the amount of adsorbed gas interacting with the frameworks, whereas *absolute adsorption* is the amount of gas both interacting with the frameworks and staying in pores in the absence of gas–solid interaction. The majority of the experimental adsorption data reported in the literature are excess adsorption isotherms. The absolute adsorbed amount can be estimated for systems with known crystal structure. From the viewpoint of hydrogen storage, the total amount that a material can store or its absolute adsorption is another informative value for real application [50].

Hydrogen Saturation at 77 K Hydrogen sorption behavior at saturation is a critical parameter for judging the practicality of porous MOF materials [51]. Hydrogen saturation is very difficult to achieve at room temperature, due to the rapid thermal motion of dihydrogen molecules. Current research is focusing on investigating hydrogen saturation uptake at 77 K. Existing studies revealed that hydrogen saturation uptake at 77 K scales up with surface area [27,51]. As shown in Figure 12.3a, some of them can reach or even pass the 2010 DOE gravimetric adsorption target of 6 wt%, albeit at 77 K, and the DOE targets are "system goals." The excess gravimetric uptake of MOF-177 is as high as 7.5 wt%, while the absolute gravimetric uptake can reach 11.3 wt% at 77 K and 70 bar [50].

Most porous MOFs are very light. Generally speaking, the higher the surface area is, the lower the crystal density. In most cases the low density decreases the volumetric hydrogen uptake of the MOF material despite its high gravimetric uptake. As indicated in Figure 12.3b, very few porous MOFs can reach the 2010 DOE volumetric hydrogen uptake target. A typical example is MOF-177, which has an excess gravimetric uptake of 7.5 wt%, passing the 2010 DOE goal of 6 wt% at 77 K. Its low crystallographic density of 0.427 g cm^{-3}, however, leads to the excess volumetric uptake of 32 g L^{-1}, which is far below the 2010 DOE volumetric goal of 45 g L^{-1} [51]. A compromise between the surface area and crystal density should be met in search of porous MOF material with both high gravimetric and volumetric hydrogen uptake. Nevertheless, in terms of absolute adsorption, there exist some MOF materials with

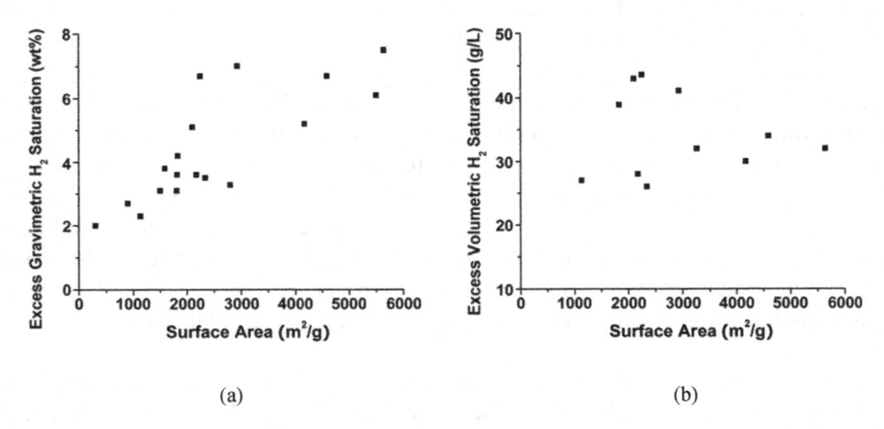

FIGURE 12.3 Correlation between surface area and (a) excess gravimetric and (b) excess volumetric hydrogen saturation uptake at 77 K.

both gravimetric and volumetric hydrogen uptake that are comparable to the 2010 DOE targets for hydrogen storage, albeit at 77 K. For example, $Mn_3[(Mn_4Cl)_3(btt)_3(CH_3OH)_{10}]_2$ (btt = 1,3,5-benzenetristetrazolate), reported by Dincă and co-workers, exhibits absolute gravimetric hydrogen uptake of 6.9 wt% and volumetric uptake of 60 g L^{-1} [52] at 77 K and high pressure. The absolute gravimetric uptake and volumetric uptake of MOF-177 can reach 11 wt% and 48 g L^{-1}, respectively [50]. It seems very promising to store hydrogen in porous MOFs at 77 K and high pressure; however, the cost of the cryostorage vessel precludes its practical on-board application, and porous MOFs with high hydrogen uptake near ambient temperature are urgently needed.

High-Pressure Hydrogen Adsorption at Room Temperature Room-temperature hydrogen adsorption studies under high pressure have been carried out for porous MOF materials. Unfortunately, these materials have very low hydrogen uptake at room temperature (less than 1.5 wt%) [27]. High surface area alone is insufficient to achieve high-capacity ambient-temperature storage, albeit it can lead to high hydrogen saturation uptake at 77 K. Small pore size is also necessary for ambient-temperature hydrogen adsorption due to the enhanced interaction energy [27,32]. This is well illustrated in $[Cu(hfipbb)(H_2hfipbb)_{0.5}]$, a microporous MOF that contains pores of two types: small (ca. 3.5 × 3.5 Å) and large (5.1 × 5.1 Å). At room temperature and 48 atm, it can adsorb about 1 wt% of hydrogen, which is more than three times that of MOF-5 (0.28 wt%, 60 atm) which contains a pore size of above 7.7 × 7.7 Å and has a high surface area of 2300 $m^2 g^{-1}$ [53].

An effective way to reduce pore size is by utilizing interpenetration (or catenation), which has been proposed as a strategy to improve hydrogen uptake. A typical example of this method is $Zn_4O(L^1)_3$ (L^1 = 6,6′-dichloro-2,2′-diethoxy-1,1′-binaphthyl-4,4′-dibenzoate), which is fourfold interpenetrated with open channels of less than 5 Å and a BET surface area of only 502 $m^2 g^{-1}$; this material adsorbs 1.12 wt% of hydrogen at room temperature and 48 bar, among the highest of the MOF materials reported [54].

The introduction of UMCs into MOFs has been known as one of the most promising ways to improve hydrogen affinity. The ability of UMCs containing MOFs to adsorb a significant amount of hydrogen is well demonstrated by Dincă and co-workers in the porous MOF $Mn_3[(Mn_4Cl)_3(btt)_3(CH_3OH)_{10}]_2$. Upon thermal activation, this MOF can adsorb 1.4 wt% of hydrogen at 298 K and 90 bar. This high uptake capacity can be ascribed partially to exposed Mn^{2+} sites within the framework that interact strongly with H_2 molecules [52].

Despite the rapidly growing efforts to investigate hydrogen adsorption in porous MOFs, the near-ambient temperature hydrogen uptake of these materials falls far short of the 2010 DOE targets. On-board hydrogen storage using adsorptive materials for fuel cell–driven vehicles remains very much a challenge.

Hydrogen Adsorption Enthalpy The major barrier limiting hydrogen uptake at ambient temperature is the weak interaction between hydrogen molecules and the frameworks of porous MOFs. Hydrogen binding energy or hydrogen affinity can be estimated quantitatively by measuring the isosteric heats of adsorption of hydrogen or the hydrogen adsorption enthalpy, ΔH_{ads}. In the most frequently used method to determine the hydrogen adsorption enthalpy, ΔH_{ads}, the Clausius–Clapeyron equation is applied to adsorption data collected at two different temperatures (typically, 77 K and 87 K) [55]. Enhancing the hydrogen adsorption enthalpy will increase the temperatures at which porous MOFs can adsorb large amounts of hydrogen; this is essential in order to develop a storage system that meets the 2010 DOE goals at near-ambient temperatures.

Recently, an optimal enthalpy of 15 kJ mol^{-1} has theoretically been proposed for ambient-temperature, high-pressure hydrogen adsorption [56]. Various strategies, described below, have been explored to increase the hydrogen adsorption enthalpy to approach this value.

Utilizing small pore sizes comparable to the kinetic diameter of a dihydrogen molecule is efficient to enhance hydrogen adsorption enthalpy. An example of this was found in a microporous magnesium MOF; this material contains very small pores, about 3.5 Å, and exhibits high hydrogen adsorption enthalpy, 9.5 kJ mol^{-1} [57]. Catenation, an effective way to confine the pore size, has also been explored as a strategy to increase hydrogen adsorption enthalpy. Isosteric heat of adsorption studies on IRMOFs showed that the catenated IRMOF-11 surpasses other noncatenated IRMOFs in the adsorption enthalpy with a value of up to 9.1 kJ mol^{-1} at low hydrogen coverage [37]. Our recent studies on catenation isomers revealed an average enhancement of hydrogen adsorption enthalpy of about 2 kJ mol^{-1} in the catenated PCN-6 over the noncatenated PCN-6′ [38].

The creation of UMCs in porous MOFs is considered to be one of the most effective strategies to increase hydrogen adsorption enthalpy. The formation of coordinately unsaturated Cu^{2+} centers after removal of the terminal aqua ligands in HKUST-1 leads to about a 2 kJ mol^{-1} increase in adsorption enthalpy compared to MOF-5 under low-loading conditions [37]. The same phenomenon was demonstrated in the porous MOFs $Mn_3[(Mn_4Cl)_3(btt)_3(CH_3OH)_{10}]_2$ and $NaNi(sip)_2$, both of which exhibit high adsorption enthalpies of 10.1 kJ mol^{-1} at low hydrogen coverage after exposing the UMCs by removing the coordinated labile solvent ligands using thermal activation [52,58].

Instead of utilizing thermal activation to achieve coordinative unsaturation by the removal of one or more ligands from a metal center, recently we advanced a biomimetic approach to coordinatively unsaturated metal centers by creating entatic metal centers (EMCs) in the porous MOF PCN-9 [59]. In bioinorganic chemistry, protein EMCs are enforced by surrounding polypeptides; they are *imposed* into an unusual coordination geometry to enhance their reactivity in electron transfer, substrate binding, or catalysis [60]. Similarly, due to the specific geometric requirements of the ligands and SBUs in a MOF, EMCs can be created and are expected to have high hydrogen affinity. By mimicking the coordinatively unsaturated iron active center of hemoglobin, the hemoglobin-like entatic cobalt centers in PCN-9 exhibit high hydrogen affinity with adsorption enthalpy of 10.1 kJ mol^{-1} at low coverage (Fig. 12.4). The accessibility of those entatic cobalt centers was also probed by carbon monoxide molecules via infrared spectroscopy studies.

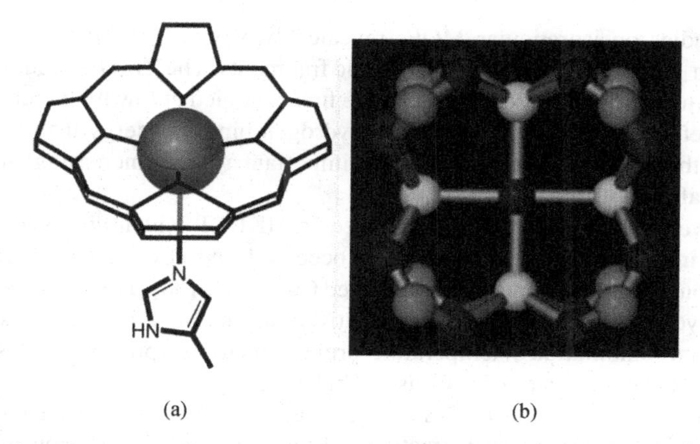

FIGURE 12.4 (a) Schematic drawing of the active center of hemoglobin; (b) $Co_4(\mu_4\text{-}O)$ (carboxylate)$_4$ SBU found in PCN-9. (From ref. 59, with permission from the American Chemical Society.)

Cation exchange in an anionic framework is yet another way to achieve coordinative unsaturation. Cation exchange on hydrogen adsorption enthalpy and capacity was recently studied in an anionic porous MOF, $Mn_3[(Mn_4Cl)_3(btt)_3(CH_3OH)_{10}]_2$. The findings revealed that the Mn^{2+}, Fe^{2+}, and Co^{2+} exchanged frameworks demonstrated the strongest H_2 binding among all the cations assessed. The Co^{2+} exchanged framework exhibited a remarkable hydrogen adsorption enthalpy of $10.5\ kJ\ mol^{-1}$ at zero coverage [61].

The introduction of naked lithium cations into porous MOF was illustrated using a chemical reduction strategy. The lithium cations reside in the porous framework to balance the negative charge of the reduced ligands. Hydrogen adsorption studies demonstrated that the inclusion of the lithium cations not only enhanced the hydrogen adsorption enthalpy significantly, but also almost doubled the hydrogen uptake [62].

Tools for the Investigation of the Interaction Between Hydrogen Molecules and a Framework The investigation of the interaction between hydrogen molecules and a framework is of great significance for future development of new porous MOF materials with high hydrogen adsorption capacities.

It is well known that neutron-based studies are the most powerful tools for hydrogen research, due to the fact that hydrogen has the largest neutron cross section [63]. One of the most useful methods for understanding hydrogen uptake in porous MOFs is inelastic neutron scattering (INS), which is extremely sensitive to the chemical environment of adsorbed H_2 and can provide important information about the sites that hydrogen molecules occupy and the order in which the sites are filled [64]. This technique has become invaluable for the understanding of hydrogen adsorption characteristics of a variety of nanoporous adsorbents, including zeolites [65] and carbon nanostructures [66], and has recently been employed to investigate hydrogen adsorption in porous MOF materials [28,48,58,67,68].

INS studies on isoreticular MOFs revealed two types of hydrogen "binding" centers: Zn_4O inorganic clusters and organic fragments. The Zn_4O inorganic clusters are the primary H_2 adsorption sites, and are first occupied at low hydrogen loadings. The Zn metal centers have much higher hydrogen binding energy than the organic linkers, although the latter also play an important role in increasing the overall hydrogen affinity of the MOF [48].

INS investigation of hydrogen adsorbed in HKUST-1 identified some specific hydrogen interacting sites, with the first occupied and strongest interacting sites located around the coordinatively unsaturated Cu^{2+} sites [68]. The strong association between hydrogen molecules and coordinatively unsaturated metal centers was also revealed by a combination of temperature-programmed desorption and INS studies on H_2-loaded $NaNi(sip)_2$ (sip = 5-sulfoisophthalate) [58].

The complementary tools to INS for the location of discrete hydrogen interacting sites are single-crystal neutron diffraction and neutron powder diffraction, which have recently been utilized for the study of hydrogen adsorption in porous MOFs [52,69–73]. Single-crystal neutron diffraction study of MOF-5 revealed two hydrogen-interacting sites, one higher-energy site over the center of the $Zn_4(\mu_4\text{-}O)(CO_2)_8$ SBU and a second site over the face of a ZnO_4 tetrahedron [69]. This is consistent with neutron powder diffraction studies on MOF-5, which also identified two additional sites at increased loading: one above the oxygen atoms of the carboxylate group and the other over the phenyl ring of the ligand [71]. Generally, these agree with an INS experiment performed on the same material, all revealing that the hydrogen molecules will occupy the sites around the inorganic $Zn_4(\mu_4\text{-}O)$ cluster first [68].

Neutron powder diffraction study on a D_2-loaded HKUST-1 sample identified six distinct D_2 interacting sites within the nanoporous structure. The first occupied and highest-energy site is located at the coordinatively unsaturated axial sites of the dinuclear Cu center, and the remaining sites are located near the benzene ring and carboxylate moieties of the ligand, which are occupied progressively. The short $C \cdots H_2$ distance of 2.39 Å at 4 K indicates appreciable interaction between a dihydrogen molecule and the d^9 coordinatively unsaturated Cu(II) center [70]. Similar association of H_2 molecules with UMCs was also observed via neutron powder diffraction studies of the porous MOF $Mn_3[(Mn_4Cl)_3(btt)_3(CH_3OH)_{10}]_2$. The 2.27 -Å $H_2 \cdots Mn^{2+}$ distance represents the first $H_2 \cdots$ metal interaction ever observed in porous MOFs (Fig. 12.5), and also accounts for the high hydrogen adsorption enthalpy of the compound [52].

As enumerated above, the primary hydrogen adsorption sites in most MOFs are metal-based. However, neutron powder diffraction studies on the zeolitic imidazolate frameworks (ZIFs), a subfamily of MOF, revealed that the strongest adsorption sites are associated directly with the imidazolate organic linkers instead of the triangular faces of the ZnN_4 tetrahedra. This observation suggests that modification of the organic linkers in ZIFs may be more important than altering the metal-containing component in order to optimize these materials for higher hydrogen adsorption capacity [72].

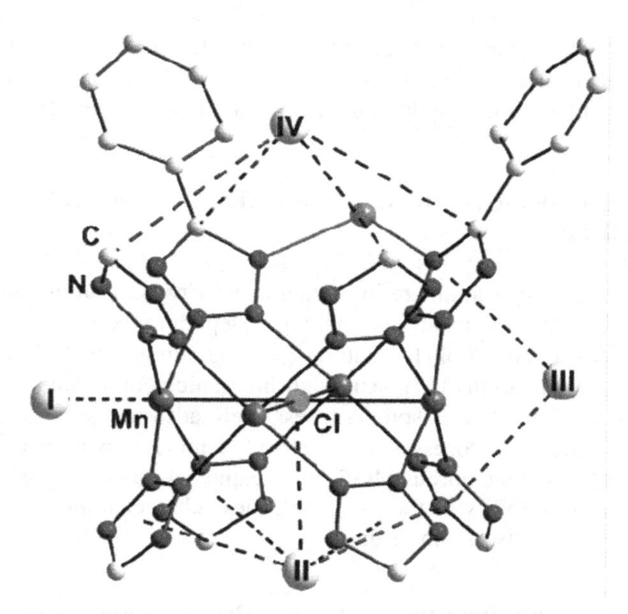

FIGURE 12.5 D_2 adsorption sites in Mn(btt), as identified by neutron powder diffraction. (From ref. 52, with permission from the American Chemical Society.)

Hydrogen Spillover One of the most intriguing methods for hydrogen adsorption enhancement is the secondary hydrogen spillover. It consists of the dissociative chemisorption of hydrogen on a metal catalyst with the subsequent migration of atomic hydrogen to the surface of a carrier contacting the metal (primary hydrogen receptor) and then to the second carrier (secondary receptor) [74]. This technique has recently been used by Li and Yang in porous MOFs [75,76]. Mechanically mixing 5% Pt on activated carbon with a MOF, followed by melting and subsequent carbonization of an amount of sucrose with the MOF, gave rise to materials with a hydrogen adsorption capacity up to 3 to 4% at 298 K and 10 MPa. In the meantime, the hydrogen adsorption enthalpy was increased to 20 to 23 kJ mol^{-1}, which is within the proposed enthalpy range 15 to 25 kJ mol^{-1}. The secondary hydrogen spillover technique shows great potential for achieving the 2010 DOE targets of hydrogen storage, but difficulties in reproducing some of the experimental results cast a shadow on the outlook of such a fascinating technique.

12.2.3 Conclusions

As a new class of porous materials, MOFs have been attracting escalating research interests. Due to their high surface area, tunable pore size, and functionalizable pore wall, MOFs are ideal candidates for adsorptive hydrogen storage materials. Although ΔH_{ads} for hydrogen adsorption in porous MOFs reported so far ranges from 6 to 12 kJ mol^{-1} currently, it is conceivable that this value will increase and eventually fall into the range 15 to 25 kJ mol^{-1}. Applying secondary hydrogen spillover to these new porous MOFs as well as existing ones will probably yield results that approach and hopefully meet the on-board hydrogen storage targets in the near future.

12.3 POROUS METAL–ORGANIC FRAMEWORKS FOR SELECTIVE GAS ADSORPTION

Gas separation and purification are important and energy consuming in industry. A few of the commercially most important gas separation challenges are: N_2/O_2 separation, N_2/CH_4 separation for natural gas upgrading, and CO removal from H_2 for fuel cell applications [77]. Although inorganic zeolites and porous carbon materials can be applied with some success, new adsorbents are still needed to optimize these separation processes to make them commercially more attractive. As a new type of zeolite analog, porous MOFs feature amenability to design, tunable pore size, and functionalizability of the pore wall; these characteristics give them great potential in selective adsorption of gases [1–9].

12.3.1 Molecular Sieving Effect in Porous Metal–Organic Frameworks

The molecular-sieving effect arises when molecules of appropriate size and shape are allowed to enter the open channels of an adsorbent, while other molecules are excluded. It accounts for the underlying principle for most selective gas adsorption processes in porous materials with uniform micropores [77].

It is essential to limit the pore size of an adsorbent for effective gas separation. To this point, apertures of porous MOFs can rationally be tuned to a certain size for selective adsorption of specific gas molecules.

Utilizing short bridging ligands is a good way to restrict pore sizes of porous MOFs for gas separation. This was exemplified by Dybtsev et al. in the microporous manganese formate MOF. The short length of the formate leads to a very small aperture size, which can discriminate H_2 from N_2, and CO_2 from CH_4 [78]. Recently, Dincă and Long confined the pore size of a microporous magnesium MOF to around 3.5 Å by using 2,6-naphthalenedicarboxylic acid. The magnesium MOF exhibited the capability of selective uptake of H_2 or O_2 over N_2 or CO [57]. A similar molecular-sieving effect was also observed in a 2,4-pyridinedicarboxylate-based cobalt MOF (CUK-1), which could separate H_2 from N_2, O_2 from N_2 and Ar, and CO_2 from CH_4 [79].

As an alternative strategy, increasing the bulkiness of the struts to constrict the apertures of porous MOFs was well illustrated by us in the construction of a zinc microporous MOF, PCN-13. PCN-13 was built from a bulky ligand 9,10-anthracenedicarboxylate (adc). Due to the bulkiness of adc, the pore size of PCN-13 is constricted to about 3.5 × 3.5 Å, which allows H_2 and O_2 to pass through the pores but excludes N_2 and CO [80].

Interpenetration is well known as an effective way to reduce the pore size of MOFs and has recently been employed to confine the pore size for selective adsorption of gas molecules. Chen et al. demonstrated that doubly interpenetrated primitive cubic nets based on bidentate pillar linkers and bicarboxylates could be rationally designed for selective gas adsorption. The microporous MOF Cu(FMA)(4,4′-Bpe)$_{0.5}$ [FMA = fumarate; 4,4′-Bpe = *trans*-bis(4-pyridyl)ethylene] was constructed by incorporation of the bicarboxylate FMA and bidentate pillar linker 4,4′-Bpe, and its pore size was tuned by double framework interpenetration to about 3.6 Å, which exhibits selective adsorption of H_2 over Ar, N_2, and CO [81]. By increasing the length of the bicarboxylates and bidentate pillar linkers, Chen et al. introduced triple interpenetration in the microporous MOF Zn(ADC) (4,4′-Bpe)$_{0.5}$ [ADC = 4,4′-azobenzene-dicarboxylate; 4,4′-Bpe = *trans*-bis(4-pyridyl)ethylene], which can distinguish H_2 from N_2 and CO [82].

12.3.2 Porous Metal–Organic Frameworks for Kinetic Separation Applications

Different from the molecular-sieving effect, kinetic separation is achieved by virtue of the differences in diffusion rates of different molecules. Kinetic separation is of great importance in industry applications, particularly chromatography applications [77].

The separation of mixed C8 alkylaromatic compounds (*p*-xylene, *o*-xylene, *m*-xylene, and ethylbenzene) is one of the most challenging separations in the chemical industry, due to the similarity of their boiling points [83]. This separation is currently performed by cation-exchanged zeolites X and Y in industry [84]; however, adsorbents with improved separation efficiency are still needed. Recently, for the first time, Alaerts et al. investigated the adsorption and separation of a mixture of C8 alkylaromatic compounds using three porous MOFs: HKUST-1, MIL-53, and MIL-47 in the liquid phase. Through chromatographic experiments, it was concluded that MIL-47 has the highest potential for the separation of C8 alkylaromatic compounds among the three MOFs investigated. Compared with zeolites used currently, MIL-47 displays high uptake capacity and high selectivity, which are advantageous for its future practical application in industry [85].

The separation of hexane isomers to boost octane ratings in gasoline represents a very important process in the petroleum industry [86]. This is achieved using the high-energy-consuming method of cryogenic distillation, albeit some alternative novel materials and technologies are now under rapid development. By making use of the pore space to capture and discriminate hexane isomers, porous MOFs have the potential to separate hexane isomers. This was well illustrated by Barcia et al. in the kinetic separation of hexane isomers by using the three-dimensional microporous MOF Zn(BDC)(Dabco)$_{0.5}$. The MOF Zn(BDC)(Dabco)$_{0.5}$ contains three-dimensional intersecting pores of about 7.5×7.5 Å along axis [100] and pores of 3.8×4.7 Å along axes [010] and [001]. By making use of the narrow channels of 3.8×4.7 Å to take up linear *n*-HEX exclusively while blocking branched hexane isomers, this MOF was used successfully in the kinetic separation of hexane isomers by fixed-bed adsorption. It exhibited extraordinary separation selectivity to separate branched hexane isomers from linear *n*-HEX. This represented the first example of using microporous MOFs for the kinetic separation of hexane isomers, demonstrating great potential for applications in the very important industrial process of hexane–isomer separation [87].

Another important process used to boost octane ratings in gasoline is the separation of alkane isomers, which is currently practiced using some narrow-pore zeolites [88]. Chen et al. recently demonstrated the application of a microporous MOF (MOF-508) packed column in the gas chromatographic (GC) separation of alkanes. MOF-508 contains 1D pores of 4.0 × 4.0 Å, which can selectively accommodate linear alkanes and discriminate branched alkanes. The subtle matching of the size and shape of the alkanes with the micropores of MOF-508 leads to different van der Waals interactions, resulting in selective GC separation of alkanes in the MOF-508 column [89].

12.3.3 Mesh-Adjustable Molecular Sieves

Inorganic zeolite molecular sieves are currently the most used adsorbents in industry for gas separations [77]. The rigidity of the bonds in zeolites affords them with fixed mesh sizes. This is advantageous when the mesh size fits the separation needs precisely. However, when the size disparity of the two gases is very small, a zeolite molecular sieve with the precise mesh size is not always readily available. In such cases, mesh-adjustable molecular sieves (MAMSs) that can always meet the separation needs are highly desirable.

Recently, we reported the first mesh-adjustable molecular sieve, MAMS-1, which was built from the amphiphilic ligand 5-*tert*-butyl-1,3-benzenedicarboxylate (BBDC) [90]. MAMS-1 is a trilayer structure and consists of hydrophilic channels and hydrophobic chambers. The hydrophilic channel and hydrophobic chamber are interconnected through a pair of BBDC ligands at their interface. Variable-temperature gas adsorption studies on MAMS-1 revealed that it exhibited a temperature-induced molecular-sieving effect, and its mesh range falls between 2.9 and 5.0 Å. When the temperature is controlled precisely, any mesh size within this range can be attained accurately. Mechanistic studies suggested that the hydrophobic chambers are the major storage room for gas molecules, which have to go through the fully activated hydrophilic channels to enter the hydrophobic chambers. The BBDC ligands at the interface of hydrophilic channels and hydrophobic chambers play the role of gates, which open linearly with increasing temperatures to let gas molecules of certain sizes gradually enter the hydrophobic chambers (Fig. 12.6). Moreover, there exists a linear relationship between mesh size and temperature, $D = D_0 + \alpha T$ (where D is the mesh size at temperature T (K), D_0 is the mesh size at 0 K, and α is a constant); D_0 and α could be tuned by ligand design. This implies the possibility of a MAMS that will be versatile in gas separation even at ambient temperatures.

12.3.4 Conclusions

Porous MOFs represent a new type of zeolite analog. Compared with inorganic zeolites, they have the advantages of being amendable to design and having controllable pore sizes and framework flexibility. Their selective gas adsorption capabilities, by either a molecular-sieving effect or kinetic separation, provide them great

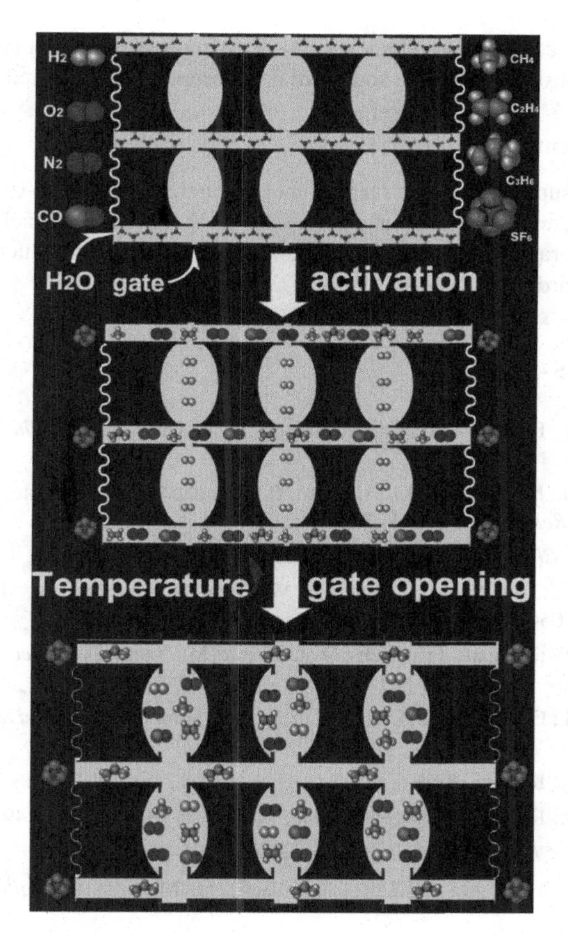

FIGURE 12.6 Schematic representation of the mechanism of the gating effect in MAMS-1. (From ref. 90, with permission from Wiley-VCH.)

promise for both industrial and chromatographic applications. The search for new MAMSs, which may be omnipotent for gas separation, deserves attention in future research.

12.4 OUTLOOK

As a new type of material, metal–organic frameworks are now attracting more and more attention from both the academia and industry. MOFs have exhibited great potential not only in hydrogen storage and gas separation applications, but also in some other applications, such as methane storage, CO_2 sequestration, and catalysis. As new adsorbents, porous MOFs hold great promise for solving the critical problem of hydrogen storage for fuel cell–driven vehicles as well as improving some current separation processes in industry, with more and more new compounds generated each year.

Acknowledgments

This work was supported by the Department of Energy (DE-FC36-07GO17033) and the National Science Foundation (CHE-0449634). H.-C. Zhou acknowledges the Research Corporation for a Cottrell Scholar Award and Air Products for a Faculty Excellence Award.

REFERENCES

1. Yaghi, O. M.; Li, H.; Davis, C.; Richardson, D. T.; Groy, L. *Acc. Chem. Res.* **1998**, *31*, 474–484.
2. Eddaoudi, M.; Moler, D. B.; Li, H.; Chen, B.; Reineke, T. M.; O'Keeffe, M.; Yaghi, O. M. *Acc. Chem. Res.* **2001**, *34*, 319–330.
3. James, S. L. *Chem. Soc. Rev.* **2003**, *32*, 276–288.
4. Yaghi, O. M.; O'Keeffe, M.; Ockwig, N. W.; Chae, H. K.; Eddaoudi, M.; Kim, J. *Nature* **2003**, *423*, 705–714.
5. Ockwig, N. W.; Delgado-Friedrichs, O.; O'Keeffe, M.; Yaghi, O. M. *Acc. Chem. Res.* **2005**, *38*, 176–182.
6. O'Keeffe, M.; Eddaoudi, M.; Li, H.; Reineke, T.; Yaghi, O. M. *J. Solid State Chem.* **2000**, *152*, 3–20.
7. Kitagawa, S.; Kitaura, R.; Noro, S.-I. *Angew. Chem., Int. Ed.* **2004**, *43*, 2334–2375.
8. Batten, S. R.; Robson, R. *Angew. Chem., Int. Ed.* **1998**, *37*, 1460–1494.
9. Janiak, C. *J. Chem. Soc., Dalton Trans.* **2003**, 2781–2804.
10. Delgado-Friedrichs, O.; O'Keeffe, M.; Yaghi, O. M. *Acta Crystallogr. A* **2003**, *59*, 515–525.
11. Seo, J. S.; Whang, D.; Lee, H.; Jun, S. I.; Oh, J.; Jeon, Y. J.; Kim, K. *Nature* **2000**, *404*, 982–986.
12. Hu, A.; Ngo, H. L.; Lin, W. *J. Am. Chem. Soc.* **2003**, *125*, 11490–11491.
13. Millward, A. R.; Yaghi, O. M. *J. Am. Chem. Soc.* **2005**, *127*, 17998–17999.
14. Bourrelly, S.; Llewellyn, P. L.; Serre, C.; Millange, F.; Loiseau, T.; Férey, G. *J. Am. Chem. Soc.* **2005**, *127*, 13519.
15. Eddaoudi, M.; Kim, J.; Rosi, N.; Vodak, D.; Wachter, J.; O'Keeffe, M.; Yaghi, O. M. *Science* **2002**, *295*, 469–472.
16. Matsuda, R.; Kitaura, R.; Kitagawa, S.; Kubota, Y.; Belosludov, R. V.; Kobayashi, T. C.; Sakamoto, H.; Chiba, T.; Takata, M.; Kawazoe, Y.; Mita, Y. *Nature* **2005**, *436*, 238–241.
17. Pan, L.; Adams, K. M.; Hernandez, H. E.; Wang, X.; Zheng, C.; Hattori, Y.; Kaneko, K. *J. Am. Chem. Soc.* **2003**, *125*, 3062–3067.
18. DOE Office of Energy Efficiency and Renewable Energy Hydrogen, Fuel Cells & Infrastructure Technologies Program Multi-Year Research, Development and Demonstration Plan, available at http://www.eere.energy.gov/hydrogenandfuelcells/mypp.

19. FY 2006 Annual Progress Report for the DOE Hydrogen Program, November 2006, available at http://www.hydrogen.energy.gov/annual_progress.html.
20. S. Satyapal et al., FY 2006 DOE Hydrogen Program Annual Merit Review and Peer Evaluation Meeting Proceedings, Plenary Session, available at: http://www.hydrogen.energy.gov/annual_review06_plenary.html.
21. Wolf, J. *MRS Bull.* **2002**, *27*, 685–687.
22. Sandrock, G. *J. Alloys Compd.* **1999**, *293–295*, 877–888.
23. Bogdanović, B.; Schwickardi, M. *J. Alloys Compd.* **1997**, *253–254*, 1–9.
24. Hynek, S.; Fuller, W.; Bentley, J. *Int. J. Hydrogen Energy* **1997**, *22*, 601–610.
25. Zrttel, A.; Sudan, P.; Mauron, P.; Kiyobayashi, T.; Emmenegger, C.; Schlapbach, L. *Int. J. Hydrogen Energy* **2001**, *27*, 203–212.
26. Nijkamp, M. G.; Raaymakers, J. E. M. J.; van Dillen, A. J.; de Jong, K. P. *Appl. Phys. A* **2001**, *72*, 619–623.
27. Collins, D. J.; Zhou, H.-C. *J. Mater. Chem.* **2007**, *17*, 3154–3160.
28. Rosi, N. L.; Eckert, J.; Eddaoudi, M.; Vodak, D. T.; Kim, J.; O'Keeffe, M.; Yaghi, O. M. *Science* **2003**, *300*, 1127–1130.
29. Chae, H. K.; Siberio-Pérez, D. Y.; Kim, J.; Go, Y. B.; Eddaoudi, M.; Matzger, A. J.; O'Keeffe, M.; Yaghi, O. M. *Nature* **2004**, *427*, 523-527.
30. Férey, G.; Mellot-Draznieks, C.; Serre, C.; Millange, F.; Dutour, J.; Surblé, S.; Margiolaki, I. *Science* **2005**, *309*, 2040–2042.
31. Rowsell, J. L. C.; Millward, A. R.; Park, K. S.; Yaghi, O. M. *J. Am. Chem. Soc.* **2004**, *126*, 5666–5667.
32. Everett, D. H.; Powl, J. C. *J. Chem. Soc., Farady Trans. 1* **1976**, *72*, 619–636.
33. Rowsell, J. L. C.; Yaghi, O. M. *Angew. Chem., Int. Ed.* **2005**, *44*, 4670–4679.
34. Isacva, V. I.; Kustov, L. M. *Russ. J. Gen. Chem.* **2007**, *50*, 56–72.
35. Lin, X.; Jia, J.; Zhao, X. K.; Thomas, M.; Blake, A. J.; Walker, G. S.; Champness, N. R.; Hubberstey, P.; Schröder, M. *Angew. Chem., Int. Ed.* **2006**, *45*, 7358–7364.
36. Chapman, K. W.; Chupas, P. J.; Maxey, E. R.; Richardson, J. W. *Chem. Commun.* **2006**, 4013–4015.
37. Rowsell, J. L. C.; Yaghi, O. M. *J. Am. Chem. Soc.* **2006**, *128*, 1304–1315.
38. Ma, S.; Sun, D.; Ambrogio, M. W.; Fillinger, J. A.; Parkin, S.; Zhou, H.-C. *J. Am. Chem. Soc.* **2007**, *129*, 1858–1859.
39. Jung, D. H.; Kim, D.; Lee, T. B.; Choi, S. B.; Yoon, J. H.; Kim, J.; Choi, K.; Choi, S.-H. *J. Phys. Chem. B* **2006**, *110*, 22987–22990.
40. Han, S. S.; Goddard, W. A., III. *J. Am. Chem. Soc.* **2007**, *129*, 8422–8423.
41. Han, S. S.; Deng, W.-Q.; Goddard, W. A., III. *Angew. Chem., Int. Ed.* **2007**, *46*, 6289–6292.
42. Ma, S.; Sun, D.; Simmons, J. M.; Collier, C. D.; Yuan, D.; Zhou, H.-C. *J. Am. Chem. Soc.* **2008**, *130*, 1012–1016.
43. Hrbner, O.; Gluss, A.; Fichtner, M.; Klopper, W. *J. Phys. Chem. A* **2004**, *108*, 3019–3023.
44. Chun, H. D.; Dybtsev, N.; Kim, H.; Kim, K. *Chem. Eur. J.* **2005**, *11*, 3521–3529.
45. Bai, X. D.; Zhong, D.; Zhang, G. Y.; Ma, X. C.; Liu, S.; Wang, E. G. *Appl. Phys. Lett.* **2001**, *79*, 1552–1554.

46. Sun, D.; Ma, S.; Ke, Y.; Collins, D. J.; Zhou, H.-C. *J. Am. Chem. Soc.* **2006**, *128*, 3896–3897.
47. Sun, D.; Ke, Y.; Mattox, T. M.; Ooro, B. A.; Zhou, H.-C. *Chem. Commun.* **2005**, 5447–5449.
48. Rowsell, J. L. C.; Eckert, J.; Yaghi, O. M. *J. Am. Chem. Soc.* **2005**, *127*, 14904–14910.
49. Chen, B.; Ockwig, N. W.; Millward, A. R.; Contreras, D. S.; Yaghi, O. M. *Angew. Chem., Int. Ed.* **2005**, *44*, 4745–4749.
50. Furukawa, H.; Miller, M. A.; Yaghi, O. M. *J. Mater. Chem.* **2007**, *17*, 3197–3204.
51. Wong-Foy, A. G.; Matzger, A. J.; Yaghi, O. M. *J. Am. Chem. Soc.* **2006**, *128*, 3494–3495.
52. Dincă, M.; Dailly, A.; Liu, Y.; Brown, C. M.; Neumann, D. A.; Long, J. R. *J. Am. Chem. Soc.* **2006**, *128*, 16876–16883.
53. Pan, L.; Sander, M. B.; Huang, X.; Li, J.; Smith, M. R., Jr.; Bittner, E. W.; Bockrath, B. C.; Johnson, J. K. *J. Am. Chem. Soc.* **2004**, *126*, 1308–1309.
54. Kesanli, B.; Cui, Y.; Smith, M. R.; Bittner, E. W.; Bockrath, B. C.; Lin, W. *Angew. Chem., Int. Ed.* **2005**, *44*, 72–75.
55. Rouquerol, J.; Rouquerol, F.; Sing, K. Adsorption by Powders and Porous Solids, Academic Press, London, **1998**.
56. Bhatia, S. K.; Myers, A. L. *Langmuir* **2006**, *22*, 1688–1700.
57. Dincă, M.; Long, J. R. *J. Am. Chem. Soc.* **2005**, *127*, 9376–9377.
58. Forster, P. M.; Eckert, J.; Heiken, B. D.; Parise, J. B.; Yoon, J. W.; Jhung, S. H.; Chang, J. S.; Cheetham, A. K. *J. Am. Chem. Soc.* **2006**, *128*, 16846–16850.
59. Ma, S.; Zhou, H.-C. *J. Am. Chem. Soc.* **2006**, *128*, 11734–11735.
60. Vallee, B. L.; Williams, R. J. *Proc. Natl. Acad. Sci. USA* **1968**, *59*, 498–505.
61. Dincă, M.; Long, J. R. *J. Am. Chem. Soc.* **2007**, *129*, 11172–11175.
62. Mulfort, K. L.; Hupp, J. T. *J. Am. Chem. Soc.* **2007**, *129*, 9604–9605.
63. Eckert, J.; Kubas, G. J. *J. Phys. Chem.* **1993**, *97*, 2378–2384.
64. Silvera, I. F. *Rev. Mod. Phys.* **1980**, *52*, 393–452.
65. Forster, P. M.; Eckert, J.; Chang, J.-S.; Park, S. E.; Férey, G.; Cheetham, A. K. *J. Am. Chem. Soc.* **2003**, *125*, 1309–1312.
66. Schimmel, H. G.; Kearley, G. J.; Nijkamp, M. G.; Visser, C. T.; de Jong, K. P.; Mulder, F. M. *Chem. Eur. J.* **2003**, *9*, 4764–4770.
67. Liu, Y.; Eubank, J. F.; Cairns, A. J.; Eckert, J.; Kravtsov, V. C.; Luebke, R.; Eddaoudi, M. *Angew. Chem., Int. Ed.* **2007**, *46*, 3278–3283.
68. Liu, Y.; Brown, C. M.; Neumann, D. A.; Peterson, V. K.; Kepert, C. J. *J. Alloys Compd.* **2007**, *446–447*, 385–388.
69. Spencer, E. C.; Howard, J. A. K.; McIntyre, G. J.; Rowsell J. L. C.; Yaghi, O. M. *Chem. Commun.* **2006**, 278–280.
70. Peterson, V. K.; Liu, Y.; Brown C. M.; Kepert, C. J. *J. Am. Chem. Soc.* **2006**, *128*, 15578–15579.
71. Yildirim, T.; Hartman, M. R. *Phys. Rev. Lett.* **2005**, *95*, 215504/1–215504/4.
72. Wu, H.; Zhou, W.; Yildirim, T. *J. Am. Chem. Soc.* **2007**, *129*, 5314–5315.

73. Dincă, M.; Han, W. S.; Liu, Y.; Dailly, A.; Brown, C. M.; Long, J. R. *Angew. Chem., Int. Ed.* **2007**, *46*, 1419–1422.

74. Conner, W. C.; Falconer, J. L. *Chem. Rev.* **1995**, *95*, 759–788.

75. Li, Y.; Yang, R. T. *J. Am. Chem. Soc.* **2006**, *128*, 726–727.

76. Li, Y.; and Yang, R. T. *J. Am. Chem. Soc.* **2006**, *128*, 8136–8137.

77. Yang, R. T. Gas Adsorption by Adsorption Processes; Butterworth, Boston, **1997**.

78. Dybtsev, D. N.; Chun, H.; Yoon, S. H.; Kim, D.; Kim, K. *J. Am. Chem. Soc.* **2004**, *126*, 32–33.

79. Humphrey, S. M.; Chang, J.-S.; Jhung, S. H.; Yoon, J. W.; Wood, P. T. *Angew. Chem., Int. Ed.* **2007**, *46*, 272–275.

80. Ma, S.; Wang, X.-S.; Collier, C. D.; Manis, E. S.; Zhou, H.-C. *Inorg. Chem.* **2007**, *46*, 8499–8501.

81. Chen, B.; Ma, S.; Zapata, F.; Fronczek, F. R.; Lobkovsky, E. B.; Zhou, H.-C. *Inorg. Chem.* **2007**, *46*, 1233–1236.

82. Chen, B.; Ma, S.; Hurtado, E. J.; Lobkovsky, E. B.; Zhou, H.-C. *Inorg. Chem.* **2007**, *46*, 8490–8492.

83. Cottier, V.; Bellat, J.-P.; Simonot-Grange, M.-H. *J. Phys. Chem. B* **1997**, *101*, 4798–4802.

84. Hulme, R.; Rosensweig, R.; Ruthven, D. *Ind. Eng. Chem. Res.* **1991**, *30*, 752–760.

85. Alaerts, L.; Kirschhock, C. E. A.; Maes, M.; van der Veen, M. A.; Finsy, V.; Depla, A.; Martens, J. A.; Baron, G. V.; Jacobs, P. A.; Denayer, J. F. M.; De Vos, D. E. *Angew. Chem., Int. Ed.* **2007**, *46*, 4293–4297.

86. Sohn, S. W. In *Handbook of Petroleum Refining Processes*, 3rd ed.; Meyers, R. A., Ed.; McGraw-Hill, New York, **2004**.

87. Barcia, P. S.; Zapata, F.; Silva, J. A. C.; Rodrigues, A. E.; Chen, B. *J. Phys. Chem. B* **2007**, *111*, 6101–6103.

88. Kulprathipanja, S.; Neuzil, R.W. U.S. patent 4,445,444, **1984**.

89. Chen, B.; Liang, C.; Yang, J.; Contreras, D. S.; Clancy, Y. L.; Lobkovsky, E. B.; Yaghi, O. M.; Dai, S. *Angew. Chem., Int. Ed.* **2006**, *45*, 1390–1393.

90. Ma, S.; Sun, D.; Wang, X.-S.; Zhou, H.-C. *Angew. Chem., Int. Ed.* **2007**, *46*, 2458–2462.

13

STRUCTURE AND ACTIVITY OF SOME BIOINORGANIC COORDINATION COMPLEXES

JIN-TAO WANG, YU-JIA WANG, PING HU, AND ZONG-WAN MAO

MOE Key Laboratory of Bioinorganic and Synthetic Chemistry, School of Chemistry and Chemical Engineering, Sun Yat-Sen University, Guangzhou, People's Republic of China

13.1 INTRODUCTION

One-third of all known enzymes are at least metalloenzymes, the unique class of enzymes in which discrete active centers are formed when metal ions or metal-containing cofactors are incorporated into proteins. These enzymes are involved in such vital biological processes as respiration, photosynthesis, nitrogen fixation, and oxygen, carbon, sulfur, and nitrogen metabolism [1]. In addition, metal centers play crucial roles in regulating gene expression and enzymatic activity.

The modeling of metalloenzymes by biomimetic metal complexes with both intellectual and aesthetic goals helps the search for useful catalysts and the understanding of their mechanisms of operation at a small-molecule level of detail through systematic and comparative studies [2]. The goal of modeling includes defining the active-site structure, understanding the active-site function, and reproducing the enzyme function as well as extending the reactivity beyond the scope of the inspiring system. First, a number of important active sites are still poorly defined, including the oxygen-evolving complex and nitrogenase. In an effort to understand these enzymes, small-molecule analogs are created and compared to the data that exist for the proteins. Second, the structure of some enzymes is very well characterized; however, the function of some components of the active site is poorly understood. This is often

Design and Construction of Coordination Polymers, Edited by Mao-Chun Hong and Ling Chen

investigated through site-directed mutagenesis. In addition, the synthesis of a model complex can suggest the function of various components. Finally, a number of enzymes are of interest since they catalyze a reaction that chemists find challenging. These reactions include the hydrolysis of the phosphodiesters in DNA and RNA by nuclease; the elimination of reactive oxygen species by superoxide dismutase, catalase, and peroxidase; the partial oxidation of a hydrocarbon by methane monooxygenase; and the oxidation and production of hydrogen by hydrogenase. "Functional" enzyme mimics or bioinspired catalysts are designed with characteristics of the enzyme in the hope of reproducing the enzymes functionality.

However, the sequence of examining specific biological reactivity, creating similar chemical architectures, and determining functional reaction conditions for model systems presents fascinating challenges to inorganic chemists. This process allows the biological code of reactivity to be deciphered and is aided greatly by a lot of modern techniques, such as x-ray crystallography, x-ray absorption, electron paramagnetic resonance (EPR), Mossbauer, resonance Raman, Fourier transform infrared (FTIR), and ultraviolet (UV)/visible/near-IR absorption and natural and magnetically induced circular dichroism (CD and MCD), as the atomic arrangement at the active site fully encodes the reactivity observed. The wide availability of such information now allows a shift in the role of synthetic modeling from structural and spectroscopic endeavors to the development of functional and catalytic models and to complete understanding of the mechanism of assembly and the role of metal centers in the molecular mechanism of catalysis or electron transfer.

13.2 BIOMIMETIC MODELING OF A METALLOENZYME WITH A CLUSTER STRUCTURE

In nature, many metalloenzymes are large protein complexes, and much of their bulk is carried out at a cluster of metals referred to as the *oxygen-evolving complex* (OEC) and hydrogenase. In this chapter we introduce some biomimetic models of these two metalloenzymes with cluster structures.

Multinuclear manganese complexes have played an important role in the search for structural analogs of the active site of the oxygen-evolving complex of photosystem II. A homogeneous manganese catalyst, $[Mn_2^{II}(mcbpen)_2(H_2O)_2](ClO_4)_2$ (**1**) [mcbpen=*N*-methyl-*N*′-carboxymethyl-*N*,*N*′-bis(2-pyridylmethyl)ethane-1,2-diamine], capable of oxidizing water to oxygen was developed by McKenzie and co-workers [3,4]. Crystal structure analysis shows that **1** is dinuclear with (μ-κO)-bridging by one oxygen atom of the mcbpen carboxylate group (Fig. 13.1). The Mn atoms are seven-coordinated by five nitrogen atoms from one ligand, a carboxylate oxygen atom from the ligand bound to the adjacent manganese ion, and a water molecule. The flexible carboxylate donor means that dinuclear manganese complexes of mcbpen show the potential to change between six- and seven-coordinate as needed, depending on the Mn oxidation state and the presence of exogenous ligands. The Mn $\cdots$ Mn separations may vary between 4.09 Å (Mn^{II}/Mn^{II}) and 2.7 Å (Mn^{IV}/Mn^{IV}). These features are a significant advance in ligand design for an OEC

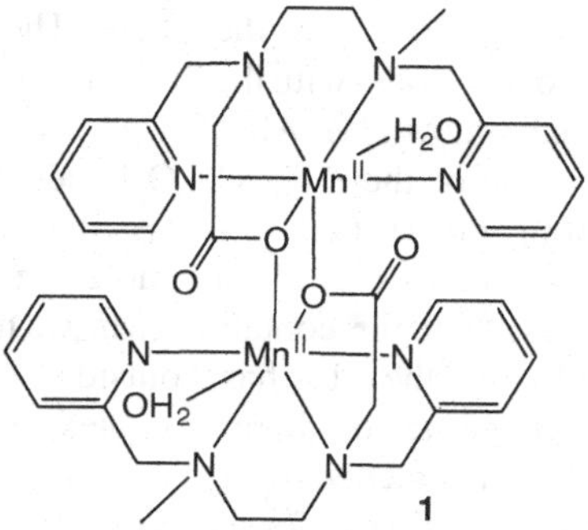

FIGURE 13.1 Schematic structure of **1**.

model and may in fact be crucial to the functionality of this system. As a result, they have directly measured the evolved O_2 concentration in solution by using the relatively new method of membrane inlet mass spectrometry (MIMS) and propose a speculative reaction mechanism. The isotopomer distribution is quantified with this method and shows that the reaction is highly specific: One oxygen atom in the evolved dioxygen comes from water, and the other is derived from the oxidant.

As the recent crystal structure resolution of the PS II of the cyanobacterium *Thermosynechococcus elongatus* at 3.5 Å that finally revealed the details of a heterometallic Mn_4CaO_4 cluster [5]. This comprises a Mn_3CaO_4 cubane, with the fourth Mn atom attached to one of its bridging O^{2-} ions. So, following is the first model complex that replicates the discrete Mn_3CaO_4 motif, with or without the fourth (dangler) manganese, which gives crystals of $[Mn_{13}Ca_2O_{10}(OH)_2(OMe)_2(O_2CPh)_{18}(H_2O)_4]\cdot 10MeCN$ (**2**) [6], which is the first high-oxidation-state manganese–calcium cluster with relevance to the water-oxidizing complex of photosynthesis. The structure (Fig. 13.2a) consists of a $[Mn_{13}Ca_2]^{42+}$ core held together by four μ_3-O^{2-}, six μ_4-O^{2-}, two μ_3-HO^-, and two μ_3-MeO^- ions. It can be described as two Mn_4O_4 cubes attached to a central planar Mn_3O_4 unit, to which are also attached two

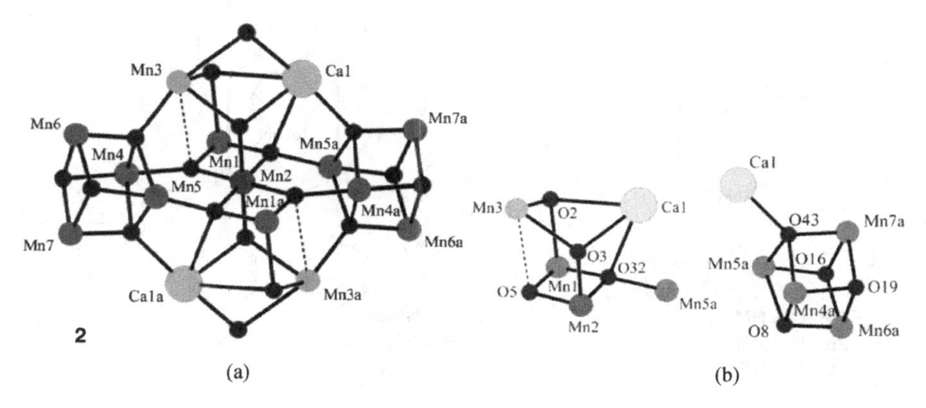

FIGURE 13.2 (a) Structure of **2**; (b) two cubane units within **2**.

Mn–Ca pairs, one above and one below the plane. The cluster is mixed-valent (one Mn^{IV}, 10 Mn^{III}, and two Mn^{II} ions) with the Mn^{IV} in the center (Mn_2) and the Mn^{II} next to the Ca^{2+} ions (Mn3 and Mn3a). All Mn and Ca^{2+} ions are six- and eight-coordinate, respectively, seven of the eight Ca–O bonds are in the range 2.297 to 2.770 Å, but the eighth is longer (Ca1–O2 = 3.039 Å). The Mn3–O5 bond is also very weak (2.998 Å), presumably due to strain in the molecule.

In fact, there are two types of cubane-containing Mn_4Ca moieties within **2**. One has the Ca^{2+} ion within the cube, and the other has it outside (see Fig. 13.2b). The former is the one relevant to the WOC (water-oxidizing complex) site. On the right is a related structure, but with the Ca^{2+} ion attached to an O atom of a Mn_4O_4 cube. Such cubes were never serious structural candidates for the WOC site, based on extended x-ray absorption fine structure (EXAFS) data.

Besides the manganese cluster compounds mentioned above, ruthenium-based catalysts also show promise for water oxidation. Zong and Thummel have designed and synthesized a new family of Ru complexes for water oxidation [7] (see Fig. 13.3). The trans-geometry for **3b** was confirmed by a single-crystal x-ray analysis. The water molecule is positioned in the Ru-5 plane with an Ru–O distance of 2.137(5) Å, which is comparable to that of other Ru(II)-aquo polypyridine complexes. The Ru–N distances fall within the normal range 2.09 to 2.10 Å, with the exception of Ru–N, which is shorter (1.92 Å) as expected. The torsion angles about the pyridine–naphthyridine bonds are only about 1.5°. The water molecule is hydrogen-bonded to a uncomplexed naphthyridine nitrogen on one side with an N···H distance of only about 2.27 Å.

Furthermore, the catalytic activities of **3a–c** and **4a–c** toward water oxidation were evaluated. At pH 1.0 the dimethylamino group is protonated and thus behaves more as an electron acceptor than as a donor group. For both mono- and dinuclear complexes, the highest turnover numbers are observed with 4-methylpyridine as the axial ligand, and these numbers indicate a robust catalyst compared to other systems reported previously. The mononuclear system **3a** produces only a small amount of dioxygen,

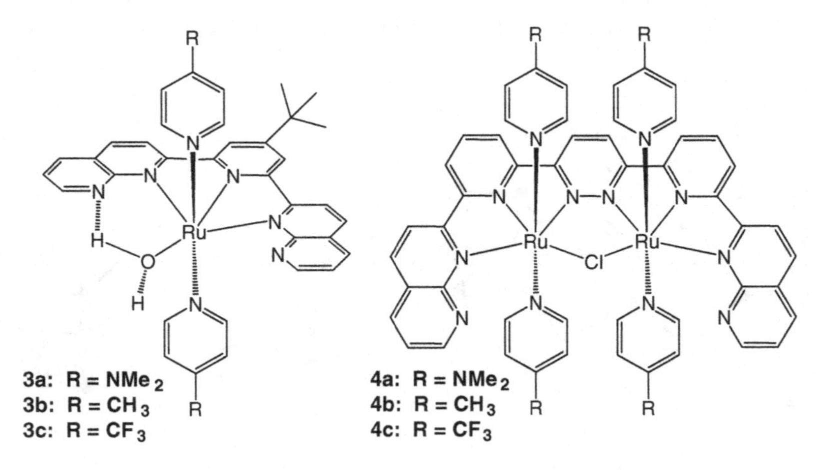

FIGURE 13.3 Structures of mononuclear and dinuclear Ru complexes for water oxidation.

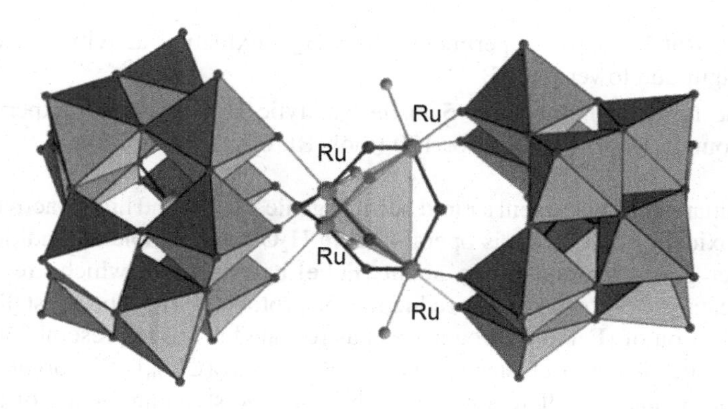

FIGURE 13.4 Structure of the polyanion in **5**, highlighting the central $\{Ru_4(\mu\text{-}O)_4(\mu\text{-}OH)_2(H_2O)_4\}^{6+}$ core (ball-and-stick representation, hydrogen atoms omitted for clarity) and the slightly distorted $\{Ru_4\}$ tetrahedron.

but does so at a fast rate. Anyway, Ru(II)-aquo systems in previous reports found no evidence for O_2 formation, indicating that complexes **3a–c** are unique in this regard.

Very recently, another functional model-based ruthenium, $Rb_8K_2[\{Ru_4O_4(OH)_2(H_2O)_4\}(\gamma\text{-}SiW_{10}O_{36})_2]\cdot 25H_2O$ (**5**), has also been synthesized by Geletii and co-workers [8]. The crystal structure of **5** reveals the "out-of-pocket" d-metal coordination polyhedra observed in water-soluble γ-diiron derivatives; namely, the ruthenium centers are corner-sharing and not ligated to the central SiO_4 unit (Fig. 13.4). The two out-of-pocket $\gamma\text{-}\{SiW_{10}Ru_2\}$ monomeric units are rotated by 90° around the vertical C_2 axis relative to one another, defining overall D_{2d} symmetry for the polyanion. The staggered structure facilitates incorporation of a $\{Ru_4(\mu\text{-}O)_4(\mu\text{-}OH)_2(H_2O)_4\}^{6+}$ core, in which the four ruthenium centers span a slightly distorted tetrahedron with Ru ··· Ru distances of 3.47 to 3.66 Å. It is noteworthy that during the synthesis of **5**, Ru^{III} ($RuCl_3\cdot H_2O$ reactant) is oxidized by O_2 to give a $\{Ru^{IV}{}_4\}$ complex, and several techniques are used to further characterize oxidation states and potentials of the ruthenium centers and the protonation states of the $\{Ru_4(\mu\text{-}O)_4(\mu\text{-}OH)_2(H_2O)_4\}^{6+}$ core.

Then they investigate the catalysis of the reaction $4[Ru(bpy)_3]^{3+} + 2H_2O \rightarrow 4[Ru(bpy)_3]^{2+} + O_2 + 4H^+$ by **5**, monitored by both accumulated $[Ru(bpy)_3]^{2+}$ spectrophotometrically and O_2 chromatographically. The result shows that at the equivalent ruthenium concentration of **5** (2.5 μm), the rate of the equation was about 100-fold higher. That is, with the help of the complex, water oxidation by $[Ru(bpy)_3]^{3+}$ became much easier. Although the kinetics of H_2O oxidation catalyzed by **5** is complicated, which might indicate involvement of multiple intermediates, four lines of evidence indicate that **5** remains intact in many oxidation states in neutral aqueous media:

(1) The acid–base titrations are fully reversible up to pH 7.5.
(2) The cyclic voltammograms are reproducible and show reversible oxidation and reduction of multiple ruthenium centers.

(3) The $RuCl_3$ control experiment shows H_2O oxidation activity two orders of magnitude lower than **5**.

(4) The turnover number for **5** in the catalytic H_2O oxidation experiments is about 18 (mol O_2/mol **5**) or 120 (mol $[Ru(bpy)_3]^{3+}$/mol **5**).

In summary, they document a rapid, all-inorganic, stable, and highly active catalyst for H_2O oxidation to O_2 that is operational in H_2O under ambient conditions.

Another group of models are about [FeFe]-hydrogenase, which are the most efficient catalysts known for the reduction of protons to H_2. The crystallographic characterization of [FeFe]-hydrogenases has revealed a striking resemblance of the di-iron subsite of the H-cluster to known $[Fe_2(\mu\text{-}SR)_2(CO)_6]$ (R = organic group) complexes. A major challenge now is to build a free-standing analog of the entire H-cluster, as this offers the prospect of understanding the interplay of the conjoined di-iron and cubane units that form the enzymic catalytic machinery and of the development of new catalytic materials.

The chemical synthesis of the H-cluster framework of [FeFe]-hydrogenase has been achieved through linking of a di-iron subsite to a [4Fe–4S] cluster [9]. They synthesized analytically pure $[Fe_4S_4(L)_3\{Fe_2(CH_3C\ (CH_2S)_3)(CO)_5\}][NBu_4]_2$ (**6**) [L = 1,3,5-tris(4,6-dimethyl-3-mercaptophenylthio)-2,4,6-tris(*p*-tolyl-thio)benzene]. Then full geometric optimization of an *in silico* model of **6**, $[Fe_4S_4(SCH_3)_3\{Fe_2(CH_3C\text{-}(CH_2S)_3)(CO)_5\}]^{2-}$, was carried out in a density functional theory (DFT) framework using the BP86 pure functional and an all-electron valence triple-ζ basis set with polarization functions on all atoms (TZVP) (see Fig. 13.5). In particular, one of the thiolate groups bridges almost symmetrically the {4Fe4S} and the {2Fe2S} units, with an Fe–Fe distance (2.6 Å) in the binuclear cluster that is indicative of a metal–metal bond. Analysis of the electronic properties of the *in silico* structure reveals that the redox state of the binuclear moiety can be described as Fe(I)Fe(I). Furthermore, the function of the artificial H-clusters as an electrocatalyst for proton reduction has also been studied.

Through linking of a di-iron subsite to a {4Fe4S} cluster, they achieve the first synthesis of a metallosulfur cluster core involved in small-molecule catalysis. In addition to advancing understanding of the natural biological system, the availability of an active, free-standing analog of the H-cluster may enable us to develop useful

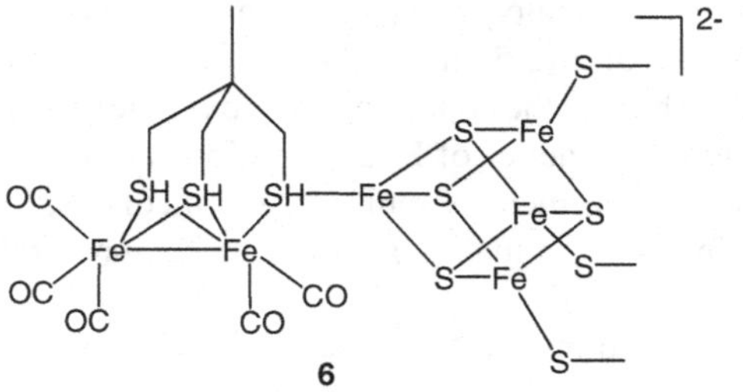

FIGURE 13.5 Structure of **6**.

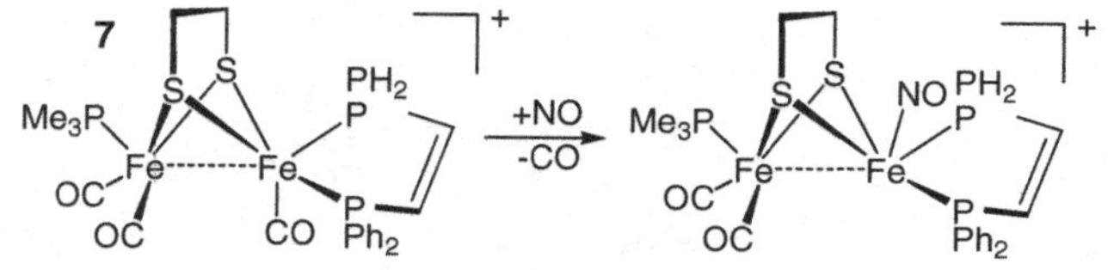

SCHEME 13.1 Structure and irreversibly bound NO of **7**.

electrocatalytic materials for application in, for example, reversible hydrogen fuel cells.

The active site of [FeFe]-hydrogenase exists in two functional states, H_{red} and H_{ox}. Here, a model for the H_{ox} state of the [FeFe]-hydrogenase is described [10]. The mixed-valence salt $[Fe_2(S_2C_2H_4)(CO)_3(PMe_3)(dppv)]$ [**7**,dppv = *cis*-1,2-$C_2H_2(PPh_2)_2$] (Scheme 13.1) is attractive, because like the active site, this di-iron framework bears three donor ligands and three CO ligands. Crystals of **7** were obtained from CH_2Cl_2/hexanes at −20°C. Crystallographic analysis revealed an unsolvated 1 : 1 salt. In general, the $[Fe_2(SR)_2L_6]$ framework approaches that proposed for the H_{ox} state of the active site. Of specific interest, the oxidation has caused the {Fe(CO)(dppv)} center to rotate around the Fe–Fe axis. The apical site on the {Fe(dppv)} center is vacant, and the $[BF_4]^-$ anion is approximately 5 Å away from this "distal" Fe center. The "proximal" $\{Fe(CO)_2(PMe_3)\}$ moiety remains relatively unperturbed, and overall, the bond lengths match those in **7** within 0.02 Å. The unique CO ligand is only partially bridging, with a relatively long Fe_{prox}–C separation of 2.62 Å compared to the Fe_{dist}–CO bond length of 1.781(3) Å.

The radical character of **7** was demonstrated by its reactivity toward NO. At −45°C, **7** rapidly and irreversibly binds NO molecule. The product, which forms with loss of CO, is the diamagnetic dicarbonyl nitrosyl $[Fe_2(S_2C_2H_4)(CO)_2(NO)(PMe_3)(dppv)]BF_4$ ($\upsilon_{NO} = 1772\ cm^{-1}$), which was isolated in analytical purity. ^1H and ^{31}P nuclear magnetic resonance (NMR) spectra show that this $\{Fe^I{}_2\}$ species is C_s-symmetric, which uniquely requires that the NO ligand occupies the apical site on the {Fe(dppv)} center (Scheme 13.1).

Indeed, the mixed-valence salt $[Fe_2(S_2C_2H_4)(CO)_3(PMe_3)(dppv)]BF_4$ displays the structure and reactivity proposed for H_{ox}, one of the two functional states of the [FeFe]-hydrogenases. It is a somewhat effective method that links a porphyrin-type photosensitizer to the simple biomimetic model for [FeFe]-hydrogenase in order to prepare a new type of light-driven model compound. Not long ago, Song and co-workers synthesized one such model compound [11] (see Fig. 13.6) in which a photosensitizing tetraphenylporphyrin group (TPP) is covalently linked to the N atom of the di-iron azadithiolate (ADT) moiety $[\{(\mu\text{-}SCH_2)_2N\}Fe_2(CO)_6]$ (**9**). The molecular structures of model **9** and its precursor **8** were confirmed by x-ray crystallography. Both **8** and **9** contain a di-iron/ADT moiety in which the six-membered ring Fe1–S1–C7–N1–C8–S2 has a boat conformation and the other six-membered ring, Fe2–S1–C7–N1–C8–S2, adopts a chair conformation. The bond between the disubstituted benzene ring and the N1 atom is axially oriented in both **8** and **9**, and all the iron nuclei in **8** and **9** have the expected square-pyramidal

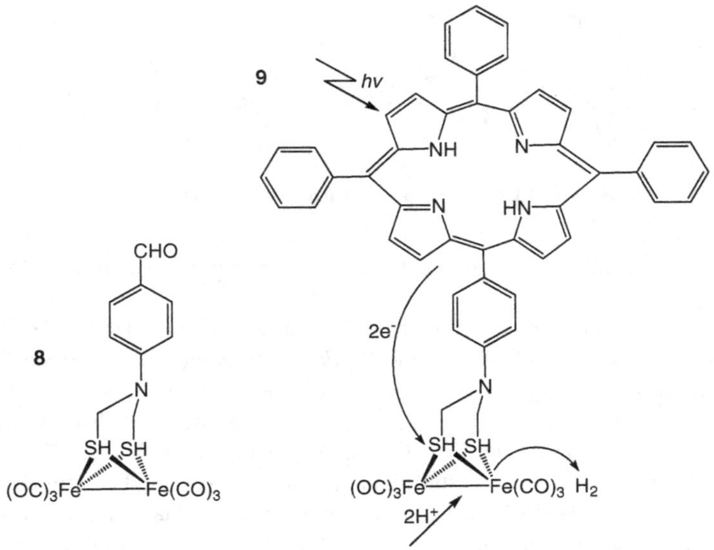

FIGURE 13.6 Structures of **8** and **9** and the light-driven process of **9**.

coordination sphere. The Fe–Fe bond lengths of 2.4956(9) Å for **8** and 2.5000(9) Å for **9** are within a reasonable range. The sum of the C–N–C angles around N1 is 356.8° for **8** and 355.3° for **9**, which implies somewhat interrupted π conjugation between the disubstituted benzene ring and the p orbital of the nitrogen atom.

Whereas **8** is a simple model compound, the target compound **9** is a novel light-driven-type model with a porphyrin photosensitizer. The C–N bond lengths in the pyrrole rings of the porphyrin unit of **9** are nearly the same [1.364(5) to 1.378(5) Å] and lie between those of normal single and double C–N bonds. The four benzene rings attached to the porphyrin unit of **9** are planar and twisted relative to the porphyrin plane in order to reduce the steric interactions between the phenyl hydrogen atoms proximal to the porphyrin unit and the pyrrole rings.

In the expected light-driven process, model **9** should first absorb a photon at the photosensitizer TPP. This photoexcited TPP is then oxidatively quenched by the di-iron unit to give a reduced iron species. After regeneration of the TPP by transfer of an electron from an external donor, this process is repeated to produce a doubly reduced di-iron species that should be able to drive the reduction of protons to hydrogen (Fig. 13.6). In view of the strong coordination ability of porphyrin macrocycles with a wide variety of metal ions and the facile substitution of the Fe-bound CO by other ligands, it is supposed that we will be able to further modify **9** to facilitate electron transfer from the photoexcited TPP to the di-iron/ADT moiety and thus finally accomplish the light-driven reduction of proton to hydrogen.

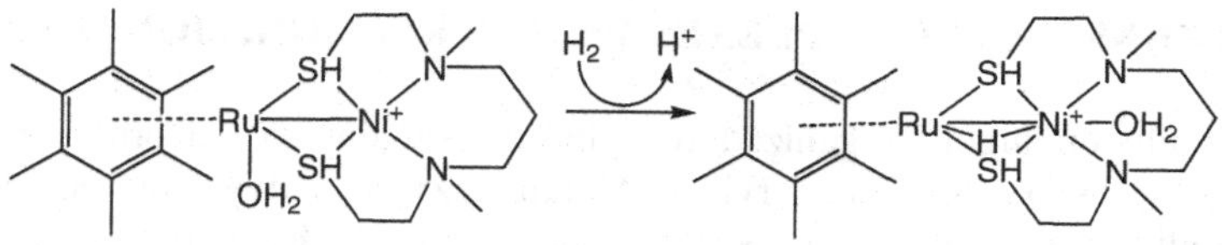

SCHEME 13.2 Structures of complexes **10** and **11**.

Recent work from Ogo and co-workers has described a combined structural and functional model for the [NiFe]-hydrogenases, the most pervasive family of biocatalysts for the production and oxidation of H_2 [12]. In examining the active site of the [NiFe]-hydrogenase, three structural criteria come to mind: a nickel–iron core, a pair of bridging thiolate ligands, and most important, a bridging hydride (Scheme 13.2). The biomimetics of the enzyme have been pursued even before structural data were available, with emphasis mainly on the first two structural criteria. Even the most realistic synthetic reproductions of the active site have not yet evolved into functional models. This situation may change in light of Ogo and co-workers' complex, which meets nearly all the structural criteria and is functional. To achieve and fully characterize their functional model, Ogo et al. successful isolation and crystal structure of the paramagnetic Ni(μ-H)Ru complex $[(Ni^{II}L)(H_2O)(\mu\text{-}H)Ru^{II}(\eta^6\text{-}C_6Me_6)](NO_3)$ [**11**, L = *N,N*′-dimethyl-*N,N*′-bis(2-mercaptoethyl)-1,3-propanediamine], which was synthesized by reaction of a diamagnetic dinuclear NiRu aqua complex $[(Ni^{II}L)Ru^{II}(H_2O)(\eta^6\text{-}C_6Me_6)](NO_3)_2$ (**10**) with H_2 in water under ambient conditions.

The molecular structures of the target model compounds **10** and **11** have been confirmed by x-ray diffraction techniques. The Ni center of **10** sits in the pocket of the tetradentate ligand in a square-planar arrangement. The Ru–O (aqua ligand) bond length is 2.154(3) Å, whereas the Ni···O distance is 2.858(3) Å. In **11**, the Ni atom adopts a distorted octahedral coordination geometry, with a tetradentate ligand, one aqua ligand, and one hydrido ligand (Scheme 13.2). The bridging H atom is closer to the Ru atom [Ru–H = 1.676(8) Å, Ni–H = 1.859(7) Å] in the Ni–H–Ru moiety. The Ni···Ru distance [2.739(3) Å] of **11** is shorter than that of **10** [3.1611(6) Å]. A similar tendency has been observed in [NiFe]-hydrogenase; that is, EXAFS studies on the [NiFe]-hydrogenase have shown that the Ni···Fe distance in the active form is 2.512(7) Å.

This was a smart move that replaced the $\{Fe(CN)_2(CO)\}$ unit with $\{Ru\text{-}(C_6Me_6)\}^{2+}$, as ruthenium forms more stable dihydrogen complexes than any other metal, whereas similar charge-neutral iron species are rare because of their intrinsic lability. The chemical and structural features of the two models allow planning for improved theoretical modeling of the [NiFe]-hydrogenases, which can now be based more on facts than on assumptions. The pseudo-octahedral Ru^{II} and square-planar Ni^{II} centers of precursor **10** are electronically saturated and are thus in principle inactive. Owing to the flexible hinge of the thiolato bridges, the ruthenium-coordinated water molecule finds enough space between the two metals without perturbing the nickel center.

13.3 BIOINSPIRED COMPLEXES WITH A RECOGNITION DOMAIN

Proteins rely on highly organized recognition domains to perform the various functions necessary for cell survival. Metalloenzyme active sites contain both metal-binding pockets (primary coordination sphere), which are composed of amino acid ligands (e.g., histidine), as well as noncoordinating residues (second coordination sphere), which regulate important structural and chemical properties of the protein. This second coordination sphere is important to maintaining the metastable structure of the active site and to selectively recognizing substrate and is regulated by modulation of hydrophobic–hydrophilic interactions in the vicinity of the metal-binding pocket, and donor–acceptor interactions between proximal amino acid residues and targeted substrate molecules. To overcome these shortcomings, mimetic proteins are being extensively pursued. A conceptually simple solution would be to replace the unstable parts of a protein with a synthetic scaffold. The more exciting and fascinating application of the catalysts is in obtaining artificial metalloenzymes with a higher substrate affinity or different substrate specificity than those of natural systems [13,14].

The first example is a series of ligands derived from the bis-2-pyridinylmethylamine structure bearing aromatic amino groups [15], which were prepared and their Zn(II) complexes **12**, **13**, and **14** (see Fig. 13.7) studied as catalysts for the cleavage of bis-*p*-nitrophenyl phosphate (BNPP) diesters. The resulting second-order rate constants were 4.2×10^{-4}, 3.8×10^{-3} and $9.7 \times 10^{-2}\,M^{-1}\,s^{-1}$ for complexes with increasing amino groups at 25°C, respectively, indicating that activation of the substrate by means of one or two hydrogen bonds produces a 9- and 230-fold

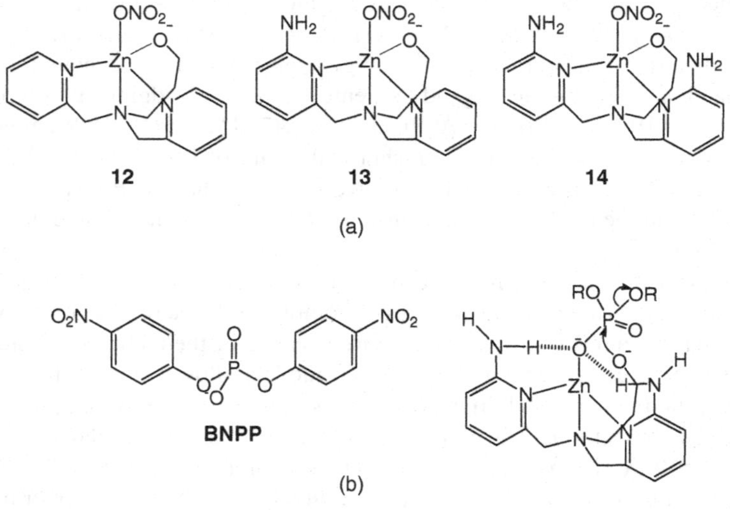

FIGURE 13.7 (a) Structures of **12**, **13**, and **14**; (b) substrate and proposed intermediate for BNPP cleavage by **14**.

reactivity increase, respectively. The comparative kinetic study indicated that the insertion of double amino groups in **14**, capable of acting as hydrogen-bond donors, substantially increases the hydrolytic activity of the metal complex. The mechanism of the reactions and the structure of the complexes were investigated in detail by means of kinetic analysis, NMR spectroscopy experiments, and theoretical calculations. The implementation of the structure of the reactive complex with groups capable of forming intramolecular hydrogen bonds with the substrate emerges as the most important feature of the system.

Relative to an amino group, a guanidinium group that has six potential hydrogen-bond donors available is one of the most representative. It contains two amines and an iminium in a plane forming a Y shape to establish its great stability as an ion in aqueous environment. These structural features make it an extremely advantageous functional group for the binding of carboxylates or phosphates in enzymes and antibodies via forming strong ion pairs, which play a key role in many biological activities, such as molecular recognition and catalysis. Anslyn and co-workers observed that the Zn(II) complex of a terpyridine ligand bearing two guanidinium groups (Fig. 13.8) leads to rate acceleration for the hydrolysis of the RNA dinucleotide ApA at pH 7.4 at 37°C of three orders of magnitude with respect to the complex without guanidinium groups [16]. In complex **15**, the cooperativity between the catalytic groups is comparable to natural staphylococcal nuclease containing the Ca(II) center.

The x-ray crystal structure of **15** shows that the Zn(II) complex exhibits distorted trigonal bypyramidal geometry about the Zn(II) center. The guanidinium auxiliary groups would be expected to have greater hydrogen-bonding ability and less ordered orientation. The position of the auxiliaries is dictated primarily by intermolecular hydrogen bonding as opposed to intramolecular electrostatic repulsions, and as a result, **15** adopts a less symmetric solid-state geometry. In the **15** structure, the

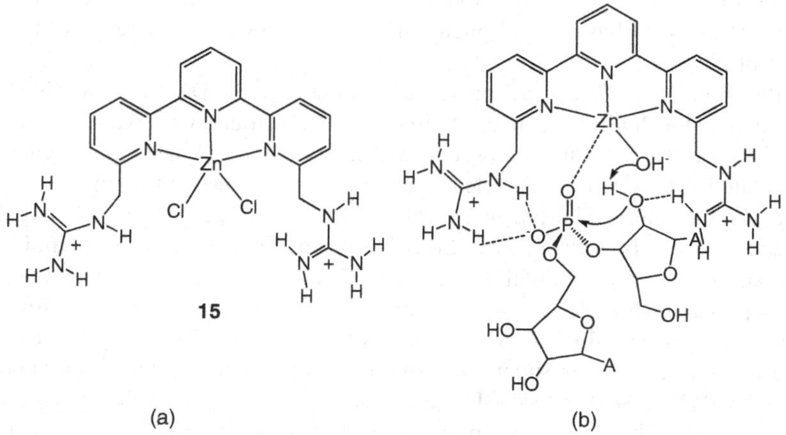

FIGURE 13.8 (a) Structure of **15**; (b) suggested intermediate for ApA cleavage.

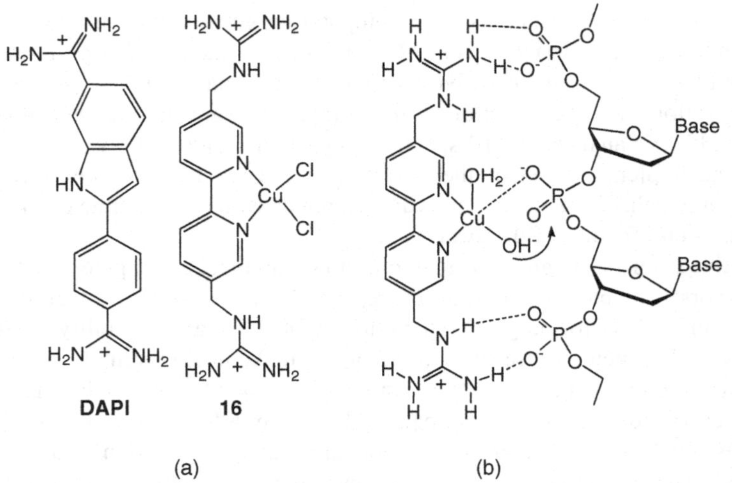

FIGURE 13.9 (a) Structures of **16** and DAPI; (b) suggested mechanism for DNA cleavage by **16**.

integrity of the Zn(II) coordination geometry is not compromised by the presence of cationic auxiliary groups.

Another example is the copper complex of a 2,2′-dipyridyl with double guanidinium pendants reported by He and co-workers [17]. A prospective view of the cationic structure of **16** shows that the geometry around the copper(II) ion is square-planar based, with the basal plane formed by the two nitrogen atoms of the ligand and two chloride atoms (Fig. 13.9). Since the guanidimium plane atoms adopts an extended conformation in the crystal stack, the distance between the Cu and central N atoms is 6.5408(5) Å, which is coincident with that of adjacent phosphodiesters in B-form DNA (ca. 6 Å). In addition, the structure of the complex is very similar to that of DAPI (4′-6-diamidino-2-phenylindole), which especially targets the minor groove of natural DNA.

In the absence of reducing agent, the supercoiled plasmid DNA cleavage by **16** was performed and its hydrolytic mechanisms was demonstrated with radical scavengers and T4 ligase. As a result, the pseudo-Michaelis–Menten kinetic parameter (k_{cat}) is calculated to be 4.42 h^{-1}. The cleavage efficiency is about 10-fold higher than that of the simple analog [Cu(bipy)Cl_2] ($k_{cat} = 0.50\ h^{-1}$). The highly effective DNA cleavage ability of **16** is attributed to the effective cooperation of the metal moiety and two guanidinium pendants with the phosphodiester backbone of nucleic acid.

In peptides, guanidine, a residue of arginine, exists in the protonated form as a guanidinium cation, which functions as an efficient recognition moiety of anionic functionalities such as the superoxide anion. Mutant studies of the copper and zinc superoxide dismutase (Cu,Zn-SOD) have revealed that Arg141 in the enzyme, about 5.9 Å away from the copper ion, promoting electrostatic steering of the superoxide substrate to and from the copper ion in the active site. Recently, two inclusion systems

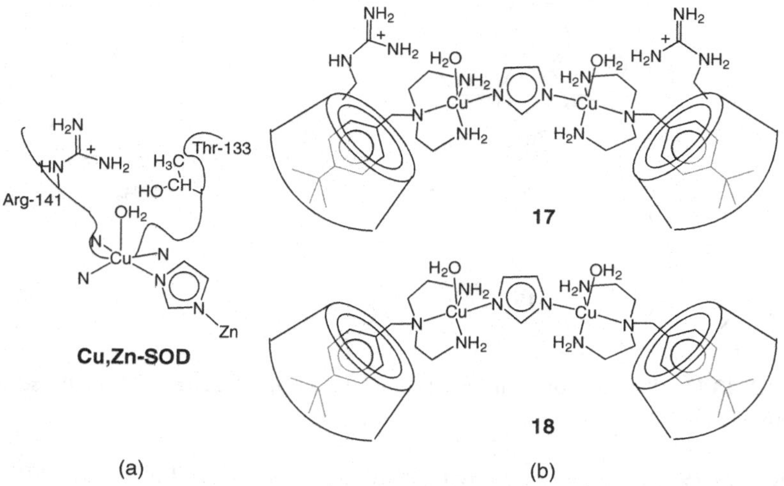

FIGURE 13.10 (a) Cu,Zn-SOD active site; (b) its supramolecular mimics **17** and **18**.

of guanidinium-modified β-cyclodextrin derivative with mono- or binuclear copper complexes were reported as new supramolecular mimics for Cu,Zn-SOD [18,19].

Interestingly, the structure of the simple analog of $\{[Cu(L)(H_2O)(\beta GCD)]_2(im)\}$-$(ClO_4)_3$ (**17**), $\{[Cu(L)(H_2O)(\beta CD)]_2(im)\}(ClO_4)_3$ (**18**), was characterized by single-crystal x-ray crystallography, where L is 4-(4′-*tert*-butyl)benzyldiethylenetriamine, βCD is β-cyclodextrin, βGCD is mono-6-deoxy-6-guanidinocycloheptaamylose, and im is imidazolate. As is shown in Figure 13.10, the two *tert*-butylbenzyl groups of the imidazolated-bridged dinuclear Cu(II) moiety insert into cavities of the two βCD molecules along the primary hydroxyl side to form the supramolecular unit $\{[Cu(L)(H_2O)(\beta CD)]_2(im)\}^{3+}$. Each Cu(II) ion has a distorted square-pyramidal geometry with four nitrogen atoms and a H_2O molecule, which is very similar to N_4O_w, the coordination sphere of Cu(II) ion in the natural enzyme. Interestingly, the two βCD molecules form a head-to-head dimer though hydrogen bonding of secondary hydroxyl groups (bond distances of 2.7 to 3.0 Å), resulting in a novel helical chiral chain.

In such a model, the copper moiety of the guest and the guanidinium cation of the host mimic the active site and Arg141 of Cu,Zn-SOD, respectively. Their SOD-like activities were investigated and obtained IC_{50} values showed that the mimic with guanidyl cation caused a 43% addition in SOD activity relative to the mimic without guanidyl cation. The result indicates that the guanidyl cation of the mimic can surely enhance SOD activity effectively.

Since βCD has an appropriate cavity size that can mimic the hydrophobic environment around the metal ion center in a metalloenzyme and weak interactions after complexation, metal complexes of β-cyclodextrin have been used for the synthesis of metalloprotein model. A typical example is that of two phen-Zn(II) complexes positioned alternately on the secondary and primary sides of βCD

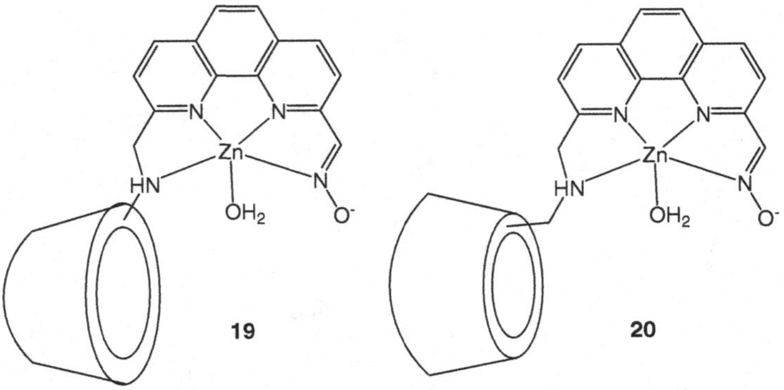

FIGURE 13.11 Structure of phen-Zn(II) complexes positioned alternately on the secondary and primary sides of βCD.

(complexes **19** and **20**), reported by Breslow and Nesnas (see Fig. 13.11) [20]. They were examined as catalysts for the formation of *p*-nitrophenoxide ion from 4-nitrophenyl acetate (pNA) in the burst kinetic region in 10 mM HEPES buffer (pH 7.0). Kinetic saturation was seen, from which a Lineweaver–Burk double reciprocal plot gave $K_m = 0.4$ mM and $k_{cat} = 0.060\ s^{-1}$ for **19** and $K_m = 0.6$ mM and $k_{cat} = 0.0039\ s^{-1}$ for **20**, respectively. In the absence of the catalyst, k_{uncat} for pNA was $2.65 \times 10^{-6}\ s^{-1}$, so k_{cat}/k_{uncat} is 22,600 and 1500 for the initial fast deacylation of **19** and **20**, respectively. The preference for the secondary face catalyst may reflect preferential binding geometry, and probably also the greater undesirable flexibility of the primary face catalyst.

Further, a cyclodextrin dimer with the Zn(II)-phenanthroline linking group (**21**) was reported and used as a catalyst for diester hydrolysis [21]. In complex **21**, the Zn(II)-phenanthroline links two CDs on their primary-sides (see Fig. 13.12). Hydrolytic kinetics of carboxylic acid esters were performed with bis(4-nitrophenyl) carbonate (BNPC) and pNA as substrates. The hydrolysis rate constants obtained

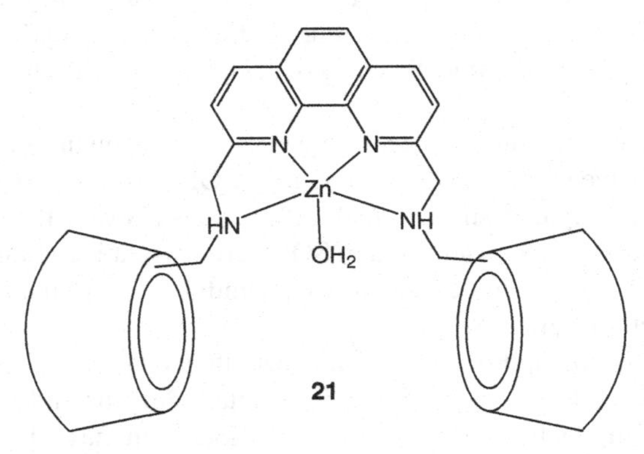

FIGURE 13.12 Structure of a cyclodextrin dimer with the Zn(II)-phenanthroline linking group.

showed that **21** has a very high rate of catalysis for BNPC hydrolysis, giving a 3.89×10^4-fold rate enhancement over uncatalyzed hydrolysis in 50 mM Tris at pH 7.01and 25°C, relative to only a 42-fold rate enhancement for pNA hydrolysis.

As phospholipase C and P1 nuclease which use three Zn(II) ions to catalyze the hydrolytic cleavage of phosphate diester bonds in phosphatidylcholine and in nucleotides such as RNA and DNA, respectively. Unlik the mononuclear complexes mentioned aboved, binuclear and trinuclear complexes especially facilitated by low-energy conformational changes of the flexible calix[4]arene backbone are another kind of artificial nucleases. Molenveld and co-workers [22] used calix[4] arenes modified with 1-3 Zn(II)-2,6 -bis(aminomethyl)pyridyl groups, **22**, **23**, and **24** (Fig. 13.13a), respectively, were investigated as models for dinuclear and trinuclear metallo-enzymes that catalyze the cleavage of phosphate diesters.

Under neutral conditions, 0.48 mM of **23** causes a 23,000-fold rate acceleration in the transesterification of the RNA model substrate 2-hydroxyproyl-*p*-nitrophenyl phosphate (HPNP, 0.19 mM). Comparison with the activities of a mononuclear complex **22** and the complex lacking the calix[4]arene backbone shows that the catalysis is due to the cooperative action of the Zn(II) centers and indicates that

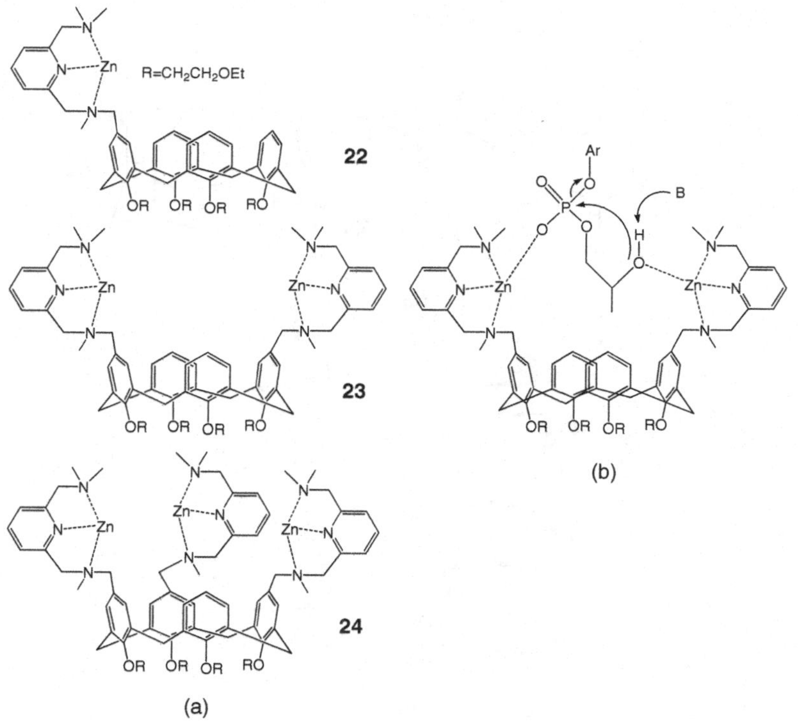

FIGURE 13.13 (a) Mono-, di-, and trinuclear Zn(II)-calixarene complexes; (b) possible bifunctional mechanism to rationalize the rate enhancement for the dinuclear system.

hydrophobic effects contribute to the catalysis (Fig. 13.13b). Saturation kinetics and pH variation studies demonstrate that the high catalytic activity of the flexible complex **23** originates from a very high substrate binding affinity, affording a Michaelis–Menten complex in which the substrate is converted with a relatively moderate rate. The trinuclear complex **24** induces a 32,000-fold rate acceleration and shows decreased substrate binding and an increased catalytic rate compared to its dinuclear analog **23**. In a possible mechanism, two Zn(II) ions activate the phosphoryl group and another activates the β-hydroxyl group of HPNP.

Besides ammonium, guanidinium, and calixarene groups, conjugation with other DNA-affine subunits and sequence-selective elements are also reported. For example, a series of *cis-cis-* triaminocyclohexane Zn(II) complex–anthraquinone intercalator conjugates, linked by alkyl spacers of different length (**25**, **26**, **27**) reported by Boseggia and co-workers (Fig. 13.14a) [23]. At a concentration of 5 μM, complex **26** with a C_8 alkyl spacer cleaves supercoiled DNA at a rate of $4.6 \times 10^{-6}\,s^{-1}$ at pH 7.0 and 37°C. Saturation kinetics have been observed with a binding constant (K_a) of about $1.0 \times 10^{-4}\,M^{-1}$, in agreement with the reported DNA affinity of anthraquinone. Thus, conjugation of the metal complex with the intercalating group led to a 15-fold increase in the cleavage efficiency compared with the Zn-triaminocyclohexane complex lacking the anthraquinone moiety.

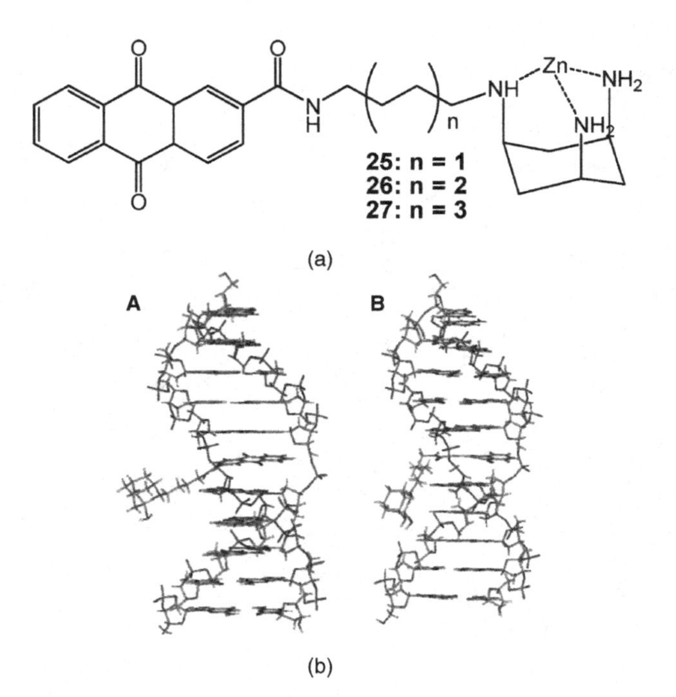

FIGURE 13.14 (a) Structure of **25**, **26**, and **27**; (b) calculated structures of the intercalation complexes between complexes **25** ($n = 1$, **A**) and **27** ($n = 3$, **B**) and DNA.

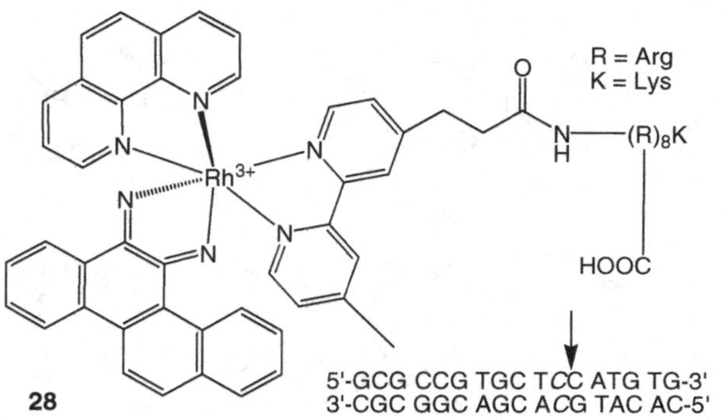

FIGURE 13.15 Structure of **28** and mismatched DNA with marked mismatch and cleavage site.

On the other hand, the reactivity trend of complexes **25** to **27** highlights the remarkable effect due to the length of the alkyl spacer. While complex **25**, with a C_4 spacer, is virtually ineffective at every concentration, the reactivity increases progressively using complexes **26** and **27** with C_6 and C_8 spacers. Such behavior indicates that the length and flexibility of the spacer play a fundamental role in the interaction between the catalytic subunit and the phosphate backbone. After intercalation of the anthraquinone moiety, the spacer must be free to fold in such a way as to position the metal complex close to a phosphate group, and a too-short spacer may prevent correct folding. This is confirmed by molecular mechanics calculations (Fig. 13.14b).

When common metal complexes interact with targeted nucleic acid, it is difficult for them to have selectivity simultaneously. To enhance the selectivity, the introduction of biological ligands (e.g., nucleic acid bases, amino acids, peptides) to the structure of the complexes is a useful strategy. Following that, a Rh complex modified by peptide is introduced. The Rh complex **28** (Fig. 13.15) was synthesized by Brunner and Barton [24] by coupling the parent [Rh-(phen)(bpy3C)chrysi]$^{3+}$, containing a pendant carboxylate (bpy3C = 4-propionic acid-4′-methyl-2,2′-bpy), to the N-(Arg)$_8$Lys-C peptide on the solid phase after activation with HOBT/HBTU.

Photocleavage was performed by irradiation after incubation with **28** of an oligomeric DNA duplex (2 μM) containing a single base-pair mismatch. After desalting the reaction mixture by the ziptip procedure, MALDI-TOF mass spectra and gel electrophoresis analysis were measured for the cleavage reaction. As a result, complex **28** of Rh containing a pendant peptide for targeting DNA mismatches site-specifically.

13.4 FUNCTIONAL COMPLEXES AS THERAPEUTIC AGENTS

Anticancer drugs based on platinum; such as cisplatin (**29**) [CDDP, *cis*-diamminedichloroplatinum(II)], which was introduced into clinical practice in 1971, are used frequently and widely today. Cisplatin is effective in treating a variety of cancers, especially testicular cancer, for which it has a greater than 90% cure rate. However, the

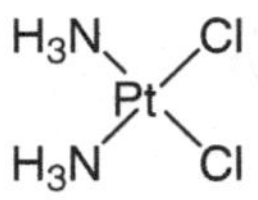

29

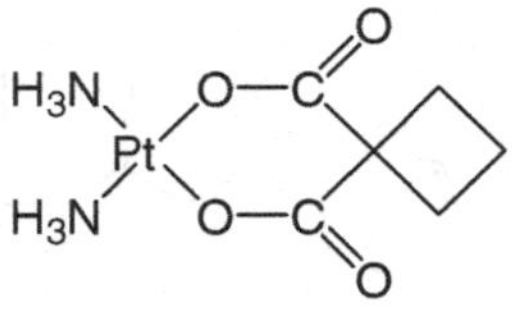

30

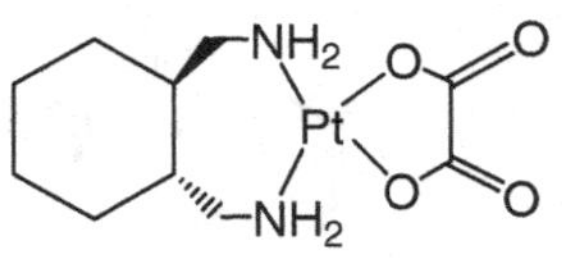

31

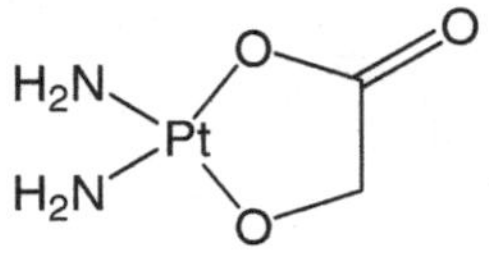

32

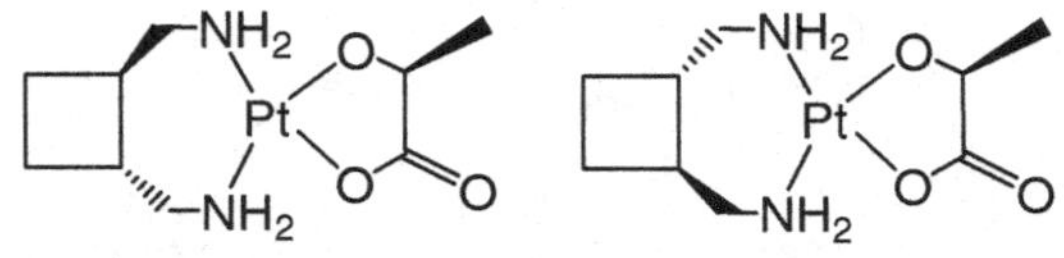

33

FIGURE 13.16 Structures of typical platinum drugs.

drug's resistance and side effects hampered the treatment efficacy of cisplatin. During the lasting years, four complexes were approved for clinical use: carboplatin (**30**), oxaliplatin (**31**), nedaplatin (**32**), and lobaplatin (**33**) (Fig. 13.16). Because their structure is similar to that of cisplatin, with two ammine or amine donor groups and two anionic leaving groups in a *cis* geometry, some of the drawbacks of cisplatin are consequently present.

Platinum exists in two main oxidation states, Pt(II) and Pt(IV). In Pt(II) complexes such as cisplatin and carboplatin, the platinum atom has four bonds directed to the corners of a square plane where four ligand atoms are located. In contrast, in Pt(IV) complexes, there are six bonds and ligands: four in a square-planar configuration and two located axially, directly above and below the platinum, thus producing an octahedral configuration [25]. A number of novel cisplatin analogs have been prepared over the years. The design and development of these cisplatin analogs have revealed common requirements that are necessary in its use as an anticancer drug.

Ruiz et al. obtained Pt(II) complexes (Fig. 13.17) with the anions of the model nucleobases [26] established by x-ray diffraction and explored the DNA adduct formation and values of IC_{50}. From the data in Table 13.1, the values of IC_{50} are apparently lower than those of cisplatin. Atomic force microscopy images of the modifications caused by the platinum complexes on plasmid DNA pBR322 were

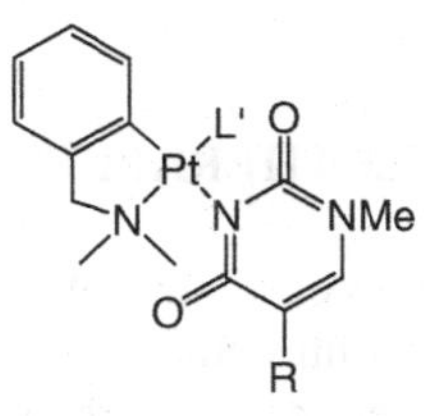

	L'	R
34	PPH_3	CH_3
35	PPH_3	H
36	DMSO	CH_3
37	DMSO	H

FIGURE 13.17 Structures of Pt(II) complexes with bmda.

Table 13.1 IC_{50} Values for Cisplatin and Complexes 34 to 37 for the Tumor Cell Line HL-60

Complex	IC_{50}/24 h (μM)	IC_{50}/72 h (μM)
Cisplatin	15.61	2.15
34	0.692	0.729
35	0.674	0.643
36	1.614	0.935
37	1.537	0.782

also obtained. From the images we observe that the complexes seem to modify the morphology of the pBR322 DNA in a mode similar to that of cisplatin. They both cause kinks and cross-linking in the plasmid forms.

The original empirical structure–activity relationships considered the *trans* isomer of antitumor cisplatin and other transplatin analogs to be inactive. However, a series of bifunctional *trans*-platinum(II) complexes have been synthesized that show anticancer activity distinct from cisplatin (sometimes even more efficient than cisplatin itself and its analogs) and bind to DNA in a manner distinctly different from that of cisplatin. *Trans, trans, trans*-[$PtCl_2(CH_3COO)_2(NH_3)$(1-adamantylamine)] [*trans*-adamplatin(IV)] (**38**) and its reduced analog *trans*-[$PtCl_2(NH_3)$(1-adamantylamine)] [*trans*-adamplatin(II)] (**39**) were examined [27] (see Fig. 13.18) and their cytotoxicities and mutagenicities were investigated. Interestingly, *trans*-adamplatin(IV) was considerably less mutagenic than cisplatin. However, IC_{50} shows that complex **39** has better anticancer activity than **38**. Consistently with the lipophilic character of *trans*-adamplatin complexes, their DNA binding mode markedly different from that of ineffective transplatin.

Multinuclear platinum complexes have shown great potential for cancer chemotherapy [28]. These complexes contain two, three, or four platinum centers, with both *cis* and/or *trans* configurations, and bind to DNA in a manner different from that of cisplatin. They react with DNA more rapidly than cisplatin does and produce characteristic long-range inter- and intrastrand cross-linked DNA adducts [29]. The interstrand cross-links are insensitive to the repair of cellular extracts, thus enhancing the cytotoxicity of multinuclear complexes. Multinuclear platinum complexes are structurally very diverse, and the trinuclear complex [(*trans*-PtCl $(NH_3)_2)_2$\{μ-*trans*-Pt-$(NH_3)_2(H_2N(CH_2)_6NH_2)_2$\}]$^{4+}$ (BBR3464) has undergone phase II clinical trials for treatment of a variety of cancers and shown IC_{50} values at least 20-fold lower than that of cisplatin.

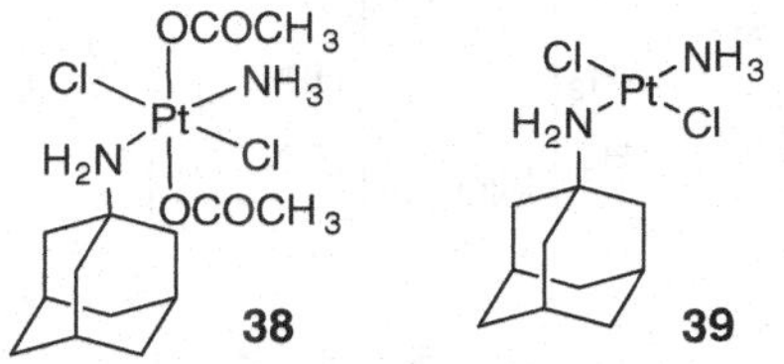

FIGURE 13.18 Structures of *trans*-adamplatin(IV) (**38**) and *trans*-adamplatin(II) (**39**).

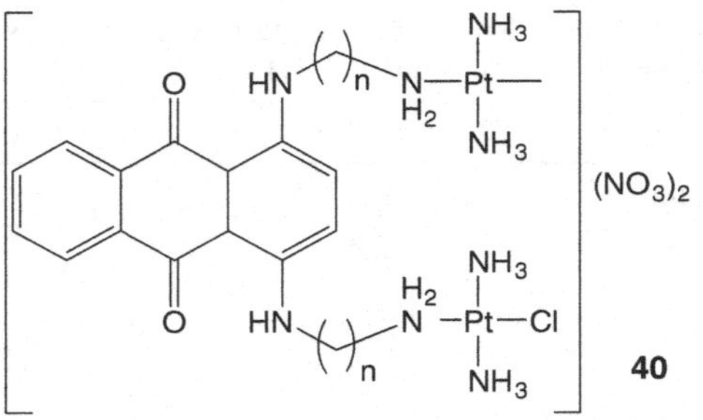

FIGURE 13.19 Structure of dinuclear platinum complex **40** with anthraquinone bridge.

We mentioned earlier that one of the drawbacks of some *cis*-platinum drugs is their resistance. Now there is dinuclear platinum with *N,N′*-bis(aminoalkyl)-1,4-diaminoanthraquinones (**40**), reported to be cisplatin-resistant over a U2-OS/Pt subline [30] (Fig. 13.19). The platinum complex is excreted from the cell via the Golgi apparatus, while the weakly basic anthraquinone ligand accumulates in the Golgi complex, where it is taken up by lysosomes and then transported to the cell surface.

Three new derivatives of the cytotoxic azole-bridged dinuclear platinum(II) complex (Fig. 13.20) has been reported and characterized structurally [31]. A cytotoxicity assay of these dinuclear platinum(II) compounds on human tumor cell lines were performed, and much higher cytotoxicity than cisplatin on several human tumor cell lines (Table 13.2) was shown.

Pt(III) is an unusual oxidation state. Only a few mononuclear complexes have been reported, and in some cases they have been proposed as intermediates in the reductive elimination and oxidative addition reactions of Pt(IV) and Pt(II), respectively. A series of trinuclear platinum(III) complexes were reported and the chemical relationship among the different complexes were investigated [31].

Gold drugs have well-documented anti-inflammatory activity; they also show their antitumor and anti-HIV activities. Tetrahedral Au(I) complexes with 1,2-bis (diphenylphosphino)ethane and 1,2-bis(dipyridylphosphino)ethane ligands display a wide spectrum of antitumor activity in vivo, especially in some cisplatin-resistant cell lines. Chrysotherapy is well established as a treatment of rheumatoid arthritis with gold-based drugs. Auranofin, an organogold compound classified as an antirheumatic

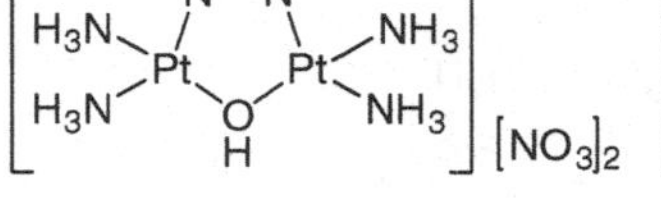

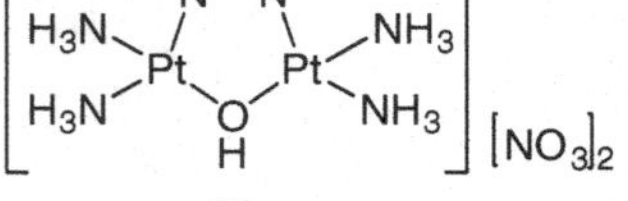

FIGURE 13.20 [{*cis*-Pt$(NH_3)_2$}$_2$(μ-pz)]$(NO_3)_2$ (**41**) and [{*cis*-Pt$(NH_3)_2$}$_2$(μ-1,2,3-ta)] $(NO_3)_2$ (**42**).

Table 13.2 In Vitro Cytotoxicity Assay of the Dinuclear Platinum(II) Complexes and Cisplatin on Human Tumor Cell Lines[a]

	IC_{50} (μM)						
Test Compound	MCF7	EVSA-T	WIDR	IGROV	M19	A498	H226
41	0.06	0.15	0.12	0.59	0.05	0.53	0.68
42	0.09	0.32	0.40	0.13	0.19	1.24	2.72
Cisplatin	2.33	1.41	3.22	0.56	1.86	7.51	10.9

[a] Cell lines: MCF7, EVSA-T, breast cancer; WIDR, colon cancer; IGROV, ovarian cancer; M19, melanoma; A498, renal cancer; H226, non-small cell lung cancer.

agent, first reported in 1979, has the advantage of being orally absorbed. It contains coordinated triethylphosphine and 2,3,4,6-tetra-*O*-acetyl-β-1-D-thioglucose ligands, which have been evaluated for cytotoxicity or antitumor activity.

Gold can exist in seven oxidation states: −I, 0, I, II, III, IV, and V. Only gold(I) and gold(III) are known to form compounds that are stable in aqueous media, and hence in a biological milieu. A common rationale for the design of potential gold-based antitumor agents is to attach gold(I) or gold(III) to a compound that has antitumor potency and good metal-ligating donor atoms.

Gold(I) phosphine complexes have two distinct classes: linear two-coordinate and tetrahedral four-coordinate geometries, which display antitumor properties. Caruso and his co-workers directed the in vitro activities of two gold(I) tetrahedral complexes, [Au(DPPP)(PPh_3)Cl][**43**](Fig. 13.21) and [{AuCl(PPh_3)}$_2$(μ_2-DIPHOS)], the result showed a dose-dependent loss of mitochondrial membrane potential (Table 13.3) [32]. By shortening the distance between both PPh_2 groups (by using DIPHOS instead of DPPP), a different Au complex is stabilized; this complex has a unique triangular Au coordination and a nonchelating structure with Au–P bonds weaker than those of the related nonchelate Au–DIPHOS compounds. Considering that the confirmation of chelating diphosphines is a necessary feature for effective

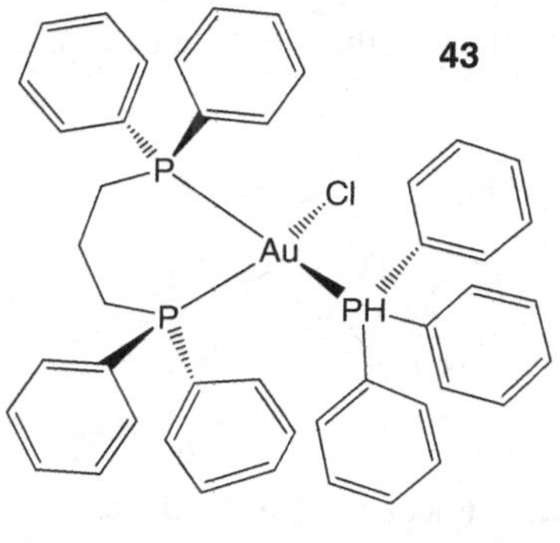

FIGURE 13.21 Structure of Au(DPPP)(PPh_3)Cl (**43**).

Table 13.3 Cytotoxic Activity of Au Complexes in Human Melanoma Cell Lines

	IC_{50} (μM)[a]	
	$[\{AuCl(PPh_3)\}_2(\mu_2\text{-DIPHOS})]$	$Au(DPPP)(PPh_3)Cl$
JR8	28 ± 0.33	0.8 ± 0.11
SK-MEL-5	2.7 ± 0.11	1.0 ± 0.56
2/60	5.5 ± 0.085	1.7 ± 0.47

[a] IC_{50} values were determined graphically from the growth inhibition curves obtained after a 48-h exposure of the cells to each drug. Data represent mean values (SD of three independent experiments). Values for $[Au(DPPP)(PPh_3)Cl]$ are taken from the literature.

antitumor activity, these results confirm that the chelate complex $[Au(DPPP)(PPh_3)Cl]$ is more effective than the nonchelate complex $[\{AuCl(PPh_3)\}_2(\mu_2\text{-DIPHOS})]$.

Gold(III) species have also demonstrated potential as antitumor agents. Gold(III) is a d^8 metal ion, a configuration it shares with Pt(II). A variety of ligands form stable complexes with this oxidation state; as a result, its complexes have a wide range of physical and chemical properties. As cisplatin $[(H_3N)_2PtCl_2]$ has been developed into a clinically important chemotherepy agent, many inorganic chemists were attracted to the structural similarities of gold(III) and Pt(II), both are d^8 metal ions and form spuare-planar, four-coordinate structures.

DNA is among the most suspected target molecules for gold(III) compounds. Messori reported the potential of bipyridyl gold(III) compounds as antitumor drugs, and their DNA binding properties. They reported a series of six oxobridged binuclear gold(III) complexes [33], all having a common structural feature: the presence of an extended, roughly planar system containing the Au_2O_2 "diamond core" and the aromatic rings of the bipyridine ligands. In particular, they have found that the 6,6′-dimethyl-2,2′-bipyridine derivative, which showed the largest structural deviations with respect to the model compound $[Au_2(\mu\text{-}O)_2(bipy)_2](PF_6)_2$ (**44**, see Fig. 13.22), had the greatest cytotoxic activity. Turn to the identification of possible structure–function relationships, this compound is the one that most greatly deviates from the "average" chemical and biological behavior within the series; in particular, it is nearly 10 times more effective than its structural analogs in terms of cytotoxic potency.

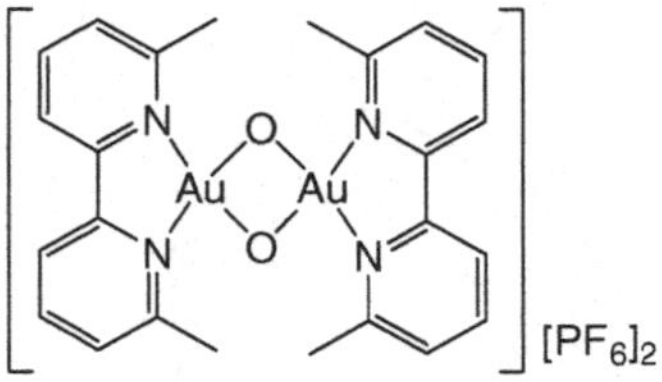

44

FIGURE 13.22 Structure of $[Au_2(\mu\text{-}O)_2(6,6'\text{-}Me_2bipy)_2][PF_6]_2$.

Overall, these observations can provide the opportunity to design new compounds of this series that bear new and specific structural features (e.g., different types of substituents) and might display increasing chemical reactivity and enhancing cytotoxic effects.

Moreover, there is no direct evidence for the formation of Au(III)-DNA adducts in living cells. Several cytotoxic gold(III) complexes have been shown to be efficient inhibitors of mitochondrial thioredoxin reductase [34]. In Alberto Bindoli's work [35], both gold(I) and gold(III) complexes are extremely efficient inhibitors of thioredoxin reductase, showing IC_{50} ranging from 0.020 to 1.42 μM, whereas metal ions and complexes not containing gold are less effective, exhibiting IC_{50} going from 11.8 to 76.0 μM. At variance with thioredoxin reductase, auranofin is completely ineffective in inhibiting glutathione peroxidase and glutathione reductase, while gold (III) compounds show some effect on glutathione peroxidase.

Generally speaking, gold(III) complexes are not very stable under physiological conditions because of their high reduction potential and fast hydrolysis rate. Therefore, a suitable ligand to stabilize the complex becomes a foremost challenge. A multidentate ligand that can provide chelating interaction to a correlated gold(III) center can help achieve this aim [36].

Vanadium has also demonstrated its insulin-mimetic properties in vivo. After being orally absorbed, vanadium is rapidly distributed in tissues (lungs, kidneys, spleen, and muscle), and it is stored primarily in bone. An important aspect of this renewed interest is the potential for improved clinical efficacy of vanadium at very low concentrations by complexation with oppropriate ligands. Vanadium compounds mimic most of the metabolic effects of insulin on different cellular types [37,38]. Besides the antidiabetic effects for which it is now so well known, vanadium also exhibits a number of other therapeutic effects, including antitumor and anti-inflammatory activities [39]. Vanadium has various oxidation states (+3, +4, +5, etc.), all of which can interact with biomolecules containing negatively charged oxygen donors such as carboxylate, phenolate, phosphate, phosphonate, catecholate, and hydroxamate groups [40].

Even though simple vanadium salts such as $VOSO_4$ and $NaVO_3$ are effective for normalization of the serum glucose levels of STZ-rats, synthetic vanadium complexes have been investigated to reduce toxicity and establish long-term efficaciousness. The possibility of the formation of a cyclic tetranuclear complex by means of solution speciation is also suggested. Here is an example that the mononuclear complex $[VO(Hhpic\text{-}O,O)(Hhpic\text{-}O,N)(H_2O)]\cdot 3H_2O$ (**45**) (see Scheme 13.3) exhibits relatively higher insulin mimetic activity than $VOSO_4$, while the tetranuclear complex, $[(VO)_4(l\text{-}(hpic\text{-}O,O',N))\ _4(H_2O)_4]\cdot 8H_2O$ (**46**) (Scheme 13.3) shows significantly lower activity [41]. Complex **45**, a six-coordinated asymmetric V center $[VO_5N]$ with a distorted octahedral shape, is readily transformed into **46**, which is the first example of a vanadium molecular square.

While **45** exerted higher insulinomimetic activity than $VOSO_4$, the activity of **46** was significantly lower than that of $VOSO_4$. This may indicate that **46** appears to retain its cyclic structure during the in vitro test (Table 13.4). In addition, the

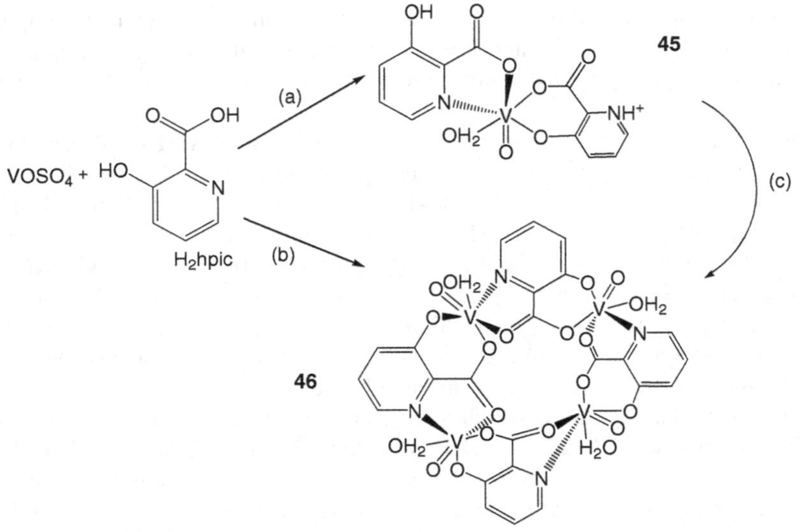

SCHEME 13.3 Strutures of oxovanadium(IV) complexes **45** and **46**.

insulinomimetic activity of VO(IV) complexes depends not only on the individual components, but also on the assembly. Hence, rational ligand design for the VO complex might be an promising approach to enhance the insulinomimetic activity and plausibly to reduce its toxicity.

Recent studies showed that oxovanadium complexes exhibit potent anti-HIV properties toward infected immortalized T-cells. However, the instability of vanadium(IV) complexes under physiological conditions has been encountering frequently. Porphyrins have a rigid square-planar scaffold which could prohibit the demetalation reaction. To address the problem of instability, porphyrinato was employed as a ligand to stabilize VO^{2+}. The water-soluble oxovanadium(IV) tetraarylporphyrin complex **47** (Fig. 13.23) has demonstrated excellent solution stability against glutathione reduction and high potency (5 μM, 97% inhibition) in inhibiting HIV-1 replication in Hut/CCR5 cells [42]. Chi-Ming Che and his co-workers reason that the porphyrin ligand should stabilize vanadium(IV) and carry

Table 13.4 Apparent IC_{50} Value for the Free Fatty Acid Release from Isolated Rat Adipocytesa

Complex	IC_{50} (μM)
$VOSO_4$	1.00 ± 0.02
45	0.78 ± 0.05
46	5.42 ± 0.29

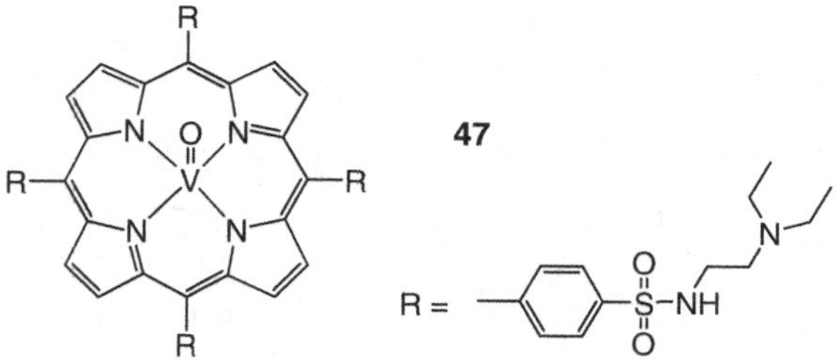

FIGURE 13.23 Oxovanadium(IV) porphyrin with anti-HIV activity.

the VO^{2+} unit to the biological target. Complex **47** shows dose-dependent anti-HIV activities in Hut/CCR5 cells (Fig. 13.24).

Tetrahedral bis(cyclopentadienyl)vanadium(IV) complexes (Fig. 13.25) were first reported for their dual potent anti-HIV (Table 13.5) and rapid spermicidal activities [43]. The oxovanadium–thiourea [OVT] nonnucleoside inhibitors [NNIs] $C_{29}H_{27}Br_2Cl_2N_6O_2S_2V$ (**48**) and $C_{31}H_{35}N_6O_4S_2V$ (**49**) were synthesized by reacting $VOSO_4$, a V^{IV} compound, with two corresponding deprotonated thiourea NNI compounds as ligands. The existence of the V=O bond ($968\,cm^{-1}$) was confirmed by IR spectroscopy. This dual function may be useful for the development of an effective and safe vaginal anti-HIV spermicide for women who are at high risk of acquiring HIV/AIDS by heterosexual transmission.

Besides the antidiabetic effects for which it is now so well known, vanadium also exhibits a number of other therapeutic effects, including antitumor activities. The present work shows that vanadium complexes can inhibit some parameters related

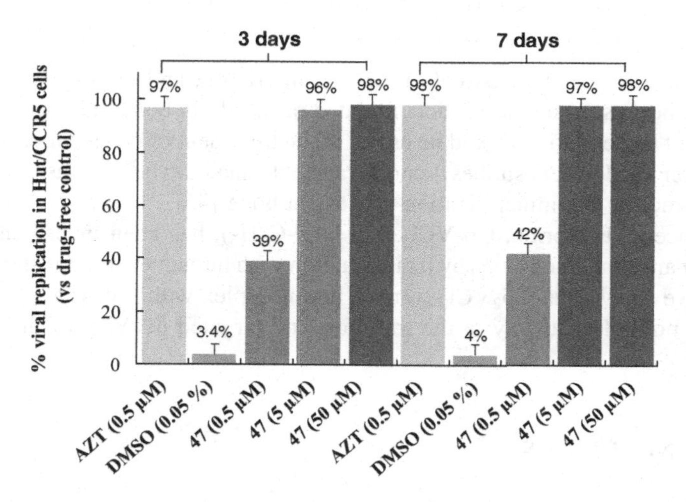

FIGURE 13.24 Percentage inhibition of HIV-1(BaL) replication in Hut/CCR5 cells (7 days) by oxovanadium(IV) porphyrins (**47**) and relative complexes.

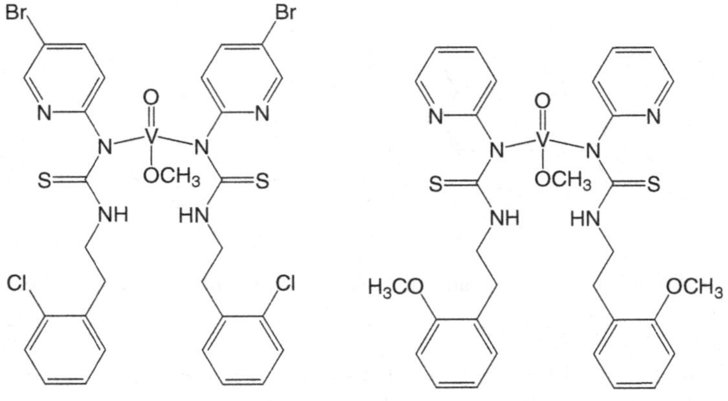

48

49

FIGURE 13.25 Structures of oxovanadium(V)–thiourea NNIs.

Table 13.5 Comparative Anti-HIV Activity of Thiourea-NNIs and OVT-NNIs Against HIV-1 $HTLV_{IIIB}$

	Anti-HIV Activity IC_{50} (μM)	
Compound	$IC_{50[p24]}$ [a]	$IC_{50[rRT]}$ [b]
48	0.08	2.1
49	0.12	0.87

[a] $IC_{50[p24]}$, drug concentration inhibiting HIV-p24 antigen production by 50%.
[b] $IC_{50[rRT]}$, drug concentration inhibiting HIV-rRt activity by 50%.

to cancer metastasis, such as cell adhesion, migration, and clonogenicity [44]. In particular, because vanadium is accumulated primarily in bone, it seems reasonable to think that its derivatives could be useful in the treatment of bone-tumor metastasis. On the other hand, recent studies demonstrate that vanadium behaves as an osteogenic agent, promoting the mineralization process in bone [45].

Vanadocene dichloride, Cp_2VCl_2 ($Cp = \eta^5\text{-}C_5H_5$), has been investigated for its promising antitumor activities by its interaction with human serum transfer [46]. The results have shown that Cp_2VCl_2 forms a new complex with transferrin, which may provide a possible pathway in the transport and targeted delivery of the antitumor agent.

13.5 CONCLUSIONS

A lot of model compounds for various metalloenzymes have been reported in the past few decades. The research described in many reviews has highlighted the advances that have been made in the study of synthetic analogs of metalloenzymes. The future of synthetic bioinorganic modeling will involve functional and catalytic models capable of reproducing or extending biological reactivity. The synthetic variability of ligands coupled with the ability to examine reactions in such fine detail provides an approach to dissecting the biological reactivity that is complementary to mutagenesis studies of the enzymes. Many significant challenges remain in modeling metalloenzymes. For example, many structural models lack functional equivalence, while many functional models have little structural equivalence. Further advances should be directed toward the synthesis of analogs that combine these properties and have both excellent structural and functional equivalence to the enzymes; this will include the incorporation of groups that mimic protein residues in the vicinity of the active site, which are postulated to play a role in the mechanism of action. In particular, the importance of careful system design and choice of the appropriate position for a highly organized recognition domain need iterated series. Developing synthetic modeling systems that are easy to synthesize and assure monomeric species, yet allow substrate accessibility, require new strategies to create site-isolated model complexes and to reach optimum efficiency.

REFERENCES

1. Bertini, I.; Gray, H. B.; Stiefel, E. I.; Valentine, J. S., Biological Inorganic Chemistry: Structure and Reactivity, University Science Books, Sausalito, CA, **2007**.
2. Parkin, G. *Chem. Rev.* **2004**, *104*, 699.
3. Baffert, C.; Collomb, M. N.; Deronzier, A.; Kjærgaard-Knudsen, S.; Latour, J. M.; Lund, K. H.; McKenzie, C. J.; Mortensen, M.; Nielsen, L. P.; Thorup, N. *Dalton Trans.* **2003**, 1765.
4. Poulsen, A.K.; Rompel, A.; McKenzie, C. J. *Angew. Chem., Int. Ed.* **2005**, *44*, 6916.
5. Ferreira, K. N.; Iverson, T. M.; Maghlaoui, K.; Barber, J.; Iwata, S. *Science* **2004**, *303*, 1831.
6. Mishra, A.; Wernsdorfer, W.; Abbouda, K. A.; Christou, G. *Chem. Commun.* **2005**, 54.
7. Zong, R.; Thummel, R. P. *J. Am. Chem. Soc.* **2005**, *12*, 12802.
8. Geletii, Y. V.; Botar, B.; Kögerler, P.; Hillesheim, D. A.; Musaev, D. G.; Hill, C. L. *Angew. Chem., Int. Ed.* **2008**, *47*, 3896.
9. Tard, C.; Liu, X. M.; Ibrahim, S. K.; Bruschi, M.; Gioia, L. D.; Davies, S. C.; Yang, X.; Wang, L. S.; Sawers, G.; Pickett, C. J. *Nature* **2005**, *433*, 610.
10. Justice, A. K.; Rauchfuss, T. B.; Wilson, S. R. *Angew. Chem. Int. Ed.* **2007**, *46*, 6152.
11. Song, L. C.; Tang, M. Y.; Su, F. H.; Hu, Q. M. *Angew. Chem. Int. Ed.* **2006**, *45*, 1130.
12. Ogo, S.; Kabe, R.; Uehara, K.; Kure, B.; Nishimura, T.; Menon, S. C.; Harada, R.; Fukuzumi, S.; Higuchi, Y.; Ohhara, T.; Tamada, T.; Kuroki R. *Science* **2007**, *316*, 585.
13. Ko1vari, E.; Kra1mer, R. *J. Am. Chem. Soc.* **1996**, *118*, 12704–12709.
14. Best, M. D.; Tobey, S. L.; Anslyn, E. V. *Coord. Chem. Rev.* **2003**, *240*, 3.
15. Livieri, M.; Mancin, F.; Saielli, G.; Chin, J.; Tonellato U. *Chem.–Eur. J.* **2007**, *13*, 2246.

16. Ait-Haddou, H.; Sumaoka, J.; Wiskur, S. L.; Folmer-Andersen, J. F.; Anslyn, E. V. *Angew. Chem., Int. Ed.* **2002**, *41*, 4014.
17. He, J.; Hu, P.; Wang, Y.-J.; Tong, M.-L.; Sun, H.; Mao, Z.-W.; Ji, L.-N. *Dalton Trans.* **2008**, 3207.
18. Fu, H.; Zhou Y.-H.; Chen, W.-L.; Deqing, Z.-G.; Tong, M.-L.; Ji, L.-N.; Mao, Z.-W. *J. Am. Chem. Soc.* **2006**, *128*, 4924.
19. Zhou, Y.-H.; Fu, H.; Zhao, W.-X.; Chen, W.-L.; Su, C.-Y.; Sun, H.-Z.; Ji, L.-N.; Mao, Z.-W. *Inorg. Chem.* **2007**, *46*, 734.
20. Breslow, R.; Nesnas, N. *Tetrahedron Lett.* **1999**, *40*, 3335.
21. Zhou, Y.-H.; Zhao, M.; Mao, Z.-W.; Ji, L.-N.; *Chem.–Eur. J.* **2008**, *14*, 7193.
22. Molenveld, P.; Stikvoort, W. M. G.; Kooijman, H.; Spek, A. L.; Engbersen, J. F. J.; Reinhoudt, D. N. *J. Org. Chem.* **1999**, *64*, 3896.
23. Boseggia, E.; Gatos, M.; Lucatello, L.; Mancin, F.; Moro, S.; Palumbo, M.; Sissi, C.; Tecilla, P.; Tonellato, U.; Zagotto, G. *J. Am. Chem. Soc.* **2004**, *126*, 4543.
24. Brunner, J.; Barton, J. K. *J. Am. Chem. Soc.* **2006**, *128*, 6772.
25. Farrell, N. P. Uses of Inorganic *Chemistry in Medicine*, Royal Society of Chemistry, Cambridge, UK, 1999.
26. Ruiz, J.; Lorenzo, J.; Sanglas, L.; Cutillas, N.; Vicente, C.; Villa, M. D.; Avilés, F. X.; Lōpez, G.; Moreno, V.; Pérez, J.; Bautista, D. *Inorg. Chem.* **2006**, *45*, 6347.
27. Halāmikovā, A.; Heringovā, P.; Kašpārkovā, J.; Intini, F. P.; Natile, G.; Nemirovski, A.; Gibson, D.; Brabec, V. *J. Inorg. Biochem.* **2008**, *102*, 1077.
28. Wheate, N. J.; Collins, J. G. *Coord. Chem. Rev.* **2003**, *241*, 133.
29. Hegmans, A.; Berners-Price, S. J.; Davies, M. S.; Thomas, D. S.; Humphreys A. S.; Farrell, N. *J. Am. Chem. Soc.* **2004**, *126*, 2166.
30. Kalayda, G. V.; Jansen, B. A. J.; Wielaard, P.; Tanke, H. J.; Reedijk, J. *J. Biol. Inorg. Chem.* **2005**, *10*, 305.
31. Forniés, J.; Fortun, C.; Ibāñez, S.; Martìn, A. *Inorg. Chem.* **2006**, *45*, 4850.
32. Caruso, F.; Pettinari, C.; Paduano, F.; Villa, R.; Marchetti, F.; Monti, E.; Rossi, M. *J. Med. Chem.* **2008**, *51*, 1584.
33. Gabbiani, C.; Casini, A.; Messori, L.; Guerri, A.; Cinellu, M.A.; Minghetti, G.; Corsini, M.; Rosani, C.; Zanello, P.; Arca, M. *Inorg. Chem.* **2008**, *47*, 2368.
34. Coronnello, M.; Mini, E.; Caciagli, B.; Cinellu, M. A.; Bindoli, A.; Gabbiani C.; Messori, L. *J. Med. Chem*, **2005**, *48*, 6761.
35. Rigobello, M. P.; Messori, L.; Marcon, G.; Cinellu, M. A.; Bragadin, M.; Folda, A.; Scutari, G.; Bindoli, A. *J. Inorg. Biochem.* **2004**, *98*, 1634.
36. Yang, T.; Zhang, J. Y.; Tu, C.; Lin, J.; Liu Q.; Guo, Z. J.; *Chin. J. Inorg. Chem.* **2003**, *19*, 45.
37. Shechter, Y.; Goldwaser, I.; Mironchik, M.; Frikin, M.; Gefel, D. *Coord. Chem. Rev.* **2003**, *237*, 3.
38. Srivastava, A.K.; Mehdi, M.Z.; *Diabetic Med.* **2005**, *22*, 2.
39. Noblia, P.; Vieites, M.; Parajon-Costa, B. S.; Baran, E. J.; Cerecetto, H.; Draper, P.; Gonzalez, M.; Piro, O. E.; Castellano, E. E.; Azqueta, A.; Cerain, A. L. de; Monge-Vega, A.; Gambino, D. *J. Inorg. Biochem.* **2005**, *99*, 443.

40. Xing, Y. H.; Aoki, K.; Bai, F. Y. *J. Coord. Chem.* **2004**, *57*, 157.
41. Nakai, M.; Obata, M.; Sekiguchi, F.; Kato, M. *J. Inorg. Biochem.* **2004**, *98*, 105.
42. Wong, S. Y.; Sun, R. W.; Chung, N. P. Y.; Lin C. L.; Che, C. M. *Chem. Commun.* **2005**, 3544.
43. Cruz, O. J. D.; Dong, Y.; Uckun, F.M. *Biochem. Biophy. Res. Commun.* **2003**, *302*, 253.
44. Molinuevo, M. S.; Cortizo, A. M.; Etcheverry, S. B. *Cancer Chemother. Pharm.* **2008**, *61*, 767.
45. Facchini, D. M.; Yuen, V. G.; Battell, M. L.; McNeill, J. H.; Grynpas, M. D. *Bone* **2006**, *38*, 368.
46. Du, H. Y.; Xiang, J. F.; Zhang, Y. Z.; Tang, Y. L.; Xu, G. Z. *J. Inorg. Biochem.* **2008**, *102*, 146.

INDEX

Design and Construction of Coordination Polymers, Edited by Mao-Chun Hong and Ling Chen

Printed in the United States
By Bookmasters